KU-168-358

Gastrophysics

The New Science of Eating

PROFESSOR CHARLES SPENCE

Foreword by Heston Blumenthal

VIKING

an imprint of

PENGUIN BOOKS

VIKING

UK | USA | Canada | Ireland | Australia
India | New Zealand | South Africa

Viking is part of the Penguin Random House group of companies
whose addresses can be found at global.penguinrandomhouse.com.

First published 2017
001

Copyright © Charles Spence, 2017
Foreword copyright © Heston Blumenthal, 2017

The moral right of the author has been asserted

Set in 12.54/15.31 pt Bembo Book MT Std
Typeset by Jouve (UK), Milton Keynes
Printed in Great Britain by Clays Ltd, St Ives plc

A CIP catalogue record for this book is available from the British Library

HARDBACK ISBN: 978–0–241–27008–0
TRADE PAPERBACK ISBN: 978–0–241–27009–7

www.greenpenguin.co.uk

MIX
Paper from
responsible sources
FSC® C018179

Penguin Random House is committed to a
sustainable future for our business, our readers
and our planet. This book is made from Forest
Stewardship Council® certified paper.

To Norah Spence, who knew implicitly the value of a good education without ever having had the opportunity to have one.

And Barbara Spence, who had to read more about the legendary F. T. than any loving wife should ever have to.

Contents

Acknowledgements

I would never have ended up in the world of gastrophysics if it hadn't been for the enduring support and mentorship of Prof. Francis McGlone then at Unilever Research, for which I will always remain grateful. As will become clear from the main text, though, it was really the introduction to Heston Blumenthal by Tony Blake of Firmenich that led to my growing interest in gastronomy, rather than food science! In recent years, I owe an especial debt of gratitude to Rupert Ponsonby (R&R), Christophe Cauvy (then of JWT), and Steve Keller (iV Audio Branding) for having believed in the multisensory approach to gastrophysics and all things fun. To Prof. Barry Smith, for helping make the Baz 'n' Chaz wine roadshow so enjoyable. Long may it continue! It has, though, really been the enthusiastic support and collaboration of the next generation of young chefs, including Jozef Youssef, of Kitchen Theory, and Charles Michel, Crossmodalist extraordinaire, that has made the latest gastrophysics research such fun to do. You will read about a number of their dishes and designs in the pages that follow.

I would also like to thank the many chefs and culinary schools for their support, and opening up their kitchens and restaurants to the 'Mad Professor': I have been lucky enough to conduct gastrophysics research over the last fifteen years together with a number of world-leading chefs including Heston Blumenthal and all the team at The Fat Duck Research Kitchen and restaurant; Chef Sriram Aylur, Quilon, London; Chef Jesse Dunford Woods, Parlour, London; Ben Reade, Nordic Food Lab; Dominique Persoone, The Chocolate Line; Chef Albert Landgraf from Epice, São Paolo; Chef Xavier Gamez, of Xavier260, Porto Allegre, Brazil; Chef Andoni and Dani Lasa from Mugaritz, San Sebastián; Chef Joel Braham, of The Good Egg, London; Chef Debs Paquette, of Etch, Nashville; and not forgetting Chef Paul Fraemohs, of Somerville College, Oxford. I have also been lucky enough to conduct research together with Ferran Adrià's Alicía Foundation in Spain, The Paul Bocuse Cookery School, Lyon, France, and Westminster Kingsway College, London. I would also like to thank Jelly & Gin, Blanch & Shock, Caroline Hobkinson, Sam Bompas, and all the students, past and present, who have done most of the research here at the Crossmodal Research Laboratory.

Finally, I would like to thank Tony Conigliaro from 69 Colbrooke Row, London, Ryan Chetiyawardana, aka Mr Lyan, Neil Perry (of Rockpool, Sydney), and Maxwell Colonna-Dashwood, of Colonna & Small's, Bath. All masters of their art. And, finally, Fergus Henderson, for the memorable evening onstage at the Cheltenham Science Festival back in 2007 (along with a bucket of tripe oh so gallantly displayed by my vegan then graduate student, Maya Shankar).

Foreword

There was a time when – apart from the late, great Nicholas Kurti – scientists didn't consider the science of food a serious or worthwhile subject for study. I'd talk with them, offering up theories based on what I'd observed and carefully tested in The Fat Duck kitchen, and get an indulgent smile that seemed to say, 'You stick to cooking and let us get on with the rest.' Admittedly, chefs were no better, insisting that cooking had little to do with science, as though the eggs they were busy scrambling weren't in fact undergoing the technical process of coagulation.

Charles, though, wasn't like this. One of his strengths is that he has a curiosity that crosses disciplines and, for all his scientific rigour, isn't confined to a narrow academic viewpoint. Upon meeting him, I discovered that many of the ideas I was exploring in my kitchen, he was also exploring in his lab. And so, as you'll see in this book, he and I began doing research together on how we react to the food we see, hear, smell, touch, and put in our mouths. We eat with our eyes, ears, nose, memory, imagination and our gut. Every human being has a relationship with food, some of it positive, some of it negative, but ultimately it's all about emotion and feeling.

To me, this is at the very heart of how we respond to food: much more than the tongue (which detects at least five tastes); more even than the nose (which detects countless aromas), it's the conversation between our brain and our gut, mediated by our heart, that tells us whether we like a food or not. It's the brain that governs our emotional response.

It's a hugely rewarding subject (and an essential one for us, as humans, to understand), but it's undoubtedly a complex one, too. Charles is the perfect guide to introduce us to this world and to investigate with us – in a truly accessible, entertaining and informative way – how it works. On every page there are ideas to set you thinking and widen your horizons, from the notion that we all of us live in separate and completely different taste worlds, to questions like, 'Is cutlery the best way to move the food from plate to mouth?'

What I take away from *Gastrophysics* is that, as Charles says, in the mouth very little is as it seems. The pleasure we get from food depends, far more than we could possibly imagine, on our subjectivity – on our memories, associations and emotions. It's a fascinating topic into which you can take your first steps through the door by reading *Gastrophysics*.

Heston Blumenthal

Amuse Bouche

'Open wide!' she said, in her most seductive French accent, and so I did. And in it went. In that one moment, in that one movement, and in that one mouthful, I was taken back to the haziest memories of being spoon-fed as a baby (or at least my imagining of what that must have been like). That dish, or rather the way in which it was served, also foreshadowed what my last meals may well be like as the darkness draws in. So, if you want just one example to illustrate how food is so much more than merely a matter of nutrition, then that was it – that mouthful of lime *gelée* at The Fat Duck restaurant in Bray, many, many years ago. It was an incredibly powerful experience, shocking, disturbing even.[1] But why? Well, I guess in part because no one had fed me that way, at least not in the last forty-five years or so.* Yet there I was, at what was soon to become the world's top restaurant, being spoon-fed my three-Michelin-starred dinner. Well, one course of it, at least. Just enough to make the point that dining is about much more than merely what we eat.

The pleasures of the table reside in the mind, not in the mouth.[2] Get that straight and it soon becomes clear why cooking, no matter how exquisitely executed, can only take you so far. One needs to understand the role of 'the everything else' in order to determine what really makes food and drink so enjoyable, stimulating and, most importantly, memorable. Even something as simple as biting into a

---

* The only thing missing was for the waitress to have me sit on her lap before serving me! I doubt that Heston and the gang would dare to repeat this little interlude at The Duck today. It is just a little too provocative, a little too 'out there' for those gastrotourists who can afford the £295 price of admission, now that the restaurant has firmly established itself as one of the grand temples of modernist cuisine. But others have since taken up the baton from Heston and his ilk, such as Dabiz Muñoz, the 'bad boy' of modernist cuisine, at DiverXo, in Madrid.

fresh ripe peach turns out, on closer inspection, to be an incredibly complex multisensory experience. Just think about it for a moment: your brain has to bind together the aromatic smell, the taste, the texture, the colour, the sound as your teeth bite through the juicy flesh, not to mention the furry feeling of the peach fuzz in your hand and mouth. All of these sensory cues, together with our memories, contribute much more than you would believe to the flavour itself. And it all comes together in your brain.[3]

It is the growing awareness that tasting is fundamentally a cerebral activity that is leading some of the world's top chefs to take a fresh look at the experiences that they deliver to their diners. Just take Denis Martin's modernist restaurant in Switzerland (see Figure 0.1). The chef realized that some of his guests were not enjoying the food as much as he thought they should, given how much effort he was putting into preparing the dishes.[4] Too often his diners were stiff and buttoned-up – 'Suits on account', as he put it. How could anyone who walked in the door sporting such a sour expression be expected to enjoy his food? The solution was brilliantly simple, and involved putting a cow on each and every table.

Nothing happens at the start of service until one of the diners, curious as to whether what they see before them on the table is a Swiss take on a salt shaker or pepper grinder, picks up their cow. When they tilt it to look underneath, it lets out a mournful moo. Diners often laugh in surprise. Then, within a few moments, the dining room erupts into a chorus of mooing cows, and the restaurant is full of chortling diners. The mood has been lifted and that is when the first course comes out from the kitchen.* This wonderfully intuitive *mental* palate cleanser is far more effective than any acidic sorbet – the traditional means of cleansing the palate – at enhancing the diners' enjoyment of the food to come. After all, our mood is one of the most important factors influencing our dining experience, so best try to optimize it.[5]

It turns out that modernist chefs are especially interested in the

---

* Notice also how this helps bring the diners in the restaurant together in a shared sonic experience (see 'Social Dining').

Figure 0.1. The only item of tableware to greet the expectant diner at Denis Martin's two-Michelin-starred restaurant in Vevey, Switzerland. But what exactly are you looking at, and why has the chef placed one on each and every table?

new sciences of eating (what I will here call gastrophysics), given their habit of recombining ingredients in new and unusual ways, not to mention their desire to play with diners' expectations. How exactly they are using this emerging knowledge to enhance the experience of eating constitutes the subject matter of this book. Many of the food and drinks companies are also becoming increasingly curious about the science of multisensory flavour perception. The aspirations of the latter, though, tend to be somewhat different from those of the chefs. Their hope is that the new gastrophysics insights may help them to use the so-called 'tricks of the mind' in order to reduce some of the unhealthy ingredients in their branded food products without having to compromise on taste.

Gastrophysics: The new sciences of eating

Many factors influence our experience of food and drink, whether we are eating something as simple as a luscious ripe peach or a fancy dish at one of the world's top restaurants. However, none of the existing approaches provides a complete answer as to why food tastes the way it does and why we crave some dishes but not others. After all, the focus of modernist cuisine is primarily on food and its preparation – often described as the new science of the kitchen.[6] Sensory science, meanwhile, tells us about people's perception of the sensory attributes of what they eat and drink in the lab, how sweet the taste, how intense the flavour, how much they like the dish. And then there is neuro-gastronomy – basically, the study of how the brain processes sensory information relating to flavour. This new discipline helps shed light on the brain networks that are involved when people taste liquidized food pumped into their mouth via a tube while lying flat on their back with their head clamped in a brain scanner. Do I have any vol-unteers?[7] Interestingly, you now find mention of the diner's brain on the menu at top restaurants like Mugaritz in San Sebastián in Spain, and at The Fat Duck restaurant in Bray. In fact, many of the science-inspired trends one now sees coursing through restaurants across the globe can be traced back to Bray, where Heston Blumenthal and his research team, together with their many collaborators, have been pushing the bound-aries of what dining could be for more than two decades now.

However, neither modernist cooking, nor sensory science nor even neurogastronomy offers a satisfactory explanation as to why our food experiences, be they special occasion or mundane everyday meal, appear to us as they do. What is needed is a new approach to measuring and understanding those factors that influence the responses of *real* people to *real* food and drink products, ideally under as *naturalistic* conditions as possible. Gastrophysics builds on the strengths of a number of disci-plines, including experimental psychology, cognitive neuroscience, sensory science, neurogastronomy, marketing, design and behavioural economics, each subject contributing a part of the story with specific techniques designed to answer particular questions.

As an experimental psychologist, I have always been interested in the senses, and in applying the latest insights from cognitive neuroscience to help improve our everyday experience. While I started out investigating sight and sound, over the years I have been slowly adding more senses to my research. Eventually this led me to the study of flavour, which is, after all, one of the most multisensory of our experiences. Given that my parents never went to school (they were constantly moving around the country, as they grew up on the fairground), I have always had a clear sense that research findings need ultimately to have real-world application. In 1997, I started my lab, the Crossmodal Research Laboratory, which is nowadays funded largely by the food and beverage industry. There are psychologists, obviously, but also marketers, the occasional product designer, musicians, and we even have a Chef in Residence. (Guess who has the tastiest lab parties in Oxford!) I have also been lucky enough to work with leading chefs, mixologists and baristas, and for my tastes the most exciting gastrophysics research lies at the intersection of these three areas – the food and beverage industry, the culinary experience designers and the gastrophysicists. I believe that gastrophysics research will come to play a dominant role in understanding and improving all of our food and drink experiences in the years to come.

What is 'gastrophysics'?

Gastrophysics can be defined as the scientific study of those factors that influence our multisensory experience while tasting food and drink.[8] The term itself comes from the merging of 'gastronomy' and 'psychophysics'[9]: gastronomy here emphasizes the fine culinary experiences that are the source of inspiration for much of the research in this area,[10] while psychophysics references the scientific study of perception. Psychophysicists like to treat the human observer much like a machine. By systematically observing how people respond to carefully calibrated sets of sensory inputs, the psychophysicist hopes to measure what their participants (or observers) perceive,[11] and then to figure out what really matters in terms of influencing people's behaviour.

Generally speaking, gastrophysicists aren't interested in simply asking people what they think. Better to focus on what people actually do, and how they respond to specific targeted questions and ratings scales, such as: How sweet is the dessert (give me a number from 1 to 7)? How much did you enjoy the food? How much would you pay for a dish like the one you have just eaten? They tend to be sceptical of much of what people say in unconstrained free report, given the many examples where people have been documented to say one thing but to do another (see 'The Atmospheric Meal' chapter for some great examples of this).

Importantly, the findings of the gastrophysics research do not apply only to high-end food and beverage offerings. If they did, they would still be interesting, certainly, but perhaps just not all that relevant in the grand scheme of things. How often do most of us get to dine at a Michelin-starred restaurant anyway? But many of the modernist chefs are incredibly creative. What is more, they have the authority and capacity to instigate change. If they are intrigued by the latest findings from the gastrophysics lab, they can probably figure out a way of putting a dish inspired by the new science on the menu next week. The large food and beverage companies, by contrast, often find it harder to engage in rapid, not to mention radical, innovation, much though they would like to. In the food industry, everything just tends to happen at a much slower pace!

In the best-case scenario, some of the most inventive ideas first trialled in the modernist restaurant provide genuine insights that can subsequently be used to enhance the experience of whatever we might be eating or drinking, whether we are on an aeroplane or in hospital, at home or in a chain restaurant. The multisensory dishes and experiences first dreamed up in some of these top dining venues provide the proof-of-principle support that gives others the confidence to innovate for the mainstream. So when the collaboration works well, it can lead to emerging gastrophysics insights being turned into amazing food and drink experiences that people really want to talk about and share. Get it right and it can result in dishes that are more sensational, more memorable and possibly healthier than anything that has gone before.

For example, just take the research that we conducted together with Unilever fifteen years ago.[12] We demonstrated that if we boosted the sound of the crunch when people bit into a potato crisp we could enhance

Figure 0.2. Chef Heston Blumenthal gets to grip with the 'sonic chip' in the golden booth at the Crossmodal Research Laboratory in Oxford, *c.*2004.

their perception of its crunchiness and freshness. Research, I am proud to say, that led to our being awarded the Ig Nobel Prize for Nutrition. This isn't the same as the Nobel Prize, but a rather more tongue-in-cheek award for science that first makes you laugh, and then makes you think. It was around this time that the chef Heston Blumenthal started coming up to the lab in Oxford, having been introduced by Anthony Blake of the Swiss flavour house Firmenich. As soon as we stuck the headphones on Heston and locked him away in the booth, he got it (see Figure 0.2)!

In fact, when interviewed on a BBC Radio 4 show at the time the chef stated: 'I would consider sound as an ingredient available to the chef.' This realization, in turn, provided the original impetus that led to the 'Sound of the Sea' seafood dish, at The Fat Duck, which became the signature dish at one of the world's top restaurants.[13] Other restaurants and brands then started working on adding a sonic element to their dishes, often facilitated by technology.

Subsequently, we worked together with The Fat Duck Research Kitchens on sonic seasoning – basically, a way of systematically modifying the taste of foods by playing specific kinds of sound or music.[14] These insights eventually made their way on to the menu at The House of Wolf restaurant in north London, courtesy of culinary artist Caroline Hobkinson. Culinary artists are more artist than chef, but use food and food installations to express themselves and their ideas. And it was on the basis of such research that British Airways launched their 'Sound Bite' menu in 2014, providing the option of sonic seasoning for their long-haul passengers.[15] More recently still, a number of health authorities have started to research whether they can generate 'sweet-sounding' playlists to help, for example, those diabetic patients who need to control their sugar intake – the idea being that if you can 'trick' the brain into thinking that the food is sweeter than it actually is, you get better-tasting food without the harmful side effects of consuming too much sugar. From the gastrophysics lab to the modernist restaurant, and on to the mainstream (though I worry that the follow-up studies have yet to be done to check just how long-lasting the effects of music and soundscapes are). And it may be that the direction of travel is reversed, with some of what is already going on in the top restaurants providing the impetus for the basic research back in the lab.[16]

What's the difference between 'crossmodal' and 'multisensory'?

Many of the insights of gastrophysics are built on the latest findings coming out of crossmodal and multisensory science. Now, these complex-sounding terms describe the fact that there is much more interplay between our senses than previously thought. While scientists used to think that what we see goes to the visual brain, what we hear to the auditory brain, and so on, it turns out that there are far more connections between the senses than we ever realized. So changing what a person sees can radically alter what they hear, changing what they hear may influence what they feel, and altering what they feel can modify what they taste. Hence the term 'crossmodal', implying that what is going on in one sense influences what we experience in the others (as, for example, when someone puts on some red lighting and suddenly the wine in your black glass tastes sweeter and fruitier).

The term 'multisensory', by contrast, is more often used to explain what happens when, say, I change the sound of the crunch you hear as you bite into a crisp. In the latter case, what you hear and feel are integrated in the brain into a multisensory perception of freshness and crispness, with both senses intrinsic to your experience of one and the same food item. Don't worry if the distinction sounds like a subtle one – it is. Nevertheless, is just the this sort of thing that gets my academic colleagues fired up.

I would certainly like to take issue with the conceit of one recent BBC TV show (*Chef vs Science: The Ultimate Kitchen Challenge*) in the UK, in which chef was set against scientist. Ridiculous, if you ask me. For no matter whether the competition is between Pierre Gagnaire and Hervé This (one of the godfathers of molecular gastronomy), or Michelin-starred *MasterChef* regular Marcus Wareing versus materials scientist Mark Miodownik, the answer isn't really in any doubt – stick with the chef.[17] What is much more interesting, at least to me, is how much of a lift the chef, molecular mixologist or barista can get by working together with the gastrophysicist.[18] In the chapters that follow, I hope to convince you that, more often than not, the combination will win out. Not only that but the fruits of this collaboration are starting to percolate down to influence our food and drink experiences no matter where we eat and regardless of what we choose to consume.

Not everyone is happy about what they see happening in the world of gastronomy, though. *MasterChef* judge William Sitwell, for instance, promised to destroy any square plates you brought to him.[19] He absolutely hates the new fashion in plating. Don't get me wrong, I understand where he is coming from. There are undoubtedly some practitioners out there who have definitely lost the plot. You know what I mean – when the dish you ordered arrives at the table served in a mini frying pan, atop a plank suspended between a couple of bricks. But let's be clear about this: the mere fact that some people take things too far does not invalidate the more general claim that our perception of, and our behaviour around, food is influenced by the way in which it is plated and what it comes served on. What is particularly exciting to me is that one can take some of the latest trends in plating from the high end of gastronomy and translate them into actionable insights that hold the promise of enhancing the food service offering in, for instance, a hospital setting.[20]

Is cutlery the best way to move the food from plate to mouth?

How much do you really like the idea of sticking something into your mouth that has been inserted into who knows how many other mouths beforehand? Think about it carefully – is a cold, smooth

Figure 0.3. Will the tableware of the future look like this? A selection of utensils created by silversmith Andreas Fabian in collaboration with Franco-Colombian chef Charles Michel, as displayed at the 'Cravings' exhibition at London's Science Museum.

stainless steel knife, fork and/or spoon really the best way to transfer food from table to mouth? Why not eat with your fingers instead? Is it mere coincidence that this is how one of the world's most popular foods – the burger – is typically eaten? Given what we now know about the workings of the human mouth and the integration of the senses that give rise to multisensory flavour perception, shouldn't we all think about designing things a little differently, moving forward? Why not give spoons a texture to caress the tongue and lips? After all, the latter are amongst the body's most sensitive skin sites (at least of those that are accessible while seated at the dining table).[21]

Why not cover the handles of one's cutlery with fur, much like the Italian Futurists might have done at their tactile dinner parties[22] in the 1930s? We have tried both here in Oxford (see Figure 0.3). There is inertia to change, certainly.[23] But since we have (mostly) accepted such radical innovations to our plateware in recent years, why not do the same with our cutlery? This question holds true no matter whether your implements of choice happen to be Western cutlery or chopsticks. Excitingly, gastrophysicists are now working with cutlery makers, industrial designers and chefs in order to deliver a better offering to the table.[24]

I am convinced that change really is possible in the world of food and drink, and that progress will come at the interface between modernist cuisine, art and design, technology and gastrophysics. Thereafter, the best ideas will be disseminated out to the mainstream by the food and beverage industry. And by chefs . . . and eventually by you.

Testing intuitions

What the gastrophysics research often does, then, is assess people's intuitions. Typically, the results provide empirical support concerning the relative importance of various different factors that people already suspected were somehow relevant. However, on occasion, the research can turn up a surprise result, one that may, for instance, show that some age-old kitchen folklore is just plain wrong. Let me give you a concrete example to illustrate the point: many chefs are taught in cookery school to place an odd rather than even number of elements on the plate (i.e., serve three scallops or five, rather than four). However, when we tested this practice by showing several thousand people pairs of plates of food and asking them which they preferred (see Figure 0.4 for an example), it really didn't matter. Instead, people's choices correlated to the total amount of food that was on the plate. The more food, the better![25] Of course, even when the gastrophysics research simply backs up people's intuitions, it can nevertheless help put a monetary value on something, which often aids in decision-making (i.e., is the extra effort/cost of doing things a particular way really worth the effort?).

In the remainder of this introduction, I want to focus on some of the questions that gastrophysicists are currently thinking about, and

Figure 0.4. Which plate of seared scallops do you prefer? The latest research shows that we care more about how much food there is than whether there happens to be an odd or an even number of elements on the plate.

bringing to the public's attention. These are some of the key themes that will be discussed in the chapters that follow.

Just how much influence does the atmosphere really have?

Now, whenever we eat, be it in a dine-in-the-dark or Michelin-starred restaurant, the atmosphere, the sights, the sounds, the smell, even the feel of the chair we happen to be sitting on (not to mention the size and shape of the table itself), all influence our perception and/or our behaviour, however subtly. From what we choose to order in the first place to what we think about the taste of the food when it comes, the speed at which we eat and the duration of our stay, the atmosphere affects everything. People will tell you that they were always going to choose what they ordered and to eat and drink as much as they actually did. However, the emerging gastrophysics research shows that this is simply not the case.[26]

In our research with the food and beverage industry, we have been quantifying just how much of an impact the atmosphere really has on people's ratings of taste, flavour and preference. We found, for example, that people's ratings of one and the same drink may vary by 20% or more as a function of the sensory backdrop where it is served. No wonder, then, that – as we will see later – top chefs and restaurateurs are increasingly recognizing the importance of such environmental effects. In some cases, they have sought to match the atmosphere to the food they serve, the image that they wish to create or the emotion that they wish to provoke. In the 'Airline Food' chapter, for instance, we will take a look at how our growing understanding of the impact of the atmosphere on multisensory taste perception is now enabling some of the world's most forward-thinking airlines to improve their food offering at 35,000 feet.

Have you heard of off-the-plate dining?

One of the trends that has been sweeping high-end modernist dining in recent years is the growing focus on off-the-plate dining (see 'The Experiential Meal'). This term is used to describe the more theatrical, magical, emotional, storytelling elements that one increasingly finds in contemporary haute cuisine. Nowadays, it all seems to be about delivering meaningful, memorable and stimulating multisensory experiences (or journeys); selling 'the experience', the *total* product and not just the *tangible* product in Philip Kotler's marketing termin-ology.[27] Better still if those experiences also happen to be shareable (e.g., for the millennials on their social media).

And while the tops chefs fight over who should get the credit for first coming up with the idea of multisensory experiential theatrical dining, the irony is that the Italian Futurists were already matching meals to sounds eighty years ago, not to mention adding scents and textures to their dinners, and they were amongst the first to experi-ment with miscolouring the foods that they served. We'll take a closer look at whether modernist cuisine really was invented back in the 1930s in the final chapter ('Back to the Futurists').[28]

Doesn't good food speak for itself?

Some commentators, including a few Michelin-starred chefs, dismiss gastrophysics as nothing more than 'sensory trickery'. 'Good food,' you hear them proclaim oh-so-earnestly, 'should speak for itself.' To them, a great meal is all about the local sourcing, the seasonality of the ingredients, the detail and technique in the preparation, and the beautiful cooking. Don't mess with the food; keep it simple, keep it slow, even. This was certainly the line I heard from Michael Caines MBE, then the Michelin-starred chef at Gidleigh Park in Devon, when I met him in 2015.[29] He'd have you believe that none of this other stuff matters, that the world would – heaven forbid – perhaps be a better place without gastrophysics.

According to the likes of Caines,* the honest chef lets their dishes do the talking. They don't need to worry about the weight of the cutlery to make *their* food taste great. And yet I don't need to go to Gidleigh Park to know that the cutlery will be heavy. There is just no way that any self-respecting chef would ever serve their food with a plastic or aluminium knife and fork. It would spoil the experience! Tell me, am I wrong? And, hold on a minute, let's take a look at the decor and context. Gidleigh Park just so happens to be a beautiful manor house set in the heart of the Devonshire countryside. I am sure that you don't need a gastrophysicist to tell you that the chef's dishes are going to taste better there than if exactly the same food were to be served in a noisy aeroplane cabin or in a hospital canteen. In other words, you cannot avoid 'the everything else', however much you might wish to.

My point, then, is that wherever food and drink is served, sold or consumed there is always a multisensory atmosphere. And that environment impacts both what we think about what we are tasting and,

---

* Caines is very much at the slow food rather than the molecular gastronomy end of the spectrum; and, truth be told, there are far worse places to practise slow food than rural Devon. My problem with the slow food movement, though, is that most of those who advocate it tend to have the luxury of living close to the verdant countryside.

more importantly, how much we enjoy the experience. Ultimately, there is just no such thing as a neutral context or backdrop. It is time to accept the growing body of gastrophysics evidence demonstrating that the environment, not to mention the plateware, dish-naming, cutlery and so on, *all* exert an influence over the tasting experience. Once you have got that straight, then surely it makes sense to try and optimize 'the everything else', along with whatever you happen to be serving on the plate. And this holds true no matter what one is trying to achieve, be it a more memorable, a more stimulating or a healthier meal. Or, I suppose, you can simply stick your head in the sand and pretend that none of this other stuff really matters. To me, the choice is clear. (And my advice for those who choose to ignore all that the emerging science of gastrophysics has to offer is to simply make sure that you are serving your food in a fancy venue with your diners holding heavy cutlery!)

So, without further ado, having polished off the amuse bouche (not to mention the naysayers), let's move on to the first course!

1. Taste

Can you list all of the basic tastes? There is sweet, sour, salty and bitter, for sure. But anything else? Nowadays, most researchers would
include umami as the fifth taste. Umami, meaning 'delicious taste',
was first discovered back in 1908 by Japanese researcher Kikunae
Ikeda. This taste is imparted by glutamic acid, an amino acid, and is
most commonly associated with monosodium glutamate, itself a
derivative of glutamic acid. Some would be tempted to throw
metallic, fatty acid, kokumi and as many as fifteen other basic tastes
into the mix as well – though even I haven't heard of most of them.[1]
And some researchers query whether there are even any 'basic' tastes
at all![2]

The mistake that many people make, though, when talking about
food and drink is to mention things like fruity, meaty, herbal, citrusy,
burnt, smoky and even earthy as tastes. But these are not tastes. Strictly
speaking, they are flavours.[3] Don't worry, most people are unaware of
this distinction. But how do you tell the difference? Well, hold your
nose closed – and what is left is taste (at least assuming that you are not
tasting something with a trigeminal hit, like chilli or menthol, which
activate the trigeminal nerve). So if we struggle to get the basics
straight, what hope is there when it comes to some of the more complex interactions taking place between the senses? Taste would be
simple, if it weren't so complicated!

Do you mean taste or do you mean flavour
(and does it really matter)?

Most of what people call taste is actually flavour, and many of the
things that they describe as flavours turn out, on closer inspection, to
be tastes.[4] Some languages manage to sidestep the issue by using the

same word for both taste and flavour.[5] In fact, in English, what we really need is to create a new word – and that neologism is 'flave'. 'I love the *flave* of that Roquefort' would do the trick. Let's see whether it catches on. There are also challenges here from those stimuli that lie on the periphery. Just take menthol, the minty note you get when chewing gum: is it a taste, a smell or a flavour? Well, all three, in fact; and it also gives rise to a distinctive mouth-cooling sensation.[6] The metallic sensation we get when we taste blood also has the researchers scratching their heads in terms of whether it should be classified as a basic taste, an aroma, a flavour or some combination of the above.[7]

Most people have heard of the 'tongue map'. In fact, pretty much every textbook on the senses published over the last seventy-five years or so includes mention of it. The basic idea is that we all taste sweet at the front of the tongue, bitter only at the back, sour at the side, etc. However, the textbooks are wrong: your tongue does not work like that! This widespread misconception resulted from a mis-translation of the findings of an early German PhD thesis that appeared in a popular North American psychology textbook written by Edwin Boring in 1942.[8] So now we have got that cleared up, let me ask, do you actually have any idea how the receptors are laid out on your tongue? No, I didn't think so. Something so fundamental, so important to our survival, and yet none of us really has a clue about how it all works. Shocking, no?

The taste receptors are not evenly distributed, but neither are they perfectly segmented as the oft-cited tongue map would have us believe. The answer, as is so often the case, lies somewhere in between. Each taste bud is responsive to all five of the basic tastes. But these taste buds are primarily found on the front part of the tongue, on the sides towards the rear of the tongue and on the back of the tongue. There are no taste buds in the middle of the tongue.[9] Interestingly, though, many people (including chefs) tend to say that they experi-ence sweetness more towards the tip of the tongue, they feel the sourness on the sides of the tongue and bitterness/astringency often seems more noticeable towards the back of the tongue.[10] And for me, a pure umami solution has a mouth-filling quality to it that none of the other tastes can quite match.

The *real* question, though, is how have so many people been so wrong for so long? Part of the reason may be due to the general neglect of the 'lower' senses by research scientists. Another factor probably relates to the 'tricks' that our mind plays on us when constructing flavour percepts, things like 'oral referral' and 'smelled sweetness' (about which more later). As we will see time and again throughout this and the following chapter, in the mouth, very little is as it seems.

Managing expectations

Why, you might well ask, does a cook – be they a modernist chef working in a high-end Michelin-starred restaurant or you slaving away in the kitchen preparing for your next dinner party – need to know about what is going on in the mind of the diners they serve? Why not simply rely on the skills that are taught in the culinary schools or picked up from watching those endless cookery shows on TV? Why not focus on the seasonality, the sourcing, the preparation, and possibly also the presentation of the ingredients on the plate? That is all you need, isn't it?[11] As a gastrophysicist, I know just how important it is to get inside the mind of the diner in order to understand and manage their expectations about food. It is only by combining the best food with the right expectations that any of us can hope to deliver truly great tasting experiences.

It really excites me to see a growing number of young chefs starting to think more carefully about feeding their diners' *minds* and not just their *mouths*. I'm sure this is largely down to the influential role of star chefs like Ferran Adrià and Heston Blumenthal, both of whom I have been lucky enough to work with. Where they lead, others surely follow. But that still doesn't answer the more fundamental question of what got the top chefs interested in the minds of their diners in the first place. After all, this certainly isn't something that they teach you at cookery school.[12]

In Heston's case, it all started with an ice cream. In the late 90s, Heston created a crab ice cream to accompany a crab risotto. The top chef liked the taste and, after a little tinkering, believed it to be perfectly seasoned. But what would the diners say? (Typically, any new dish is trialled in the research kitchen across the road from the restaurant. Then, once it has met with Heston's approval – a slow and exacting process – the next step is to try the new dish out on a few of the regulars and see how they like it. Only if a dish passes all of these hurdles will it stand a chance of making its way on to the restaurant's tasting menu.)

Imagine the scene: just like in one of the chef's TV shows, you can almost see Heston looking on expectantly from the kitchens, waiting for the diners' approval as his latest culinary creation is brought out to the guinea pigs sitting at the tables. Surely the diners will think it tastes great, given who made it. But, in this case at least, the response was not what was expected. 'Urrrggghhh! That's disgusting. It's way too salty.' Well, maybe I exaggerate a little – but trust me, the response wasn't good.

What had gone wrong? How could one of the world's top chefs consider a dish to taste just right only to have some of his regular guests find it far too salty? The answer, I think, tells us a lot about the importance of expectations in our experiences of food and drink. In other words, it is as much a matter of what is in the *mind* of the person doing the tasting as what is in their *mouth* or on the plate.[13] When the diners saw that pinkish-red ice cream (this was also evaluated in the lab with a smoked salmon ice cream), their minds immediately made a prediction about what they had been given to eat. Tell me, what would *you* expect to taste were such a dish to be placed before you?

For most Westerners, pinkish-red in what looks like a frozen dessert is associated with a sweet fruity ice cream, probably strawberry flavour. 'Sweet, fruity, I like it, but it isn't so good for me' – all that goes through a diner's mind in the blink of the eye. After all, one of our brain's primary jobs is to try to figure out which foods are nutritious and worth paying attention to (and perhaps climbing a tree for), and which are potentially poisonous and hence best avoided. However, on the rare occasions when our predictions turn out to be wrong, the surprise, or 'disconfirmation of expectation', that follows can come as quite a shock. It can, in fact, be rather unpleasant.[14] The diners in Heston's restaurant presumably thought that they were going to taste something *sweet*, but what was brought out from the kitchen was actually a *savoury* frozen ice. In other words, they were expecting strawberry and got frozen crab bisque instead! The savoury ice may have been popular in England a century ago, but it has very much fallen out of favour these days.[15]

In a great series of gastrophysics experiments, Martin Yeomans and his team at the University of Sussex, together with Heston, showed that it was possible to radically influence people's perception and liking of the frozen pink treat simply by changing the name of the dish.[16] All it took to modify the participants' expectations in the lab setting was to tell them that this was a savoury ice, or else give the dish the mysterious title 'Food 386'. The expectations that go with the name or description of the dish led people to enjoy the ice cream significantly more than those who had not been told anything about the dish before tasting it. Crucially, they no longer found it too salty either.

Research suggests that our first exposure to a flavour affects those that come thereafter, even once we know exactly what it is that we are tasting. And though the effects may not always be as dramatic as in the case of Heston's pink savoury ice cream, we have probably all had our own experience of this. I still remember, on my first trip to Japan, fifteen years ago, buying a pale-green ice cream from a street vendor. It was a hot spring day and everyone seemed to have one of these refreshing-looking ices in their hands. I had absolutely no doubt that it was mint-flavoured, just as it would be back in the UK. But I recoiled in shock on tasting what turned out to be something most unexpected; it was, in fact, green-tea-flavoured ice cream. Delicious in its way, but I must confess that I have somehow never been able to quite get over that initial surprise whenever I am served a bowl in Japan.

Whatever the name and/or description of a dish, and no matter what it looks like, these cues are always there, helping to set our expectations.[17] And those expectations influence our judgements and perception, however subtly. Even when cooking at home, how those you serve experience your food is as much a matter of what is going on in their minds as it is a matter of what they put in their mouths.

However, it is not just the colour and other visual properties of food that set our expectations.[18]

What's in a name?

Imagine yourself in a fancy restaurant, scanning the menu for something to eat. You already know that you want fish, but which one? Now, let's suppose you came across Patagonian toothfish. Would you order it? No, I didn't think so. Nor, for that matter, did anyone else. Sales of this veritable 'monster of the deep' had been disappointing for years. No matter how chefs prepared it, diners just turned their noses up and chose something else instead. Their eyes would continue scanning down the menu, looking for something that sounded, how shall I put it, a little more enticing.

Would the response be different, do you think, if they were to come across Chilean sea bass on the menu? It certainly sounds a lot more appealing, doesn't it? The thing is, though, that these two names refer to one and the same fish! Sales of this currently sustainable fish have increased by well over 1,000% – yes, that's three zeros – in a number of markets around the world (including North America, the UK and Australia). The trick was simple: just change the name. This is one of the most impressive examples of 'nudging by naming', as the behavioural economists like to call it.[19] In fact, in no time at all, this fish started appearing on the menus of all the best restaurants, a trend that, even today, shows no signs of letting up. Once again, it is what is in the diner's mind, and the associations they make with different labels or descriptions that are crucial here.

The frozen crab bisque/smoked salmon ice cream and Patagonian toothfish cases are exceptional: they have, in fact, been chosen to make a particular point – about the importance of naming to our experience of food. Nevertheless, look around and one finds many everyday examples demonstrating much the same point. Have you ever wondered, for instance, why golden rainbow trout is so much more popular than regular brown trout? The traditionally trained chef's mind may immediately start to ponder differences in taste or texture, or perhaps to consider how the fish was dispatched.[20] But why stop there? When was the last time you ate an *ugli* fruit (the result of the hybridization of a grapefruit – or pomelo – an orange and a tangerine)? Exactly. You have to wonder how much more popular this member of the citrus family could be had it been given another name. The decline in popularity of everything from faggots to pollack and Spotted Dick in recent years can, at least in part, be put down to their unfortunate names.[21]

Great expectations

Some of you may already be wondering whether you could use the same naming 'tricks' to enhance people's perception of whatever food or beverage product you happen to be serving. Unfortunately, though, I very much doubt that you will be able to increase sales of most everyday foods by anything like as much as the Patagonian toothfish – sorry, Chilean seabass – example might lead one to believe. Nor, unless you have had your head very firmly stuck in the modernist cookbooks, will the colours of the dishes you prepare at home give as misleading an impression of the actual taste or flavour to come as the pinkish-red hue of Heston's frozen treat undoubtedly did.[22] No one will, I presume, get the wrong end of the stick on seeing whatever you might be thinking of serving at your next dinner party. The colours of the foods we prepare normally give a pretty reliable indicator of the probable tasting experience. It is mostly in the modernist restaurant or when in parts foreign that things start to go awry. So relax!

Getting the name and/or description of a dish right is definitely worth investing some time in, even for those of you cooking at home. Take the following examples: simply referring to a pasta salad as a salad with pasta (i.e., just reversing the order in which the same words appear) makes people think of the dish as being that little bit healthier. And adding more descriptive elements, as when a restaurateur describes a dish as 'Neapolitan pasta with crispy fresh organic garden salad', is likely to lead to the number of positive comments that a dish garners increasing.[23]

The topic of expectation management is just as important in the setting of the supermarket. Why else, after all, have the supermarkets started to create *phoney* farms to use in the labelling of their food packaging?[24] Here I am thinking of farms like Rosedene and Nightingale; these names may well conjure up images of some rural idyll, but they do not actually exist. So why are the supermarkets doing it? Well, it turns out that we will pay more for exactly the same food, let's say a ploughman's sandwich, if we are told that the cheese inside was produced by farmer John Biggs, from Duxfield Farms in Cumbria. Obviously neither you nor I have any idea what this particular farmer's cheese tastes like, because I just made him up. And yet this kind of description adds value to the food offering or, in marketing-speak, it increases the consumer's willingness to pay. It may even make your sandwich taste different, perhaps better, as a result. These, then, are precisely the sorts of experiments that the gastrophysicist is interested in conducting, and the kind of results they want to share.

Others, though, have used the naming of the dish as an opportunity to capture people's attention. Heston Blumenthal received a phenomenal amount of press when he decided to call one of his new dishes 'Snail Porridge'; Had he given this dish a French name ('*Escargots à la Something*'), no one would have batted an eyelid; the dish would probably have tasted much more authentically French too. Over in Bror in Denmark, two ex-Noma chefs have decided to call one of their dishes simply 'Balls'. They come to the table breadcrumbed, fried to a reddish brown and dusted with sea salt. Delicious, apparently.[25]

Paul Pairet, the chef at Ultraviolet, a multisensory experiential restaurant in Shanghai, has this to say on his restaurant's website: 'What is the "psycho taste"? The psycho taste is everything about the taste but the taste. It is the expectation and the memory, the before and the after, the mind over the palate. It is all the factors that influence our perception of taste.' So here's another of the world's top chefs explicitly recognizing the importance of 'the everything else' to the mind-blowing dining experiences that he provides.

Of course, we do not just have expectations about the taste and flavour of food and drink, and how much we will like it. We also have expectations about the kinds of food served by specific chefs or in specific venues; the same dish will taste very different to us as a function of whether it is served in a modernist restaurant, at your friend's house or up in an aeroplane.[26] And then there is the anticipation, the booking of your meal.[27] This is undoubtedly all part of pleasure too. You know, finding a great restaurant, even getting there, in some cases. Believe it or not, some chefs, with their minds squarely on the design of the experience, even consider how the diners will arrive at their restaurant. Just take Mugaritz, in Spain. As chef Andoni says: 'Mugaritz is not only the restaurant but also the road leading up to it, the countryside that you can see from the car and that, bend after bend, stokes the anticipation of everyone who visits us. Mugaritz is also its setting.'[28]

Or take Fäviken, the restaurant set in the wilds of the Swedish countryside. No one will doubt your credentials as a proper gastrotourist if you manage to make your way to this remote location! The approach to El Celler de Can Roca, consistently rated first or second in lists of the world's top restaurants, is also chosen to discombobulate diners, situated as it is at the far side of an industrial park in Girona. So, should you be inviting any friends from afar to visit for dinner, be sure to recommend the scenic route.

'Tell me what you eat, and I will tell you what you are.' So said Jean Anthelme Brillat-Savarin in his much-quoted classic text *The Physiology of Taste*, first published back in the 1820s.[29] Perhaps, but I would be tempted to put it rather differently: 'Tell me what a person expects to eat, and I'll tell you what they taste. I'll also estimate how much they'll enjoy the experience.'[30] Expectations are key. Rare, after all, is the occasion on which we put something into our mouth without having first being informed, or having at least made a prediction, about what it is and whether or not we are going to like it. Our response to food – both the decision about what we choose to buy, order or eat, and what we think about it once we do – is nearly always affected by our beliefs (our expectations, in other words). It is the latter that subsequently anchor, and hence disproportionately influence, our tasting experience.[31]

Do pricing, branding, naming and labelling influence taste?

Typically, we are aware of the brand and/or price of whatever it is that we eat and drink. On many occasions, the food will also be accompanied by some form of label or description. Such product-extrinsic cues, as they are known, all exert a profound influence over what people say about the taste, flavour and/or aroma of a food, not to mention how much they enjoy it.[32] While we have known for years that pricing, branding and other kinds of product description can influence what people say about food and drink, until recently we had no real idea whether and how such factors affected the way in which the brain processes taste.

However, the latest neurogastronomy research demonstrates that the changes in brain activity resulting from the provision of such information can be dramatic. Differences are seen both in terms of the network of brain areas that are activated and the amount of activation that is seen there. What is more, these effects have, on occasion, been shown to affect the neural activity at some of the earliest (i.e., primary) sensory areas in the human brain. For example, in what has become one of the classic studies of branding, people had their brains scanned while one of two famous colas was periodically squirted into their mouth. Different patterns of brain activation were seen depending on which brand the participants thought that they were tasting.[33] The fact that branding has such a marked effect on flavour perception presumably helps to explain why blind taste tests are such a common feature of commercial product testing.[34] There is, though, a question about what such tests actually tell us. Think about it for a moment. How often do you put something in your mouth without knowing what it is that you are tasting? While it may be a worthwhile exercise when it comes to the detection of flaws in food and beverage products, I suspect we should be doing more of our testing in the presence of all the other cues that normally go along with consumption. In that way, we will stand a much better chance of re-creating the more naturalistic conditions of everyday life.

Does food and drink taste better if you pay more for it? Not always, for sure, but more often than not. In support of such an intuition, neuroscientists in California investigated what happened in the brains of social wine drinkers (aka students) when given different, and some-times misleading, information about the price of a red wine. A \$5 bottle of wine was either correctly described or else mislabelled as a \$45 bottle; meanwhile, a \$90 bottle of wine was presented as costing either \$10 or \$90, and a third was correctly labelled as costing \$35 a bottle. The price was displayed on a monitor whenever a small amount of the wine was squirted into the participant's mouth. In some trials, participants had to rate the intensity of the wine's taste, whilst in other trials they judged its pleasantness.[35]

Everyone reported liking the expensive wine more than the cheap wine. Crucially, analysis of the brain scans revealed increases in blood flow in the reward centre of the brain associated with the price cue (see Figure 1.1). Telling people that the wine was more expensive (regardless of what wine they were actually tasting) led to an increase

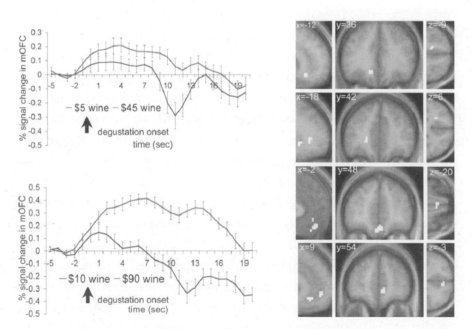

Figure 1.1. Images showing the percentage signal change in brain activation in the medial orbitofrontal cortex (mOFC; the brain's reward centre) over time (seconds are indicated on the *x* axis) as a function of the price associated with a wine that the participants in the scanner are tasting.

in activation in the medial orbitofrontal cortex (mOFC), a small part of the brain located just behind the eyes. By contrast, no change in blood flow was observed in the primary taste cortex, the part of the brain that processes the sensory-discriminative attributes of taste (e.g., when judging how sweet, sour, etc. something is). Intriguingly, though, when the same wines were presented eight weeks later, now without any indication as to their price (and away from the confines of the brain scanner), no significant differences in pleasantness were reported. And the latest evidence suggests that the effects of misleading pricing may work better in the mid-price range. So I am afraid that no matter what you say, you have little chance of convincing people that the Two-buck Chuck (i.e., cheap plonk) you are serving them is premier cru.[36]

Imagine being given a clear solution to drink. You are told either that you are about to taste something very bitter or else something that is much less bitter. Should you happen to be lying in a brain scanner at the time, changes are likely to be seen in some of the earliest sites in the brain after the taste and smell signals are initially coded by the sensory receptors. In particular, researchers have shown that activity in the middle and posterior insula, an area deep in the cerebral cortex, can be modulated by the verbal description people had been given regarding the intensity of the to-be-delivered taste. The response in the brain's reward centre, the OFC, also varies systematically as a function of people's expectations concerning the drink's bitterness.[37] Elsewhere, researchers have tried verbally describing an odour as 'smelly cheese' and found that people rated it as more pleasant than when exactly the same odour was labelled as 'sweaty socks' instead. Once again, the brain's response was modified as a result of the provision of a product-extrinsic cue.[38]

While such neuroimaging results are undoubtedly fascinating, it is perhaps worth bearing in mind here just how unnatural the situation in which the participants find themselves really is. How often, after all, do you go out on a Friday night and find yourself lying flat on your back, inserted several feet into a narrow tube, with your head clamped still. The latter precaution is needed in the brain scanner in order to minimize any head movements that can make it difficult to analyse the neuroimaging data. And that is not all. You will have a tube held between your teeth as a few millilitres of wine is periodically squirted into your mouth. You are told to evaluate its taste, without swallowing. Eventually you are allowed to swallow, then your mouth is washed out with artificial saliva (yes, really). Then the whole process starts again.

People's beliefs about the origins of their food also impact on how they perceive taste. For instance, in a recent study illustrating this point, students from the US were given identical samples of meat (e.g., beef jerky or ham) and told either that it was factory farmed or that it was free range. Those who were told that the meat was factory farmed rated it as tasting less pleasant, saltier and greasier. What is more, the students ate less of it too, and said that they would have been willing to pay less for the meat. Crucially, the same pattern of results was obtained across three separate studies. One finds that describing a food as organic or free range has much the same effect – despite the fact that in blind taste tests consumers mostly cannot tell the difference. So what this means in practice is that if you shell out for some organic, free-range, hand-fed food, you should be sure to let your guests know its provenance if you want them to be able to taste the difference.[39]

One of the many challenges facing food and drink companies in this area is that even though they may be making real and sustained progress in terms of reducing the less healthy ingredients in their branded products, they are often best advised not to state 'low fat' or 'reduced sugar' on their labels, because doing so is likely to cause the consumer to say that it tastes different. Keep quiet about it, and they may not detect that anything has changed. Health by stealth, that's the way to do it! It is worth stressing here that the interests of the food and beverage industry tend to be quite different from those of the modernist chef. The latter is trying to create unusual, surprising and sometimes spectacular results. The majority of the diners at the top restaurants tend not to care too much about the health/nutritional content of the meals they are served (given that it is likely to be a one-off occasion[40]). Rather, they want surprise and novelty. The former, by contrast, are typically more interested in trying to keep their successful branded products tasting the same to consumers as they always did, whilst gradually making their products less un-healthy.[41]

Once you understand just how important naming, labelling, branding and pricing can be, you might start to wonder how much, if anything, is actually happening at the level of the taste buds. Ulti-mately it is the interaction between what is in the mouth and what is in the mind that determines what the final tasting experience is like, and how much we enjoy it. Master both the food and the gastrophys-ics and you'll be in a good place to impress, whoever you are and whomever you are cooking for.

Taste worlds

Tell me, what does coriander (or cilantro) taste like to you? Do you love it or loathe it? The majority of people, it has to be said, like its fresh, fragrant or citrusy characteristics. Others, by contrast, are con-vinced that it tastes soapy (some even describe spinach as soapy too). It reminds them of dirt, bugs or mould, they say. Those in the latter camp will typically avoid *any* food containing what John Gerard, writing back in 1597, called a 'very stinking herbe' with leaves of 'venemous quality'.[42] So who is correct? What does coriander *really* taste of?

Both sides are right, though more of the population fall into the former category. Most of us – 80% or more – are likers, the exact fig-ure depending on the ethno-cultural group tested.[43] Are those on the soapy side of the spectrum simply unable to detect one of the many compounds that make up the distinctive flavour of coriander? Or per-haps those on the citrusy side are anosmic to something ('anosmia' being the technical name for being unable to smell some volatile chemical or other). No one knows for sure! What is more, there is even uncertainty about whether that soapy sensation should itself be characterized as a taste, an aroma or something else entirely. What-ever it is, it doesn't seem to fit any of the commonly recognized basic tastes.[44]

Though this may be more of a topic for the next chapter, it is worth noting here that something like one in every two people can't smell androstenone,[45] an odorous steroid derived from testosterone. They are anosmic to this particular volatile organic molecule. Meanwhile 35% of the population find that it has a very powerful – and deeply unpleasant – stale, sweaty, urine smell. (This is the reason why male pigs are castrated, i.e., to minimize the unpleasant aroma known as 'boar taint'.) Worse still, the individuals in this group tend to be exquisitely sensitive to this compound; some can detect it at concentrations of less than 200 parts per trillion. The remaining 15% or so of the population, well, they say that it smells sweetly floral, musky and/or woody. Some people (like me) experience the smell simply as chemical-like. Same molecule: completely different experience!

The prevalence of these genetic differences in the worlds of taste/flavour perception varies by region and culture. So, if you had to guess, in which part of the world do you think the likelihood of people perceiving the urinous note in their uncastrated pork meat would be highest? I have heard that it's the Middle East – i.e., exactly the place where religion bans pork as a legitimate source of food. Just mere coincidence, you think? Seems unlikely, doesn't it?

Coriander and androstenone are just the tip of the iceberg as far as genetically determined differences are concerned. That is to say, every one of us is anosmic to some number of compounds, many of which are associated with food.[46] So, for instance, our sensitivity to isovaleric acid (a distinctive sweaty note in cheese), ß-ionone (a pleasant floral note added to many food and drink products; think of the fragrance of violets), isobutyraldehyde (which smells of malt) and cis-3-hexen-1-ol (which gives food and drink a grassy note) all show a significant degree of genetic variation,[47] and roughly 1% of the population are unable to smell vanilla. What this means, in practice, is that there are some pretty profound individual differences in people's ability to perceive these compounds.

Who knows, then, how many of the disputes between wine experts can be put down to such genetic variability? Just take the famous disagreement between Robert M. Parker Jnr, the influential American wine critic, and the British Master of Wine Jancis Robinson regarding the 2003 Château Pavie. The former absolutely loved this wine, whereas the latter slammed it, giving the *en primeur* wine a score of 12/20. Robinson had the following to say: 'Completely unappetising overripe aromas. Why? Porty sweet. Port is best from the Douro not St Emilion. Ridiculous wine more reminiscent of a late-harvest Zinfandel than a red Bordeaux with its unappetising green notes.' Parker responded by saying that the Pavie 'does not taste at all (for my palate) as described by Jancis'. So were these two international experts tasting the same wine differently? Did they perceive the same attributes, which one writer appreciated and the other disliked? Or did the wine really taste different to the two star wine writers?[48]

I myself am totally anosmic to tri-chloro-anisol (TCA for short), the chemical that gives rise to cork taint in wine.[49] This form of 'smell blindness' is something that my wine colleagues find most amusing, as I am sure you can imagine. When a corked bottle comes to the table, they will order a replacement and put two glasses of the same wine down in front of me, one from either bottle. I will normally find them to be identical, whereas my friends will not be able to drink from one of the two glasses. Once again, TCA is one of those chemicals to which people show huge differences in terms of sensitivity. I normally have the last laugh, though, since when the uncorked wine runs out, there is still plenty more wine left that only I enjoy!

What I'm hinting at is that we all live in very different taste worlds (see Figure 1.2).[50] Some people are able to detect bitterness in food and drink where others taste nothing (the former group are commonly referred to as supertasters). Supertasters may have as many as sixteen times more papillae on the front of their tongue as others (known as non-tasters). Not only do people vary in terms of their sensitivity to bitterness but also – to a less pronounced degree – in terms of their perception of saltiness, sweetness, sourness and oral texture.[51] Taster status, like odour sensitivity, is largely heritable (i.e., genetically determined).[52] In fact, back in the 1930s, scientists were thinking of using this taste test as a paternity test. And beyond these individual differences in sensitivity to the basic tastes, we all vary quite markedly in terms of our

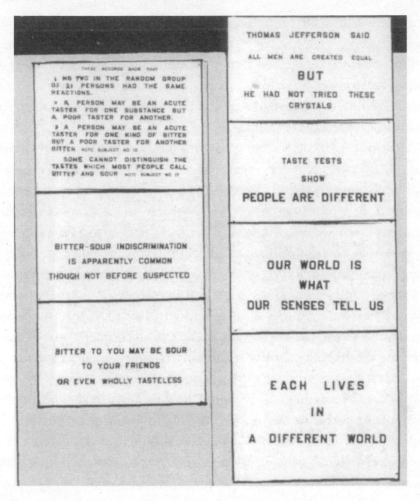

Figure 1.2. One of the original posters from a public demonstration of the different taste worlds in which we live – from the 1931 New Orleans American Association for the Advancement of Science Meeting.

hedonic responses too. So, for example, there are those who are sweet likers, whereas others (including myself) are best classified as being more ambivalent about sweetness.[53]

But why should bitterness be the taste for which individual differences are most pronounced? Why are the individual differences not so apparent for the salt, sweet or sour tastes? It is likely that individual differences in sensitivity to bitterness may have been especially important for our ancestors. In times of plenty, the supertasters would have had a competitive advantage, since they would be unlikely to ingest something bitter and hence potentially poisonous. By contrast, in lean times, it would have been the non-tasters who had a slight competitive advantage since they would have been more likely to ingest those bitter foodstuffs that happened not to be poisonous and hence less likely to starve to death. It is a little harder to make such an argument for the other tastes.[54]

However, a liking for bitter-tasting foods (associated with super-taster status) also correlates with psychopathic tendencies! Or, as the authors of one recent study put it: 'General bitter taste preferences emerged as a robust predictor for Machiavellianism, psychopathy, narcissism and everyday sadism.'[55] Though, of course, it is important to note that correlation is not causation – you are not necessarily a psychopath should you be one of those who likes bitter-tasting food and drink. Intriguingly, the latest research shows that tasting something bitter can give rise to increased hostility. By contrast, tasting something sweet can apparently make you feel more romantic and increase the likelihood of you agreeing to go on a date. Even more remarkably, those who are thinking about love will rate water as tasting sweet. Meanwhile, men whose hockey team had won rated a lemon-lime sorbet as tasting sweeter than did those supporting the losing side. And going a step further, marketing professor Baba Shiv and his colleagues in California have reported that handling large amounts of money can change people's taste thresholds.[56] Once again, the sense of taste turns out to be so much more than merely a matter of taste.[57]

Some global food companies have already taken advantage of this distinction by launching two versions of a particular product into the marketplace, one targeted at supertasters, another for all the non-tasters out there. Not that it will say this on the label; the company will just let the market segment itself. Remember: taster status runs in families. As it happens, my mother, brother, sister and nieces are all supertasters, whereas my father can taste none of the bitterness in broccoli that the rest of the family can. This, I believe, helps to explain why my father would always make us kids eat all of our vegetables before leaving the dinner table. He never understood, or so we like to think, how terrible these green vegetables tasted to the rest of us.[58] If only we had known then that people live in different taste worlds.

*'There is more to taste'**

Taste is crucial to our survival. In a way, one might think of it as the most important of our senses – helping us to distinguish between that which is nutritious and that which may be poisonous.[59] And yet, on closer inspection, it turns out not to be so important, at least not in terms of perception. You get a sense of this from looking at how much of the cerebral cortex is given over to each of the senses. While more than half of the brain is involved in processing what we see, only something like 1% of the cortex is directly involved in taste perception.[60] The reason for this is that our brains pick up on the statistical regularities of the environment, and so we learn to predict the likely taste and nutritional properties of potential foodstuffs on the basis of other sensory cues, such as colour and smell. We may learn, for example, to expect pinkish-red foods to be sweet.[61] This allows us to assess the likely consequences of ingesting a whole host of different foods without necessarily having to stick them into our mouths first in order to determine what they taste like.

Ultimately, if you know about the expectations set up by the other senses, then you are in a much better position to modify people's perception of taste.[62] It may even help all those exasperated parents out there wondering how to get their offspring to eat more vegetables.[63] So, no matter how you define (or think about) taste, it is clear that the other senses play a far bigger role in determining what we think we are tasting and how much we enjoy the experience than we generally realize. So, in closing, let me introduce you to Eleanor Freeman, senior snack inventor at online health food company Graze. Her taste buds have been insured for £3 million. Gennaro Pelliccia, an Italian coffee master for the Costa coffee chain in the UK, has his

---

* This is the strapline from a recent advert for Lavazza coffee.

insured for £10 million, while the taste buds of Hayleigh Curtis, who
works for Cadbury chocolate, are insured for a measly £1 million.[64]
To me, though, this just sounds like a headline-grabbing stunt, for as
we will see in the next chapter, what a top taster really needs to worry
about is their nose.

2.   Smell

Think about the last time you had a head cold and your nose was blocked. Food and drink didn't have much of a taste, did they? Ever wonder why that is? What is missing in such circumstances is not taste – trust me, your taste buds are working just fine – but rather aroma. Assuming that you don't have a cold at the moment, try holding your nose pinched tightly closed, and get a friend to feed you something without letting you know what it is. Unless they pick something really pungent (and if they do that, maybe they aren't such a good friend after all), you will most likely have very little idea of what you are tasting – onion or apple, red wine or cold coffee. These pairings are surprisingly hard to tell apart without a functioning sense of smell.*

It is important to distinguish here between the two different ways in which we smell. There is the 'orthonasal' route: when we sniff external aromas from the environment. And there is 'retronasal' smell: when volatile aromatic odour molecules are pulsed out of the back of the mouth into the back of the nose whenever we swallow while eating and drinking. The orthonasal sniffing of food aromas is especially important because it allows us or, rather, our brains, to form the rich flavour expectations concerning both what the experience of tasting will be like and how much we expect to enjoy it.[1] But it is the retronasal perception of aroma, on swallowing, that really provides our tasting experiences with their rich variety and interest. Most of the time, though, we remain acutely unaware of how much of the information that we think we are tasting via the tongue

---

* Or, for a mischievous version (borrowed from Sam Bompas, one colourful half of the jelly-mongers extraordinaire Bompas & Parr), why not chop up some cabbage, boil it in water and pour the liquid into a teapot, then surprise your friends when they let go of their noses with something that looks like tea but has a most unsavoury aroma!

comes in via the retronasal olfactory route. This is, in large part, because food aromas are experienced as if coming from the mouth – as if being sensed by the tongue itself. This strange phenomenon goes by the name of 'oral referral'.[2]

To illustrate the point, try eating a jelly bean with your nose closed between thumb and forefinger. What can you taste? You will most likely experience sweetness, maybe some sourness and, who knows, perhaps a hint of spiciness too (at least if you get a cinnamon one). Then, after a few bites, let go of your nostrils. You will suddenly get the fruity flavour, orange or cherry, etc. But you will experience that flavour as coming from your mouth, not from your nostrils. That is oral referral in action, the mislocalization of aroma to the mouth![3]

Does vanilla smell sweet to you?

The answer for most people is a definite 'yes'. People give exactly the same answer when it comes to the aroma of caramel and strawberries too.[4] Now, this is confusing, right? After all, didn't I just say in the last chapter that 'sweet' was a taste descriptor? So how can an aroma be said to *smell* sweet? Some have argued, I think wrongly, that this is a kind of synaesthesia.[5] Interestingly, food companies add vanilla flavouring to ice cream to bring out the sweetness.[6] They do this because at very cold temperatures your taste buds don't work so well, and hence you can no longer taste sweetness – but you can still smell it. Surely you have had the experience of drinking a warm glass of cola by mistake? Doesn't it suddenly taste sickly sweet? The composition of the drink itself hasn't changed, but the signals that your taste buds send through to your brain have changed as a function of the drink's temperature.[7] Since the drink is usually served cold, the manufacturer has added some sweetness through the nose.[8] Confused? You should be.

In the other direction, i.e., when it comes to taste's influence on aroma and flavour perception, things are very different. One of the classic studies in this area involved people tasting a solution whose sweetness had been carefully calibrated to be just below the level of awareness (in other words, it tasted like plain water). Nevertheless, when people held a small amount of that subjectively tasteless liquid in their mouth, their ability to detect the cherry-almond aroma in another drink that they were smelling suddenly increased dramatically.[9] Importantly, however, further research showed that the taste had to be *congruent* with the smell in order for this effect to occur.[10] Adding a sub-threshold dose of monosodium glutamate into the mouths of Western participants didn't have the same effect. However, the response may well be different in Japanese consumers. In other words, the research suggests that while everyone's brain uses the same rules to combine the senses, which particular combination of tastes and smells gives rise to the enhancement or suppression of flavour depends on the food culture that person has grown up in.[11]

The amazing thing is how quickly this kind of learning happens, and what's more it continues to occur throughout our lifetimes. Take a novel odorant, the smell of water chestnut (as in one study conducted on Australian adults a few years ago). Then, simply pair it with either a sweet or bitter tastant in the mouth. Believe it or not, after no more than three co-exposures, the smell starts to take on the appropriate taste qualities. More remarkably still, this can occur even when the tastant is presented at a level below perceptual awareness.[12]

Have you ever noticed how freshly ground coffee often smells wonderful, yet when you come to taste it the flavour can be a little disappointing? The same thing in reverse occurs if you take a ripe French cheese.[13] It may well smell like the inside of a jock's training shoe (please excuse the metaphor), yet if you can manage to put some in your mouth, the pleasurable experience that follows is often sublime. What is happening?[14] These changes in our hedonic ratings – basically, how much we like something – illustrate the distinction between our two ways of smelling: orthonasal (when breathing in) and retronasal (when breathing out through the back of the nose). Normally, we are remarkably good at predicting the likely retronasal flavour of a food based on nothing more than an orthonasal sniff. So good, in fact, that we simply don't know we're doing it.[15]

Scenting the scene

Look around the world of high-end modernist cuisine and molecular mixology and one sees an increasing use of scene-setting scents and mood-inducing aromas. These are being added to dishes, to drinks, to the dining table and even, on occasion, to the entire dining room (especially in those situations where the chef has the luxury of serving a single sitting, with everyone eating the same course at the same time). The aim in many cases is to create a particular atmosphere or mood, or else to trigger a specific memory in the mind of the guest, no matter what they happen to be consuming. For example, Heston Blumenthal serves a moss-scented 'Jelly of Quail with Langoustine Cream and Oak Moss' dish at his flagship restaurant, The Fat Duck. Steaming scented vapour pours out from the moss-scented carpet placed in the centre of the table (see Figure 2.1). No one, I presume, starts salivating at the thought of a mouthful of the green stuff and yet the theatrical use of scent definitely helps transport the diner to

Figure 2.1. One of the fragrant dishes served at The Fat Duck. The scent of moss covers the table and fills the diners' nostrils.

another place and in so doing enhances their experience of the dish. At Alinea in Chicago, hot water is poured over a bowl of flowers when the 'Wild Turbot, Shellfish, Water Chestnuts, and Hyacinth Vapor' dish is served. The chef, Grant Achatz, is also famous for his pheasant, served with shallot, and cider gel, which is accompanied by burning oak leaves. The idea here is to use scent to trigger happy memories of an autumnal day from childhood.[16]

Of course, one has to be careful not to overuse scene-setting scents. One diner, writing on TripAdvisor about their experience at The Fat Duck, stated: 'The final dish was "Going to Bed" ["Counting Sheep"] and I think was meant to be reminiscent of being a baby, but the smell of baby talc was overwhelming and that's not an aroma one necessarily wants while eating.' While this was certainly not my recollection of the dish, the quote does highlight the potential danger for anyone trying to use a background scent that may compete with the foreground aromas of the food itself. The challenge is, in part, made worse by the fact that even ambient smells can be mislocalized to the oral cavity and experienced as tastes or flavours in the mouth if one is not careful.[17] (Yes, oral referral strikes again!)

Fortunately, the gastrophysicist has a few tips up his sleeve to show the modernist chef/molecular mixologist how best to convince their customer's brain to segregate the background environmental smell from the foreground aroma of the food and drink (assuming, of course, that that is what the chef is trying to do). Ensuring that the various smells are first encountered at different points in time will help here, as this will make it easier for the customer's brain to localize the background aroma in something other than the food or drink. And presumably Achatz keeps his oak leaves and hyacinths very visible in order to do the same. Crucially, by using such an approach, the perceived source of the scent is likely going to be correctly localized in something other than the food.

I would like you to imagine that a sugar cube is soaked in a few drops of rose oil and placed into a glass of champagne. Imagine the drink sitting in front of you, effervescing gently. The fragrant smell of an English rose garden emanating from the glass surrounds you. Before you know it, you find yourself being transported to a pleasant scent-infused summer afternoon somewhere in your memory.[18] This is exactly what top mixologist Tony Conigliaro, of 69 Colbrooke Row, wants you to experience.

Conigliaro is using scent to prime positive memories and associations. One of the specific advantages of this approach is that smell has a much closer, more direct connection with the emotional and memory circuits in our brain than any of our other senses. It turns out that the olfactory receptors in our nose are actually an extension of our brain. In fact, it is only a couple of synapses from the cells in the olfactory epithelium lining the inside of the nose through to the limbic system, the part of the brain that controls our emotions.[19] By contrast, information from the other senses has a much longer path to travel through the brain before it hits the emotion centres, and hence it can be more easily filtered out. The challenge, though, for a multi-course, smell-enhanced tasting menu is how to clear one fragrance out before the next course arrives, one of the key practical problems that eventually scuppered early attempts at scented cinema – remember Smell-O-Vision, anyone?[20]

If smell is indeed such an important part of what we taste, and if it is such an effective means of triggering our moods, emotions and memories, then any one of the innovative approaches mentioned thus far in this chapter really makes sense from the gastrophysics perspective. However, those of you who are not lucky enough to eat or drink at one of these gastronomic hotspots might be thinking: how exactly am *I* supposed to use this knowledge? In the next section, I want to share some of the intriguing ways in which we will all be exposed to this new world of augmented flavours in the coming years. For where the modernist chefs, molecular mixologists and culinary designers lead, you can be sure that food and beverage manufacturers are never far behind.

Making sense of smell

Given how important smell is to our enjoyment of food and drink, it is surprising to realize quite how many of our everyday food – and especially beverage – experiences are not optimized to deliver the best orthonasal aroma hit. Perhaps the best (or should that be worst?) example of poor olfactory design are those plastic lids that are routinely placed over millions of paper cups of hot coffee. While these lids undoubtedly allow you to drink without having to worry about spillage, what they singularly fail to do is to allow the drinker to appreciate the orthonasal aroma of the cups' contents by sniffing them. This is especially unfortunate in the case of a freshly ground cup of coffee, given that it is one of the most universally liked smells.[21] Much the same problem occurs whenever we drink directly from a bottle or can (uncouth as it may be!). Once again, it is the orthonasal olfactory hit that is mostly missing from the experience. We can either sniff the contents or we can drink them, but there is simply no way that we can do both at the same time, no matter how hard we try. And drinking through a straw – well, that is even worse!

So, having identified the problem, what can be done about it? In terms of design, there are a number of simple solutions out there. They include the reshaping of the lid, or adding a second opening to allow the coffee (or tea) lover to sniff the aroma of their favourite beverage while sipping. This is part of the innovative solution incorporated into the ergonomic lid introduced by Viora Ltd. Their novel design allows the consumer to smell their coffee without having to take the lid off. Common sense, really. But if that's the case, you have to ask yourself why it took so long for someone to come up with this solution. My suspicion is that it is all down to oral referral again. It's not obvious that retronasal smell is involved in tasting, and hence no one bothers to factor it into their designs.

Another intriguing solution comes from Crown Packaging. The company designed a can (see Figure 2.2) with a top that lifts off completely, allowing the thirsty consumer to see and, perhaps more importantly, sniff the contents more easily than when drinking from a traditional can (i.e., one with a small tear-shaped opening).

At the opposite extreme to the traditional lid, bottle or can that resolutely prevents the consumer from orthonasally enjoying the aroma of whatever it is that they are drinking, let's take the humble pint. Back in the day, when all lager beers used to taste the same,[22] the lack of any protected headspace over the drink in the glass probably

Figure 2.2. Two examples of enhanced olfactory design: the Viora lid (*left*) and Crown's 360End™ can (*right*).[23]

didn't matter much. However, given the revolution in craft beer in recent decades,[24] there are now a host of drinks out there that many of us are willing to pay a hefty premium for (because, for a change, they really do taste of something). The problem with the traditional pint glass filled (as it always is) to the rim is the lack of any protected head-space over the beer in the glass. This, in turn, means that there is no way of concentrating the drink's aromas. So, assuming we take it as read that a more intense smell is a good thing, perhaps we should be thinking again about the design of the glass in which so many pints of beer are served every day as well. So what, one might ask, would the gastrophysicist recommend here?

You might start by considering what happens in the world of wine. After all, there is always around ten times more research into wine than into anything else that we might be tempted to drink (presum-ably because researchers like drinking wine). Firstly, notice how wine glasses are never filled to the rim. Somebody out there certainly believes, rightly or wrongly, that the empty headspace over the drink is important. It is meant to help preserve the aroma and bouquet of the contents of the glass, so as to delight the taster's nostrils.[25] In fact, the better the wine, the larger the proportion of the glass that is left empty, or at least so it would seem.

One might, of course, argue that having a full pint glass doesn't really matter; after all, as soon as the thirsty drinker has taken a few draughts from their pint, won't they have created an aromatic head-space over the drink anyway, so why worry? Bear in mind, though, that more often than not the initial sniff will set up expectations about what's coming next. It is these expectations that end up anchoring, and hence disproportionately influencing, the tasting experience that follows. Isn't that first mouthful so much more important (not to mention more enjoyable) than any of the swigs you take in the middle of your pint? And both have got to be better than the last mouthful, when all you are left with are the warm flat dregs in the bottom of the glass. So, if we value the flavour and aroma – as we should – then

Figure 2.3. Lidded stein glass: an early example of intelligent olfactory design?

maybe we should all make sure a little room is left in the glass when the beer is first served.

Of course, the danger is that the average beer drinker has become so accustomed to seeing their glass filled to the brim that they might feel short-changed were they to be given anything less. An alternative solution to this olfactory problem might well simply be to dust off those old stein glasses that once came with their very own lid attached (see Figure 2.3). The purpose, at least according to one early commentator, writing back in 1886, was to protect the gases released from the beer's surface.[26] I like to think of this as clever olfactory design from 130 years ago!

How can the delivery of flavour be enhanced?

Have you ever noticed how little aroma there is in the public areas in airports? Walk into a railway station or book store and your nostrils will almost certainly be assaulted by the smell of coffee. Airports, by contrast, appear to be olfactorily neutral spaces. That said, next time you pass through London Heathrow Airport's Terminal 2, why not stop for a bite to eat at The Perfectionist's Café. If you do get the chance, then I'd recommend the fish and chips. What may surprise you about the dish is the use of an atomizer to dispense the aroma of vinegar.[27] This, as we will see below, is just one of the ways in which creative individuals are starting to change how the more aromatic elements in a dish are delivered to the table.

Over the last couple of years, London-based chef Jozef Youssef has been experimenting with the atomized delivery of aroma in a number of the dishes he serves. During his 'Elements' dinners, for instance, the mossy-earthy smell of geosmin was sprayed over each and every diner's bowl of leek consommé, leek ash and goat's cheese cream. Meanwhile, in the chef's sell-out 'Synaesthesia' dining events, it was the scent of saffron (saffranel) that was sprayed over a butter-poached lobster with white miso velouté instead. Sounds simple, right? So, why don't you try this for yourself (i.e., 'atomizing' your guests) the next time you invite them over for dinner? You never know, do it right and they may just thank you for the way their senses are awakened and for how much better their food tastes as a result. All you need is a small, clean spray bottle in which to place your 'food perfume'.[28]

It is F. T. Marinetti and the Italian Futurists, though, who really deserve the credit for first bringing atomizers to the dining table, back in the early decades of the twentieth century. Though they were more likely to spray perfume (smelling of carnations) into their diners' faces should any of them be foolish enough to look up from their plates while eating. Quite what effect this had on the multisensory tasting experience has sadly not been recorded for posterity! But the Futurists were much more about provocation than about delivering the best multisensory dining experience possible.

Today, inventive chefs and mixologists are using smoking guns to deliver the requisite aromas to the dishes and drinks they serve (see Figure 2.4). The dry-ice-based cloud pourer allows the creative flavourist to add concentrated aromas to a dish or drink in the form of a misty vapour that can be poured tableside, or at the bar, right in front of the wide-eyed and open-mouthed customer.[29] Come on now, don't be afraid . . .

Figure 2.4. A smoking gun – the chef's best friend?

You might also not have realized that aromatic packaging has been in the marketplace for years. Just take the humble chocolate ice cream bar. While everyone loves the smell of chocolate, the real stuff lacks its delicious aroma when frozen. One manufacturer even tried adding a little synthetic chocolate aroma to the glue seal of the packaging to make up for the lost smell, so that when the customer ripped open the packaging they caught a whiff of chocolate smell which they, naturally enough, assumed must have come from the chocolate covering the ice cream itself. Not everyone uses scent-enabled packaging solutions, that's for sure. Truth be told, it can be tricky to deliver an authentic-smelling encapsulated chocolate aroma.

Then, of course, there are the reports of coffee companies injecting various aromatic substances (some, so the probably apocryphal stories go, extracted from the rear end of the skunk) into the headspace of their packaged coffee. This presumably helps to explain why the experience so many of us have on first opening a packet of coffee can be so great. Your nostrils get a sharp and pungent hit of what you undoubtedly thought was great-smelling freshly ground coffee. But why, you should ask, is the experience nearly always disappointing when the container is opened again subsequently?[30]

Another intriguing example of how design can be used to modify flavour came with the launch of The Right Cup in 2016. This glass drinking vessel includes a colourful sleeve that gives off a fruity aroma. The apple-flavoured cup is bright green, the lemon yellow, and the orange – well, what else but orange-coloured? The idea of the cup is that the consumer can drink water from it and have a tasting experience that approximates to what might be expected on actually drinking the appropriate fruit juice, or at least fruit-flavoured water. I bet that the colour cue provided by the sleeve will turn out to be pretty important to the tasting experience.[31] Along somewhat similar lines, in 2013, PepsiCo applied for a patent for the use of encapsulated aroma in their drinks packaging.[32] These scented capsules would only be broken, and the aroma released, when the consumer unscrewed the lid. The thinking was that a better aroma experience could be delivered by scenting the packaging rather than the product itself.

Canadian company Molecule-R sells an Aromafork kit: for

around US$50, you get a set of four metal forks, with a bag of circular blotting papers to insert into the end of the fork, and twenty small phials of different aromas intended to augment the flavour of food with each and every mouthful. I have yet to try The Right Cup, but my experience with the Aromafork is that unless one is very careful the aroma can all too easily end up striking one as synthetic. This was certainly the response of the guests on the BBC Radio 4 show *The Kitchen Cabinet* when I tried the scent-enabled fork on them. This is not to say that we can always distinguish synthetic from natural aromas; more often than not, we can't.[33] The problem here is just that many of the aromas that are currently sold with the kit smell cheap and artificial, and most of us certainly don't like our food to taste artificial.[34]

I remain to be convinced by Molecule-R's suggestion that their innovative fork is ideal for the home chef who has forgotten to add a particular ingredient to a dish. Its most beneficial use, as far as I can see, would be to replace some particularly expensive ingredient. I can, for example, imagine dribbling a few drops of quality truffle oil, say, on to my fork, thus delivering far more culinary pleasure than if the same amount of oil were simply to be drizzled haphazardly over the food itself. The atomized scent of saffranel could be used in much the same way – saffron being, it is said, gram for gram, more expensive than gold. So are you game to try this at home? Simply put a few drops of something fragrant on to the middle of a wooden spoon or fork before using it to serve your guests. You'll be guaranteed to deliver a dining experience with a difference!

Slow food this most definitely is not. But should the consumer realize that they can get a better experience, or else the same flavour, at a fraction of the price (at least in the case of truffle, saffron and other similarly expensive ingredients), then who knows, the Aromafork – or some more aesthetically pleasing successor – might just revolutionize how we eat in years to come.*

---

* Assuming, that is, that our desire for those pricey ingredients is really about the taste, and isn't just functioning as a Veblen good – i.e., as something used to signal our wealth to others.

Ultimately, the chances of such augmented approaches succeeding in the long term are going to depend on the delivery of quality scents at a reasonable price. Remember the perceived synthetic (rather than natural) nature of the scents in the Aromafork. Once the sniffer realizes that what they are sniffing doesn't originate from their food or drink but instead comes from the cutlery, glassware or packaging, they might well start to believe that it smells artificial. Rightly or wrongly, there are frequent scares about our exposure to synthetic flavours and scents in everything from fragranced candles through to processed foods.[35] It is that belief or concern that the smell is artificial or 'chemical' – though, of course, *all* aromas are chemical – as much as the evidence before our nostrils that will, I predict, reduce hedonic ratings when sampling these products.

Have you noticed how often the modernist chef stresses, either explicitly or otherwise, the *natural* origins of their off-the-plate aromas? Be it the scent that is released when hot water is poured over the hyacinths at Alinea, or the recently deceased Homaro Cantu's use of fresh sprigs of herbs in the curly handles of his cutlery at Moto, also in Chicago. We will just have to wait and see how the consumer of tomorrow responds to this all-new world of olfactorily enhanced food and beverage packaging, glassware and cutlery. No doubt the food and beverage companies, together with the flavour houses, will take a page out of the chefs' and mixologists' book, and figure out how best to emphasize the naturalness of their novel olfactorily enhanced design solutions.

The olfactory dinner party

So far, we have focused on the use of scent to deliver both food and non-food aromas in a more effective manner, so as to enhance the multisensory flavour experience, or else to provoke a certain mood, memory or emotion. But can flavourful smells also be used to promote healthier eating behaviours? Go back to the 1930s and one finds the Italian Futurists (yes, them again) suggesting that 'in the ideal Futuristic meal, served dishes will be passed beneath the nose of the diner in order to excite his curiosity or to provide a suitable contrast, and such supplementary courses will not be eaten at all'.[36] The same notion makes an appearance in Evelyn Waugh's 1930 novel *Vile Bodies*: 'He lay back for a little in his bed thinking about the smells of food, of the greasy horror of fried fish and the deeply moving smell that came from it; of the intoxicating breath of bakeries and the dullness of buns . . . He planned dinners, of enchanting aromatic foods that should be carried under the nose, snuffed and then thrown to the dogs . . . endless dinners, in which one could alternate flavour with flavour from sunset to dawn without satiety, while one breathed great draughts of the bouquet of old brandy . . .'[37]

Smell (olfaction), then, is more important to our tasting experiences than any of us realize. That being the case, you might be thinking, why not simply enjoy the aromas of great-tasting food without the calories associated with actual consumption? That, at least, is one of the ideas behind the olfactory dinner party. However, you don't need a gastrophysicist to tell you that your cravings are never really going to be satisfied by smell alone.

And yet there are a growing number of companies out there who are starting to deliver food aroma as an end in itself. Just take the coffee inhaler: wherever you might be, you can get a caffeine hit without having to find a coffee shop. You can inhale chocolate aroma too. Culinary artists Sam Bompas and Harry Parr fit right in here, with their 'Alcoholic Architecture' installations. These UK-based jelly-mongers have experimented with a series of 'cloud bar' installations, where punters could spend fifteen minutes or so in a space infused with a gin & tonic mist. Then there is the Vaportini, a bit of kit that gently heats your drink and concentrates the aromas, so that you can just inhale them and supposedly enhance your experience as a result.[38]

This all sounds intriguing enough, but I really don't see the idea of the olfactory dinner party gaining popularity any time soon. For I doubt that our brains can be truly satisfied without our actually consuming something. As Lockhart Steele, owner/creator of *Eater.com* puts it: 'Novelty is everything in a certain corner of the dining world, no matter how fleeting . . . Dining in the dark, dining without talking – all that's left is eating without eating.'[39] And as we will see in the 'Touch' chapter, oral-somatosensory (i.e., mouth touch) and gustatory (or taste) stimulation appear key to driving our brains towards satiation (i.e., fullness).

That is not to say that enhancing the delivery of food aromas while *actually* eating and drinking (what I am calling 'augmented flavour') might not be a good idea. The female participants in one recent laboratory study became fuller sooner following the enhanced delivery of food aroma while they were eating a tomato soup. In this case, simply ramping up the olfactory component of the dish reduced people's consumption by almost 10%.[40] Thus, all of us might be able to reach satiety sooner if only we knew how to stimulate our senses more effectively. Such findings support the idea that augmenting the orthonasal aroma of food and drink could (say, through food and beverage packaging or via smelly cutlery) lead to greater enjoyment, not to mention possibly smaller waistlines.

Scents and sensibility

Have you been to a Hilton Doubletree hotel recently? If so, you will no doubt be familiar with the deliciously sweet and aromatic cookie scent that fills the lobby at check-in. Smile at the person behind the counter and they are likely to give you a freshly baked cookie. Once again, we see a desirable food aroma being combined with an unexpected (at least on a first visit) gift.[41] Definitely a good idea from the sensory marketing perspective. I must confess that I am a regular visitor, and I can't help but worry that my exposure to the high-energy sweet food scent encourages me to consume a cookie that I might otherwise not have eaten.

My grandfather, who had a grocer's shop in the north of England, would sprinkle quality coffee beans behind the counter (see Figure 2.5). When a customer came into the shop he would crush the beans underfoot, releasing the coffee aroma that would hopefully nudge them into buying some of the real stuff. So, given what I have just told you, you will understand why it does not come as any surprise to me to see food stores now releasing the enticing scent of their

Figure 2.5. My grandfather's shop in Idle, Bradford, where olfactory sensory marketing was being used intuitively half a century ago.

products in order to try to lure customers in. But is all this scent-based marketing actually provoking our desire to eat in many situations where we might not otherwise have thought about consumption?

We really do need to be more aware of the consequences of the looming commercialization of food-based scent and flavour market-ing. Those exposed to food aromas exhibit an increased appetite not only for any food that happens to be associated with that specific aroma but also for other foods and beverages that are similar in terms of their macronutrient profile. That is, exposure to one sweet high-energy food leads to our appetite for other foods with a similar aroma increasing.[42] Have you noticed how food chains tend to locate their stores in those positions in a mall that ensure the optimal disper-sion of their signature smell, and often combine this strategy with the use of the least effective extractor fan permitted, so that more of the aroma hits the consumer's nostrils.[43]

Let me leave you for now, though, with an even scarier thought. Have you ever considered what is going to happen when food runs out as a result of global warming, overfishing and food blights? Well, artists Miriam Simun and Miriam Songster have imagined how three foods that are currently threatened, namely chocolate, cod and pea-nut butter, might be enjoyed in the future. To illustrate their take on the future, they drove *The Ghost Food Truck* from Philadelphia to

New York in 2013. Those who visited this most unusual piece were given a mask to wear that allowed them to smell what it was that they were supposed to be eating. According to one commentator: 'When you get your sample, it comes with something that looks a bit like a medical breathing tube. Fit it around your face, and it holds a small bulb soaked in synthetic chocolate, cod, or peanut butter scent right next to your nose. Once you finish eating [vegetable protein and algae], the attendant pops off the bulb and takes it away, cleaning the frame for the next guest.'[44] Should this dystopian view of the future one day come to pass, then F. T. Marinetti's conception of plates of food that you would be allowed to smell but not eat from more than a century ago (see the final chapter for more on the Futurists) may turn out to be closer to the truth than any of us realized. In summary, then, there really is no escaping the value that understanding the nose brings to the way we connect to our plates.

In the next chapter, we are going to explore sight. We will take a closer look at the phenomenal rise of gastroporn and the recent emergence of *mukbang* in the Far East. What's that, I hear you ask. Well, I'm afraid you will just have to wait to find out . . .

3.   Sight

Your brain is your body's most blood-thirsty organ, utilizing around 25% of total blood flow (or energy) – despite the fact that it accounts for only 2% of body mass.[1] Given that our brains have evolved to find food,[2] it should perhaps come as little surprise to discover that some of the largest increases in cerebral blood flow occur when a hungry brain is exposed to images of desirable foods. Adding delicious food aromas makes this effect even more pronounced. Within little more than the blink of an eye, our brains make a judgement call about how much we like the foods we see and how nutritious they might be.[3] And so you might be starting to get the idea behind gastroporn.

No doubt we have all heard our tummies gurgling when we contemplate a tasty meal. Viewing food porn can induce salivation, not to mention the release of digestive juices as the gut prepares for what is about to come. Simply reading about delicious food can have much the same effect.[4] In terms of the brain's response to images of palatable or highly desirable foods (food porn, in other words), research shows widespread activation of a whole host (or network) of brain areas, including the taste (i.e., gustatory) and reward areas (the insula/operculum and orbitofrontal cortex respectively).[5] The magnitude of this increase in neural activity, not to mention the enhanced connectivity between different brain areas, typically depends on how hungry the viewer is, whether they are currently dieting (i.e., whether they are a restrained eater or not) and whether they are obese. (The latter, for instance, tend to show a more pronounced brain response to food images even when full.)[6]

Apicius, the first-century Roman gourmand and author,[7] is credited with coining the aphorism: 'The first taste is always with the eyes.' Nowadays, the visual appearance of a dish is just as important as, if not more important than, the taste/flavour itself. We are bombarded by food images everywhere, from adverts through to social media and TV cookery shows.[8] There is simply no getting away from them. Unfortunately, though, the foods that tend to look best (or rather, that our brains are most attracted to) are generally not the healthiest. Quite the reverse, in fact, as we will see later.

We may all face being led into less healthy food behaviours by all of the highly desirable images of foods that increasingly surround us nowadays. In 2015, just as in the year before, food was the second most searched-for category on the internet (after pornography). Thus the blame, if any, doesn't reside solely with the marketers, food companies and chefs; a growing number of us are actively seeking out images of food – a kind of 'digital foraging', if you will. How long, I wonder, before food takes the top slot?

Can you taste the colour?

What we taste is profoundly influenced by what we see. Similarly, our perception of aroma and flavour are also affected by both the hue (i.e., red, yellow, green, etc.) and the intensity, or saturation, of the colour of the food and drink we consume.[9] Change the colour of wine, for instance, and people's expectations – and hence their tasting experience – can be radically altered. Sometimes even experts can be fooled into thinking that they can smell the red wine aromas when given a glass of what is actually white wine that has just been coloured artificially to give it a dark red appearance![10]

At various points in history, scientists (some of them really rather eminent, like one of the godfathers of psychology, Hermann Ludwig Ferdinand von Helmholtz – anyone with a name that long should probably not be messed with!) have confidently asserted that there is absolutely *no* association between colour and taste. At the other extreme, there are some artists out there today who are inviting the public to 'taste the colour'.[11] I have to say that I don't think either side has got it quite right. Colours are very definitely linked with tastes, and yet I do not believe that you can create a taste out of nowhere, simply by showing the appropriate colour.

Consider the amuse bouche created by the London-based chef

Jozef Youssef as part of Kitchen Theory's 'Synaesthesia' dining events, which developed out of the latest research findings from my Crossmodal Research Laboratory. We have spent the last few years researching what tastes people around the world associate with different colours, and looking at the colours that people naturally/ spontaneously associate with the four most frequently mentioned basic tastes.[12] The results fed directly into the design of the chef's dish. Had you been a guest at the restaurant, you would have had four spoons of espherified colourful taste placed randomly down in front of you – one red, one white, one green and one browny-black. Once everyone has their four spoons, they are informed that the chef recommends starting with the salty, then the bitter spoon, next the sour, and ending on a sweet note – leaving you and the other diners a little perturbed as to which order exactly you should taste the spoons in. The idea is that the diners arrange their own spoons from left to right in front of them in the order: salty, bitter, sour and sweet. Having arranged their own spoons, the diners normally start to look around and compare notes with each other. In the restaurant, or online, we get somewhere around 75% of people ordering the spoons in the way that the chef (and the gastrophysicist) intended. So on the basis of such results, I would say that tastes are very definitely associated with specific colours.[13]

Colour can be used to modify people's perception of a taste that is already present in the mouth. I can, for example, make food or drink taste sweeter by adding a pinkish-red colour, but, as yet, I haven't come across a way to do that while serving someone a glass of water.[14] Turning water into wine – now that's still a step too far, even for the gastrophysicist at the top of their game (though, if you remember, in the previous chapter, that was pretty much what the inventors of The Right Cup were promising to do)! Nevertheless, a food or beverage company might be able to increase perceived sweetness by up to 10% by getting the colour of their product, or the packaging in which it comes, just right. Every little counts, as they say.

Some people want to know whether the effects of adding colour are like adding sugar. Surely psychologically induced sweetness must taste different from the chemically induced kind? Well, the results of side-by-side tests show that people will sometimes rate an appropriately coloured drink (imagine a pinkish-red drink) as sweeter than an inappropriately coloured (say, green) comparison drink. Such results can be obtained even if the latter drink has as much as 10% more added sugar. In other words, psychologically induced taste enhancement is indeed indistinguishable from the real thing, at least sometimes. Sweetness without the calories – now who wouldn't want that?[15]

Our responses to colour in food and drink aren't fixed but change over time. For instance, a few decades ago, marketers and cultural commentators were telling anyone who'd listen that blue foods would *never* sell.[16] Roll the clocks forward a few decades, though, and we now have cool blue Gatorade, Slush Puppy and the London Gin Company all successfully promoting blue drinks. A Spanish company even launched a blue wine in 2016.* Given the rarity of this colour in nature, it is normally introduced solely as a marketing ploy to capture the attention of consumers by standing out on the store shelf. The problem often comes later, though, when people actually get to taste that eye-catching drink. Seeing a transparent blue colour will put a certain idea in the consumer's mind about the likely taste. And if the expectation doesn't match up with the reality, the manufacturer may well have a problem on its hands.[17]

In fact, you would probably be surprised to learn how many companies have come looking for help over the years, when their consumer panels and focus groups tell them that their brands taste different, even though all that has changed is the colour of the product or pack. For example, a mouthwash manufacturer told me that their orange variant didn't taste as astringent to people as their regular blue variety, despite the formulation of the active ingredients staying the

---

* This is such a bad idea that I am pretty sure it won't still be around by the time you read this. Maybe those old marketers weren't completely wrong after all. The drink is supposedly targeted at millennials who, so we are told, like brightly coloured alcopops, etc. But I am really not sure that they will want to be seen drinking blue wine. Ironically, the Italian Futurists used to serve their guests blue wine to shock them. And to think that someone now believes it will sell!

same. It makes no sense until you learn something about the rules of multisensory integration governing how the brain combines the senses. Here, I am thinking of 'sensory dominance' – where the brain uses one sense to infer what is going on in the others.

The impact of colour depends on the food. In the context of meat and fish, blue causes a distinctively aversive response. In one of my favourite 'evil' experiments – the sort of thing that ethics panels were surely brought in to put a stop to – a marketer by the name of Wheatley served a dinner of steak, chips and peas to a group of friends. At the start of the meal, the only thing that might have struck anyone as odd was how dim the lighting was. This manipulation was designed to help hide the food's true colour. For when the lights were turned up part-way through the meal, Wheatley's guests suddenly realized that they had been tucking into a steak that was blue, chips that were green and peas that were bright red. A number of them apparently started to feel decidedly ill, with several heading straight for the bathroom![18]

Can you taste the shape?

This is another of those questions to which the intuitive answer has to be: 'But of course not.' And yet, over the last decade I have had lots of fun at food and science festivals around the world, giving people different foods to taste while asking them to tell me whether some aspect of the tasting experience is more 'bouba' or 'kiki'? If you don't have the faintest idea what I am talking about, don't worry. Just take a look at the shapes shown at either end of the scale in Figure 3.1, and ask yourself which one would you give each of those made-up names. Most people say that 'kiki' must be the angular shape while 'bouba' is obviously the rounded blob.

Next, I would like you to imagine the taste of dark chocolate, Cheddar cheese or sparkling water. Place an imaginary mark on the scale that, in some sense, 'matches' the sensory properties of the experience of each of those foods. Sounds silly, right? But just try it. Do the same thing for milk chocolate, a nice ripe piece of Brie or still

Figure 3.1. Do tastes, aromas and flavours have shapes? This figure illustrates the simple shape symbolism scale that we often use in our research. The crayon indicates the mid-point of the scale.

water – the choice is yours. My gastrophysics research indicates that you will probably have marked those last three more towards the left (bouba) end of the scale, whereas the other three will be more towards the right.[19]

What's fascinating about this kind of research is just how consistent people's answers are (given that there are no objectively correct answers to these questions). Most people will place carbonated, bitter, salty and sour-tasting foods and drinks towards the angular end of the scale. Sweet and creamy sensations, by contrast, are nearly always paired with rounder shapes.[20] In other words, we all seem to show a generalized tendency to want to match specific shapes (or contour) with particular tastes, aromas, flavours and even food textures.[21]

Why is it that people associate shapes with tastes at all? One evolutionary account for this is that angular shapes, bitterness and carbonation are all somehow linked with danger or threat. An angular shape could be a weapon, bitterness might indicate poison, while sourness and carbonation would, once upon a time at least, have been an avoidance cue signalling overripe or spoiled foods. By contrast, sweet and round both have positive connotations. So maybe we just choose to put together those stimuli that we feel the same way about, even if we have never experienced them together before.[22] Alternatively, there may be something in the environment – some correlation between shape and taste properties – that we (or rather, our brains) are picking up on. I sometimes wonder whether the acidity in cheese might be correlated with its texture. Could it be that harder cheeses (i.e., those that are more likely to retain their angular form on cutting) are, on average, more acidic than soft and runny cheeses? Does dark chocolate tend to break into more angular shapes than milk chocolate?[23]

Next time you are in the supermarket, take a look at the beer or bottled water shelves. The majority of beer and carbonated water brand logos are angular, not round. There are exceptions, of course, but it is remarkable how often you see a red star or triangle on the front of a bottle or can – the many red stars that adorn the San Pellegrino sparkling water bottle, or the prominent red star on the Heineken label, for example.[24] See how the food and beverage industry are communicating with you in ways that are functionally subliminal? But apart from the marketing angle, just how important is knowing about shape symbolism? Well, I have always been most interested in what happens to taste when you start manipulating the shape of food itself. Back in 2013, Cadbury decided to update the shape of the iconic Dairy Milk chocolate bar by rounding off the corners, reducing the weight of the chocolate by a few grams in the process. Consumers wrote in and called in droves to complain. They were convinced that the company had changed the formula – that their favourite chocolate now tasted sweeter and creamier than before. But a spokesperson from Mondelez International (the American owner of Cadbury) stated: 'We have been very clear and consistent that we have not changed the recipe of the much-loved Cadbury Dairy Milk, although it's certainly true that we changed the chunk last year from the old, angular shape to one that's curved.'[25] So you can imagine how companies could (if sugar weren't so cheap) perhaps reduce the sugar content while making the shape rounder. The product itself would be a little less unhealthy, and it's just possible that the taste wouldn't change in the mind of the consumer either. It is a challenge, obviously, to get companies to reformulate their products effectively, but it can be done.

What is true of chocolate turns out to apply to other foods as well. When you serve food, be it a beetroot jelly or a chocolate confection, in a rounder shape, people often rate it as tasting sweeter than when exactly the same food is presented in a more angular form instead.[26]

In fact, even categorizing angular (rather than round) shapes on a piece of paper just before eating influenced the rated sharpness of a piece of Cheddar cheese in one North American study.[27] Similarly, varying the shape made by the chocolate sprinkled on top of a caffè latte can be all it takes to modify the drinker's taste *expectations*. We have been working with baristas in Australia to show that people expect a caffè latte with a star shape sprinkled on top to taste a little more bitter than a latte with a rounded, bouba-esque shape sprinkled on top.[28] Though whether that change in expectations is sufficient to alter the perceived taste depends on how close the taste of the drink is to what was predicted by the customer.

When the taste is pretty much as expected, the shape may well influence what people say about the taste. However, if there is too much of a divergence, the brain seems to discount the shape cue altogether.[29] That, at least, is our current hypothesis in the lab. And, of course, even when the shape does change the taste, whether that results in increased liking or not is going to depend on the taste of the food itself, and on the taster. While sweetness is normally liked, we conducted one study in a Scottish hotel restaurant where an overly sweet dessert was actually liked less when it was served on a 'sweeter'-looking plate.[30]

Can you taste the plate?

Most people will simply shake their heads when you ask them whether changing the colour of a plate would influence the taste of the food on it. However, in research conducted together with the Alicía Foundation in Spain, exactly the same frozen strawberry mousse was rated as tasting 10% sweeter and 15% more flavourful, and was liked significantly more, when eaten from a white plate rather than from a black plate instead (see Figure 3.2). Remarkably, follow-up research by scientists working in Greenland obtained even more striking results by varying both the colour and the shape of the plate. Strange though it may seem, round plates are just 'sweeter' than angular plates![31]

In other research, we have been able to modify the taste of everything from a vended cup of hot chocolate to a caffè latte by varying

Figure 3.2. White and black plates with red frozen strawberry dessert. The funny thing is that people rated the dessert from the white plate as tasting significantly sweeter and more flavourful than exactly the same food served on the black plate. Such results certainly make me wonder about the sense of using different coloured charger plates, as in Denis Martin's namesake restaurant in Vevey, Switzerland (remember the cow?).

the colour of the cup. Hot chocolate tastes more, well, chocolatey, and is liked significantly more, when served from an orange plastic cup (rather than a white one). Meanwhile, caffè latte is judged more intense while the sweetness is somewhat suppressed when served in a white porcelain rather than a clear glass mug.[32]

Intriguingly, gastrophysics research also shows that enhancing the visual contrast on the plate can lead to a substantial increase in food and liquid intake in those suffering from advanced Alzheimer's disease. In one study conducted at a long-term-care facility in the US, for instance, switching to high-contrast coloured plates and glasses led to a 25% increase in food consumption and liquid intake going up by as much as 84%![33] The results of another hospital study were equally dramatic: average consumption amongst the older and more vulnerable patients, including those suffering from dementia, went up by 30%, just by changing the plate colour! So impressed were they with the results in this case that the hospital went on to replace their standard-issue white plates with blue crockery.[34]

But hold on a minute, I hear some of you say. There is something that doesn't quite make sense here. What about the 'blue plate special'? As the wonderfully named Bunny Crumpacker put it in her book *The Sex Life of Food*: 'the term blue plate special became popular during the Great Depression because restaurant owners found that diners were satisfied with smaller portions of food if it was served on blue plates.'[35] So how can it be that blue plates reduced consumption back in the 1920s but significantly increase how much patients eat today? One explanation might be that the majority of the food served in hospitals tends to be bland in both taste and colour, so, it simply fails to stand out against a white plate. By contrast (boom-boom), served on a blue plate it is suddenly much easier to see what one is eating. This is precisely why, at home, I used to like serving my signature Thai green chicken curry (greenish-white in colour) with white rice on a black plate: the visual contrast was much more striking. Or as one article I read put it, serving steak on a white plate is fine, but porridge never should be. Though, sadly, the delightful Mrs Spence isn't of the same opinion. She thought the angular black plates were a little too batchelor-ish, so out they went. So much for gastrophysics!

Even to a non-gastrophysicist, presenting hospital food on a red plate or tray must seem intuitively wrong. There is a justification for this approach, of course. It is supposed to help the relevant healthcare professionals more easily identify those patients needing special nutritional attention. I suspect that it is a bad idea, though. Why? Well, because the colour red tends to elicit avoidance motivation. What this means in practice is that people eat significantly less when served food on a red plate than when offered exactly the same food from plateware of another colour. And the effects aren't small either.

In one study, people consumed almost twice as many pretzels (under lab conditions) when they ate from a white plate than from a red plate.[36] There seems no good reason to imagine that serving food on a red tray, rather than from a tray of a different colour, wouldn't trigger exactly the same kind of avoidance motivation. So red plates and trays might be recommended for anyone who wishes to lose weight, but this is simply not the situation that most hospital patients find themselves in.[37]

So, while you can't literally taste the plateware on which food is served, its colour (not to mention its size) is likely to modify your behaviour. Perhaps it might bias you to eat more (or less) than you otherwise might have done.[38] What is more, it is also likely to influence your experience of whatever is served, possibly making it seem more delicious, sweeter or more flavourful.

The effect of background colour on taste and flavour perception causes all manner of problems for the food and beverage industry, as a number of companies have found to their cost. Simply altering the can colour – say, adding a little more yellow to the side of 7Up (as was done in the 1950s), or launching a white Christmas Coke can (as happened in 2011) – can change what consumers think about the taste.[39] Of course, modifying the colour of the packaging shouldn't affect the taste of the contents, especially for a brand that people are already familiar with. And yet the evidence suggests that it most certainly does! The phenomenon provides an example of visual dominance (of the packaging colour) – which may, in fact, be the only product-relevant sensory cue that the consumer has to go on before they start drinking directly from the can.

Now, take a look at the plate shown in Figure 3.3. What do you think serving dinner from such an odd bit of plateware will do to the experience? Given the emerging gastrophysics research, no one should be surprised by the explosion of interest in enhancing people's enjoyment of food by breaking away from the rigid tyranny of the large, round, white plate (what some refer to as the American plate). No matter whether you are a diner or a chef, you can benefit from paying a little more attention to the latest gastrophysics research and to the plateware itself, not just what is on it.

It used to be that only the world's top restaurants served each of the dishes on their multi-course tasting menus from dedicated plateware but this is now slowly percolating down through chain restaurants all the way to the adventurous home chef. Are we reaching the point where the fact that a dish looks good (or photographs well) is becoming more important to its success than how it tastes? This growing trend towards gastroporn has been parodied by Canadian chef Carolyn Flynn, aka 'Jacques La Merde'. La Merde has an Instagram site with over 100,000 followers which shows food picked up from the

Figure 3.3. A fabulous piece of plateware, as used at the Montbar and Tickets restaurants in Barcelona. Serving food on one of these plates may well change what people say about the taste. Ice cream, for instance, might taste that little bit sweeter when served on this than if the same product were to be served on an angular black slate.

convenience store or fast-food restaurant – i.e., junk food like crumbled Oreos and Doritos – plated up so as to look like it came from a top Michelin-starred restaurant.[40]

Ideally, of course, you want both. Doesn't everyone want food that not only looks beautiful but tastes great too? There are a couple of important challenges here for the gastrophysicist. One is to develop robust experimental tests to assess the impact of changes in plating on people's perception (and how much they would be willing to pay). But the gastrophysicist can also provide theoretical insights into the aesthetics of plating (i.e., not just analysing what the chefs have done to see what people like but also making predictions/recommendations about the sorts of configurations that the diner ought to like, based on what we already know about the brain). Taken together, these will hopefully enable those who are preparing really great-tasting food to come up with plating arrangements that have maximum visual appeal as well. Delivering on that promise is easier when the chef combines their skills in the kitchen with the latest gastrophysics testing techniques and insights.[41]

When did '*food porn*' first appear and where is it heading?

People have been preparing beautiful-looking foods for feasts and celebrations for centuries.[42] (And, of course, artists have for centuries captured them in all those still life paintings too.[43]) However, for anything other than an extravagant feast, the likelihood is that meals in the past would have been served without any real concern for how they looked. That they tasted good, or even just that they provided some sustenance, was all that mattered. Now, this was true even of famous French chefs, as highlighted by the following quote from Sebastian Lepinoy, Executive Chef at L'Atelier de Joel Robuchon, describing the state of affairs before the emergence of nouvelle cuisine: 'French presentation was virtually non-existent. If you ordered a coq au vin at a restaurant, it would be served just as if you had made it at home. The dishes were what they were. Presentation was very basic.' [44]

Everything changed, though, when East met West in the kitchens of the French cookery schools in the 1960s. It was this meeting of culinary minds that led to nouvelle cuisine and, with it, gastroporn. The latter term dates back to a witty review written in 1977, describing Paul Bocuse's *French Cookery* cookbook as a 'costly ($20.00) exercise in gastro-porn'.[45] The term stuck, and has now made its way into the *Collins English Dictionary*, which defines it as 'the representation of food in a highly sensual manner'. Others prefer the term 'food porn'.[46] Make no mistake, though, they are all talking about the same thing.

These days, more and more chefs are becoming concerned (obsessed, even) by how their food photographs. And not only for the glorious full-colour, full-page pictures that will adorn the pages of their next glossy cookbook. As one restaurant consultant put it: 'I'm sure some restaurants are preparing food now that is going to look good on Instagram.'[47] Certainly, chefs who serve something visually stunning and plated beautifully, or else plated on or in something most unusual, like a brick, trowel or flat cap, can give their dishes some real eye-appeal, and hence hopefully help grow their digital presence.

Figure 3.4. Foodography – it's all about getting the perfect shot. The curved plate and smartphone stand make taking a great photo a cinch.

Some chefs have been struggling with how to deal with the growing trend for diners to photograph their food and share it on social media. Their much publicized responses include everything from limiting their diners' opportunities to photograph the food during the meal through to banning photography inside the establishments they preside over completely. The latter approach seems doomed to failure, for you just can't fight the tides of change. Best to get with the program and figure out how to adapt your food offering for the growing number of experience-hungry millennials out there who want to share their every waking moment with their social networks. It would, however, seem as though the chefs have now, mostly, embraced the trend, acknowledging that it is all part of 'the experience'. As Alain Ducasse, chef at London's three-Michelin-starred Dorchester Hotel says: 'Cuisine is a feast for the eyes, and I understand that our guests wish to share these instants of emotion through social media.'[48]

When thinking of a more constructive approach to resolving this looming technology-driven crisis in high-end cuisine, just consider those cutting-edge restaurants that have now started to incorporate specially shaped plates into their service in order to provide the perfect backdrop for the food in their diners' photos. Others, like Foodography, at Catit, a restaurant in Tel Aviv, Israel, offer their diners camera stands at the table, or else serve their food on plates that spin 360 degrees, thus promising their customers the perfect shot (see Figure 3.4).[49]

Spinning plates might well strike you as a bit over the top. But then again, haven't we all subtly, almost unconsciously, rotated the plate once or twice after a waiter has placed a dish in front of us in a restaurant? Have a look at the two plates of food shown in Figure 3.5 and tell me which one you prefer. The left one, right? Yet all that differs is the orientation! Our latest online research shows that people's preference for a plate of food can vary quite dramatically simply as a function of the orientation in which it is presented.[50] In our online experiments, for instance, we find that people are willing to pay significantly more for exactly the same food when shown in one orientation as compared to another. Now that you know this, don't you want to optimize the orientation of the plate every time you serve food? If such a simple step can genuinely enhance the

Figure 3.5. The same plate of food served two ways. Several thousand people have been spinning this particular plate, the signature dish from Brazilian chef Albert Landgraf, into the perfect orientation (over the internet, in a massive citizen science experiment). Based on the results, we now know that the majority of people prefer the plate shown on the left. Ideally, the onions should be plated so that their tips point 3.4 degrees past 12 o'clock. The chef had decided to plate it at 12. So the chef intuitively got it more or less spot on. No need for the gastrophysicist, other than for confirmation, in this case!

perceived value and enjoyment of the food you serve, no matter who you are and no matter whom you are serving, you'd be crazy not to.

Beyond that, there is a sense in which the visual appeal of the meal has become an end in and of itself, with a growing number of people taking images of what they eat.[51] Certainly, these days the press is full of tips on how to make your food photography more visually attractive, offering, as one newspaper did recently, to help you 'turn your dull food images into Instagram food porn in 12 simple steps'.[52] Researchers and food companies have, in recent years, begun to establish which tricks and techniques work best in terms of increasing the eye-appeal of a dish, including, for instance, showing food, and especially protein, in motion (even if it is just implied motion) to attract the viewer's attention and convey notions of freshness. The gastrophysicist knows only too well just how important this is: a beautifully plated dish is likely to taste better than it might if the elements were randomly plonked on the plate. At the same time, I and many of my colleagues who work as chefs worry that this increase in eye-appeal sometimes comes at the cost of the optimization of the actual flavour of the dish.[53]

Have you heard of 'yolk-porn'?

What do you get if you show protein (e.g., oozing egg yolk) in motion? Answer: yolk-porn. Seriously! This is the all-new trend in addictive food images (see Figure 3.6).[54] I came across an example recently in a London underground station. There were a whole host of video advertising screens along the wall as I ascended the escalator. All I could see, out of the corner of my eye, was a steaming slice of lasagne being lifted slowly from a dish, dripping with hot melted cheese, on screen after screen. As the marketers know only too well, such 'protein in motion' shots are very attention-grabbing; our eyes (or rather our brains) find them almost irresistible. Images of food (or more specifically, energy-dense foods) capture our visual awareness, as does anything that moves.[55] 'Protein in motion' is therefore precisely the kind of energetic food stimulus that our brains have evolved to detect, track and concentrate on visually.

Figure 3.6. Just a toast soldier being harmlessly dipped into an egg yolk, do you think, or something rather more pernicious?

The British retailer Marks & Spencer have acquired something of a reputation for their food porn with much of their advertising over the last decade or so including highly stylized and gorgeously presented visual imagery. Look closely at their ads and one finds plenty of protein in motion (both implied and real).[56] Their most famous ad, from 2005, was for a chocolate pudding shown with an extravagant melting centre. A sultry voiceover came out with the iconic – though subsequently much parodied – line: 'This is not just chocolate pudding, this is a Marks & Spencer chocolate pudding.' But do you have any idea what this one ad did to sales? They skyrocketed by around 3500%!

What was especially noticeable about M&S's 2014 campaign (which paid homage to the original gooey pudding) was that *all* of the food was shown in motion. In fact, one of the most widely commented on images was of a Scotch egg being sliced in half, with the yolk oozing out slowly. According to one company executive, the ad showcased 'the sensual and surprising aspects of food – like its textures and movement – in a modern, stylish and precise format'. And M&S certainly aren't the only ones to have adopted this approach. An informal analysis of the food commercials shown in the highly valued advertising slots during the US Super Bowl revealed that in the years 2012–14 two-thirds showed food in motion.[57] The attraction that we all feel towards images of food in motion might also help to explain the viral explosion of interest in one particular video of a melting chocolate dessert that swept the internet recently, described by a number of journalists as simply 'hypnotic'.[58]

One other reason to show food in motion (real or implied) is that it looks more desirable, in part because it is perceived to be fresher. For instance, food psychology and marketing researcher Brian Wansink and his colleagues at Cornell University conducted research showing that we rate a picture of a glass of orange juice as looking significantly more appealing when juice can be seen being poured into the glass than when the image is of a glass that has already been filled. Both are static images but one *implies* motion. That was enough to increase the product's appeal.[59] (For those of you at home, who may not be able to guarantee that your food moves, another strategy is simply to leave the leaves and/or stems on fruit and vegetables, to help cue freshness.)

Mukbang

Now let me tell you, dear reader, about one of the strangest trends relating to food porn that I have come across in recent years; it's called *mukbang*.[60] A growing number of South Koreans are using their mobile phones and laptops to watch other people consuming and talking about eating food online. Millions of viewers engage in this voyeuristic habit, which first appeared back in 2011, every day. Interestingly, the 'porn' stars, or 'broadcast jockeys' (BJs) as they prefer to be known, are not top chefs, TV personalities or restaurateurs but rather just regular (albeit generally photogenic) 'online eaters' (see Figure 3.7). One can think of this as yet another example of food in motion; it's just that the

Figure 3.7. *Mukbang* – it roughly translates as 'food porn'. The live-streaming of people eating food started in South Korea in 2011 and now attracts millions of viewers.

person interacting with the food happens to be more visible than in many examples of dynamic food advertising in the West, where all you see is the food moving – just think of those M&S ads. I also get the sense, though, that some of those who, for whatever reason, often end up eating alone are tuning in for a dose of *mukbang* at mealtimes to get some virtual company (see the 'Social Dining' chapter).

It would be interesting to see whether those who eat while tuning in to one of these increasingly popular BJs end up consuming more than they would were they really eating alone (i.e., without any virtual dinner guests). One might also wonder if *mukbang* is as distracting as regular television, which has been shown to dramatically increase the amount consumed.[61] If so, one might expect not only that the viewer's immediate intake of food will increase but also that the amount of time that passes before they get hungry again ought to be reduced.

With *mukbang*, viewers can imagine themselves having a meal with whoever they see on the screen. However, research shows that food imagery is most visually appealing when the viewer's brain finds it easy to simulate the act of eating, i.e., when the food is seen from a first-person perspective. This is rated more highly than viewing food from a third-person view (as is typically the case with *mukbang*).[62] Marketers, at least the smarter ones, know only too well that we will rate what we see in food advertisements more highly if it's easier for our brains to mentally simulate the act of eating that which we see. So, for example, imagine a packet of soup. A bowl of soup is shown on the front of the packaging. Adding a spoon approaching the bowl from the right will result in people being around 15% more willing to purchase the product than if the spoon approaches from the left instead. Why so? Well, it's because the majority of us are right-handed, and so we normally see ourselves holding a spoon in our right hand. Simply showing what looks like a right-handed person's spoon approaching the soup makes it easier for our brains to imagine eating. Now, for all those lefties out there saying, 'What about me?' – it may not be too long before the food ads on your mobile device might be reversed to show the left-handed perspective. The idea is that this will help maximize the adverts' appeal to you (assuming, that is, that your technology can figure out your handedness implicitly).[63]

How worried should we be by the rise of food porn?

Is there anything to be worried about here, or is it all just a storm in a teacup (or should that be Twitter-feed)? Why shouldn't people indulge their desire to view all those delectable gastroporn images. Surely there is no harm done? After all, food images don't contain any calories, do they? Well, it turns out that there are a number of problems that gastrophysicists have proven and that I think we should be concerned about:

1. *Food porn increases hunger.* One thing that we know for certain is that viewing images of desirable foods provokes appetite. For example, in one study, simply watching a 7-minute restaurant review showing pancakes, waffles, hamburgers, eggs, etc. led to increased hunger ratings not only in those participants who hadn't eaten for a while but also in those who had just polished off a meal.[64] As Italian researchers put it: 'Eating is not only triggered by hunger but also by the sight of foods. Viewing appetizing foods alone can induce food craving and eating.'[65]

2. *Food porn promotes unhealthy food.* Some of Nigella Lawson's delicious-looking cakes, the ones you see her making on TV, contain in excess of 7000 calories. In fact, many of the recipes that top chefs make on television cookery shows are incredibly calorific or unhealthy (despite chefs' pronouncements about healthy eating in the press!). Indeed, those who have systematically analysed the TV chefs' recipes find that they tend to be much higher in fat, saturated fat and sodium than is recommended by the World Health Organization's nutritional guidelines.[66] This is not only a problem for those viewers who go on to make the foods that they see their idols making. (Surprisingly few people actually do this; maybe it is not such a problem after all. According to a recent survey of 2000 foodies, fewer than half had ever cooked even one of the dishes that they had seen prepared on food shows.) Rather, the bigger concern here is that the foods we see being made, and the food portions we see being served on TV may set implicit norms for what we consider it appropriate to eat at home or in a restaurant.[67]

3. *The more food porn you view, the higher your Body Mass Index (BMI).* While the link is only correlational, not causal, the fact that people who watch more food TV have a higher BMI might nevertheless cause you to raise your eyebrows. They might, of course, be watching more television generally, not just food programmes – after all, the term 'couch potato' has been around for longer than the term 'food porn'. The key question, though, from the gastrophysics perspective, is whether those who watch more food television have a higher BMI than those who view an equivalent amount of non-food TV. It would certainly seem likely to be the case, given all the evidence out there showing that food advertising biases subsequent consumption, especially in kids.[68]

4. *Food porn drains mental resources.* Whenever we view images of food – on the side of product packaging, in cookery books,

or increasingly on TV shows or social media sites like
Instagram's 'Art of Plating' – our brains can't help but
engage in a spot of embodied mental simulation. That is,
they simulate what it would be like to eat the food. At some
level, it is almost as if our brains can't distinguish between
images of food and real meals. We therefore need to expend
some mental resources, silly though it may sound, to resist
all of these *virtual* temptations. So what happens when we
are subsequently faced with an actual food choice? Imagine
yourself watching a TV cookery show (such as on The Food
Network) and then arriving at a train station; the smell of
coffee wafting through the air leads you by the nose into
buying a cup. While at the counter, you see the sugary snack
bars and fruit laid out in front of you. Should you go for a
bar of chocolate, or pick a healthy banana instead? In one
laboratory study, participants who had been shown
appealing food images tended to make worse (i.e., more
impulsive) food choices afterwards than those who had been
pre-exposed to a smaller number of food images.[69] All of
this (increased and increasing) exposure to desirable food
images results in involuntary embodied mental simulations.
That is, our brains imagine what it would be like to
consume the foods that we see, even if those foods are only
on the TV or our smartphones, and we (or our brains) then
have to try to resist the temptation to eat. One recent study,
conducted in three snack shops located in train stations,
investigated whether people could be nudged towards
healthier food choices simply by moving the fruit closer to
the till than the snacks – the reverse normally being the
case. The 'nudge' worked in the sense that people were
indeed more likely to buy a piece of fruit or a muesli bar.
Unfortunately, though, people continued to purchase snacks
like crisps, cookies and chocolate *as well*. In other words, an
intervention that had been designed to *cut* people's
consumption actually resulted in their consuming *more*
calories (assuming that what was purchased was consumed)![70]

Can I engage in a spot of gastroporn at home?

The good news is that labs like mine have been researching ways to turn this visual attraction to food into a positive, so that we can learn to make healthier meals at home. Any one of us can make the food we serve more attractive by plating for the eye. In one of the studies conducted here in Oxford, for example, we served a salad in a college dining room to 160 diners either just as a regular tossed salad, or with the elements arranged to look a little like one of Kandinsky's paintings. The results showed that the diners were willing to pay more than twice as much for exactly the same food when it looked more visually attractive (see Figure 3.8). The complete Kandinsky salad might look daunting to try to create at home (given that it has more than thirty elements). But even a typical UK at-home meal of steak and chips and a garden salad (with only three ingredients) can be made significantly more desirable by paying a little more attention to the presentation.[71]

Figure 3.8. Same food, alternative plating. Diners were willing to pay twice as much for the food shown on the left despite it having exactly the same ingredients as the less artfully arranged salad on the right. Once you realize how much difference even simple changes to the eye-appeal of a dish can make, it seems crazy not to take a few tips from the gastrophysicists and serve something that is optimally visually attractive. What have you got to lose?

Another trick here is to use the visual appearance of food to help yourself eat less. One idea is to serve it off a smaller plate (thus making it look like there is more food). One should avoid serving food on a plate that has a wider rim too.[72] Even looking at the picture of the bowl on the front of a packet of breakfast cereal can influence how much we think it appropriate to serve ourselves (i.e., it acts as a serving/consumption norm). Show the cereal in a bowl with a rim and it will look like a smaller portion than if shown in a rimless bowl. In some of our latest research, we have been able to demonstrate that showing a given amount of cereal in a bowl with no rim versus a wide rim really can make a difference. According to researchers from Oxford and Cambridge, reducing the size of the plate or bowl we eat from reduces average calorie intake by roughly 10% (or 160 calories).[73]

Imagined consumption

Imagine yourself eating lots of M&Ms. Do you think that this could affect how much you would end up eating if offered a bowl of the brightly coloured sweets afterwards? Research shows that you would eat significantly fewer of them if they were subsequently offered to you. Thus, the *imagined* consumption of food can reduce actual consumption.[74] Those of you on a diet, take note! Simply by repeatedly imagining the act of consumption, you are likely to then eat a little less. This habituation effect is, however, food specific. Getting people to imagine eating a different food, like cheese, doesn't, I'm afraid, suppress the desire to eat, say, chocolate.

Similarly, getting people to remember the last meal that they ate also suppresses subsequent snacking behaviour. It is, of course, theoretically possible that just seeing a food once could result in the viewer imagining the act of eating it, and hence give rise to reduced consumption due to habituation. However, in practice this does not seem to be the case, presumably because no one in their right mind would spend their time imagining eating the same food sixty times in a row! (Or just thirty times in one of the other studies.) Instead, the exposure to palatable food images (in the absence of instructions to repeatedly imagine eating the seen food) generally, though not always, leads to increased consumption of whatever food is made available thereafter. Though I do wonder whether, in much the same way that watching pornography may reduce a viewer's sexual activity, there might be situations under which the same could be true for at least some of those individuals who watch large amounts of food porn – i.e., they use it to in some sense control their cravings for the real stuff.

Ugly fruit

Those who are really serious about losing some weight could do worse than to make their food look as ugly as possible.* At the opposite extreme we have enhanced 3D virtual reality food blogs (like Perception Fixe, by Matheus DePaula-Santos of Myo Studios) whose aim is to make food look as desirable as possible. According to one journalist: 'Myo Studios is banking on the notion that providing an enhanced visual experience through virtual reality will markedly up its food blog's ante. Users will be able to "sit down in front of a steak from some restaurant, even though there's no reservation for three months." DePaula-Santos told me, "One of my hopes is to not just take photographs of food, but also be able to animate it. If you see a sizzling steak in front of you, that's just one way of stimulating more senses." '[75]

So, has there, one might ask, been any reaction to all this focusing on the beautiful in food? Well, I suppose there has, though it only started recently and is still small-scale. You may have heard that various supermarket chains are now selling boxes of wonky fruit. This is undoubtedly a good idea, both to help avoid all that food waste and because, in general, the more beautiful a source of food looks, the less aromatic it tends to be. It is a bandwagon that celebrity chefs like Jamie Oliver are jumping on to as well.[76] Of course, the fact that most

---

* If you need some inspiration in this regard, there is a Tumblr site called 'Someone Ate This', dedicated to awful-looking food. Why don't you check it out, and see if you are still feeling hungry after viewing some of the truly ghastly meals shown there.

consumers reject bruised-looking fruits and unusually shaped vegetables highlights, once again, the importance of eye-appeal.

Speaking for myself, as the son of a one-time greengrocer I know
only too well that the best banana cakes are always made with the
ugliest, blackest fruit. The family joke is that us kids never realized
that bananas came in any colour other than black, as my father always
brought home the leftover produce that no one else wanted to buy,
the irony here being that the blackened fruit tend to be the most
flavourful.[77]

Seeing sense

Our brains have evolved to find sources of nutrition in food-scarce
environments. Unfortunately, however, nowadays, we are surrounded
by more images of energy-dense, high-fat foods than ever before.
While there is undoubtedly an increasing desire to view images of
food, not to mention take pictures of it, and more is known about
what aspects of these images attract us than ever before, we should, I
think, be concerned about just what consequences such exposure is
having on us all. I, for one, am growing increasingly concerned that
all this 'digital grazing' of images of unhealthy energy-dense foods
may be encouraging us to eat more than we realize and nudging us all
towards unhealthier food behaviours in the long term.

Describing desirable images of food as gastroporn, or food porn, is
undoubtedly pejorative. However, I am convinced that the link with
actual pornography is more appropriate than we think.[78] And given
the oft-discussed concerns expressed about the latter, perhaps we
really should be thinking seriously about moving those food magazines bursting with images of highly calorific and unhealthy food up
on to the newsagents' top shelf? Or preventing cookery shows from
being aired on TV before the watershed?[79] While such suggestions
are, of course, a little tongue-in-cheek, there is a very serious issue
here. What is more, the explosion of mobile technologies means
that we are all being exposed to more images of food than ever
before, presented with foods that have been designed to look good, or

photograph well, more than for their taste or balanced nutritional content.

Let me end with a quote from Max Ehrlich's 1972 book *The Edict*, set in a future where the strictly calorie-rationed populace can go to the movies (the Vistarama Theater) to see a 'Foodie': 'To those watching, what they saw was almost unbearable, both in its pain and ecstasy. Mouths dropped half open, saliva drooling at their corners. People licked their lips, staring at the screen lasciviously, their eyes glazed, as though undergoing some kind of deep sexual experience. The man in the film had finished his carving and now he held a thick slice of beef pinioned on his fork and raised it to his mouth. As his mouth engulfed it, the mouths of the entire audience opened and closed in symbiotic unison with the man on the screen . . . The Foodie had been designed to titillate, and it did. What the audience saw now was not simple greed. It was pornographic. Close-ups of mouths were shown, teeth grinding, juice dribbling down chins.'[80]

But I don't want to leave this chapter on a pessimistic note. In the coming years, gastrophysicists will surely continue to examine the crucial part the foods we are exposed to visually play in our food perception and eating behaviours. There seems little chance that the impact of sight will decline any time soon, especially given how much time the majority of us spend gazing lovingly at our mobile devices and computer screens. So my hope is that by understanding more about the importance of sight to our perception of, and behaviour around, food and drink we will all be in a better position to optimize our food experiences in the future.

4.   Sound

Ask yourself which sense is the most important when it comes to determining your experience of food and drink. Most people will mention taste first. Smell will rank pretty highly too, of course. Some might talk about what a food looks like, and maybe even the mouth-feel and oral texture. But virtually no one, be they sensory scientist, chef or regular consumer, talks about sound. However, as you will see in this chapter, what we hear when we eat and drink – even the noises of food preparation, the rattling sounds of product packaging or loud background music – plays a much more important role than any of us realize. Sound, in other words, is the forgotten flavour sense.[1]

The sounds of preparation

What would go through your mind if you were sitting in a fancy restaurant and you suddenly heard the unmistakeable 'ding' of the microwave? It would be pretty disconcerting! I would argue that the sounds of food and beverage preparation are important precisely because they help set our expectations. No wonder, then, that so many people deliberately try to disable the microwave's distinctive sound because of the negative impression that it conveys to all who hear it (especially in a restaurant setting). Go online and you'll be amazed how many blogs and discussion groups there are complaining about this sound and requesting help to turn it off. Large electronics manufacturers such as GE have, in recent years, been working on redesigning it. (Perhaps, though, attitudes are changing, at least in the home environment; according to the results of a recent survey, a third of those questioned said that they wouldn't mind if served a micro-waved meal at a dinner party.[2]) Of course, the sounds of food cooking can also capture our attention. Just think about all of the desirable food preparation sounds that may have you salivating before you can say 'dinner time'.[3] In fact, one of the classic observations in the field of psychology made by the Russian scientist Ivan Pavlov, way back in the 1920s, was that the dogs he was studying started salivating in response to the sound of the bell used to alert them to the arrival of food. The dogs had soon come to associate the sound of the bell with the delivery of food.[4]

Just take the grinding, gurgling, spluttering and sizzling noises emanating from a coffee machine. These sounds are diagnostic, i.e., they are rich in clues about the probable tasting experience that is to follow. Even the screeching and squealing of the hot-air bubbles that make the milk froth provide information, at least for those who know how to listen. It's the change in pitch that tells the barista when the milk in the jug has reached the right temperature. And if you think that's impressive, what about the guy who says that he can distinguish a hundred different brands of beer based simply on the sound of the bubbles when a drink is poured into a glass![5]

Klemens Knöferle, a former post-doc in my Oxford lab, conducted a study in which he systematically influenced what people said about a cup of Nespresso coffee simply by filtering the sounds made by the machine as it turned those colourful pods into cups of coffee. As he enhanced the harsher, higher-pitched noises, people said that the coffee didn't taste so good. When he cut them, suddenly taste ratings went up.[6] So it's no wonder that so many manufacturers are now trying to engineer the 'right' noises into their machines. They are, in other words, slowly catching up with all the car companies out there who, for decades, have been modifying the design of everything from the sound the car door makes when it closes through to the noise made by the engine as heard by the driver inside the vehicle. Do you remember the iconic Volkswagen 'Sounds just like a Golf' adverts?[7]

Some innovative chefs have started to work creatively with the sounds of preparation. This is what you would have experienced had you been lucky enough to score a table at Mugaritz in San Sebastián in 2015: at one point during the meal, mortars and pestles were brought out from the kitchen, and diners were encouraged to grind their own spices prior to a hot broth being poured into each and every mortar. Just imagine a roomful of diners sitting in a prestigious two-Michelin-starred restaurant, all grinding their spices in synchrony, creating a resonant sound that fills the room. All of the guests sitting at their separate tables united, at least for a moment, by the playful sounds of food preparation.

One Swedish composer, Per Samuelsson, has made something of a career composing with these sorts of food preparation sounds. He is often to be found recording the peeling, chopping, slicing, dicing, grinding, shaking and stirring noises in a busy kitchen. The young Swede then turns those kitchen sounds into a musical composition that is played back to diners while they eat the fruits of the chefs' labour.[8] These compositions are intended to highlight the often unacknowledged effort involved in creating the food that we consume, while at the same time delivering an immersive multisensory environment, one that is designed to enhance the meal experience. Meanwhile, Massimo Bottura, voted the world's top chef in 2016, was recently recorded in an anechoic chamber. The aim? To capture all of the sounds of him making lasagne, the favourite comfort food of his childhood.[9]

It started with a crisp

Back in 2008, Max Zampini and I were awarded the Ig Nobel Prize for Nutrition for our groundbreaking work on the 'sonic chip' (see Figure 4.1). Yes, I know what you are thinking: completely preposterous. Ten of these prizes are awarded every year to a select bunch of international scientists for research that, on the face of it, sounds crazy, ridiculous or preferably both. But the point is that this work, which hopefully makes you laugh first, is actually serious. And getting the prize garners a huge amount of publicity for the winning researchers. Believe it or not, a decade after receiving the award, I am still fielding enquiries on a monthly basis. (This despite my family's incredulity – 'Not the crisps again,' they moan whenever they see the story resurface in the press.) Film crews from around the world

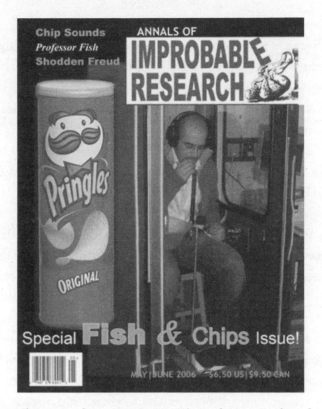

Figure 4.1. My former student Max Zampini (now the esteemed Prof. Zampini) demonstrating the 'sonic chip' experiment on the front cover of *The Annals of Improbable Research*.

periodically descend on the lab, wanting to recreate the magic moment when, simply by changing the sound of the crunch, we were able to change people's perception of the crunchiness and freshness of the crisps (or potato chips, if you are reading this in North America). Truly, my life has not been the same since.

When we first published our results in what, even by academic standards, was a fairly obscure outlet, we thought that they were important, certainly, but nothing out of the ordinary. While Unilever, who funded the study, were interested in our findings, it was hard to see who else might get excited by them.[10] And, if you are wondering why exactly Unilever would fund a project using one of their competitor's products (at the time, Pringles were owned by P&G),

well, the answer is that Pringles are ideal for gastrophysics research. Why so? Because they are *all* exactly the same size and shape. So you can be sure that any change in people's responses must be attributable to the sound manipulation that you have introduced, not to any individual differences in the crisps themselves. And Pringles have another practical advantage: they are large enough that people don't normally put them in their mouths in one go. (You don't, do you?) Hence the relative contribution of air-conducted to bone-conducted sounds in the overall multisensory tasting experience is enhanced.

Basically, we discovered that, solely by boosting the high-frequency sounds that people hear when they bite into a Pringle, we could make them seem around 15% crunchier and fresher than when we cut those sounds. Of course, you might reasonably expect that chefs would be less influenced by such superficial sound modifications of the food they eat. But that turns out not to be the case, at least not if the trainee chefs at the Leith cookery school in London were anything to go by. When we tested them for a BBC TV show a few years ago, they were so busy concentrating on the texture of the crisps that they were just as easily fooled as the Oxford undergraduates[11] who provided the subject matter for our original study.

You can play exactly the same sonic tricks with apples, celery, carrots or, in fact, with any other noisy food, be it dry, like crisps and crackers, or moist, like fruit and vegetables. In one recent study, this time conducted in northern Italy, ratings of the crispness and hardness of three varieties of apple were systematically modified by changing people's biting sounds.[12] This crossmodal illusion is important for a couple of reasons. For one thing, it provides one of the most robust demonstrations that what we hear really does influence what we taste. And it turns out that this particular crossmodal effect works just as well even when you know exactly what is going on. It continues to work no matter how many sonically enhanced chips you've bitten into too. I should know, having crunched more than most – all in the name of science. In other words, the sonic chip illusion is an automatic multisensory effect, as my colleagues in the cognitive neurosciences would say.

Can the sonic chip change product innovation?

But perhaps the more important consequence of our 'ground-breaking' research is that this kind of neuroscience-inspired testing protocol has now been adopted by many of the world's largest food companies. The virtual prototyping approach developed here in Oxford, where we assess how consumers respond to augmented reality products (rather than real product prototypes) allows these companies to figure out how people would respond if their products were to have added crunch or crackle, say. Importantly, the food and beverage companies can do this without having to go into the development kitchens and make a whole host of new products that actually do sound noticeably different from each other. This, the traditional approach, tends to make for a much slower and more effortful development process, especially when the feedback from the tasting panel, as is often the case, is that they didn't like *any* of the new variants that the researchers worked so hard to create. On hearing such news, they have to go back to the kitchens, shoulders slumped and heads drooping, and prepare a whole new set of samples. This is product innovation, yes. But it can be painfully slow!

By contrast, having the testing panel evaluate virtual product sounds first allows one to assess a whole range of alternatives, and find out what, if anything, really makes a difference. So the process is inverted: first, you try to figure out what sounds people like their products to make, and only then do you go into the kitchens to determine whether the chefs and culinary scientists can actually create foods with the requisite sonic properties. Sometimes, those working in the kitchens will shake their heads and laugh, replying that what they are being asked to do is physically impossible. Other times, though, they will know just what is required. But whatever they say, at least everyone knows in which direction they should be heading, sonically speaking. And as a result, product innovation occurs much more rapidly.

The sound of food

Many of the food properties that we all find highly desirable – think crispy, crackly, crunchy, carbonated, creamy★ and, of course, squeaky (like halloumi cheese) – depend, at least in part, on what we hear. Most of us are convinced subjectively that we 'feel' the crunch of the crisp. However, this is simply not the case. Introspection, after all, often leads us astray and, based on the results of the gastrophysics research, I can assure you nowhere is this more true than in the world of flavour. (Take, for instance, the experience of carbonation. Most people, if you ask them, will swear blind that they enjoy the 'feel' of the bubbles bursting or exploding in their mouths. It turns out, though, that the sensation is actually mediated largely by the sour receptors on the tongue; i.e., by the sense of *taste*, not by the sense of *touch* at all.)[13]

Given that we don't have touch receptors on our teeth, any feeling we get as we bite into or chew (masticate) a food is largely mediated by what is felt by the sensors located in the jaw and the rest of the mouth. The latter, removed as they are from the action, do not provide any especially precise information about the texture of a food. By contrast, the sounds that we hear when a food fractures or is crushed between our teeth generally provide a much more accurate sense of what is going on in our mouths. So it makes sense that we have come to rely on this rich array of auditory cues whenever we evaluate the textural properties of food.

Some of these sounds are conducted via the jaw-bones to the inner ear, while others are transmitted through the air. Our brains integrate all of these sounds with what we feel and in the case of the sonic chip this happens both immediately and automatically. And so if you change the sound, the perceived alteration in food texture that follows is experienced as originating from the mouth itself, not as a funny sound coming from your ears.[14] This means that most of us are

---

★ 'Creamy?' I hear you ask. Yes, even creamy foods subtly change the sound your tongue makes when you run it around your mouth.

oblivious to just how important the sound of the crunch is to our overall enjoyment of food! And this isn't just about the crunch: the same goes for crispy, crackly, creamy and carbonated, though the relative importance of sonic cues to our perception of texture and mouthfeel probably varies for each particular attribute. My suspicion is that what we hear is probably more relevant, and hence influential, in the case of crackly, crunchy and crispy foods than in the case of carbonation and creaminess perception. Nevertheless, research points to the conclusion that what we hear plays at least some role in delivering *all* of these desirable mouth sensations, and more.[15]

Crispy and crunchy

One of the major problems associated with working in this area stems from the fact that despite all of the research that has been conducted over the years, it is still not altogether clear just how distinctive 'crisp' and 'crunchy' are as concepts to many food scientists, not to mention to the consumers whom they spend their life studying. Certainly, judgements of the crispness, crunchiness and hardness of food turn out to be highly correlated, thus suggesting that they are indeed very similar concepts to most of us. But matters start to get more complicated when it comes to languages other than English. Some use different terms; others simply don't have any relevant descriptors with which to capture these textural distinctions, if such there be. The French, for example, describe the texture of lettuce as *'craquante'* ('crackly') or *'croquante'* ('crunchy') but not as *'croustillant'*, which would be the direct translation of 'crispy'. Meanwhile, the Italians use just a single word, *'croccante'*, to describe both crisp and crunchy sensations.

Things become really confusing when it comes to Spanish. For Spanish-speakers don't really have words for 'crispy' and 'crunchy' or, if they do, they certainly don't use them. Colombians, for instance (and, I imagine, the Spanish-speakers from many other South American countries), describe lettuce in terms of its freshness (*'frescura'*) rather than as crispy. And when a Spanish-speaking Colombian wants to describe the texture of a dry food product, they will either borrow the English word 'crispy' or else use *'crocante'* (the equivalent of the French word *'croquante'*). This confusion apparently extends to mainland Spain, for when questioned 38% of consumers there did not even know that the term for 'crunchy' was *'crocante'*. What is more, 17% of the consumers in one study thought that 'crispy' and 'crunchy' meant the same thing. This is kind of bizarre when you think how important noisy foods are to our experience and enjoyment of eating. Given that we still can't seem to agree on the definitions and differences between the various textural attributes of food, it is perhaps little wonder that research on the sound of food hasn't progressed quite as rapidly as one might have hoped. This is unfortunate for, as top chef Mario Batali has noted: 'The single word "crispy" sells more food than a barrage of adjectives describing the ingredients or cooking techniques.'[16]

'Why is a soggy potato chip unappetizing?' This was the title of a commentary in a top science journal called (you guessed it) *Science* a few years ago.[17] The nutrient content doesn't change as a crisp becomes stale but, for whatever reason, none of us seems to like the soggy variant. And yet, no one was, I suspect, born liking noisy food. It is on this point that I have to disagree with Mario Batali when he says: 'There is something innately appealing about crispy food.'[18] No, there isn't. Indeed, most of what we think of as innate is, in fact, learnt. In other words, we all *learn* to like specific food sensory cues, in large part because of what they signal to our brains about what we are consuming (and what physiological rewards are to come). Crisp and crunchy – well, they signal fresh, new and maybe seasonal too.

Perhaps the more fundamental question that we should concern ourselves with is why exactly crispy, crunchy and crackly have come to constitute such universally desirable food attributes? These sounds don't directly signal the nutritional properties of a food – or do they? Let's examine crunchiness, which undeniably provides a pretty good (i.e., reliable) cue to freshness in many fruits and vegetables. This information would have been important for our ancestors, since fresher foods preserve more of their nutrient content and hence are better to eat.

Cooking induces the Maillard reaction, the non-enzymic browning that results from strong heat being applied to nitrogen- and carbohydrate-containing compounds. In his book *The Omnivorous Mind*, John S. Allen points to the fact that cooking by fire simultaneously makes foods both more nutritious (or, more accurately, easier to digest) and crispier.[19] Think, for instance, of the delicious crustiness of freshly baked bread. Hence this may be why, in evolutionary terms, the sounds of crispy, crackly, crusty and crunchy are so important. Part of the answer to the question of why soggy foods are so unappealing may also relate to the latest research showing that as the crunchiness of a food increases, so too does its perceived flavour.[20] No surprise then that we all want more crunch. In fact, consumers the world over demand it!

Our strong attraction to fat may also be relevant here. The latter, after all, is a highly nutritive substance, perhaps explaining why we have taste receptors in the oral cavity sensitive to the presence of fatty acids. However, it can still be a challenge for our brains to detect fat directly in food and drink.[21] Why so? Well, often its presence is masked by other tastants like sweet and salt. While cream, oil, butter and cheese are all associated with a distinctively pleasing and desirable mouthfeel, my suspicion, at least as far as dry snack foods are concerned, is that our brains may have learnt, as a result of prior experience (i.e., exposure), that these sonic cues are suggestive of the presence of fat. That is, the louder the crunch, crackle, etc., the higher the fat content of the food we are biting into is likely to be. So we all come to enjoy those foods that make more noise because they probably have more of the rewarding stuff in them than other, quieter foods. Now you know why you find it so hard to resist the sound of crunch!

How do you feel about eating insects?

I'm sure I know the answer to that question without you having to say a word. But, knowing what we now know about sound, we should all find the idea rather more desirable than, in fact, we do. Many insects are, after all, pretty crunchy, at least those with a hard

Figure 4.2. A crunchy and protein-rich snack. Deep-fried crickets should tick all the right boxes as a highly desirable and moreish treat.

carapace or exoskeleton (see Figure 4.2). What's more, they provide an excellent source of protein and fat. It would be good for the planet, too, if we all ate more of these little critters (and less red meat). And yet, no matter what you or I say, most Western consumers are yet to be convinced.[22] Marketing insects so that people – by which I mean us in the West – find them desirable is, I feel sure, going to be one of the ultimate challenges for gastrophysics in the coming years. How exactly do we take everything we know about the mind of the consumer and make this currently most undesirable food source truly delicious? Or, at the very least, make sure that insect matter constitutes a larger proportion of our diets. Playing on the sound of crunch might offer one way in to the popularization of entomophagy (that's the fancy term for eating insects).

So what gastrophysics insights might help here? Well, one option is simply to surreptitiously increase the amount of insect matter that people already eat. (If you are a peanut butter fan, you might want to skip the rest of this paragraph!) I bet you didn't know that there can be as many as 100 insect parts, for example, in every jar of peanut butter before the producer has to declare this on the label). The same must presumably also be true for jam (due to the difficulty of keeping tiny creatures out), and who knows what else is in your ground coffee. So why not slowly increase that number (while at the same time reducing other ingredients that are in short supply, or are unhealthy)? I bet that in the future insect-based matter will become a more substantial part of our diets without the consumer ever noticing. This is like the health-by-stealth strategy that worked so well for the cereal companies when they decreased the amount of salt in their breakfast cereals. They reduced the salt content by as much as 25% by doing it gradually. Consequently, each successive reduction was essentially imperceptible to the consumer; and over the long term, levels of unhealthy ingredients have dropped substantially.[23]

Alternatively, we might work on distinguishing between more and less disgusting critters. For, if you think about it, we already eat lots of bee-based products, everything from honey (sometimes mistakenly called 'bee vomit') through to royal jelly and propolis. So eating bee brood (i.e., baby bee) ice cream shouldn't be too much of a leap, should it? We don't seem to find ladybirds so disgusting either – certainly if one lands in my beer, I will happily flick it out and keep on drinking.

It may be, though, that the sensory strategy that works best in the long term is one that builds on the sound of crunch – something, after all, that we know most consumers really like. All the gastrophysicists have to figure out now is which insects, and which method of preparation, would make the loudest crunch of all?[24] Then away we all go, to a crispier, crunchier and more sustainable future.

Why do crisps come in such noisy packets?

As well as the sounds of preparation and the noises associated with our consumption of food and drink, the sounds of product packaging also have a pronounced impact on our tasting experiences. Do you think that it is an accident that crisps come in such noisy packets? Of course it isn't! From the very beginning, marketers intuited that it would make sense to have the sound of the packaging be congruent with the sensory properties of the contents. This is as true today as it was back in the 1920s when crisps were first packaged for fresh, portion-controlled delivery direct to the consumer.[25] Even Pringles, whose packets typically make less noise than most other snacks, have done something to enhance the sound of their foil seal. You don't have to take my word for it – try running your fingers over it next time you come across a tube and just listen to the difference.

But just how much influence does the sound of the packaging *really* have on our judgements of the product within? Well, a few years ago, we tackled this very question. Together with Oxford undergraduate Amanda Wong, we conducted a study showing that the louder the rattling packaging sound that people heard as they ate, the crunchier-seeming were the crisps they had been asked to rate. While the effects were nothing like as dramatic as those we saw when we modified the sound of the crunch itself, they were still significant.[26] In other words, in terms of perception, our brains appear to have a remarkably hard time distinguishing the product from the packaging.

Frito-Lay may have taken these findings a little too seriously when they came out with their all-new biodegradable packaging for SunChips. This, let me tell you, is probably the loudest packaging ever created (see Figure 4.3). When my colleague Barb Stuckey sent me a couple of packs from California, we got the sound meter out to determine in the lab how much noise they gave off when gently agitated in the hands. The answer: in excess of 100 dB! To put that number into perspective, this is the background noise level that you will find in the very loudest of restaurants. Moreover, it is the sort of noise level that, if one is exposed to it for long periods of time, can lead to permanent hearing damage!

The packaging was so loud that many consumers wrote in to complain. It was so loud that the company was forced to offer ear plugs to try and quell the growing consumer backlash. The idea, I suppose, was that you would buy a packet of SunChips, get them home, then insert the ear plugs and enjoy your snack in peace. You'd have to feel sorry for anyone who was sitting nearby. Even worse if they suffered from misophonia (sufferers find the sound of other people masticating unbearable). Eventually, of course, this damage-limitation exercise failed miserably, and the company withdrew their

Figure 4.3. Sonic warfare? Quite possibly the noisiest packaging on earth, creating in excess of 100 dB of rattling noise when the packaging is gently agitated in the hands. Who thought that was a good idea?

sonically supercharged packaging from the shelves altogether,[27] never, I suspect, to be seen nor, more importantly, heard again.

You can almost hear the marketing executives rubbing their hands together with glee, though, at the ingeniousness of their original idea. We had shown at the Crossmodal Research Laboratory that noisy packaging helps bring out the crunch of the crisp, so surely selling your noisy product in even noisier packaging ought to make your product seem crunchier still. (Assuming, that is, that people eat direct from the pack, rather than pouring the contents out into a bowl or plate. This, though, is a reasonable assumption, given claims that a third of all food is eaten direct from the packaging; if anything, I would imagine that this figure is likely to be much higher in the case of crisps.) The noisy packaging also had the advantage of being extremely effective at capturing the attention of consumers. For as soon as anyone picked a pack off the shelf in the supermarket, you could guarantee that everyone else in the aisle would be looking around to see what the hell was going on. While I doubt there will be anything as noisy ever again, you may be sure that many other companies are thinking about or even actually subtly modifying the sounds that their packaging makes.[28] But the real moral of the SunChips story is 'everything in moderation'. Just because more sound is good, that doesn't mean that even more sound will necessarily be better!

'Snap, crackle and pop'

Product sounds can help set our expectations regarding the product category, or even the specific brand.[29] Some years ago, Kellogg's even tried to patent their crunch. They wanted to trademark the distinctive sound that their breakfast cereals made when the milk was added. They hired a Danish music lab to create 'a highly distinctive crunch uniquely designed for Kellogg's, with only one very important difference from traditional music in commercials. The particular sound and feel of the crunch was identifiably Kellogg's, and anyone who happened to help himself to some cornflakes from a glass bowl at a breakfast buffet would be able to recognize those anonymous cornflakes as Kellogg's.'[30]

Packaging sounds can help set our expectations too. According to Snapple (the beverage company owned by Dr Pepper Snapple Group, Inc.), the distinctive (or signature) sound that the consumer hears on unscrewing the cap from an unopened bottle provides a cue to freshness. 'The company calls it the "Snapple Pop" and says it builds anticipation and offers a sense of security, because the consumer knows the drink hasn't been opened before or tampered with. Snapple was so confident about the pop's safety message that in 2009 it eliminated the plastic wrapping used to encircle the lid. It saved on packaging costs and eliminated an estimated 180 million linear feet of plastic waste, the company says. "We were a lot more comfortable making that decision because we knew there was this iconic pop," says Andrew Springate, senior vice-president of marketing.'[31]

Given how much money the food and beverage companies spend on developing and protecting their visual identity, it is surprising how few of them seem to give much thought to their sonic identity. Interestingly, though, as far as I am aware, Snapple hasn't protected the 'Snapple Pop'. This may be because it has proved hard to trademark specific product sounds (as Harley-Davidson found to their cost when they tried to protect the rich, low-pitched 'potato-potato-potato' sound of their motorcycle exhausts).[32]

Many advertisers have picked up on the potential of sound – not much escapes the ears of marketers. Often they attempt to draw attention to the noises that products make when opened, poured and/ or consumed on screen. The JWT ad agency, for example, worked on a campaign in Brazil to emphasize the sound that Coca-Cola makes when poured into an ice-filled glass.[33] Or think of the loud crack you hear on the Magnum ice cream ads, or the iconic sound of someone running their fingers along the foil seal of the old-fashioned Kit-Kat.* What other distinctive food packaging sounds can you think of?

What does your food sound like at home?

All right, I hear you say, I can see why big companies or chefs might be interested in sound or in sonically mediated food textures, but how does this affect us *mere mortals*? Well, the latest findings from gastrophysics highlighting the importance of sound also provide insights that you can take advantage of at home. For instance, the next time you throw a dinner party, be sure to ask yourself where the sonic interest lies in the dishes you serve. If it isn't crunchy, crackly, crispy or creamy, are you stimulating your guests' senses as effectively as you might? The solution can be quite simple: just sprinkling

---

* This latter is just one of those rituals, as targeted by questions like 'How do you eat your Oreo cookie?', that has been chucked out by the bean-counters who, wanting to save money on packaging costs, changed Kit-Kat wrappers from foil and paper to plastic. Shame!

some toasted seeds over your salad, or adding some crispy croutons to your soups at the last minute. This presumably explains the ubiquitous presence of the gherkin and Batavia lettuce (also known as French crisp lettuce) in your burger bun too – they add a sonic element that makes you enjoy the experience of eating the burger that much more.[34]

Those of you who are a little more adventurous might want to try sprinkling some popping candy into your chocolate mousse, or even into the potato topping of your shepherd's pie. These are both approaches that top chefs have incorporated into their dishes over the years.[35] And if you want to make the sonic surprise all the more memorable, 'hide' it. Your guests will be taken aback when, several mouthfuls into that mostly silent chocolate mousse, say, they suddenly experience an explosion of sound in their mouths. Something that they won't soon forget, I can assure you! This is a good thing, as we will see in the 'Meal Remembered' chapter.

Have you ever wondered why the pairing of Melba toast with pâté works so well? Is this not a classic example of taking a great-tasting but silent food (that's the pâté), and adding to it a burst of noise (when biting into the crisp toast)? Sure, there is texture contrast here (and that is important too). But fundamentally, is it not really about injecting some sonic interest into the dish? More often than not, in fact, our perception of flavour in a meal is enhanced when sound is introduced. Indeed, as we saw earlier, some of the most interesting recent research has shown that as the crunchiness of a food increases, so too does its flavour. In *The Omnivorous Mind*, J. S. Allen also suggests that loud foods might be more resistant to habituation than silent ones. This, he believes, might be part of their universal appeal.[36] So there you have it. Whatever you do, make sure you add some sonic excitement to mealtimes. Better just first check that there aren't any misophones at the table . . .

If sound is as important to the perception of crunchiness and crispness as we think, then maybe there is a solution here if you find that you only have stale crisps in the cupboard next time you throw a party. According to research, your guests probably won't realize just as long as you turn the background music up loud enough to mask the missing sound of the crunch.[37] For as soon as you introduce some loud background noise, your guests' brains will fill in the missing sound of the crunch that they can no longer hear properly. Be warned, though: all that loud noise is also likely to impair your guests' ability to determine how alcoholic your punch is. And if any of them should be impertinent enough to ask you why the music is so loud, just tell them that all the best chefs do it nowadays (about which more below).[38]

'Pardon?' Taking the din out of dinnertime

When was the last time you went out to eat, only to find it difficult to hear the conversation of your companions? My guess is that it probably wasn't long ago. The problem of overly loud restaurants (not to mention bars) has become far more widespread in recent years. Public spaces can be so noisy that many of us can no longer hear ourselves think, never mind get our orders across. Noise is currently the second most common complaint amongst restaurant-goers, behind poor service. In fact, over the last decade or two, many restaurants have become so loud that some critics now report on the noise levels alongside the quality of the food in the venues they judge.[39]

The blame for this growing cacophony has been laid, in part, at the doors of the New York chefs, famous for listening to very loud music while preparing for service.[40] No one knows quite when it started, but at some point one of them had the 'bright' idea that maybe the diners would like it as well. Bad move! For as one journalist astutely notes: 'No matter how elegant the food at a restaurant, the music that plays as it's prepared is likely to be less refined. No one is listening to Vivaldi as he buffs baby vegetables and dismembers ducks.' Charles Michel, the former chef-in-residence at the lab, told me that when he worked at the Louis XV restaurant in Monaco's Hôtel de Paris, the chef de cuisine, Frank Cerutti, would put on some heavy metal during the *mise en place* in order to make the kitchen staff go faster![41]

Part of the responsibility must lie with those designers who have insisted on the restaurants they advise ditching their soft furnishings, leaving spaces filled with hard reflective surfaces.[42] The new Nordic look – you know, all that bare wood, and with carpets, upholstered chairs and tablecloths removed – has a lot to answer for, sonically speaking! There is nothing left to absorb all the noise. That said, though, the chefs might not be entirely blameless. After all, part of the reason for removing the tablecloths in Grant Achatz's Michelin-starred restaurant Alinea, in Chicago, was to give the dishes a little more sonic interest when first placed on the table.[43]

The backlash against restaurants and bars that are overly noisy has been growing steadily.[44] According to a recent press report: '[A] group of Michelin-starred chefs have started a campaign to reduce noise levels in Spanish restaurants amid concerns that it is spoiling the gastronomic experience for some guests.' Others have gone further. Ramón Freixa, chef at the two-Michelin-starred Hotel Único in Madrid had this to say recently: 'Gastronomy is a sensual experience and noise prejudices this pleasure. A good conversation with the people you are dining with should be the only sound in a restaurant.'[45] So how is the din to be taken out of dinnertime? While a successful restaurant full of customers can do without music, it is obviously going to be a much harder trick to pull off in a lightly pop-ulated venue. Furthermore, it is also worth remembering here that background music serves to insulate one table's conversation from the prying ears of those seated nearby. A better aim, then, might be to set the volume level so that the music is 'very present but it's never overpowering.'[46]

A number of chefs and culinary artists have spotted a gap in the market and have been arranging silent dining events.[47] If you eat in silence, then you can really concentrate on the food (and no, texting isn't allowed either, before you ask). This should enhance the sensory pleasure of the experience. Much the same logic underpins all those dine-in-the-dark restaurants out there. Such mindful dining might even lead to reduced consumption.[48] That said, these events have not been a commercial success, I suspect because while this strategy can help to build anticipation by emphasizing the sounds of preparation coming from the kitchen (at least if managed appropriately), insisting on silence also precludes the main activity when eating and drinking, namely the social aspect of communicating with those who you are with.[49]

Across a number of studies we, and a growing number of other researchers, have been able to demonstrate that what we listen to (and how much we like what we hear) influences the taste, texture and aroma of a diverse range of foods. People's hedonic ratings (i.e., how much they like the food or drink) are also affected. Often – but, interestingly, not always – the more we like the music, the more we enjoy the taste of the food or drink consumed while listening to that music.[50] For instance, one recent study revealed that liked music brought out sweetness in gelati, whereas disliked music tended to bring out bitterness in a group of trained panellists.[51] While some have wanted to dismiss such results as little more than a party trick,[52] I would argue that this research may actually have some profound implications for the way in which we think about the senses and food. Furthermore, such results will, I believe, affect the future design of multisensory tasting experiences, no matter where you may happen to find yourself. In fact, we will come across a number of examples in the chapters that follow. So my prediction is that you are going to hear a lot more about sonic seasoning in the years to come.

Sonically enhanced food and drink

When it comes to consumption, many of the food attributes that we find most pleasurable – think here of the crispy, crunchy, crackly, creamy and carbonated – are all influenced, to a greater or lesser extent, by the sounds we hear during consumption. While this is undoubtedly important for younger consumers, in the future, I suspect, boosting the sonic interest of one's dishes is going to become even more important for the rapidly growing ageing population. After all, the number of us living past seventy is steadily increasing, and this is the age at which one starts to see some really dramatic decline in the ability to taste and smell (much though my octogenarian parents deny it!).[53] Now, for any of you out there feeling relieved that this doesn't apply to you, I'm afraid that I've got some bad news.

According to the sensory scientists, if you are out of your teens, the chances are that the decline has already started. Not feeling quite so

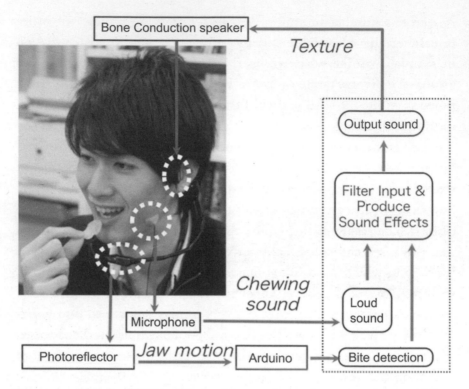

Figure 4.4. The playful 'Mouth Jockey' detects the user's jaw movements and then plays back a specific pre-recorded sound.

smug? I didn't think so. Research clearly shows that the majority of older adults are pretty much anosmic: i.e., they can no longer smell anything much at all. Unfortunately, there is currently absolutely nothing that can be done to bring back the sense of taste or smell once they have begun their inevitable decline (unlike the boost we can give failing vision or hearing with spectacles or hearing aids). But one thing that can be done is to make sure that dishes served to the elderly have lots of crunch and crackle – more sonic interest, in other words. This should help to stimulate the mind and palate of whoever is eating.[54]

So what have younger people, whose taste buds and, more importantly, noses are still in good working order, got to look forward to? Researchers in Japan have developed a playful headset called the 'Mouth Jockey' (see Figure 4.4), which detects the user's jaw movements and plays back pre-recorded sounds while the user eats. Just imagine hearing the sound of screaming as you bite into a gummy bear, say. Others have been working on augmented straws that can recreate the sounds and feeling associated with sucking liquidized food through a straw: place the straw over a mat showing a picture of the desired food and then suck. It is amazing how realistic and how much fun the experience can be.[55]

The EverCrisp sonic app allowed you to 'freshen up' your stale crisps by using your mobile device to add a little more crunch. While it is tempting to imagine that technology will come to play an increasingly important role in augmenting our experience of food in the years to come (see the 'Digital Dining' chapter), I believe that there will also be a crucial role for design. I want to leave you, therefore, with the Krug Shell (see Figure 4.5), as an example of what is possible here. This Bernardaud Limoges porcelain listening device, which was released as a limited edition offering back in 2014, sits snugly on top of a custom Riedel 'Joseph' champagne glass. If you can somehow get your hands on one, I'd encourage you to try it. You will hear how the sounds of all those bubbles popping in your glass have been pleasantly amplified.[56] Then sit back and contemplate whether you are really happy to let sound remain 'the forgotten flavour sense'.

Figure 4.5. The Krug shell, designed by French artist Ionna Vautrin to amplify the sound of the bubbles popping in a glass of champagne.

5. Touch

At a number of the world's top Michelin-starred restaurants these days, the first three, four or even five courses are all eaten with the fingers. This is happening at Noma in Copenhagen, at Mugaritz in San Sebastián, Spain, and even at The Fat Duck in Bray.[1] Any self-respecting gastrotourist will know what I am talking about. But just think about it for a moment. This would have been unheard of in two- or three-Michelin-starred restaurants even a few years ago. Sure, we probably always eat with our hands while snacking in the car, and we mostly take the bread with our fingers, even in the fanciest of restaurants, and shellfish too for that matter, but to eat other dishes with our fingers? That is something new. And, going one stage further, Mugaritz declared in 2016 that they were no longer going to use traditional cutlery at all.

In this chapter, I want to show you just how important what we feel is to our experience and enjoyment of food and drink. Not only what we feel in our mouths but also what we feel in the hand. Mark my words, dining and drinking are slowly but surely becoming much more tactile activities. And I am not just talking about the increasing range of textures and mouthfeels that you might be exposed to while eating the latest creation of some famous modernist chef or other; rather, I am talking about our total tactile interaction with food and drink. Touch is, after all, both the largest and earliest developing of our senses, with the skin accounting for about 16–18% of body mass.[2] Ignore it at your peril!

Does texture influence taste/flavour?

The answer is a very definite yes,[3] although this is a hard subject to study empirically, as it is difficult to manipulate the sensory cues independently. So, for example, increasing the viscosity of a liquid will reduce the number of volatile aroma molecules released from the drink's surface. So while you may be certain that the oral-somatosensory properties of a food or drink (basically the mouthfeel) will affect the taste/flavour of food and drink, determining the cause of that interaction can be a little harder to discern.[4]

Not so long ago, though, sensory scientists finally managed to figure out a way of varying texture and aroma independently. Specifically, they delivered the latter by means of a tube placed into the mouth of their unfortunate subjects and were able to demonstrate that the addition of certain fatty aromas modified the 'mouthfeel', and perceived thickness, of liquids. On the other hand, increasing the oral viscosity of the liquid in the mouth (just think about the different mouthfeel associated with cream versus water) also affected the perceived aroma/flavour. This is true even when the aroma is delivered via a tube, and so one can be sure that the aroma release hasn't changed as a result of the change in viscosity.[5]

Why not try eating something like a strawberry or a cookie, or whatever you have at hand? Ask yourself where the taste/flavour originates. The answer, most likely, is that it appears to come from the food that you can feel moving around in your mouth. Am I right? However, as a food breaks down and is transported around the oral cavity via the saliva as a result of chewing (what my colleagues like to call mastication), you are probably tasting with *all* your mouth – and every time you swallow, with your nose too (via the retronasal olfactory route). Your brain does an amazing job of putting all of these sensory cues back together again and associating them in your mind with their presumed source, i.e., the food you feel in your mouth. Such a good job, normally, that we rarely, if ever, stop to think twice about it.[6]

When you watch a film at the cinema, your brain ventriloquizes the voices from the loudspeakers located around the auditorium so that they appear to be coming from the lips that you see moving on the screen. Much the same thing happens whenever we eat and drink. The simplest way to illustrate this phenomenon is to take a teaspoon of a salty or sweet solution into your mouth and then run a tasteless

Q-tip across your tongue. You should feel the taste coming from the tactile stimulus moving across your tongue, despite the fact that the swab itself has no taste. What this shows is how tactile stimulation in the mouth 'captures' where we perceive taste to originate from. In fact, knowing that the brain does this has allowed us (in some of our more bizarre research) to mislocalize taste out of the mouth and on to a butcher's tongue (or a rubber imitation of a human tongue)![7] Some people are convinced that they can taste what they see being applied (e.g., a drop of lemon juice) to a fake tongue outside of their body when, in fact, what is dropped on to their own tongue is water.

'Oral referral' describes the phenomenon of experiencing food attributes like fruitiness, meatiness and smokiness in the mouth, rather than in the nose, where such aromas are first sensed. For more than a century, it was thought that the tactile stimulation experienced in the mouth while eating and drinking explained the oral referral of odours to the mouth. However, this turns out not to be the case.[8] The fact that the multisensory integration of flavour cues occurs so effortlessly, and so automatically, shouldn't hide how incredibly complex it all is. So, in response to the question: 'Can you feel the taste?', the answer is: 'No.' That said, what we feel, in the mouth or out, most certainly does affect both the taste and flavour of what we eat and drink.

Marinetti's tactile dinner party

F. T. Marinetti, the founding father of the Italian Futurists, had a great interest in touch. He produced a tactile manifesto called *Il tattilismo* in 1921 and organized the very first tactile dinner parties back in the 1930s. Unfortunately, however, there was just one problem: the Futurists couldn't cook. They were dismissed as 'a fart from the kitchen' by the Italian press.[9] Their approach to cuisine was never meant to last (their questionable political views didn't help their cause much either [10]). No wonder, then, that they disappeared without trace in the 1940s. Though, as we will see later in this chapter, contemporary chefs and culinary artists are starting to revive a number of Marinetti's marvellously mad ideas about experience design (forget the food), with some surprising results.

Not being able to cook certainly didn't stop the Futurists from throwing some of the most influential dinner parties. According to a description of one such dinner, the meals were to be eaten to the accompaniment of perfumes which were to be sprayed over the diners, who, holding their fork in their dominant hand, would stroke with the other some suitable substance – velvet, silk or emery paper. So, if you want to give your dinner guests the full Marinetti experience, insist that they turn up wearing pyjamas made of different materials, such as velvet or silk. Then, when the food is placed on the table, encourage them to eat with some kind of textured implement, while feeling their next-door neighbour's jim-jams with their other hand![11] And, if that seems a tad risqué for the suburbs, I have a couple of more modest solutions that all you budding Futurists out there could try instead.

You could, for instance, take inspiration from London chef Jozef Youssef, of Kitchen Theory. He served a dish entitled 'Marinetti's Vegetable Patch' as part of his sell-out 'Synaesthesia' dining events in 2015. There were various different textures on the plate itself. But there were also a number of black cubes (designed to fit in the hand) scattered across the dining table, each surface, or opposing pair of surfaces, of which was covered with a different material: Velcro, velvet, sandpaper, that sort of thing. The diners were encouraged to taste the various elements of the dish while feeling the textured cubes. They were instructed to look out for any correspondence between what they were feeling in their hands and the textures in their mouths. Now, it has to be said, some people had no idea what was going on. This certainly wasn't one of those dishes/experiments that works on absolutely everyone. But there were some diners (roughly a third, I'd say) who volunteered that their experience of the food had been altered by changing the surface they touched.[12] Weird, almost synaesthetic, some would say. So maybe the Futurists were on to something after all!

Together with Professor Barry Smith of the University of London, we do something very similar when we run tutored multisensory wine-tastings. We get a selection of swatches cut from different materials and hand them out to everyone. We then serve a couple of reds and have people rate how well the different textures go with each of the different wines. This is something so simple, and yet it intrigues many people. And it is something that anyone can try at home. Why not try it yourself next time you open a bottle with some friends? It will, at the very least, get your guests to pay a little more attention to the tasting experience. It may also help to explain why all those adverts for red wine invoke textural metaphors: velvety, silky, and so on.

Is the first taste really with the hands?

In many parts of the world – think Africa, the Middle East and the Indian sub-continent – people mostly eat with their hands. However, in the restaurant setting, especially in Westernized countries, we nearly always eat with the aid of cutlery, be it cold, smooth knife and fork in the West or chopsticks in the East. And whenever we drink, we always pick up the cup, glass, can or bottle first.* In a very real sense, then, the first taste is with the hands. According to sensory scientists and flavour chemists, the feel of the cutlery or drinking vessel shouldn't exert any influence over the taste of the food and drink.[13] Nor should it really affect how much you enjoy the tasting experience. After all, everyone – chefs, food critics and regular consumers alike – thinks that we can simply ignore 'the everything else' and concentrate squarely on the taste of the food on our plate, or the flavour of the drink in our glass. But we cannot! By this stage of the book, I hope you are convinced that 'the everything else' really does matter. And what we feel is no different. In truth, it has more of an influence on our experience of food and drink than many of us would credit, or even, perhaps, be willing to believe.

A growing body of research from the emerging field of gastrophysics now demonstrates how what we feel really can influence our

---

* Except if we drink through a straw. But that is a very bad idea, given that it minimizes the orthonasal olfactory hit! Just see the 'Smell' chapter.

tasting experiences. Some of the world's top chefs, molecular mixologists, culinary artists and even packaging and cutlery designers are starting to pay much closer attention to what we feel when we eat and drink. They are playing around with changing everything from the texture and weight, through to the temperature and firmness of whatever it is that we happen to be holding in the hand while consuming. And take note: they are not stopping at the hands! The most creative experience designers out there are also thinking about how best to stimulate your lips, and even your tongue, more effectively.

Do you really like the feel of cold smooth metal?

We did not evolve to find the cold smoothness of stainless steel or silver cutlery particularly appealing to the touch. Rather, we were always meant to eat with our hands. So why are so many of our interactions with food mediated by metal cutlery? As top interior designer Isla Crawford once put it: 'Surfaces made from natural materials are often preferable, as irregularity is far more sensual than clinically perfect surfaces.'[14] To my way of thinking, it's bizarre: so many of the world's top chefs are doing such amazing things on the plate (assuming there is a plate – not always guaranteed these days), expressing culinary genius and creativity in ways we have never seen, nor could even have imagined back in the 1970s. And yet those self-same chefs have their diners eat with the traditional combination of knife, fork and spoon. It really isn't very imaginative now, is it?

There are, after all, few other objects that you would put in your mouth after they had already been in who knows how many other people's mouths beforehand. How would you feel if I suggested you use someone else's toothbrush? So what, exactly, is so different about cutlery?[15]

In the years to come, I believe, we are going to see some radical innovation in the way in which we move food from plate or bowl to mouth. I hope that open-minded cutlery makers out there will take the scientific insights about the receptor profile of the average human mouth, and all the latest gastrophysics research, and translate this knowledge into aesthetically pleasing cutlery designs that will enhance our tasting experiences. The likelihood is that the results of their labours will first be found within the confines of the modernist restaurant. And from there, signature cutlery will slowly start to appear in the marketplace, perhaps under the brand name of one top chef or another.

Would you enjoy eating with a textured spoon?

So, to get us started on this tactile journey, take the example shown in Figure 5.1. How do you think your experience would be if you ate something with the aid of one of these spectacular-looking utensils? More memorable, probably. More stimulating, absolutely.

Unfortunately, the spoon shown in Figure 5.1 is one of a kind. I very much doubt that you will find this designer's work for sale on Amazon any time soon. A shame, really, given how boring the texture of most cutlery is currently. But at least one mainstream cutlery manufacturer has recently brought out a commercial range of sensorial spoons with added textural interest (see Figure 5.2). These four textured spoons caress your tongue in unusual ways, but we are still

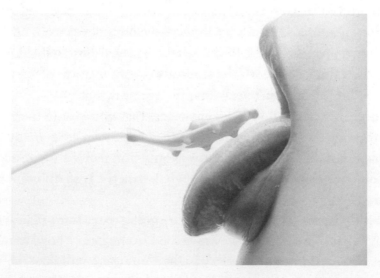

Figure 5.1. 'Tableware as sensorial stimuli cutlery' from the wonderful designer Jinhyun Jeon.

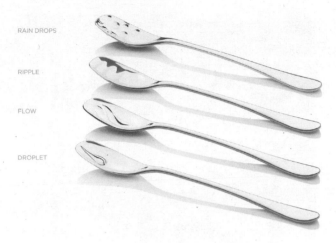

RAIN DROPS

RIPPLE

FLOW

DROPLET

Figure 5.2. A set of four textured spoons from Studio William.

researching, together with chef Jozef Youssef and the top cutlery designer William Welch, into whether any of the spoons do an especially good job of enhancing a specific taste, flavour or texture of food.[16]

For those of you who haven't yet acquired your own set of sensorial spoons, there are nevertheless some very simple ways to stimulate your guest's tongues more effectively without having to invest in a whole new set of cutlery. Next time you invite them round for dinner, why not surprise them? Wet some spoons in lemon juice (best not try this with the silver ones, though, if you don't want to get into trouble). Next, dip them into something crystalline or gritty, like sugar or a little ground coffee, say, and allow them time to dry. Then, just before you are about to serve, place a dollop of something tasty on top and hand them to your guests. This, anyway, is the technique that some of my culinary artist friends, like Caroline Hobkinson, use to tickle their guests' tongues in ways that they might not have been stimulated previously. Even top restaurants have used this approach, for example, Alinea's 'Osetra' dish. At the very least, the unusual texture will surprise the diner and hence make them a little more mindful about what they are eating.[17]

Another way in which to alter your guests' experience of their food is to change the material properties of the cutlery itself. One cheap solution here involves laying the table with some wooden

Figure 5.3. High-end wooden cutlery, as served at Noma in Copenhagen. An unusual texture, yes, but too light for some.

picnic cutlery rather than your normal knives and forks – it'll probably save on the washing-up too! But bear in mind how one diner responded when Noma, currently one of the world's top restaurants, tried something similar. They introduced some high-end wooden utensils into the service at their restaurant in Copenhagen (see Figure 5.3). While I have not yet had the opportunity to try them for myself, one colleague who visited the restaurant in 2015 was certainly disappointed. As she wrote to me after her return: 'It was like eating with a spork from a takeaway.'[18] It would be really interesting to conduct the appropriate study in a restaurant setting, to see whether this response would be true more generally.

Weight, what is it good for?

I cannot emphasize enough just how important weight is to the design of cutlery. You definitely want it to have a good heft in the hand, not to mention the right balance between one end and the other. One of the first things I noticed at Heston Blumenthal's The Fat Duck restaurant was how heavy the cutlery is (really hefty pieces, made from wood and steel, in the French Laguiole style).* The cutlery maker William Welch (the designer of the textured spoons we saw in Figure 5.2) knows intuitively that making sure his cutlery feels good in the hand is really important. As important, he told me, for most people as its looks.

In contrast, it is amazing to me how many young chefs skimp on their cutlery. It is understandable, of course. Imagine some Young Turks opening their first gastropub out in the countryside somewhere. They have poured their life savings into the endeavour, and are likely to be running short of cash when they open. Heavy cutlery might seem like more of a luxury than a necessity, right? But if they cut that corner, they'll end up serving beautifully prepared food to be eaten with what feels like light canteen cutlery. It really detracts from the overall experience, as I am sure you'd intuitively agree. But what does the gastrophysics research say?

Given the differing priorities out there, it was obviously time for the gastrophysicist to step in and conduct an appropriate study. Of course, our first job was to check the scientific journals to see what had already been done. The really surprising thing was that the literature was pretty much silent on the topic (i.e., on the impact of cutlery on the experience of food and drink). How could something so fundamental have been ignored for so long? So, in our own research, we wanted to determine once and for all just how important the weight of cutlery in the hand really is to the experience of food in the mouth (or mind). We had already conducted a series of studies here at the Crossmodal Research Laboratory, demonstrating that if people tasted food with a heavier spoon they generally had better things to say about it than when exactly the same food was eaten with a lighter spoon instead.[19] But evaluating store-bought yoghurt from a heavy plastic spoon in the lab is a long way from the setting of a high-end restaurant. Would the same results also apply *there*?

In many of our highly controlled laboratory experiments, the same people taste putatively different foods with spoons varying markedly in terms of their weight. One of the benefits of using the same participants is that we can be sure that the results we obtain are

---

* At least it was before the restaurant closed to refurbish their kitchens in 2015.

due to our experimental manipulation and not to individual differences between people. However, on the negative side, it's possible that the format of our studies might have focused the participants' attention unnaturally on weight. Imagine yourself being asked to taste food over and over again, for up to an hour, when the most salient thing that varies is the weight of the cutlery that you are given to use. With nothing else to occupy your mind, there is a very real danger that the cutlery's weight might capture your attention, and hence influence your behaviour, in a way that it simply wouldn't do in a restaurant, say.

Given such concerns (which, it should be said, apply to much laboratory-based research), I was looking out for an appropriate opportunity to test the cutlery idea out in the wild.[20] Of course, I would love to carry out a study with the diners holding that oh-so-heavy cutlery in The Fat Duck. But that is just never going to happen. Why not? Well, which funding body would agree to pick up the tab for all my *subjects* (yes, those well-fed guinea pigs) at the end of the night? Currently the price at the restaurant is nearly £300 a head, and that's before wine or service.

Luckily enough, though, at around this time, I was invited to give a talk at an International Egg Confederation conference. The organizers wondered whether I would mind awfully running a three-course experimental lunch for the delegates, so that they could get an idea of what real gastrophysics research was like. I couldn't believe my luck! They had just offered me the perfect opportunity to test out the theory that weight in the hand really does matter to our enjoyment of food in the mouth. It was the moment that I had been waiting for. But would the experiment actually work, out there in an ecologically valid testing situation?

Imagine the scene: 150 international conference delegates in a swanky hotel restaurant somewhere in the centre of Edinburgh. The diners were randomly dispersed across different tables. There was a scorecard by each place setting and pencils on the tables so that the diners could record their responses. They were asked how much they liked the food, how artistically they thought it had been plated, and how much they would have been willing to pay for the dish in a restaurant like the one they happened to be sitting in. While the conference delegates were very much aware of the fact that they were taking part in an experiment, they didn't know what the specific research questions being addressed with each dish were. For the main course, a piece of Loch Etive salmon, the service staff had laid light canteen cutlery at half of the tables, and heavy, expensive cutlery at the rest.* But still, we weren't asking people about the cutlery; we were only asking them about the food.

The results were unequivocal. Those eating with heavier cutlery thought that their food had been plated more artistically. And crucially, they were willing to pay significantly more for it than those eating the same food, on the same day, in the same dining room, but who just so happened to be holding lighter cutlery instead.[21] So, it really is that straightforward: Adding weight to your guests' hands will most likely make them think that you are a better chef! With that in mind, why not reach into the cutlery drawer right now and feel up some of your own. Are you sure that you are creating the right impression? But you don't want to take things *too* far: I have heard rumours of one restaurant where the diners started to complain because the cutlery was just too heavy to lift comfortably.†

Furry cutlery, anyone?

Periodically, we have lab dinner parties at my house here in Oxford. On one such occasion, Charles Michel, the then chef-in-residence,

---

* The eagle-eyed reader may have noticed that *both* the weight and the quality of the cutlery were varied. However, the fact of the matter is that it is really hard to get heavy poor-quality cutlery. And even if we had found some, there was no way anyone was going to splash out on seventy sets of it just so we could run the study. As is so often the case, then, we had to work within the confines of what was on hand – in this case, canteen versus restaurant cutlery.

† There was a great Pugh cartoon in the *Daily Mail* based on our heavy cutlery findings. It showed a couple dining at the table, with an older women slaving away in the kitchen. Underneath, the punchline read: 'I've changed the cutlery twice and your mother's cooking still tastes horrible.'

Figure 5.4. *Left:* Furry cutlery *à la* Charles Michel. Best served with rabbit; *Right:* Meret Oppenheim's 1936 *Object*, a fur-covered cup, saucer and spoon. How would you feel about putting the cup to your lip to drink? This art piece was deeply subversive at the time, given its sexual connotations. There is just something about fur that you don't want coming too close to your lips. One can only imagine what Freud would have to say on the matter!★

decided to get rabbit from the market and prepare a stew. And it was delicious, I can assure you. But the most memorable thing about the evening, and about that dish in particular, was what the Paul Bocuse-cookery-school-trained chef had done to my wife's cutlery. He had asked at the butcher's for the cleaned rabbit pelts, the stuff that normally just gets thrown away. And in a stroke of *genius* he wrapped them around the handles of the spoon (see Figure 5.4, *left*). In an instant, the cutlery became a truly multisensory dining implement – F. T. Marinetti would have been proud! Sitting around the dining table, we all tentatively held the soft, furry skin in our 'paws', the faint

---

★ Now, when the chef was working on a short placement at The Fat Duck, I went to dine, and the kitchens came out with something really interesting. Take a tomato and boil to loosen the skin. Try to remove pieces of skin that are as large as possible. Then take the skin and press it in strips on to the lip of a glass, then serve in it something like a gazpacho soup. What you get, as a diner, is a most peculiar sensation, a little animalistic, especially when the inverted tomato skin comes into contact with your lower lip.

aroma of the animal emanating from our hands. There was no doubt – straight away everyone had a much greater awareness of where our dinner had really come from (see Figure 5.4, right, for another famous example).

Imagine my surprise, then, when just a few months later the over-weight furry white spoon shown in Figure 5.5 greeted me when I got to the last course of The Fat Duck's revamped tasting menu. Now, I am not sure that this spoon was necessarily the ideal choice. The dish itself is all white, light and airy. So I could imagine a surprisingly light spoon working well here. Instead, what you get is a spoon that is noticeably heavier than you thought it was going to be. (The distinctive baby-powder aroma of heliotropin originates from the handle.) Though perhaps that is the idea: to create even more of a contrast with what you might have expected that you were going to feel. Diners may come away from the experience thinking a little more carefully about the weight of the cutlery, and the influence that it may have on their experience of food.

However, while you want the cutlery to enhance the taste of the food, you don't necessarily want it to become the centre of attention, to distract the diner from whatever they are eating.[22] There is, I think,

Figure 5.5. 'Counting Sheep', the final dish on the tasting menu at The Fat Duck. If you saw a spoon like this, just how heavy would you expect it to be? (It is actually much heavier than that, believe me.)

a very real danger of this happening with the 'Counting Sheep' dish. At least there would be if it weren't for the fact that your dessert is actually spinning on a pillow that is magically floating in mid-air! Yes, really: magnetic levitation, the likes of which you have probably never seen (unless, that is, you have stopped in at the Artesian bar at The Langham Hotel in London, where a cocktail with a balloon floating over it has been on the menu for a couple of years now). This, then, is the aim of much of the research from my lab: to figure out how to use the latest insights from gastrophysics to create better, more memorable tasting experiences.

Eating with your hands

Have you ever thought about the fact that the hamburger, one of the world's most popular foods, is usually eaten with the hands? I would go so far as to say that it tastes better when grasped between one's thumbs and forefingers than when eaten daintily with a knife and fork from a plate or plank. The same goes for fish and chips straight from the newspaper at the seaside too (at least it used to, before someone banned newspaper wrapping as unhygienic). Now, I am the first to admit that there is a lot more going on in the latter case to explain the pleasure of the experience than merely using one's hands. Still, think about it carefully, and it is surprising how many foods really do seem to taste better when eaten like this. No wonder, then, that popular US chef Zachary Pelaccio titled one of his cookbooks *Eat With Your Hands*.[23] He was very much 'on trend'.

And it is not just fast food that people are eating with their fingers nowadays; it is also haute cuisine. For, as we saw at the beginning of this chapter, a growing number of top Michelin-starred restaurants have been incorporating dishes that are to be eaten without cutlery, or else with totally new forms of cutlery.* Interestingly, though – and

---

* Going the whole hog is probably doomed to failure, at least in the context of fine dining. Restaurants like Il Giambellino in Milan, where the diners were not provided with any cutlery at all, don't tend to last long, especially not in the capital of risotto!

I am still trying to figure out the reason(s) behind this – finger food tends to make its appearance at the start of the meal rather than later on. If any of you have any ideas about quite why this should be so, please do let me know.

Many people write to me saying that, for them, food really does taste better when eaten with the hands.[24] This seems to be especially true for those from India, say, who have grown up using their fingers for this purpose. A number of them report that food just appears to lose its taste whenever they have to eat using cutlery. The following, from the Indian narrator in Yann Martel's book *The Life of Pi*, illustrates the point: 'The first time I went to an Indian restaurant in Canada I used my fingers. The waiter looked at me critically and said, "Fresh off the boat, are you?" I blanched. My fingers, which a second before had been taste buds savouring the food a little ahead of my mouth, became dirty under his gaze. They froze like criminals caught in the act. I didn't dare lick them. I wiped them guiltily on my napkin. He had no idea how deeply those words wounded me. They were like nails being driven into my flesh. I picked up the knife and fork. I had hardly ever used such instruments. My hands trembled. My sambar lost its taste.'[25]

I have always wanted to compare what people say when eating and rating a range of foods with either their fingers or with a knife and fork, or even chopsticks. Of course, the answer is likely to depend on the food being served and on the context, not to mention on the individual diner and what they are used to, or have grown up with.[26] Nevertheless, there are some intriguing results already out there showing that the food we feel in the hands really does influence our perception in the mouth. For instance, my colleague Michael Barnett-Cowan conducted a study in Canada in which he somehow managed to glue together two half-pretzels with the opposing ends sometimes having the same texture, either both fresh or both stale, and sometimes having different textures. Just imagine the situation: you might be holding a stale pretzel in your hand but munching on a soft one, or vice versa. The results revealed that the feel of the food in the hand really did influence what people had to say about their in-mouth experience.[27]

Once again, this is a simple change that any one of us could try out the next time that we invite some friends round to eat: withhold the cutlery. A few of you readers may be worrying what Debrett's *Guide to Etiquette* would have to say on the subject. Good news: the 2012 version of the guide finally acknowledged that finger food was acceptable in polite society, at least for certain foods such as pizza, calzone and ice-cream cones. Whatever you do, though, be sure not to lick your fingers afterwards![28]

And finally, eating with the hands can also be a good idea on a first date, at least if you believe the results of a survey of 2,000 people reported in the papers recently. Men apparently find women making a mess while eating with their hands a big turn-on. So now you know! (And if you are a man hoping to make a good first impression on the ladies, the top tip was to make sure not to order a salad for your main course.)[29]

How much would you enjoy food moving in your mouth?

My colleague Sam Bompas describes a dinner he attended in Korea where the tentacles of a live squid were cut directly on to his plate. His hosts earnestly recommended that Sam chew vigorously in order to prevent the still-squirming suckers from sticking to his throat on the way down! Disgusting, right? So it is ironic, then, that while our *visual* attention is undoubtedly drawn to food in motion (see the 'Sight' chapter), once that food enters the oral cavity, movement is the last thing we want. This deep-seated aversion to things wiggling around in our mouths (and worse still, in our throats) was presumably partly what caused such a stir when Noma served live ants a few years back. (The entomophagy angle not helping much here either.[30])

It was also the thought of movement in the mouth that made everyone so squeamish when I was at school in Canada for a year, a long, long time ago. The prefects there challenged themselves to eat a live goldfish fresh from the bowl every time the home ice-hockey team scored (which was, thankfully, pretty rare).

Evolutionarily speaking, this dislike of movement in the mouth is probably an old mechanism that helped our ancestors to avoid the risk of choking.* That said, note how the language of menus often tries to give the impression that there is still life in whatever is being served. There's a great line in Steve Coogan and Rob Brydon's TV series *The Trip* where, after a dish is introduced as 'resting' by the waiter, Rob Brydon points out: 'Rather optimistic to say they're resting. Their days of resting have been and gone. They are dead.'[31] True, but 'dead' just feels like a word that should never appear on a menu.

More generally, the texture of food in the mouth (even when it is not moving) seems to be a particularly strong driver of our food likes and dislikes. So, for example, many Asian consumers find the texture of rice pudding to be more off-putting than its taste (or flavour). By contrast, for the Westerner breakfasting in Japan, fermented black natto has a texture and consistency that won't soon be forgotten. And take the oyster – it's this shellfish's slippery, slimy texture, not its taste or flavour, that people typically find so objectionable, many agreeing with the late British food critic A. A. Gill's memorable description of them as 'sea-snot on a half-shell'.[32]

Of course, the textural (oral-somatosensory) properties can also constitute a key part of what we find so pleasing about the foods that we love. Indeed, a number of researchers have argued that this is a key part of the appeal of chocolate, one of the few foods to melt at mouth temperature. (Try eating a very cold versus a warm piece of chocolate to experience this difference.) Texture, then, plays a crucial role in determining our perception of a food's quality, its acceptability and ultimately our food and beverage preferences. Just think about it: comfort foods typically have a soft texture (e.g., mashed potatoes, apple sauce and many puddings). In fact, it has been argued that foods having this texture tend to be thought of as both comforting and nurturing. By contrast, many snack foods are crispy, like chips and pretzels. Texture contrast is something that many chefs and food developers work with and, more generally, it is known to be something

---

* Even food that moves on the plate can itself be both hypnotic and disturbing. The bonito flakes served in Japanese restaurants sometimes do this, not to mention the squirming of the live abalone and hagfish that one sometimes sees at Korean BBQs.

that consumers value in food. As Barb Stuckey puts it in her book *Taste What You're Missing*: 'Good chefs go to great lengths to add texture contrast to their plates, utilizing four different approaches: within a meal, on the plate, within a complex food, and within a simple food.'[33]

What's so special about bowl food?

Hard though it is to believe, five – yes *five* – books were published on the subject of bowl food in 2016. Why, you have to ask yourself, should serving food in a bowl matter? Well, apparently the main appeal is that it makes everything taste better. Even Gwyneth Paltrow thinks so. So it must be true.[34] Of course, it isn't just about serving the same old food in a new receptacle. Part of the appeal to the bowl foodies is filling their receptacles with foods that are wholesome, nutritious and, well, filling.[35]

Serving hot food in a bowl allows, maybe even encourages, the diner to take a hearty sniff of the steaming contents. Most of us are less likely to do this if exactly the same food is served on a plate when it is placed on the table in front of us, say. And as we saw earlier, anything that enhances the olfactory hit associated with a dish is likely to lead to improved flavour perception and possibly also increased feelings of satiety. Holding the bowl in your hands means that you feel its weight too. And the evidence here shows that the heavier the bowl, the more satiated (i.e., fuller) you will expect to feel.[36] This is a problem, of course, for the food and beverage companies, who are being advised by government to make their packaging lighter, especially those trying to promote a filling snack (a yoghurt, say). Time and again in our research, we find that adding weight to a soft-drink can, to a box of chocolates or to a carton of yoghurt leads people to rate the product, no matter what it is, more highly.[37]

Holding a bowl, you feel the warmth of the contents, and possibly the texture and the reassuring roundness of the underside of the bowl. Note here that the texture of plateware has also been shown to modify people's experience of food. In one of our recent studies, for example, we found that people rated ginger biscuits as tasting significantly more spicy when served from a rough plate than from a more traditional smooth plate.[38] Holding a warm cup or bowl in your hand can even make those around you appear a little friendlier too.[39] And, as if all that wasn't enough, serving food from bowls without rims can trick our brains into thinking that there is more than when exactly the same amount is served from a bowl with a wide rim. It is supposedly more photogenic too. Ultimately, then, from a gastrophysics perspective, bowl food is indeed likely to work especially well for those looking for a filling, healthy meal.

Affective ventriloquism

But why exactly should what we feel have such an effect on our taste experiences, especially when what we are interacting with is not the food itself? One possible answer relates to the notion of 'affective ventriloquism'. My colleague Alberto Gallace and I noted a few years ago that people appear to transfer the affective response generated by whatever they touch to what they think about the food or drink itself.[40] That is, we find it hard to maintain separate impressions of the food or drink on the one hand, and of the cutlery, glassware or plateware on the other. Instead, what we think about one can all too easily bias our judgements about the other.

Given that we consume as much as a third of our food and drink direct from the packaging, it should come as no surprise to hear that product designers and marketers are interested in optimizing the feel of their product packaging. In fact, this may be how most of us will be exposed to the whole new world of tactile design. In some cases, the aim is to prime or convey notions of fruitiness, by treating the packaging surface to give it the same feel. The good old Jif lemon juice container is a classic example. In this case, the product itself imitates the size, colour and even feel of a lemon (see Figure 5.6).[41]

However, my absolute favourite comes from the high-end Japanese designer Naoto Fukasawa, a fabulous range of hyper-realistic packaging prototypes, e.g., a drinks container that perfectly captures the experience of touching the fruit – which makes the Jif version of a

Figure 5.6. From left to right: Granini glass bottle; hyper-realistic juice drink packaging by Naoto Fukasawa; and Jif lemon juice container. These examples of multisensory packaging probably enhance the consumption experience by mimicking the feel of the fruit they contain.

lemon feel cheap by comparison. Amazingly, the designer has perfectly rendered the surface of a banana, of a strawberry and, most impressive of all, the hairy skin of a kiwi fruit.

It is more than fifteen years ago now since we started working with Unilever on exactly this topic. Our idea then was to try to enhance the fruity notes in Lipton peach-flavoured iced tea by treating the packaging to give it something of a furry-peach feel. At that time, the solution was just too expensive to be practicable, but – good news for food and beverage companies – giving packaging surfaces a unique and realistic feel is now becoming much cheaper. And I am more convinced than ever, given the research evidence of the intervening years, that getting the feel in the hand right is an important way in which to improve the consumer's experience of food and drink in the future.

No matter whether it be cutlery, glassware, plates or bowls, the gastrophysics approach is providing the evidence and insights to support the creative tactile designs that are now emerging for the dining table. The most extreme and intriguing examples will come from designers, modernist chefs and molecular mixologists. However, my prediction is that the majority of us will be exposed to this new approach through food and beverage packaging – everything from the textured paint on a Heincken can (brought out in 2010, the special cans were intended to have a 'signature' feel) to the silky surface of some high-end boxes of chocolates.

6. The Atmospheric Meal

When we imagine an eating experience, we can't ignore the setting. If you are far from home, in some foreign city, deciding where to eat, don't you always end up gravitating towards the restaurant that has a buzz, a vibe, in other words, the one that has 'atmosphere'? And don't we all mostly stay away from those places where there aren't any diners – you know, restaurants that look stone dead – no matter how highly they come recommended?

Can the atmosphere in a restaurant determine how much we eat, not to mention how much we spend? Many restaurateurs certainly believe so. As the owner of Pier Four in Boston said, back in 1965 when it was one of North America's most successful restaurants: 'If it weren't for the atmosphere, I couldn't do nearly the business I do.'[1] But beyond its effect on table turnover, not to mention maximizing the bottom line, can you really enhance the perceived taste and/or enjoyment of a meal just by picking out the right sort of background music? Does the *same* food actually taste *different* when the atmosphere, or environment, in which it is served changes? As I will show you in this chapter, emerging gastrophysics research shows that the answer to these questions is very often yes.[2]

Environmental attributes, from the music through to the lighting, and from the ambient scent through to the feel of the chair you are sitting on, can – and in many cases do – influence the dining experience. Marketers have long been aware of the profound influence of the environment. The famous North American marketer Philip Kotler, for instance, in his seminal early paper on 'atmospherics', emphasized how a key part of the total product offering is the atmosphere in which a product or service is presented, which is itself multisensory. He drew a distinction between the *tangible product* and the *total product*. He is, I think, worth quoting at length given how influential this work has been: 'The *tangible product* – a pair of shoes, a refrigerator, a haircut, or a meal – is only a small part of the total con-sumption package. Buyers respond to the *total product*. [. . .] One of the more significant features of the total product is the *place* where it is bought or consumed. In some cases, the place, more specifically, the *atmosphere* of the place, is more influential than the product itself in the purchase decision. In some cases, the atmosphere is the primary product.'[3]

To date, the majority of the research on atmospherics has tended to focus on music – the easiest aspect of the environment to change. So, to begin with, let's take a look at the evidence concerning the effect of background music on our dining behaviour.

Moving to the beat

Do you think that you would eat and drink more rapidly if the tempo or loudness of the music in a restaurant increased? Would you end up spending more if they were to play classical rather than top-forty music, say? And would you be more likely to choose something French if accordion music was playing in the background? Sounds unlikely, right? Yet in one of the most impressive demonstrations of background music's impact on consumer purchasing behaviour, this is exactly what was found. In particular, researchers alternated the type of music in the wine section of a British supermarket. When French music was played, the majority of people bought French wine; however, when distinctively German (Bierkeller) music was played instead, the majority of wines sold were German. The numbers have to be seen to be believed (see Figure 6.1).

Most people, when told about such results, are convinced that *they* wouldn't be so easily influenced. So too, in fact, were the custom-ers who were questioned on leaving the tills in the study itself, the majority of whom resolutely denied that the music playing in the background had swayed their purchasing decisions that day. They confidently asserted that they had always intended to buy French wine, as the accordion music played in the background. However, the sales figures tell a very different story. Given results like these, you

Background Music

	French accordian music	German Bierkeller music
Bottles of French wine sold	**40** (77%)	**12** (23%)
Bottles of German wine sold	**8** (27%)	**22** (73%)

Figure 6.1. Number (and percentage in brackets) of bottles of French versus German wine sold as a function of the background music in one of the most oft-cited marketing studies on the impact of ambient music on people's behaviour.[4]

can probably better understand why gastrophysicists are often so sceptical of people's subjective reports. Better to look at what people do, rather than merely relying on what they say.[5]

Do you think that your food preferences would be affected by a restaurant deciding to change its decor? Well, one study that went some way towards answering this question was conducted at the Grill Room at Bournemouth University, in the UK, back in the early 1990s.[6] Researchers wanted to know whether they could alter the perceived ethnicity of a range of Italian dishes without changing the food offering. To that end, a selection of Italian and British foods was offered up over four days. On the first two days, the restaurant was decorated as normal (e.g., with white tablecloths and the walls and ceiling unadorned). For the other two days, the restaurant was given an Italian feel: Italian flags and posters were mounted on the walls and ceiling, the tables were covered with red and white chequered tablecloths. Oh, and a wine bottle was placed on each table for good measure.

The diners (138 of them, to be precise) were invited to complete a questionnaire once they had finished their meal. They were asked how ethnic their meal had been, as well as how acceptable they found the food overall. Giving the restaurant an Italian theme resulted in diners choosing more pasta and Italian desserts such as ice cream and zabaglione, and significantly fewer fish dishes. Adding an Italian feel also resulted in the pasta items being rated as more authentic. The perceived ethnicity of the meal as a whole went up too, with 76% of the diners describing the meal as Italian, as compared to only 37% in the baseline condition. Such results illustrate how what people think about a meal can be influenced by changing nothing more than the visual attributes of the environment in which that dish happens to be served.[7] And given what we have already seen, had Italian music been played too, who knows how much more pronounced the multisensory atmospheric effects would have been?[8]

So at home, you could perhaps make your pizza and pasta taste more authentic by playing a bit of Italian opera. The film director Francis Ford Coppola, for one, insisted on 'musical accompaniments matched to menus – accordion players for an Italian *pranzo*, mariachi for Mexican *comida*' whenever he was filming.[9]

The question remains, what might work best, practically speaking, in terms of music to accompany your takeaway pizza? Well, you'll be glad to hear that here at the Crossmodal Research Laboratory we have done the relevant study in order to determine just that (even if Italian music might make Italian food seem more authentic, it still might not give rise to the best experience for the diner). In a recent project conducted on behalf of Just Eat (an online food-ordering company, like Seamless in the US), we asked more than 700 consumers which of 20 musical tracks worked best with the 5 most common types of takeaway food in the UK: Italian, Indian, Thai, Chinese and sushi. The music spanned several genres, from R'n'B and hip-hop, pop and rock, through to classical and jazz. Pavarotti's 'Nessun Dorma' came out as the top match for takeaway Italian food. Across the board, Nina Simone's 'Feeling Good' and Frank Sinatra's 'One for My Baby' were always ranked in the top three tracks, regardless of the type of takeaway that our participants were evaluating. So they too would seem to be a safe bet for anyone without a compendious music collection. But the really big surprise was that Justin Bieber's 'Baby' came pretty much bottom. So that one is a big no-no – you have been warned. (Sorry, all you Beliebers out there . . . but you can't argue with the data!) As to why the results turned out this way, well, that is a question that we are still working to answer.[10]

Playing classical music in the background tends to result in people splashing out more. This turns out to be true no matter whether one is looking at diners' willingness to pay for the food served in a student cafeteria or customers' actual spending behaviour in a restaurant setting. In fact, a 10% increase in the average bill is not unheard of. For instance, in one study conducted by Professor Adrian North, at Softley's restaurant in Market Bosworth, Leicestershire, the diners spent £2 a head more, on average, when classical music was playing in the background rather than pop.[11] In another study, the customers in a wine store were shown to spend more when classical music rather than top-forty music was played.[12]

Background music can impact how pleasant we rate the food itself.[13] No prize for guessing that the more discomfiting the music, the less time people spend in a given environment, while the more they like the music, the longer they stay.[14] And, generally speaking, the more you like the music, or the environment, the more you like the food and drink.[15] That said, I always suggest to clients that they should conduct their own research in order to know what kind of musical backdrop will work best for them. There may well be some significant cross-cultural differences here in terms of the appropriateness of music (and/or conversation) at mealtimes too.[16] In Korea and Japan, for instance, it is far more common to eat a restaurant meal in silence without people talking much to one another, and with no music in the background.

That said, it is important to bear in mind the congruency between the restaurant concept, the clientele and the type of music that is played.[17] It is hard to believe that classical music would necessarily be the right choice in one of those grungy burger bars that are popping up all over the place these days. It just wouldn't seem appropriate, would it? But for a high-end hotel restaurant, where the mark-up on the French wine is much higher than for any of the other stuff that they have in the cellars, the implications are clear. What we can infer here is that classical music is more likely to prime notions of class, and/or that those who are drawn to classical music may, on average, be a little more affluent.

Next, let's take a look at tempo, i.e., the number of beats per minute (bpm), and loudness. Is the speed at which you eat or drink influenced by the speed of the music playing in the background? By now, I am sure you can guess what the answer to this question will be. And indeed, playing faster music has been shown, across a number of studies, to result in people eating and drinking more rapidly. In what is perhaps the classic study in this area, this time conducted by R. E. Milliman, a North American professor of marketing, back in 1986, the tempo of the music playing in a medium-sized restaurant was manipulated. The 1,400 North American diners whose behaviour was assessed ate much more quickly when fast (as compared to slow) instrumental music was played. When slow music was played, diners spent more than 10 minutes longer eating, bringing the total duration of their restaurant stay up to almost an hour. Although the musical tempo had no effect on how much money people spent on their food, there was a marked difference on the final bar tab, with those exposed to the slow music spending around a third more! Slowing the music down increased the restaurant's gross margin by almost 15%, no doubt a sensible idea at quiet times. However, when the customers are queuing out of the door, the restaurateur might be better off playing some fast-paced music instead.[18]

But would a restaurant chain really go to such lengths to control the flow of customers? Absolutely! Just listen to Chris Golub, the man responsible for selecting the music that plays in all 1,500 Chipotle restaurants in the US: 'The lunch and dinner rush have songs with higher bpms because they need to keep the customers moving.'[19] In fact, Golub is often to be found sitting in his local New York branch of Chipotle, observing people's behaviour in response to the different music that he is thinking of adding to the playlist. Then, depending on the customers' responses, he will fine-tune it both for tempo and style before it is beamed out to branches across the land.[20] All that is missing here is the statistical analysis and this would be fully-fledged gastrophysics research!

Ultimately, of course, restaurateurs and bar managers are primarily interested in increasing their profits. The Hard Rock Café chain, for instance, plays loud music in its venues because of the positive effect on sales. Just take the following, from a piece that appeared in *The New York Times*: '[T]he Hard Rock Café had the practice down to a science, ever since its founders realized that by playing loud, fast music, patrons talked less, consumed more and left quickly, a technique documented in the International Directory of Company Histories.'[21] And according to another report: 'When music in a bar gets 22 per cent louder, patrons drink 26 per cent faster.'[22] This shines a whole new light on why so many restaurants and bars are getting louder than ever before. Put simply, it makes us spend more!

That said, much of the underpinning research was conducted years ago, in a different time and place; the results may no longer apply. My recommendation to the restaurateur is just to be aware of the importance of the atmosphere to the food offering you provide. This should at the very least help you to avoid the situation whereby the chef who obviously cares passionately about their food allows the front-of-house manager to put their iPod on random shuffle. You know the sort of thing: you suddenly find yourself listening to Frank Sinatra singing 'Jingle Bells' in the middle of July when eating in a Thai restaurant, say! This really shouldn't happen. But I think we all know that it does, and more often than it ought to. If you have the opportunity to experiment, why not try French music this week and American Rock next, or classical music today and the top-forty tomorrow, and see for yourself the impact on what people say (or, more importantly, on sales). Gastrophysics research provides a number of suggestions as to what might happen, but you will have to check for yourself to be sure of getting the effect you want.[23] Broadly speaking, though, when the atmosphere matches (or is congruent with) the food offering, people will tend to enjoy the experience more.[24]

Do you care about being comfortable?

Have you ever wondered why most trendy coffee shops have such hard and uncomfortable seating? Well, put simply, they just don't want you to linger. I know of a number of baristas who deliberately chose hard, uncomfortable furniture in order to discourage their customers from loitering and hogging the tables all day long. You don't need a gastrophysicist to tell you that the less comfortable the chair, the shorter your stay will be. McDonald's has been doing this for years; as one commentator put it: 'The rule written into the design of the seats [in McDonald's] is that 10 minutes is the appropriate length of one's stay before they become uncomfortable.'[25] At the high end, though, where the duration of the customer's stay isn't an issue, restaurants are increasingly thinking about how to augment the feel of the space in which their food is served. A few innovative chefs, such as Joshua Skenes, chef-owner of Saison in San Francisco, have started to play with giving their restaurants a distinctive feel. According to the chef: 'You need great food, great service, great wine, great comfort. And comfort means everything. It means the materials you touch, the plates, the whole idea that the silverware was the right weight. We put throws on the back of the chairs.'[26] Take a look at Figure 6.2; it seems Noma in Copenhagen are doing much the same thing.

Would you rather sit at a round or a square table? Generally speaking, people prefer round (or curvilinear) to angular forms, a preference that extends all the way from everyday objects through to architectural spaces and even furniture. Some evolutionary psychologists put

Figure 6.2. A chair with texture at Noma in Copenhagen.

this seemingly ubiquitous preference down to angular forms being associated with danger (think sharp and dangerous weapons). Of course, due to practical constraints, the majority of traditional restaurant floor plans are angular. But a square space can be the frame for much rounder forms, in the decoration or the furniture.

In one recent study, a group of North American university students was shown pictures of interior environments containing either angular or rounded furniture. The results highlighted a preference for rounder furniture, with the latter tending to elicit greater feelings of pleasure as well. Interestingly, in this case, the participants reported a greater desire to approach the curvilinear rather than the rectilinear furniture. As one of the participants said: '[R]ounded furniture seems to give off that calming feeling.'[27] Round tables can be used to help make the interior of a restaurant appear more welcoming. But they also reduce capacity – presumably why many restaurant consultants recommend a mixture of round and angular tables, aiming for a balance between approachability and profitability.

How would you like to dine in a white cube?

There are some traditional establishments where no attempt has been made to augment the atmosphere whatsoever: Just think of those temples to haute cuisine with their unadorned white walls, diners sitting before a starched white tablecloth (or, as is currently fashionable, just a starched napkin), eating in hushed and respectful silence.[28] No one could claim that such venues were trying to distract their diners from the food, right? The idea of releasing ambient fragrance or changing the temperature of the space to match the dishes being served would, one imagines, be complete anathema to such traditional restaurateurs.[29] I guess that there will always be a place for such austere dining rooms. But my sense is that it is hard to make this kind of approach seem contemporary or exciting, at least in the current climate. More often than not, such venues are being replaced (in the San Pellegrino listing of the world's fifty best restaurants, say) by the more experiential dining concepts.

What is more, bear in mind that by removing atmospheric cues one is still making a statement. Here, I am reminded of a quote from one commentator: 'The modern restaurant is an experience in codes. The architecture, the foods served, and even the customers are codes built up to the total consumable image. Restaurants then do not just serve food. They serve an experience.'[30] So while the decor may be minimalist, the atmosphere is very definitely never 'neutral'. Make no mistake, the dishes served in the 'white cube' environment will be rated differently by diners than in any other environment that one might care to choose. Based on the evidence, the food would probably be rated as better in quality and more expensive, though perhaps less memorable. The key point is that there is always an atmosphere wherever food is served and consumed.

The same also goes for those healthy, natural, organic stores and restaurants, the places with the baskets of fresh produce on display as you walk in (many of Jamie Oliver's restaurants fall into this category). Make no mistake about it, this kind of atmosphere is itself priming notions of healthy and natural in the mind of the diner.[31] It may look casual, but it most certainly isn't – the creation of the display is itself artifice. Often, a great deal of thought has gone into constructing that 'natural' environment to be just so. That is part of the conceit; in fact, I bet it affects the experience just as much as in other restaurants where the atmosphere changes on a course-by-course basis. They are still creating an impression, an expectation that will colour the whole encounter between the diner and their food and drink.[32]

Over the years, some restaurants have really gone overboard in terms of delivering a multisensory atmosphere. One of the most famous early examples here is the Tonga Room & Hurricane Bar, which opened down in the basement of the Fairmont hotel in San Francisco back in 1945.[33] I still remember visiting as a young graduate student, long before my current interest in multisensory dining had taken root. A spectacular tropical thunderstorm, with simulated thunder and lightning, would unfold every thirty minutes or so during opening hours. Good idea though this was, presenting the same old sound and light show for so many years had taken its toll on the place – it was looking a little tired. Furthermore, it is easy to imagine how the customers might well habituate to the repeating multisensory scene that unfolds.

A little over five decades after The Tonga Room first opened its doors to the public, and on the other side of the pond, as it were, one finds The Rainforest Café. This well-known London restaurant also delivers an experience that tries to stimulate *all* of the customers' senses. Every half an hour or so, the restaurant goes dark while the guests are 'treated' to all the rumbling and flashing of a thunderstorm in the tropics. While The Tonga Room targets a more mature clientele, the latter venue obviously has its sights set on a much younger market (or rather, on the pockets of those who have been charged with looking after them). As the self-styled engineers of the experience economy (see the 'Experiential Meal'), B. J. Pine, II, and J. H. Gilmore, say: 'The mist at the Rainforest Café appeals serially to all five senses. It is first apparent as a sound: Sss-sss-zzz. Then you see the mist rising from the rocks and feel it soft and cool against your skin. Finally, you smell its tropical essence, and you taste (or imagine that you do) its freshness. What you can't be is unaffected by the mist.'[34]

No matter whether the grown-ups like it or not, there can be no doubting how incredibly successful the experience has been amongst the target audience, namely children. For a few years, my nieces were huge fans, though I suspect that they have rather grown out of it now. And what is absolutely clear is just how successful the venture has been, commercially speaking.[35] In other words, atmospherics sells, or at least it does when done well.[36]

There has sometimes been, it has to be said, a lingering suspicion that restaurateurs are interested in the atmosphere only insofar as it differentiates them from the competition and increases their bottom line. Of course, it is all too easy to get a little sniffy about the grubby financial side of things, but who isn't ultimately interested in at least breaking even? As the influential British chef Marco Pierre White once put it: 'Any chef who says he does it for love is a liar. At the end of the day it's all about money. I never thought I would ever think like that but I do now. I don't enjoy it. I don't enjoy having to kill myself six days a week to pay the bank . . . If you've got no money you can't do anything; you're a prisoner of society. At the end of the day it's just another job. It's all sweat and toil and dirt: it's misery.'[37]

One could also mention the popular dine-in-the-dark restaurants here (e.g., Dark Restaurant Nocti Vagus in Berlin) – they clearly fit into the atmospherics framework, with a sensory input removed, rather than added. Nevertheless, going to one of these restaurants is very definitely an experience, but not necessarily one that is centred on great-tasting food.[38]

To summarize, then, the atmosphere affects our food behaviours in a number of ways: everything from influencing where and what we 'choose' to eat, through how long we stay, not to mention what we think of the overall experience (see Figure 6.3).[39] But it is worth noting that we haven't so far addressed the more fundamental question of

Figure 6.3. All of the senses play a role in controlling our behaviour while drinking and dining. The intelligent restaurateur knows how to work with the senses to create the right environment. The scientific approach to multisensory atmosphere design has led to a number of restaurant chains increasing their profitability.[40]

whether changing the environment really does modify what people perceive on the plate or in their glass. This is the kind of question that is of most interest to the gastrophysicist.

Atmospheric tasting

Listen to the chefs and you'll hear conflicting views. When interviewed a couple of years ago, French chef Paul Pairet was quoted as saying that he didn't believe that all the multisensory atmospherics in his restaurant Ultraviolet in Shanghai made any of his dishes taste better. Rather, he thought simply that 'the memory of the dish is stronger'. A worthy enough aim, but is that all there is to it? Ironically, the press report in which Pairet is quoted itself seems to come to a different conclusion. For according to the journalist: 'each dish is accompanied by a carefully choreographed set of sounds, visuals and even scents, all intended to create a specific ambience to enhance the flavors of the meal.'[41] Pairet is not alone in his views, though. Others include French chef Alain Senderens, who once complained about the Michelin man's penchant for fancy fittings. 'I was spending hundreds of thousands of euros a year on the dining room – on flowers, on glasses,' he said, 'but it didn't make the food taste any better.'[42]

In the other camp there are those, like Heston Blumenthal, who have latched on to the fact that the atmospherics really can modify the tasting experience. We first demonstrated this with Heston at the 'Art and the Senses' conference held in Oxford back in 2007. The lucky participants at this event got to eat oysters while listening to the sounds of the sea and to taste bacon-and-egg ice cream to either the sound of sizzling bacon or the clucking of farmyard chickens.[43] We demonstrated that people rate bacon-and-egg ice cream as tasting significantly eggier when listening to the sound of farmyard chickens clucking in the yard, but when we played the sounds of sizzling bacon instead, suddenly the bacon flavour became rather more intense. Changing the atmospheric sounds altered people's perception of the test foods. Playing the sounds of the sea also made the oysters more pleasurable (but no saltier).[44]

In the years since, I have been lucky enough to team up with some of the world's leading drinks brands to conduct various large-scale multisensory tasting events with members of the public. Typically, these events have built on the belief that changing the atmosphere will influence the tasting experience. And rather than manipulating just the sonic environment, we have been working to modify the visual and olfactory aspects of the environment too. Let me share a couple of these experiences with you.

'The Singleton Sensorium'

Typical of this gastrophysics approach was 'The Singleton Sensorium', which took place over three evenings, in the heart of Soho, London, in 2013. My colleagues from Condiment Junkie, a UK-based sound agency, decorated three rooms in an old gun-maker's studio in very different styles. One room aimed to recreate an English summer afternoon, another was designed to prime nations of sweetness, while the third room had a distinctly woody theme. The Condie boys also generated some atmospheric soundscapes to play in the background in each room. Take the sweet room. It was decorated in a pinky-red hue, chosen because that is the colour that most people generally associate with sweetness. There was nothing angular in the room; everything was round – the pouf, the table, even the floor plan and the window frames. Why? Well, because our research had shown that people associate rounder shapes with sweetness. There was also the sweet-smelling but non-food-related fragrance and the high-pitched tinkling of what sounded like wind chimes coming from a ceiling-mounted loudspeaker. The latter choice was again based on our laboratory research showing that people associate such sounds with sweetness.[45] So every sensory cue had been chosen on the basis of the latest gastrophysics research to help prime, consciously or otherwise, notions of sweetness on the palate. The first room, by contrast, had been designed to prime grassiness on the nose. The final 'woody' room was meant to prime a textured finish, or aftertaste, in the mouth.

Over three evenings, nearly 500 people were escorted in groups of

10 to 15 through an experience lasting no more than 15 minutes. At the outset, everyone was given a glass of whisky, a scorecard and a pencil. They filled in one section of the scorecard while standing in each room. The punters were asked about the grassiness of the whisky on the nose, the sweetness of its taste, and the woody aftertaste. They indicated how much they liked the whisky, and what they thought of the decoration in the room that they were standing in. I was one of the tour guides, and let me tell you, it was an exhausting experience. It was the first time that anything like this had been tried on this scale. Would the experiment work as planned, or would people simply walk away saying that the whisky obviously tasted the same in all three rooms because it was, after all, the same whisky?

It was a huge relief to find, once the results were analysed, that, as a group, people rated the grassiness of the nose of the whisky as significantly more intense in the grassy room. Meanwhile, the second room brought out the sweetness on the palate (as expected), and the final (woody) room really did accentuate the textured finish of the whisky. As a psychologist, one always worries about so-called 'experimenter expectancy effects',[46] i.e., that your subjects may say what they think you want to hear, rather than tell you what they actually experienced or thought. In fact, at the end of the 'Sensorium', one or two people did approach me and say something of this sort: 'We knew what you were up to. You wanted us to say the whisky tasted grassier in the green room, right? So we did the opposite!'

Note, though, that even these truculent individuals were not *unaffected* by the multisensory environment (at least in a manner of speaking). And crucially, the group analysis revealed that such individuals were clearly in the minority. What is more, people enjoyed the whisky most in the woody environment. So, manipulating the multisensory atmosphere in this scientifically inspired manner really did affect what people had to say about the drink in their hand. Depending on the room, the change in people's ratings of the nose, taste and finish of the whisky were in the order of 10 to 20%.[47]

Would whisky experts have been equally affected by 'The Singleton Sensorium'? It is difficult to say for sure. However, it is worth noting that neither the whisky expert nor, for that matter, the wine aficionado can necessarily do all of the things that they think (or say) that they can when it comes to blind-tasting.[48] What is perhaps more relevant is that the experience was powerful enough for a number of chefs, restaurateurs and designers to go away and change the way in which they delivered some of their food and beverage offerings. For example, staff at one famous restaurant in the Lake District, in north-west England, started serving guests whisky from a wooden tray, matching the environment in which they themselves had most enjoyed the drink while taking part in the event.

'The Colour Lab'

What do you think would be the best colours to bring out the fruitiness and freshness of a wine? And could you achieve the same effect by playing sweet or sour music (sour music tends to be dissonant, high-pitched, rough, sharp and staccato)? We set out to answer these questions in what may well have been the biggest ever tasting event of its kind, known as 'The Colour Lab'. More than 3,000 people were tested over an unseasonally warm May Bank Holiday weekend on the banks of the River Thames in London, as part of 'The Streets of Spain' festival. Each person was given a glass of Spanish Rioja in a black glass. They had to rate the wine first under regular white lighting (to get a baseline measure), then under red illumination, then again in a green environment with 'sour' music. Finally, they tasted the wine under red lighting again, but this time accompanied by 'sweet' music. Once again, a 15 to 20% change in people's ratings was observed on switching from one audiovisual atmospheric combination to another.[49] The red lights and sweeter music (consonant, high-pitched, neither rough nor sharp but smooth and flowing) were found to accentuate the fruitiness in the wine, while the green colour and sourer music brought out its fresher notes instead.

While previous gastrophysics research had already demonstrated (albeit on a much smaller scale) that changing the colour of the light bulbs or the music playing in the background can change what people say about the wine they are tasting, we were the first to combine the senses in a multisensorially congruent manner. We were looking for what some have termed a 'superadditive' effect. Put simply, this is the idea that the various atmospheric cues might combine to deliver a multisensory effect that was bigger than the sum of its parts (i.e., greater than what you might expect merely from adding the effect of light and sound when presented individually). And as we had hoped, the sonic seasoning – sweet music with red lighting and sour music with the green lighting – did indeed enhance the effect of the lighting on the taste of the wine.[50]

One of the results (or deliverables) of such multisensory events is the statistical evidence that the environment influences people's perception. On occasion, the results may also reveal the relative importance of the senses to that experience. Often, though, what is much more powerful, at least in terms of convincing people, is the change in what they themselves felt.[51] In fact, when we trialled 'The Colour Lab' on the wine makers from Campo Viejo, they were so impressed that they left saying that they would be redesigning their own cellar-door experience as soon as they got back to Spain. Furthermore, one of the wine writers with whom I work, who was, by his own admission, initially sceptical, now uses changes in the ambient lighting as a party trick when leading informal wine tastings.[52] So next time you open a bottle of wine at home and find that it is not quite to your taste, why not just try changing the music and/or the lighting first before you reach for a different bottle? Sometimes, it really is that simple (assuming that the wine doesn't have any obvious flaws).[53] Nowadays you can buy remote-control colour-changing light bulbs for virtually nothing online. So there really is no excuse, is there?

If you are wondering what counts as sour music, you could try Nils Økland's 'Horisont'. For something sweet, look for tracks with lots of tinkling, high-pitched piano notes. I often use something like 'Poules et Coqs' from Camille Saint-Saën's *Carnival of the Animals* or tracks 6 and 7 from Mike Oldfield's *Tubular Bells* (1973). Choose something like *Carmina Burana* by Carl Orff or 'Nessun Dorma' from the third act of Puccini's *Turandot* if you want to bring out the depth in a red wine, say a Malbec.[54]

These lighting manipulations work well enough when the wine is served in a black tasting glass (as in 'The Colour Lab'). I could well imagine that the effects would be even more pronounced if the wine were to be served in a clear glass, where the colour of the wine itself may also change as a function of the ambient lighting. You do need to be a little careful here, though, because if too dramatic a change in the lighting is introduced at mealtimes, it may actually change the visual appearance of the food itself. As one commentator put it: 'A red light makes everything look red; a green light makes meat look grey and spoiled.'[55]

Of course, different people have different objectives as far as the use of ambient lighting is concerned. While some may wish to bring out the freshness in their wine, others may be wondering whether there are certain colours or, for that matter, types of music, that may help to promote healthy eating, for instance, use of red lighting to bring out sweetness (without the calories). Research shows that the colour of the ambient lighting can influence a diner's appetite. For example, yellow lighting was found to increase people's appetite, whereas red and blue lighting decreased people's motivation to eat. When the colour of the food and of the ambient lighting match, it seems to stimulate appetite, whereas complementary colours suppress it.[56] Also relevant here are the results of a recent study conducted in Sweden, in which it was found that dieting Swedish males eating their breakfasts under blue lighting felt fuller with less food.[57]

Controlling the restaurant environment

So we know what happens in experiential events. But do you think that you would eat less if the lighting and music in a restaurant were softened to create a more relaxed atmosphere? Well, researchers tested the combined impact of changing the lighting and music on the behaviour of diners in Hardee's, a fast food restaurant in Champaign, Illinois. The restaurant in question had two dining areas (ideal for gastrophysics research). In one, the lighting was set at its normal bright level, the colour scheme was also bright and the music playing in the background was loud. The other 'fine dining' environment had a much more relaxed atmosphere: there were pot plants and paintings, window blinds and indirect lighting. Oh, and did I mention the white-tablecloth-covered tables with candles on top, and soft jazz instrumental ballads in the background? Those who ate in the more relaxed side of the restaurant rated their meal as significantly more enjoyable, while at the same time consuming less (their calorie intake was reduced by an average of more than 150 calories, or 18%).[58]

The fact that the environment has such an effect on us could obviously have implications for the restaurateur. In fact, it has been suggested that this is precisely why the Hard Rock Café and Planet Hollywood chains have no windows, which gives them (just like the casinos) greater control over the environmental stimulation their customers are exposed to.[59]

The future of atmospherics

So, how might the atmospheric aspects of dining change, moving forward? As one designer put it recently: 'In the short time I have been in the business of designing restaurants, design has definitely become a major element of the dining experience. The environment and its uniqueness are becoming as important as the food, and designers and owners are becoming more sophisticated in how they use light, colour and materials.'[60] And if you want to see what the future may hold in terms of restaurant design, why not take a look at the Goji Kitchen & Bar at the Marriott Bund hotel in Shanghai. The decor in this futuristic dining space actually changes to give the restaurant one of two different feels, depending on the time of day.[61] This is undoubtedly an expensive solution, but it is also an acknowledgement that decor and atmosphere really do matter. It stands as testament to the importance of the atmospheric component of 'the everything else' to mealtimes. And while it is always difficult to figure out how much to spend on the decor, as soon as you are aware of just how much it influences the dining experience, then there really is no looking back. However, while getting the atmosphere 'right' is undoubtedly a challenge, it is also important to remember that there is no way to avoid its influence, much though one may wish to.

There is an interesting challenge here in terms of how to customize the atmospherics to individual diners, or tables.[62] Currently, most high-end multisensory dining experiences involve either a single-sitting restaurant (think Ultraviolet in Shanghai or Sublimotion in Ibiza), or headphones being brought to the table to accompany a particular dish (the 'Sound of the Sea' at The Fat Duck, for instance). However, I know of restaurateurs out there who are already thinking about whether they can use hyper-directional loudspeakers positioned over the tables in order to deliver a soundscape personalized to whatever the diners happen to be eating and drinking. Crucially, none of the diners would be able to hear what was going on at any of the other tables. This kind of solution is currently prohibitively expensive for all but the most affluent of restaurateurs. Nevertheless, looking ahead, I can well imagine that it will become more common, given the growing emphasis on personalization and customization, together with the falling cost of technology.

In this light (if you'll excuse the pun), there are now multi-coloured LEDs installed over each of the tables at The Fat Duck restaurant, following its recent refurbishment. These bulbs subtly change colour as the diners at a given table progress on a journey from night to day and on to the next evening during the course of their meal.[63] The lighting changes take place at different times on each table. Is this really the future of personalized atmospherics? I suspect that it might be a start.[64]

7. Social Dining

I don't know about you, but I don't like dining out alone. And so a human-interest story in the papers recently caught my attention. It concerned a nonagenarian British widower, Harry Scott, who had been dining alone at his local McDonald's pretty much every day, sometimes twice a day, for the last three years. The sad truth was that he simply has no one else to share his meals with since his wife died. So, on the occasion of his 93rd birthday, the staff at the branch in Workington, Cumbria, held a party for the old geezer. I have to say, looking at the picture that appeared in the papers at the time, that Harry appeared to be in much better shape than you might have expected.[1]

While this is an isolated case, I nevertheless think that it is indicative of what is happening in society at large. For the fact of the matter is that a growing number of us are dining alone (see Figure 7.1). Nearly half of all meals are now eaten alone, and more than a quarter of us eat by ourselves more often than we eat in company, according to one recent British survey. Worse still, many of us eat our main meal in isolation most days (perhaps eating lunch alone at a desk, or consuming a microwaved meal, or grabbing a bite to eat at a drive-thru).[2] Given that the figures vary by culture and age, why not count up how many meals *you* have eaten by yourself over the last week, say, and see whether or not you buck the trend.

But why, you might ask, should we even care about this change in our eating behaviour, beyond being concerned by the growing isolation that it hints at in society today? And what exactly has the company we keep got to do with the experience of the food we eat, that is, with the topic of this book, namely gastrophysics? Not everybody thinks we should be worried; here's Nell Frizzell, writing in the *Guardian* : 'Like life's other great pleasures, eating alone is something you can do one-handed, lying on your back, in nothing but an old jumper, should you so wish. It isn't lonely, it isn't distasteful, it isn't desperate: it's a celebration of existence. It keeps us alive – as simple as that.'[3] I beg to disagree. As we'll see later, much of the evidence points towards dining solo having a negative impact on people's physical health and mental well-being.[4] A recent meta-analysis of seventeen different studies, involving more than 180,000 adolescents and children, revealed that regularly sharing meals as a family reduced the odds of youngsters being overweight by 12%. It also increased the likelihood of their eating healthy foods by almost 25%. Not only that but as a gastrophysicist I would agree with American psychologist Harry Harlow, who, back in the 1930s, put it thus: '[A] good meal tastes better if we eat it in the company of friends.'[5] Gastrophysics offers a constructive framework in which to look for solutions to the growing problems associated with solo dining.

Figure 7.1. Dining solo – a growing problem in society today.

Why are so many of us eating alone?

Part of the reason is undoubtedly linked to the fact that more of us are *living* alone than ever before, due to people marrying later, divorce rates going up and people living alone for longer.[6] Another important factor here relates to our changing food habits. For one thing, there are simply far fewer meal-based family gatherings than at any point in the recent past. When was the last time you asked anyone over to your house for dinner? According to the results of one recent survey, 78% of Brits virtually never invite their friends over at mealtimes any more. When quizzed about why, the reason that people often give is that they just find it too much of a chore to prepare food from scratch, given their increasingly hectic lifestyles. Indeed, the average time spent preparing a meal has dropped from around an hour back in 1960 down to just 34 minutes today. What all of this means is that one in three of us will now go an entire week without eating dinner with another person.[7]

Is solo dining really so bad?

Dining solo is bad for us in a number of ways. On the one hand, those who eat by themselves tend to engage in poorer food behaviours. In males, for instance, eating and living alone are jointly associated with a higher prevalence of weight problems, by which I mean obesity at one end of the spectrum and being underweight and/or engaging in unhealthy eating behaviours, such as a low fruit and vegetable intake, at the other.[8] Unsurprisingly, those who dine solo are also more likely to feel lonely.[9] Many older individuals who find themselves in hospital or in long-term care also suffer from undernutrition, which is made worse by their often being forced to eat solo. *Anything* that can be done to bring the social element back to dining is likely to help improve the nutritional status of these vulnerable individuals. For instance, a couple of studies conducted in the US demonstrated that elderly hospitalized patients end up consuming significantly more food when they are encouraged to engage in more active, inter-personal behaviour with their care-givers at mealtimes.[10] Those who live by themselves, and hence regularly eat alone, also tend to generate far more food waste than those who live and dine with others, a problem exacerbated by the fact that the portion sizes that are commonly available in the supermarket often do not cater to those living alone. According to a 2013 UK Government survey, people living alone throw away 40% more of their food than those living with others.[11]

Dining with distraction

The decline of the more social aspects of dining is not restricted to the increasing proportion of us who, for whatever reason, eat alone. Technology also has its part to play. How often, after all, do *you* have the TV on at mealtimes? And how often do *you* find yourself with fork, spoon or chopsticks in one hand and your smartphone in the other? Even those of us who are physically sitting together around the dining table are too often distracted by what is on the TV, or else fid-dling with our mobile devices. In fact, according to the statistics, almost half of us watch TV while eating,[12] and many of us do so in separate rooms from our companions! One ingenious solution to counteract all this technology-induced distraction was trialled by Brazilian bar Salve Jorge back in 2013. They introduced the 'Offline Glass', a beer glass whose base is partly cut away so that it can only stand up if supported by the customer's mobile device. The idea was that this enforced 'disconnection' from the patron's technology would hopefully result in people being more sociable when out drinking.[13]

I am sure that we have all seen those unromantic couples dining together, not speaking to one another, engrossed instead in whatever happens to be on their mobile screens. Dining together alone, as it were. Of course, sometimes even those who aren't distracted by their hand-held technology don't necessarily have anything much to talk about. One idea to alleviate this situation comes from the Bocuse res-taurant at the Culinary Institute of America campus in upstate New York, where they provide a box filled with cards on every table. Each card poses a culinary question or joke. But why are they there? It is certainly not something that you would normally expect to see in a fancy restaurant, is it? When I asked, the last time I visited, the institute's director told me that these games had been strategically introduced to help break the ice for those couples who may be lost for anything to say. The hope is that the cards will improve the diners' mood, and so enhance their enjoyment of whatever they happen to order. This is another example of a mental palate cleanser, like the plastic cow we came across earlier.

Eating with the TV on is one of the worst things you can do in terms of increased consumption. Finding that people eat 15% more food with the telly on as compared to when it is off is not unusual. That said, not all shows are equally bad for our waistlines. It seems to depend on how engaging the TV show is, and whether or not we have seen it before. For instance, Dick Stevenson and his colleagues in Australia found that women who viewed the same episode of *Friends* twice consumed significantly more food than those who got to watch different episodes of the hit TV show instead. Generally speaking, the more food-related sensory cues we are aware of, the less we tend to eat. Hence, when distracted by the television (and presumably this also applies to mobile devices), the danger is that we simply fail to pay attention to the food-related stimulation and so end up consuming more before we realize that we are actually full. And there are other reasons not to watch TV while eating, as this no-nonsense advice suggests: 'Mealtimes have been noted as one of the most common times children communicate with parents, so if possible, guard your mealtimes from outside distractions. Turn off the TV and cell phones.'[14]

Do you enjoy dining alone?

I have a few colleagues, mainly, it has to be said, chefs, who say that they sometimes (only sometimes, mind you) prefer to eat alone. Why? Well, because it allows them to really pay attention to what is on their plate (i.e., to the flavour combinations and texture contrasts). If they are going to a gastronomic hotspot, they may well choose to dine solo rather than risk being distracted by the need to make conversation. Perhaps I should have used this line the time I turned up by myself for a meal at The Fat Duck. Call me unromantic, but I hadn't even registered that it was Valentine's Day!* Needless to say, Heston has enjoyed teasing me about this whenever the topic has come up since.

For me, like many others, I suspect, the experience of dining alone is never anything like as enjoyable as dining with others, no matter how enjoyable the food. After all, great food and drink are nearly always pleasures best shared, and the better the meal or wine, the more you want to share it.[15] To the extent that our mood is likely to be better when we dine in company than when sitting by ourselves, I am sure that the food and drink really does taste better to us too when in company (at least when in company we enjoy). Intriguingly, dramatic mood swings are associated with significant changes in taste and smell sensitivity. Our hedonic response to food and drink can also be affected. Who, after all, ever had a great-tasting meal when fighting with their partner?[16]

The shared meal is a universal human phenomenon, with evidence of feasting going back more than 12,000 years in the archaeological record.[17] And there is little more expressive of companionship – derived from the Latin *cum*, 'together', plus *panis* 'bread' – than the shared meal. Carolyn Steel, in her book *Hungry City*, makes the point that '[w]e are hard-wired to feel close to those with whom we share food, and to define as alien those who eat differently from us'.[18] (She also picks up on a great line from Oscar Wilde's *A Woman of No Importance*: 'After a good dinner one can forgive anybody, even one's own relatives.') The latest research shows that eating together can increase agreeableness too – thus giving a whole new slant to the topic of gastrodiplomacy.[19]

According to my Oxford psychology colleague Professor Robin

---

* In the good old days, the solo diner would distract themselves with a book. Nowadays, of course, it is increasingly the smartphone that offers companionship at the table. However, being one of the only people left on the planet without a mobile phone, that option unfortunately wasn't open to me.

Dunbar: 'The act of eating together triggers the endorphin system in the brain and endorphins play an important role in social bonding in humans. Taking the time to sit down together over a meal helps create social networks that in turn have profound effects on our physical and mental health, our happiness and wellbeing and even our sense of purpose in life.'[20] All the more worrying given the latest statistics showing that almost 70% of people have never shared a meal with their neighbours. Furthermore, when quizzed, 20% of us admitted that it had been more than six months since we last shared a meal with our parents.[21] It should never be forgotten that: 'The table is the original social network.'[22]

Gastrophysics research shows that dining with companions can, and probably does, exert a significant impact on how much we eat. Whether we eat *more* or *less* depends on who we are with, and how much we are trying to impress them. Evidence from both the laboratory and from more naturalistic dining studies shows that the amount of food consumed typically goes up when we dine with others, as compared to when we eat alone. This increase is more pronounced when dining with friends and family than when we are with those with whom we are less familiar.[23] Males, in particular, tend to eat more in groups than when alone in the restaurant setting. These social effects on consumption may, in part, relate to the longer duration of a dinner in company. However, if we are trying to impress whoever we are with (or else are nervous!), then we may well end up eating less. We also tend to consume less than we otherwise might when those around us hardly touch their food. Remarkably, such social effects on consumption have even been demonstrated in subjects who have had nothing to eat for twenty-four hours.[24]

The next time you go out for a meal you should bear the following in mind. The biggest problem with dining in company is the fact that it reduces the likelihood of you getting to order first. This matters because those who order first generally tend to enjoy their food and drink more than those who order later. The latter often end up feeling that they should have chosen something different and, as a result, they may enjoy the experience a little less than they would have done had they chosen first.[25] And given that women tend to order first in mixed company, this probably means that they generally enjoy their meals out a little more than men.

As we saw in the 'Sound' chapter, many people have been complaining in recent years about restaurants and bars that are so noisy that they simply can't hear themselves think, never mind taste the food.[26] As one commentator put it: 'You go to restaurants to be social. These days, you often come out none the wiser of what the other person has said.'[27] As we have seen, one response to this has been to go to the other extreme and introduce silent dinners, where none of the diners are allowed to speak. However, it is the fact that the meal is fundamentally a social activity/occasion that helps, I think, to explain the long-term failure of this particular concept. Putting in earplugs to drown out the noise, or donning a pair of headphones to play a dish-specific soundscape or piece of music, works well for one course, but any more than that and the social dynamics of the meal will be disrupted too much.

Notice, in contrast, that while all those dine-in-the-dark restaurants also remove one of the diner's senses, the darkness doesn't mess with the social aspects of dining. This is the key difference between dark and silent dining. If anything, diners are likely to have *more* to talk about when the lights go out – for instance, comparing their uncertainty about what exactly it is that they are eating.

Catering for the solo diner

Until recently, if you saw someone dining alone, they might have seemed a little sad, almost a social outcast. 'What, don't they have any friends?' you might have thought to yourself. This stigma, though, is starting to wane. Indeed, more people are dining out by themselves than ever before, with the number of solo diners more than doubling in the two years to 2015. In fact, a party of one is the fastest-growing size for reservations at restaurants in the UK. But what do all these solo diners do to keep themselves entertained while waiting for their order to arrive? According to one recent survey, 46% of people said that they would pass the time reading a book, while 36% played with their phone.[28]

The changing sentiment is nicely captured by the following quote from someone writing to the BBC after the publication of an article on solo dining: 'I remember a time, only a few years ago, when I found the idea of eating out alone to be a depressing prospect. I would view solo diners as sad and lonely people. Now, I eat out by myself quite often, and sometimes prefer it to the company of others. One thing more than any other has made this change from my perspective – a smartphone. I suppose solo diners really aren't solo any more at all.'[29] Some even view those individuals who dine solo in public as being confident high-achievers, out enjoying the well-deserved rewards of their labour, like the wonderfully opinionated food critic Jay Rayner: 'I'm not worried about anyone thinking I'm a sad bastard [. . .] Eating alone should be dinner with someone you love.'[30]

Part of this change in attitude may relate to the fact that more diners are now sharing every detail of what they eat, and pictures of their meals, on blogs or social media. There is even a collection of images on Tumblr called 'Dimly Lit Meals for One'. This recent trend seems to be gaining traction and goes hand in hand with our growing connection with our mobile devices, a relationship that, according to some, is best described as nothing short of a 'love affair'. Others, meanwhile, are happy to isolate themselves with their MP3 players and over-ear noise-cancelling headphones.

Many forward-thinking restaurateurs see this shifting demographic as a marketing opportunity,[31] like Ivan Flowers, the executive chef at San Diego's Top of the Market restaurant. He was brought in by the management specifically to increase the number of solo diners: '[They] felt that while the eatery already had bar seating in front of the open kitchen, it was underutilized because the chefs weren't interacting enough with the customers.' Flowers continues: '[S]olo diners sitting by the kitchen now get "to see a show", which includes cooking demos, free tastings and conversation with the chefs.' Some newspapers have now even started to make recommendations for solo diners in their restaurant reviews.[32]

Eenmaal, a pop-up restaurant in Amsterdam only features tables for one. Who would have guessed that such a venue would have been booked solid for a year since opening its doors? No surprise, then, that those behind this venture are currently planning to expand, with branches planned for London, Berlin, New York, Antwerp and beyond. In the words of Marina van Goor, the project's designer: 'I noticed that in our society, there is no room for being alone in a public space, unless you are going somewhere.'[33] My suspicion is that changing the layout of the restaurant so that it caters solely for individual diners will never be anything more than a niche offering, given how much we still rate the eating experience more highly when in company. That said, many restaurateurs could nevertheless do much more to adapt what they offer to this change in our eating habits.

Tapas-ization

Even as solo dining is on the rise, statistical analysis of the language of restaurant menus reveals a marked increase in the use of sharing terms. Nowadays, one is far more likely to see charcuterie boards, tapas and mezze plates appearing on the menu, all, note, dishes designed for sharing. They are also more informal – another popular trend in dining currently.[34] Here in Oxford, for example, at our top gastro-pub, the Magdalen Arms, many of the dishes on the menu are for two, three, four or even five diners to share.

And then there is the rise of the tasting menu – lots of smaller plates chosen by the chef. More often than not, everyone at the table has to agree to order these. As a gastrophysicist, I would advise the restaurateur to seat diners at a round table if that is what they want customers to do, because people are more likely to feel like they belong when seated at a round table than at a square one. By contrast, those at an angular table tend to show more selfish traits in group settings.[35] All of a sudden, the fact that Chinese banquets are always held around a circular table starts to make much more sense. And going even further back, don't forget about King Arthur and his Knights of the Round Table. The main reason for seating diners at angular tables is to maximize the number of diners in a space. That said, while the round table is certainly the most democratic solution, if it is too large, it can make conversing with those on the other side of the table awkward.

It is not just the plates that restaurateurs want you to share these days. You have probably come across the communal/informal dining concept whereby everyone gets to sit at large long tables. This approach constitutes a distinctive aspect of the design of restaurants like Wagamama and Busaba Eathai (both incredibly successful ventures from top restaurateur Alan Yau). The Pain Quotidiens chain does much the same. In some sense, anyone who chooses to frequent one of these informal eating establishments is sharing the table with strangers – though, as the saying goes, 'There are no strangers, just friends that we have not yet met.'[36] The physical distance from those sitting next to you might well be the same as in other popular restaurants where they cram the tables for two in against the banquette. And yet I believe that there is still something qualitatively different about 'being connected', parked at the same long table. In fact, I think that I can feel a new gastrophysics experiment coming on to get to the bottom of the matter! Just how much do we enjoy the experience when crammed at a long table with strangers sitting right next to us? As always, there is definitely more research to be done here.

Have you ever thought about how strange dining out is?

Think about it carefully, and you begin to see how strange all this eating in public in close proximity to strangers really is. How, for example, might someone from another culture view the social aspects of dining in a restaurant if they were to visit our Western twenty-first-century world? Perhaps much like Antoine Rosny, a Peruvian traveller (despite his French-sounding name), who described his experience on first visiting a Parisian restaurant back at the start of the nineteenth century (i.e., in the very early days of the restaurant): 'On arriving in the dining room, I remarked with astonishment numerous tables placed one beside another, which made me think that we were waiting for a large group, or perhaps were going to dine at a table d'hôte ('host's table'). But my surprise was at its greatest when I saw people enter without greeting each other and without seeming to know each other, seat themselves without looking at each other, and eat separately without speaking to each other, or even offering to share their food.'[37]

One intriguing performance art piece that probes the meaning of sharing a meal comes from Indonesian artist Mella Jaarsma (see Figure 7.2). Members of the public (anywhere between two and six of them) are invited to put on a bib from which a flat table surface is suspended. Diners pair up and then both order for and subsequently feed each other. This kind of intimate performance gives rise to a literally shared meal. Notice also how this wearable table temporarily binds the diners (performers), as they create a mutually supported surface from which to eat. (Though I do worry about what exactly happens when one of the diners needs to visit the toilet!) The comment of one participant who took part is interesting: 'In enacting Mella Jaarsma's piece, I experienced feeding another person and being fed by another person for the first time in my adult life. [. . . Throughout the meal] one constant remained – the way the ritual of feeding and being fed articulates power relations [. . . O]ur proximity to

Figure 7.2. Still from a performance of *I Eat You Eat Me* (2001–12).[38]

those we have power towards makes us generous; one wishes this was the case more often outside of art.'[39]

Artist Marije Vogelzang (from the Netherlands) has created a piece called *Sharing Dinner*, in which the diners are connected by cloth. In this visually striking installation, diners insert their heads and arms through slits in a white tablecloth suspended from the ceiling (see Figure 7.3). As the artist notes: 'I used a table with a tablecloth, but instead of putting the cloth on the table, I made slits in it and suspended it in the air, so that the participants sat with their heads inside the space and their bodies outside. This physically connects each person: If I pull on the cloth here, you can feel it there. Covering everyone's clothing also created a sense of equality. Initially I was concerned that people would reject the experience, particularly because the participants didn't know each other beforehand, but it actually increased their desire to relate to one another, and brought

Figure 7.3. *Sharing Dinner* by culinary artist Marije Vogelzang (Tokyo, 2008).

about a feeling of being in something together.'[40] Vogelzang also uses food to encourage sharing. She would serve one person a slice of melon on a plate that was cut in two. Meanwhile, the person sitting opposite would be given ham on a similar plate. This resulted in the participants (many of whom didn't know each other) naturally sharing their food to make the classic combination.

The telematic dinner party

Many of us who are fortunate enough to have a family to dine with nevertheless still sometimes find ourselves away from home, when travelling for work, say. As a result, we can all too easily miss out on the shared time typically centred around the dinner table. Indeed, this has been identified as a growing problem by researchers working in the field of human–computer interaction, a number of whom have started to investigate whether technology can somehow be used to reconnect people who find themselves far apart, allowing them to share meaningful (i.e., embodied) virtual mealtime experiences. Thus, the idea of the 'telematic dinner party' is born – just think of it as 'Skeating', i.e., Skyping while eating.[41]

While the concept is undoubtedly an intriguing one, there are a number of challenges for anyone wanting to make a successful technological intervention in this space. For example, what happens when those who are dining together digitally eat different foods – does that hinder their making 'a connection'? Another important issue here concerns how to make this kind of virtually shared dining experience more immersive and engaging. As a guest at a preliminary telematic dinner party said: 'I don't feel that I shared food with them. It felt like we [were] together in one room and they were eating in another room. There was no sense that we were sharing.'[42] This is definitely not what tomorrow's experience designers want to hear. Dining together involves the tight (albeit unconscious) synchronization of the diners' behaviour, so that trying to replicate the precisely coordinated choreography at a distance, where a lag can sometimes delay the signal coming from the other end, is likely to disrupt the more communal aspects of the meal.

I could, just possibly, imagine this kind of technological solution having value in extreme circumstances, for someone on a long space mission to Mars, say, who wants to reconnect to their family back on planet Earth. However, stuck down here on the ground, I really can't work up any enthusiasm for the telematic dinner party concept, even assuming it worked perfectly (i.e., with no synchronization issues). I'm tempted instead to place my bets on a different technological innovation. I am thinking of the various meal-sharing apps that have been launched over the last couple of years.[43] For a small fee, they allow those who find themselves alone in an unfamiliar place to dine with a local in their own home; so a meal is physically shared, just not with the subscriber's family. The various sites offering this kind of service each have a somewhat different feel, thus there is most likely something for every taste – and if there isn't yet, you can guarantee there will be soon! For instance, the US site EatWith has the feel of a supper club, while UK-based VizEat emphasizes the opportunity to eat with locals, to get inside a culture by eating it, as it were. According to one of VizEat's co-founders, Camille Rumani, this site had attracted more than 170,000 hosts in 115 countries within a couple of years ago of starting up (in July 2014).[44]

The question that we should probably all be asking ourselves is whether such meal-sharing apps will come to revolutionize the way in which we dine when we are away from home in much the same way that Airbnb has transformed where we stay and Uber has transformed transportation. (Note here that an UberEats app was recently launched in cities across the US, with the promise to make 'getting great food from hundreds of restaurants as easy as requesting a ride'.[45]) According to Euromonitor International, a market intelligence agency, 2015 was *the* year to look out for 'peer-to-peer dining' as one of the biggest trends. This involves direct interaction between cook and diner, without the mediation of a restaurant (chain) – just think of all those chefs who are starting to cook from home as symptomatic of this trend.[46] But, I would argue, by far the biggest potential market is those individuals who haven't left their own home, and yet still have no one to dine with. One recent start-up connecting them is Tablecrowd, which combines eating with social networking; another is Tabl, which offers essentially curated social dinners in the south of England.[47] Dining together is such a primal urge, so next time you get peckish, why not invite someone to eat with you? Most likely, you will end up having a better time than you would on your own. Just remember to order first, though, if you want to maximize your enjoyment!

8.   Airline Food

It was back in 2014 that I first got to thinking about the food and drink while in the sky. Sitting on yet another long-haul flight when my laptop battery finally died, I ended up watching the cabin attendants slowly wheeling the drinks trolley down towards the back of the plane. And it was then that it struck me – just how many people order a tomato-juice-based beverage while up in the air. Now, though you might see the odd person ordering a Bloody Mary down on the ground, it really is a pretty rare occurrence, at least in the circles in which I move. However, up above the clouds, watching the drinks flying off the trolley, it really felt like every fourth order or so involved tomato juice. But what exactly is so special about the red fruit (or is it a vegetable?) in the air, and how might the insights garnered from the drinks trolley in the sky lead to a radical redesign of airline food?[1]

Well, firstly, you shouldn't just take my word for it. Before going any further, we need to check on the veracity of the observation. Luckily, my intuitions proved correct: it turns out that tomato juice makes up 27% of the drinks ordered in the sky. What is more, there is a whole section of the population who regularly order a tomato juice from an air steward or stewardess but who would never think of doing so with their feet planted firmly on the ground. Of the more than 1,000 passengers questioned in one survey, 23% fell into the latter category.[2] So what exactly is going on here? Before answering that question, though, let's take a quick look at the history of airline food.

The way it was

Plane food hasn't always been bad. Back in the early days of commercial flight, the airlines would put on quite a spread for anyone who was rich enough to fly. Believe it or not, they used to compete on the

Figure 8.1. Airline food as it used to be! 'How many lobsters would you like today madam?' Passengers being served fresh Norwegian lobsters (with their shell still on). Note the aperitifs set on ice too.

quality of their food offering, with a carvery, lobster, prime rib, etc. available to all who wanted it (see Figure 8.1).[3] Perhaps this helps explain why the pop-up dining concept Flight BA2012, in Shoreditch, East London, was such a success. The hipsters of Hoxton were able to sample a three-course meal inspired by the airline's 1948 first-class menu. It is hard to imagine that contemporary airline food would have had quite the same appeal (see Figure 8.2).[4]

Everything changed in 1952, though, with the introduction of economy class and the associated economies of scale required once passenger numbers increased dramatically. The International Airline Transport Association (IATA) didn't help matters, either: they actually brought in guidelines limiting what food could be offered up in the skies, at least in economy class. Scandinavian Airlines were even fined

Figure 8.2: An impressive-looking French menu served on board a flight from Copenhagen to Singapore, stopping in Bangkok on the way, in the 1950s.

$20,000 for serving their transatlantic passengers a bread roll that was deemed to be just too good, following a complaint from Pan Am, one of their competitors. In recent times, though, there can be no doubt that (in real terms) less and less money is being spent on the food offering in the air, assuming, that is, that any food is provided at all.[5]

Once upon a time, the food was pretty much the only thing that kept air passengers from thoughts of their own demise, should the plane's none-too-reliable engines fail. Hence the quality of the offering was especially important, given that the passenger had little else to occupy their minds other than admiring the view from the window.[6] Nowadays, though, everything has changed. For one thing, plane travel is, thankfully, safer than ever. What is more, all manner of entertainment is available to passengers at the touch of a button.

Mind you, the cabin atmosphere up at 35,000 feet isn't all that conducive to fine dining. The lowered air pressure, together with the lack of humidity (the air, don't forget, is recycled through the cabin every 2–3 minutes), really doesn't help, with food and drink losing roughly 30% of its taste/flavour when sampled at altitude. Aware of these problems, a number of the airlines test their menus under conditions that mimic the atmosphere up there down here, if you see what I mean. In Germany, for instance, at the Fraunhofer Institute, half an old Airbus plane has been plonked in a low-pressure chamber and where they test people's reactions to the foods that they are thinking of serving up above the clouds.[7]

More often than not, though, the airlines have opted to load the food they serve with ever more sugar and salt, to enhance the flavour. No surprise, therefore, that the airline food served these days isn't the healthiest. In fact, it has been estimated that the British consume more than 3,400 calories between their check-in at the airport and their arrival at their destination.[8]

Over the years, the airlines have sought advice from chefs to help improve their food offering. Back in the day, it was French chef Raymond Oliver who was brought in by Union de Transports Aériens (the precursor of what was to become Air France). His advice fundamentally changed what was served in the air, and soon established the standard format for the meals that so many of us have now become all too used to. In fact, the chicken or fish dishes that one finds on many of today's economy-class menus can be traced directly back to the chef's early suggestions. For instance, Oliver recommended serving foods that the passengers would find familiar; not quite comfort food, exactly, but at least something reassuring. He was looking for hearty meals that were easy to prepare and heavy to digest, the idea being that then the passengers wouldn't get hungry again before the plane landed. The meals shouldn't lose too much of their flavour when reheated either. The chef's suggestions: coq au vin, beef bourguignon and veal in cream sauce. (It was 1973, after all!) These dishes also had the advantage that the meat, drenched as it was in sauce, wouldn't dry out too much when heated up in the plane.[9]

Can celebrity chefs really cut the mustard at 30,000 feet?

Today, it's much more common for airlines to bring in a chef to try to improve their food offering. And a number of them have hired celebrity chefs in order to spruce up their airline meals, Neil Perry from Australia teaming up with Qantas, for instance, or Heston Blumenthal with British Airways, or the late, great Charlie Trotter advising United Airlines. (Trotter's top tip was short ribs spiced with Thai-style barbecue sauce; the addition of spice and the inclusion of sauce are both good ideas when it comes to dining at altitude.) Meanwhile, Air France currently has so many great chefs to choose from that they rotate their affiliation on a regular basis.[10]

My guess is that even those sitting in the front of the plane would have little inkling of the star chefs' involvement in the dishes that they are tucking into were it not for their names emblazoned on the menu cards. Certainly, I have yet to see any evidence to support the claim that the chefs' interventions in any of the cases just mentioned actually led to a significant increase in passenger satisfaction. And, perhaps more tellingly, the airlines that have sought advice from the top chefs don't seem to appear any more frequently in the list of the top-ten airlines for food that is published annually. Regardless of how many Michelin stars a chef has, their food will never taste (quite) as good in the air as when served in their flagship restaurant down on the ground. What many of the frequent business passengers do appear to appreciate, though, is the change in format from the fixed-course meal service of old to more of a grazing approach, allowing them to eat what they want, more or less when they want – food on demand, as it were.

As we have already seen, to a large extent what we think about food and drink depends on the context or environment in which it happens to be consumed. And airline food is no different.[11] One of the other significant impediments to progress in the sky relates to the long-term catering contracts that many of the airlines have signed up to. Thus, even if the airline, or the chef that the airline has brought in, wants to change the culinary offering, it can prove difficult to do so in practice. Behind the scenes, a number of the catering suppliers have now started to bring in some innovative chefs to help advise them directly. The more fundamental issue here, though, is that the chefs' input is typically restricted just to the ingredients, recipes and preparation of the food. And as we are about to see, any solution that is solely focused on the food will only take you so far. It's time to bring in the gastrophysics perspective.

What's the link between the humble tomato and aircraft noise?

Let's get back to the juice! Once the plane has reached cruising altitude, the passenger's ears will be exposed to somewhere in the region of 80–85 dB of background noise, depending on how close they are to the engines. This racket suppresses our ability to taste.[12] However, it does not affect our perception of all foods equally. The really special thing about tomato juice and Worcestershire sauce (both ingredients in a good Bloody Mary) is umami, the proteinaceous taste, experienced in its purest form as monosodium glutamate or MSG. While it has long been popular in East Asian cuisine (e.g., in Japan, where the term translates as 'delicious', 'savoury' or 'yummy'), it has recently started to attract the interest of chefs from many other parts of the world too. In the West, those foods that you are likely to have come across that are rich in umami include Parmesan cheese, mushrooms, anchovies and, of course, tomatoes.[13] So, could this help to explain the mystery of why so many people order a tomato-juice-based beverage while up in the air?

Well, in 2015, researchers from Cornell University finally got around to assessing the effect of loud aeroplane noise on people's ability to taste umami. Participants sitting in the lab had to rate the strength of a series of clear drinks, each containing one of the five basic tastes presented at one of three different stimulus concentrations. Each of the solutions was tasted in silence and also while listening to pre-recorded aeroplane noise at a realistic decibel level. Intriguingly, the perceived intensity of the umami solutions was rated as significantly higher when the background noise was ramped up. By contrast, ratings of sweetness were suppressed, while ratings of the taste of the salty, sour and bitter solutions were unaffected. Given such results, British Airways' decision to introduce an umami-inspired menu on their flights back in 2013 starts to make a lot more sense.[14]

But why should loud noise affect some tastes but not others? Well, one intriguing theory is that our responsiveness to different tastes varies as a function of how stressed we are – this being just what many passengers may be feeling while flying, especially on a bumpy flight. In one older study, for example, sweet, but not salty, solutions were rated as significantly more pleasant under conditions where stress had been induced by the presentation of loud noise. One suggestion that has been put forward to try to explain this surprising result is that the energy that is signalled by sweetness might be just what an organism needs when it comes to dealing with the kinds of situation that give rise to stress in the first place. Presumably, a similar evolutionary story could be told about the increase in the perceived intensity of umami under conditions of loud noise. For, just like sweetness, umami is also a nutritive tastant, signalling, as it does, the likely presence of protein.[15] But whatever the correct explanation ultimately turns out to be, the key point remains that loud noise generally suppresses sweetness and sometimes saltiness, while at the same time enhancing the taste of umami.

What happens when people are given proper food to taste, rather than just pure tastants dissolved in solution? Well, playing loud white noise – think of the static sound on an untuned radio – leads to ratings of both the sweetness and saltiness of various snack foods such as potato chips, biscuits and cheese being suppressed.[16] Somewhat surprisingly, though, crunchiness ratings were actually higher when the background noise was turned up (as compared to silence). Perhaps, then, the airlines should be thinking about adding more crunch to the food that they serve, and adding more of the other noisy textural food attributes, such as crackly and crispy too. This is likely to have the added advantage of improving the perceived freshness and palatability of the food. Indeed, this is why having a bowl of fresh fruit available (as some airlines do for business-class passengers) is a good idea. And sprinkling the salad with sesame seeds to boost the crunch factor would be a darn sight cheaper than hiring one of the world's top chefs.

So, counterintuitive though it may sound, donning a pair of noise-cancelling headphones could actually be one of the simplest ways in which to make the food and drink taste better at altitude. But, now that we have got rid of the background noise, the next question is: what else, if anything, can you listen to, in order to make your food taste better?

Supersonic seasoning

Late in 2014, British Airways introduced 'Sound Bites' for their long-haul passengers. Once they had chosen their in-flight meal, passengers could tune in to one of the channels in the seatback entertainment system. There they would find a carefully chosen playlist of popular tunes that had been specially selected to complement the taste of the food. The musical selections were based, in part, on research findings from my lab. A number of the tracks were chosen to boost the perceived authenticity/ethnicity of the dishes, given research showing that this attribute can be enhanced by presenting matching music (or, for that matter, any other sensory cues) appropriate to the region associated by people with the food (see 'The Atmospheric Meal'). Think lasagne, or pasta, while listening to one of Verdi's arias (and if you can find a red and white checked tablecloth, even better), or perhaps Scottish salmon with The Proclaimers?[17]

Some of the first empirical evidence supporting the existence of sonic seasoning came from research conducted with The Fat Duck Research Kitchen in Bray. Together with the then head research chefs, Steffan Kosser and Jockie Petrie, we were able to demonstrate that listening to soundscapes containing lots of tinkling, high-pitched notes accentuated the sweetness of a bittersweet cinder toffee, whereas listening to low-pitched noises brought out the bitterness instead. The effects, it should be said, weren't huge (5–10%), but they were large enough to potentially make a difference to the tasting experience while up in the air.[18] So why don't you forget about adding sugar the next time you are eating at altitude, and instead just tune in to some sweet, calorie-free music? That said, while we have now got some pretty effective sweet tunes, we are still struggling to create the perfect sonic salty backdrop.

Now, I am assuming that you have taken my advice and got yourself a pair of noise-cancelling headphones, and that you are listening to the right sort of music to complement, and thus improve, the flavour of whatever you happen to be eating or drinking. So, what next? What else can be done to improve the meal experience while passengers are up in the air? Well, if the food really is worth savouring, one simple tip here would be to pause the movie. For according to the ground-based research covered in the 'Social Dining' chapter, you ought to find that you enjoy your food a little more while, at the same time, finding yourself satisfied with less of it.

Tasting under pressure

In addition to the background noise, one of the other problems with tasting at altitude is the reduced cabin air pressure. These days, planes are pressurized to create an atmosphere that is equivalent to what one would find at an altitude of approximately 6,000–8,000 feet. Under such conditions, it becomes harder to taste sweet, sour, salty and bitter.[19] No wonder then that airline food tastes so bad. However, the more profound problem is that the number of volatile aromatic molecules in the air also decreases as the cabin air pressure drops. It is this that can really suppress our flavour perception. One innovative solution might be to wear a Breathe Right nasal strip. These plasters were originally designed for athletes to place over their nostrils in order to increase the intake of air, potentially enhancing their sporting performance. Wearing one can lead to nasal airflow going up by as much as 25%. So thinking laterally here, one might consider providing the passengers with one along with their earplugs on boarding the plane, as a means of increasing their exposure to any of the volatile aromas of food and drink that are still floating around in the atmosphere at altitude. However, while this solution has yet to be tried in the skies, the findings from research conducted down on the ground have unfortunately so far been, how shall I put it, less than encouraging.[20]

Another recommendation for improving the tasting experience while up in the air (i.e., at low atmospheric pressure) comes from Professor Barry Smith, a philosopher and wine writer working out of the University of London. He has noticed that high-altitude wines (e.g., New World Malbec from Argentina) tend to be rated better in the air than one might have anticipated from tasting them on the ground.[21] Why should that be so? Well, his suggestion is that the atmospheric conditions on the side of the mountain where many of these wines are made (i.e., blended) is, in some sense, closer to those one finds in an aeroplane cabin than for some other wines. The grapes that go into Argentinian Nicolas Catena's Zapata wines, for instance, are grown at around 5,700 feet. So perhaps it is no wonder his wines taste better at altitude. So next time you get the chance to choose your wine on a plane, remember another one of Smith's suggestions and go for a fruitier number. Whatever you do, Smith argues that you'd do well to avoid those prestige wines with firm tannins, as they may well leave you with a fiercely astringent/bitter taste in your mouth.

Another problem with the atmosphere in a plane cabin is that the humidity levels are much lower than on the ground (below 20%, as compared to 30% or more in the average home). The good news again, at least for those travelling in style, is that the humidity apparently tends to be a little higher at the front of the plane. Lower humidity levels also impair our ability to taste, since they tend to result in the drying of our nose, making it harder to detect the remaining volatile odour molecules.[22] A few years ago, chef Heston Blumenthal came up with his own idiosyncratic solution to this particular problem. His recommendation was that anyone wanting to enjoy their food and drink more while flying should give themselves a nasal douche with a water spray. The idea here (perhaps a little tongue-in-cheek) was to try and increase the humidity in the nostrils to make up for the lack of humidity in the air that is recirculated every couple of minutes or

so through the cabin. With all due respect, though, while the suggestion undoubtedly makes for engaging television, it is hard to imagine
anyone actually taking the advice seriously. And anyway, before getting too carried away, just remember that should the nasal douche
work as hoped, it would also increase your ability to smell the passengers sitting close by. Are you sure that is really what you want?

Simple tips for service

For my brother, a self-confessed wine buff, the realization came
while staying in a Swiss ski chalet a few years ago. He had decided to
open a long-treasured bottle of wine, but the only thing that he could
find to pour the wine into late one night was the plastic water
cup from the bathroom. He knew exactly what the wine – a Kistler
Chardonnay – *ought* to taste like. He had, after all, liked it so much
that he had bought a couple of cases of the stuff. Somehow, though,
with the wrong drinking vessel, my brother just wasn't able to
recreate the great tasting experience that he knew he should be
having – one that, moreover, he desperately wanted to be enjoying,
given how much he had paid for the wine in the first place. On a
plane, served an unfamiliar vintage, it is going to be even harder for
any of us to discount the glassware and focus instead solely on the
quality of the contents and how much we are enjoying the experience. Haven't we all been disappointed while drinking something
expensive from a flimsy plastic container? No matter how prestigious
the contents, the cheap feel of the vessel takes away from the pleasure
of the experience.

What all of us know intuitively, I think, and the research data now
backs up, is that we like beverages more when they are served in the
appropriate receptacle than when served in something inappropriate.
Just think about it, would you enjoy drinking tea out of a wine glass?[23]
Of course not! Knowing this, one really has to wonder what on earth
many of the airlines are thinking: How, I ask you, can anyone honestly justify serving the complimentary glass of champagne that
many business-class service encounters start with from such a light

and flimsy cheap plastic glass. While serving the champagne from a plastic flute would help a little, I would suggest that anyone hoping to optimize the tasting experience should really be using glass, not plastic – for weight is crucial to the experience, no matter whether we find ourselves in the air or on the ground.

Back in the days of supersonic flight, every gram counted (much more so than it does on regular flights). At some point, designers were brought in to help develop some stylish new extra-lightweight cutlery for Concorde (plastic, while undoubtedly very light, was obviously not an option) and created some beautiful titanium cutlery: exquisite to look at, and lighter than anything metal that had gone before. Job well done! Or so one imagines they must have thought. The problem, though, was that people simply didn't like it. When trialled, it just felt too light and consequently it was never introduced on board.

Finally, I know of one innovative airline that has recently been thinking about the material properties of the cutlery they hand out. Why would they want to do that? Well, because the material from which the fork and especially the spoon (i.e., those items that enter your mouth) are made can modify the taste of the food. We conducted some research relevant to this a few years ago, together with the Institute of Making in London. We were able to demonstrate that an everyday sample of yoghurt, to which a small amount of salt had been added, was rated as tasting saltier when eaten with a stainless steel spoon that had been electroplated with copper or zinc. Such results raise the question of whether novel cutlery designs like this could be used to help season the food served in the air. Remember here that it is primarily sweet and salty tastes that are suppressed by all that loud background aeroplane noise. Unfortunately, though, while there are certain metals that can be used to bring out the salty, bitter and sour taste in food, I am not aware of any metal that can boost sweetness. Well, there is lead, I suppose; it is just that it is also poisonous. So, not quite what one wants in one's cutlery.[24]

Will multisensory experience design really take off?

All well and good, I hear you say, but will anything actually change in the near future? What will the airline meal of tomorrow look like? The good news, according to my sources, is that one of the big airlines plans to launch an in-flight food and beverage offering that will put everything that we have become accustomed to in recent years to shame. I'm afraid I can't say any more just yet. But where one airline leads, others may, sooner or later, follow, and if this is the case, then I am hopeful that we might finally see a return to the early days of flight, when the fledgling airlines competed on the quality of their food service offering (for their admittedly deep-pocketed passengers).

Does this sound unbelievable? Well, before you dismiss it, do allow me to paint a picture of air travel as it was back at the end of the 1960s. At that time, Trans-World Airlines started running themed 'foreign accent' flights between major US cities. Let me quote Alvin Toffler, former editor of *Fortune* magazine and author of the bestseller *Future Shock* directly (for otherwise, I fear, you will think I am making it up): 'The TWA passenger may now choose a jet on which the food, the music, the magazines, the movies, and the stewardess's outfits are all French. He may choose a "Roman" flight on which the girls wear togas. He may opt for a "Manhattan Penthouse" flight' – the mind boggles – 'or he may select the "Olde English" flight on which the girls are called "serving wenches" and the décor supposedly suggests that of an English pub.'

Toffler continues: 'It is clear that TWA is no longer selling transportation, as such, but a carefully designed psychological package as well. We can expect the airlines before long to make use of lights and multi-media projections to create total, but temporary, environments providing the passenger with something approaching a theatrical experience.' And before their demise TWA weren't the only ones. For a short period in the early 1970s you might even have come across a piano lounge with a fully-functioning Wurlitzer electric piano at the back of some American Airlines 747 planes. And the British Overseas

Airways Corporation (the precursor to British Airways) apparently had it in mind to provide their unmarried male passengers with a 'scientifically chosen' blind date when they touched down in London. Little surprise that the latter scheme, called 'The Beautiful Singles of London', was scrapped when the government-owned airline came under criticism from Parliament.[25] Ultimately, then, the sky really is the limit for anyone willing to recognize the power of multisensory experience design.

9. The Meal Remembered

Humour me for a moment – what would you say was your perfect meal? What exactly can you remember of the experience? What you ate? Who you were with? Perhaps even more interesting: what do you know that you have forgotten? If that is all a bit too much, why not take a more recent dinner instead, the last time you went out to a restaurant, say, and answer the same questions. My guess is that, while you can probably remember where you were and who you were with, the details of the meal itself, the specific flavours and foods you ate, will be much hazier.* Unless, that is, you always order exactly the same dishes in your favourite restaurants, as I do.

No matter how good or bad a meal, it will never last more than a few hours. The mediocre ones we forget, the marvellous occasions hopefully stay burnished into our memories, bringing pleasure whenever we think about them. The really bad ones – well, all too often they stay etched in our minds as well, much though we might wish to forget them! For my brother, the sous vide liquorice salmon that he really didn't like is the dish that he wants to forget but has been unable to, even though he was served it more than a decade ago.

Our memory of a meal, at least of an enjoyable one, is where so much of the pleasure of the experience resides. It can last for days, weeks or even years. From the perspective of anyone trying to sell us a restaurant meal, this is important because it is a major factor in our decision whether or not to return to a particular venue or chain.[1] Our flavour memories also play a crucial role in our decision to stick with one brand or switch to another while browsing the supermarket aisle;

---

* And if you are anything like me, you probably find that you have forgotten quite what your main course was by the time the dish arrives. This is why it can be so important for the waiter to explain the dish when they bring it to the table, especially if it is a non-standard offering.

once again, such a decision is often based on our recollection of the taste of the product, or what we thought about the experience the last time we encountered it.

What is a food memory?

The simplistic view here would be that our recollection of a meal is merely a weaker version of what happened at the time – 'devoid of the pungency and tang', as William James, one of the godfathers of experimental psychology (and brother of novelist Henry), once so evocatively put it.[2] However, the gastrophysicist knows only too well that our minds play tricks on us. Not only do we forget altogether certain of the things that we experienced so vividly even just a short time ago but we also misremember and confabulate others. More often than you might imagine, we recollect things that probably didn't occur at all, or at least not in the way we remember them.[3] Our recollections of meals, both good and bad, are no different in this regard.

Storing every detail of an experience (be it a meal or anything else for that matter) in memory would simply be too effortful. So our brains use a number of cognitive shortcuts to help. For instance, we tend to keep track of the high and low points (the peaks and troughs), and how meals start and end (termed 'primacy and recency effects'). As a result of another shortcut, we also tend to neglect the duration of events that don't change much over time. The latter (known, unsurprisingly, as 'duration neglect') has been demonstrated to apply to meals. Such mental heuristics are efficient in that they help us to recall the gist without necessarily having to remember every detail of our lives. However, which elements of a meal will stick (i.e., the end, the peak, etc.) appears to depend on the specifics of the situation.[4]

I would argue that knowing about such 'tricks of the mind' is absolutely crucial for anyone wanting to deliver more memorable food and beverage encounters. So, bearing these factors in mind, it's time to call in the 'experience engineers', the researchers who have made it their life's work to study what exactly sticks in our memory, and why.[5]

The main aim here, in the gastrophysics context, is to get those you serve, whoever they might be, to remember more of the good stuff about your interaction with them. Do you, for instance, still remember the lime *gelée* incident from the start of the book? One chef in Washington, DC, Byron Brown, even created a theatrical dining experience in 2011* with the specific intention of enhancing his diners' memory of the event.[6]

Common sense suggests that the best food and beverage designers – and here I am thinking of the world's top chefs, molecular mixologists and culinary artists – should be trying to create the ultimate tasting experiences for their customers. However, what those at the top of their game *really* ought to be focusing on is the creation of the most robust memories possible. Until you get that distinction clear, that our perceptions of food and drink while eating and drinking and our recollections of those consumption episodes differ, both quantitatively and qualitatively, you really can't hope to deliver the best long-term memories. The meal itself and our recollection of it are linked, obviously, but they diverge from one another in systematic ways that the gastrophysicists and experience engineers know just how to capitalize on.

One of the chefs with whom I work closely, and who has obviously been hanging around me too much, has been running his own experiment (just like a proper psychologist). What he wanted to know was just what his diners remembered of the fabulous meals he served. The chef emailed a questionnaire out to his guests a couple of weeks after they had visited his restaurant. He was in for quite a shock! For while those who chose to respond could certainly recall that they had very much enjoyed the experience, what they turned out to be very bad at was remembering precisely what it was that they had eaten. Funnily enough, the kind of thing that made the biggest impression in his diners' minds was, say, that time when the waitress sprayed some aroma or

---

* I was particularly pleased to see that it was one of my former students, Ed Cooke, a psychologist with a truly phenomenal memory, who was advising on this project. In the middle of tutorials, Ed would come out with truly amazing detail from books that he had read many years earlier.

other over their dish while they were seated at the table. In other words, it was the more theatrical, surprising and/or unusual aspects of service that stuck in people's minds, far more than the taste of the food itself. This failure to remember the food wasn't any comment on the quality of the chef's cuisine. The dishes themselves were delightful. Most were, one would have said at the time, quite memorable. However, it turned out that they weren't actually remembered that well after all, or at least not the specific ingredients and flavour combinations.

As I am sure you can imagine, this chef really had the wind knocked out of his sails when the results came in. Why, he kept muttering darkly, did he bother putting so much effort into the creation of his dishes if his diners simply couldn't recall what they had eaten, nor the unusual flavour combinations that he had created? I told the chef not to be too hard on himself, that there was a psychological explanation for all of this and that it wasn't any reflection on his culinary skills. That the main thing to concentrate on was that people remembered their hedonic response, that they had really liked 'the experience'. And, if they also happened to engineer a memory of some confabulated meal, a construction of their overactive imaginations, well, so be it. It was especially surprising to hear from those diners who were convinced that their memories were so vivid they could almost taste the dish again. They were, as likely as not, imagining the flavour of something that they hadn't tasted, at least not at this chef's table! (All of this ought to make one wonder about the value of those online reviews.)

Using my most consoling tone, I told the chef that it was no use to try to fight the foibles of recollection. What he needed to do instead was to better understand the many ways in which memories fade and our minds deceive us. We mostly do not pay attention to what we taste. Rather, our brains just do a quality check first, to ensure that there is nothing wrong with the food or drink and that it tastes pretty much as we expected (or predicted) that it should.[7] After that, once we know that we are safe, we devote our cognitive resources (what the psychologists call 'attention') to other, more interesting matters, like our dining companions, or what's on the TV or who has just sent us a text. That is, we no longer feel any need to concentrate on what we are consuming. And, as psychologists know only too well, unless you pay attention (e.g., to what you eat), you have little chance of remembering, even a few moments later, never mind after a few weeks or months.[8] Emotion may also play a role here.[9]

In fact, if you change the flavour of a food while eating (and yes, the experiment has been done),[10] people often don't notice. It is as if we are all in a constant state of 'olfactory change blindness'.[11] Intriguingly, this is something that the food companies have been trying to exploit to their, and hopefully our, advantage for a few years now. The basic idea is that you load all the tasty but unhealthy ingredients into the first and possibly last bite of a food, and reduce their concentration in the middle of the product, when the consumers are not paying so much attention to the tasting experience. Just think about a loaf of bread with the salt asymmetrically distributed towards the crust. The consumer will have a great-tasting first bite, and then their brain will 'fill in' the rest by assuming that it tastes exactly like the first mouthful did. This strategy will probably work just as long as the meal isn't high tea and the taster eating cucumber sandwiches with the crusts cut off! Or imagine something like a bar of chocolate, which most people will presumably start and finish at the ends, not in the middle. In fact, Unilever has a number of patents in just this space.[12]

This innovative product development strategy is based, on the one hand, on the phenomenon of 'change blindness' and, on the other, on the assumption our brains make that things that appear the same probably taste the same too. The promise of the latest gastrophysics research is that by understanding such 'tricks of the mind', food and beverage companies, or at least the more innovative amongst them, will be able to deliver the same great taste that the consumer has come to expect without as much of those ingredients that we should *all* be consuming less of, like sugar, salt and fat.

Choice blindness

Do you think that you would notice the difference between two jams with a similar colour and texture, or two different flavours of tea? Most people will answer these questions in the affirmative. After all, isn't the very reason we buy one kind of jam rather than another, or keep multiple types of tea in our homes, precisely because we can differentiate their flavour? And yet gastrophysics research has highlighted some pretty worrying limitations concerning our perceptual abilities. It turns out that we actually have surprisingly little recollection (or awareness) of even that which we tasted only a few moments ago. In one classic demonstration of this phenomenon, known as 'choice blindness', shoppers (nearly 200 of them) in a Swedish supermarket were asked whether they would like to take part in a taste test.[13] Those who agreed were then given two jams to evaluate. They were similar in terms of their colour and texture (e.g., blackcurrant versus blueberry). Once the shoppers had picked their favourite, they sampled it once again and said why they had chosen it, and what exactly made it so much nicer than the other jam. The shoppers were more than happy to oblige, regaling the experimenter with tales of how it was their favourite, or that it tasted especially good spread on toast, etc.[14]

What many of the shoppers failed to notice, though, was that the jams had been switched before they tasted their 'preferred' spread the second time around. The experimenter was using double-ended jam jars in order to effect this switch unnoticed. In other words, the unsuspecting customers were justifying why they liked the spread that they had just rejected. Exactly the same thing happened in another experiment with fruit teas. Overall, the deception was noted by less than a third of shoppers. Even when the jams tasted very different – think cinnamon-apple and bitter grapefruit jam, or the sweet smell of mango versus the pungent aroma of aniseed-flavoured tea – only half of the switches were detected. What these findings imply is that many of the people tested actually did not have a clear memory of the flavour of the food that they had tasted only a moment or two earlier.

Such results, surprising though they undoubtedly are, nevertheless fit well with the research on blind taste tests. Consumers are convinced that they can pick out their preferred brand when given a range of alternatives to taste blind (i.e., without being able to see the labels). Time after time, they select one product and confidently assert that it is most definitely their preferred brand (presumably by comparing the taste to their memory of the taste). Why else, after all, would they be paying more for a branded product than for a cheaper unbranded or home-brand alternative? Only the thing is, in most cases the brand that they so confidently pick out happens to be something other than the brand that they normally choose. It is not that all of the products taste the same; more often than not they don't.[15] It is just that our flavour memories aren't quite what we take them to be.

But surely this is not applicable to every product? Some of my colleagues always pipe up at this point that things are very different in the world of wine. They protest that I shouldn't believe the results of all those blind wine tasting studies in which people perform so badly, even the experts.[16] And, truth be told, there certainly are some remarkable feats of blind wine tasting out there; that I don't doubt. But it is important here to distinguish between two alternative scenarios. On the one hand, there is the case of the expert who tries a wine blind and suddenly has a flashback to some long-ago vineyard where s/he first tasted it, remembering also who they were with and even what shoes they were wearing. In the alternative scenario, on the other hand, a more measured and rational assessment of a wine's sensory properties is undertaken by the taster. In the latter case, a careful process of elimination helps the expert to determine what the likely provenance of the wine is. Both can, of course, be most impressive, but only the former can really be said to demonstrate an outstanding feat of sensory flavour memory. It is interesting in this regard to see how guides to blind wine tasting tend to focus on the second approach.[17] I suspect that, more often than not, the amazing feats of blind wine recognition are a matter of inference and cool calculation, rather than taste/flavour memory.

Have you heard of 'Sticktion'?

'In the context of experience management, [Sticktion] refers to a limited number of special clues that are sufficiently remarkable to be registered and remembered for some time, without being abrasive.

Sticktion stands out in the experience, but does not overpower it; well-designed, it is both memorable and related to the "motif" of the experience.'[18] For all of you out there who are interested in knowing how to manage your customers' (or, for that matter, your friends') food and drink experiences, the good news is that there are various strategies that can be used to create 'stickier' memories, positive recollections that will hopefully form the basis of a decision to return to a given restaurant (or for those cooking at home, that will have your friends reminiscing about what a great cook you are). One suggestion is to deliver an unexpected gift, such as an amuse bouche, a little taster from the kitchen that the diners (or your guests) hadn't been anticipating, and certainly hadn't ordered. This is just the kind of positive surprise that is likely to stick in their memories long after they leave.[19]

Similarly, the rise of the multi-course tasting menu also provides the opportunity to create stickier interactions, with the first taste of each and every dish providing a potential moment of 'flavour discovery'. Serving a large plate of the same food is, from the point of view of engineering great food memories, absolute madness. You know that people will remember the first few mouthfuls and that is it. This is the 'duration neglect' that we heard about earlier.[20] The rest of the food will be lost to recollection no sooner than the plate has been cleared from the table. A wasted opportunity if ever there was one!

Anyone who is trying to design 'great-tasting memories' should probably also be thinking about primacy (and recency) effects here. Let me explain: if I were to give you a list of items to remember, like, say, the names of the dishes on a tasting menu, then you would be more likely to retain the first few courses and the last.[21] The middle items will have to do more to stand out. No wonder, then, that so many chefs seem to really excel in their starters (not to mention their amuse bouche). Perhaps one can think of this as an intuitive example of experience engineering. If you know which dishes are most likely to end up sticking in people's memory, then working on really perfecting them can be time well spent, in terms of leaving your guests with the best impression (or memory) possible. Looking forward, it ought to be possible to figure out an ideal balance in terms of the number of courses that diners will have a reasonable chance of remembering while at the same time offering the chef enough scope to show what they are made of.

Just take the following as illustrative of the problems associated with trying to create more memorable food experiences. In one study, conducted in a restaurant, one-third of the diners questioned had no recollection whatsoever that they had eaten any bread, even minutes after having done so. This ought to make you think differently about how much effort you should put into that particular aspect of your dining experience. Indeed, it does seem like a growing number of restaurants in cities like New York no longer offer their diners bread. Intriguingly, it would also seem to argue against a primacy effect. But perhaps it is just that people simply don't think of the bread as a meaningful part of the meal. Perhaps it is treated as background, like the tablecloth, rather than the foreground experience of the dishes themselves.[22]

Those experience engineers who have studied people's memories of mainstream restaurants find that our recollections rarely revolve around food. Just take the following as a case in point: when more than 120 customers who had eaten at a branch of the UK chain Pizza Hut were questioned a week after their visit, it turned out that the enthusiasm of the opening exchange with the restaurant staff – the warmth and energy with which the employees introduced themselves – was the single most salient memory for most of them.[23] What was also important was how long it took to be acknowledged by a member of staff. Ultimately, if you know what your customers are most likely to remember, you are going to be in a better position to modify your food service offering, no matter whether you are a Michelin-starred restaurant or a gastropub. Knowledge really is power!

Do you remember what you ordered?

But, I hear you say, while the specifics of the tastes and flavours of the dishes that we have eaten previously might well fade from memory, don't we all at least remember our favourite dishes, or perhaps the chef's signature plate at the local eatery? For me, in my neighbourhood Italian, it is always fried whitebait followed by cannelloni con carne. That is something that I *never* forget.* And whenever I find myself out for an Indian, then it's chicken jalfrezi, pilau rice and peshwari naan. The regularity with which I order exactly the same dishes is one of the things that convinces my wife that I am on the autistic end of the spectrum! But I would say it's not so much a question of neophobia (versus neophilia), but rather, if you know what you like, why change?

It is here that restaurants like The House of Wolf in Islington, north London, tend to fall down. Its business model was to offer a dining space for pop-up chefs and culinary artists, each doing a 4–6-week stint. Individually, each of the chefs who came into the kitchens was great. But they were also very different, one from the next. Hence, while you may be able to remember that you liked the chef who was cooking last time, you don't really have any specific positive memories of the dishes that might encourage your return, i.e., there is nothing concrete to look forward to. The management gurus are absolutely clear on this point: any really successful restaurant needs to have a few staple dishes that they are known for, dishes that customers remember and will come back to experience time and again.[24] And indeed, The House of Wolf, just like many other restaurants with an ever-changing menu, didn't last long (the leaking roof probably didn't help much either). A shame, really, as I used to enjoy my role as professor-in-residence there. By contrast, think of those restaurant chains such as L'Entrecôte, where the menu is essentially fixed. That is precisely why the customer looks forward to returning – to eat exactly the same food that they remember so well from every previous visit. There are even those, like my wife's family, who have been dining at L'Entrecôte for more than four decades now, despite the perennially long queue, as you can't book a table (see also 'The Personalized Meal'). Presumably the queuing is also part of 'the experience', lending what appears to be some scarcity value.

---

* Though thinking about it carefully, maybe it is actually difficult to say whether I know what those dishes are generally supposed to taste like or whether I am remembering the specific taste of the dish the last time I went.

Do you remember what you ate?

Another technique that can help one's diners to create memorable dining experiences is to tell stories around the food. Eating at The Fat
Duck restaurant is a good example. The experience starts with the diners being presented with a map and magnifying glass complete with
duck's feet (see Figure 9.1). The idea that you are on some kind of journey has been ramped up, even while many of the dishes have stayed
pretty much the same. Storytelling can, I think, help the diner to make
sense of a multi-course tasting menu – a series of dishes that might otherwise seem like a random sequence from the chef's own personal hall
of fame (a selection that might seem sentimental, even). By providing a
storyline, a narrative framework, not only will the diner be better
placed to parse or 'chunk' (that is, to group together items so that they
can be processed or memorized more easily) the whole experience, but
once 'chunked', the experience will be easier to remember too.[25] This

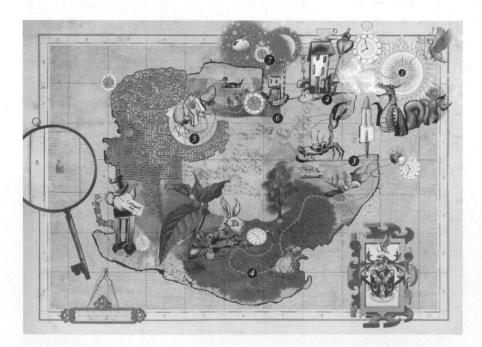

Figure 9.1. The map that diners are given to inspect (with the aid of a magnifying
glass) on arrival at The Fat Duck.

becomes all the more important once we have lost the more traditional structure of the three- or five-course meal.

For special meals, it can also be a good idea to give the diner a copy of the menu to take away with them.[26] The walls of the kitchen at home are adorned with framed examples of fabulous meals, like the menu from my first visit to Heston Blumenthal's restaurant in Bray. Even though it is now nearly fifteen years later, reading the descriptions of the dishes on the wall is all it takes to trigger so many pleasant memories (not necessarily of flavour, you understand, but rather recollections of the meal itself, and what I imagine the dishes could have tasted like). And, in this case, presumably, the more descriptive the menu, the better.[27]

Even better, at least in terms of creating a lasting (or 'sticky') impression, was what happened when the menu was placed on the table midway through that meal. I still remember, just as vividly as if it was yesterday, picking up what looked like a regular vellum envelope, hoping to remind myself of the dishes that I had already polished off (and which I was already starting to forget – there had, after all, been so many) and to see what was coming next. I was shocked to feel something under my fingertips that had the texture of skin (the envelope had been specially treated). That was not at all what I had been expecting, and that moment of 'hidden surprise' has stayed with me ever since.[28] Indeed, as noted earlier, more often than not it is the unusual or surprising experiences, the ones that capture our attention, that really force us to stop whatever else we may be doing (or concentrating on). It is the moments where we have to figure out what it is, exactly, that is going on that are remembered best. It is those events (and dishes) that need to be processed more thoroughly in order to be understood that will really stick in your memory. This is what the psychologists describe as 'depth of processing'; the deeper the processing, the better our recollection will be.

A final suggestion here, in terms of creating better food memories, relates to what is known as the 'end effect'. Our recollections of experiences tend to be dominated by what happened at the end, and food is no exception.[29] Consequently, ending a meal on a high note can lead to greater remembered enjoyment. In one simple demonstration of this effect, researchers gave eighty people an oat cookie followed by a chocolate cookie. Another eighty people ate the cookies in the reverse order. Those who finished their snack with the more enjoyable chocolate cookie remembered the food as having tasted better when quizzed thirty minutes later.[30] The end effect is presumably also what explains why 'all you can eat' meals are unlikely to be remembered too fondly. The end in such cases, at least in my personal experience (as a student, I hasten to add), is that feeling of being unpleasantly stuffed and knowing that you have eaten more than you should have. By contrast, those Italian restaurants where the meal ends with a surprise shot of limoncello might well be engineering a more positive memory, by ending the encounter with a mood-inducing unexpected gift. So, why not think about what you can do to surprise your dinner guests before they leave the table?

What is so special about mindful dining?

The next time you find yourself eating in front of the TV or computer, think carefully about what you are doing. As we have seen already, mindful (or attentive) eating and drinking is important, and anything that we can do to make ourselves more aware of what we are consuming is going to help in terms of increased enjoyment, enhanced delivery of multisensory stimulation, and quite possibly increased satiety too.[31] But does mindful dining and drinking also lead to better memories of food and beverage experiences? It certainly feels like this ought to be the case, and that eating mindfully would also lead to reduced consumption subsequently (be it later in the meal or in subsequent meals).[32] However, I think that we need to await further gastrophysics research in order to know for sure. That said, there is, I suppose, a sense in which the many diners taking all those pictures of their food and posting them on social media are creating an external memory of the event, and of the dishes themselves. They are an aide-memoire, if you will.[33] It is those images, more than anything else, that will help people to remember what they probably know they will otherwise forget.

Away from the restaurant, though, one might ask what exactly we remember of our favourite foods? Consumers often complain whenever too drastic a change is introduced into the formulation of their favourite brands – the consumer backlash that was triggered by the introduction of New Coke or by the change in the shape of the Cadbury Dairy Milk bar, for example. Such behaviour could certainly be taken to imply that we must retain the taste of our favourite foods in memory. However, it turns out that our recollection of branded foods and beverages may only be as good as the last consumption episodes. This, after all, is what enables the food and beverage manufacturers to engage in their health-by-stealth strategies, gradually cutting the amounts of unhealthy ingredients, like sugar or salt, in their products without consumers ever realizing that anything has changed.[34] Do it too fast and you'll have the consumer writing, calling and emailing in to complain that their favourite brands no longer taste the way they did! The challenge for the food companies is made all the more difficult, though, by the fact that even extrinsic product changes, such as to the colour of the can or the shape of the chocolate, can affect the perceived taste of the product (and give rise to an increase in consumer complaints), even if the formulation itself hasn't changed (see the 'Sight' chapter).[35]

The gastrophysicist will tell you that as soon as we are alerted to a change or a potential modification (e.g., when we read 'low fat' or 'reduced salt' on the label) we start to pay more attention to what we are tasting.[36] Thus it is normally a bad idea for a company to market a new reduced-fat, -sugar, or -salt product as such because that is likely to set expectations about the taste in the mind of the consumer (as discussed earlier), who will then pay more attention to the taste, looking for any differences. This is rarely a good thing. For as the Dutch sensory scientist Ep Köster notes, as far as the senses of smell, taste and mouthfeel are concerned, memory seems to be focused on detecting change rather than on identification and precise recognition of the food stimuli that we have encountered previously.[37]

So, how are we to make sense of all this – failure to identify our favourite brands in all those blind taste tests on the one hand, and consumers' vociferous complaints when they perceive that their favourite brands have changed on the other? It doesn't seem to quite add up – or does it? Well, maybe we remain essentially blind to taste, i.e., we pay no attention unless our brains happen to detect that something is not quite as we expected it to be. Only then do we really start to concentrate. So if you are trying to nudge your family towards slightly healthier eating behaviours, the implications are clear: introduce changes to your cooking (such as reducing the salt) gradually, and whatever you do, don't let them know what you are up to.

The meal forgotten

I wouldn't want you to think that the 'meal remembered' is only of interest to those who, like in the movie *Total Recall*, want to insert particular memories into your mind (be it for financial or commercial gain). There is also an important element to this research, relevant to all of those unfortunate individuals who are losing or have totally lost their ability to recollect recent events, those who can forget, just as soon as the dishes have been cleared away from the table, that they have just polished off a three-course meal. For instance, amnesic patients suffering from Korsakoff's Syndrome (normally resulting from extreme alcoholism) may retain no memory whatsoever of having just eaten, and will happily start a second and even a third meal, if it is placed before them, providing they are momentarily distracted after the last meal has finished.[38] One solution here is to leave visual cues of a recently finished meal lying around to provide an external reminder of the meal that has just been eaten.

In a related vein, a few years ago, I was a consultant on a project with a London-based scent expert and a design agency to help develop an intervention for early-stage Alzheimer's/dementia patients who forget to eat. The basic idea was that if these patients could be reminded to eat, they could maintain a semi-independent existence for a little longer than might otherwise be the case. The solution that my colleagues came up with is a plug-in that releases the scent of breakfast in the morning, of lunch at – you guessed it – lunchtime and enticing dinner aromas in the evening. The product has been available commercially for a few years now. In this case, one of the challenges for the gastrophysicist was to figure out which aromas to use, for while the smell of sizzling bacon might work as a breakfast cue for some, it clearly isn't an option for those who abide by certain religions. Furthermore, the foods we eat also change as the decades go by. So we had to try to find those food aromas that would be meaningful to individuals who were mostly expected to be of retirement age.

The importance of developing such sensory interventions becomes clear once you realize that close to 50 million people worldwide are currently suffering from dementia. In a small test of the designed solution, called 'Ode', fifty people living with dementia (along with their families) used the device for almost three months. The weight of more than half of those who took part either stabilized or increased, leading to an average weight gain of two kilograms. No wonder that it was voted the most innovative British Business Idea of 2013 by Small Business Cup.[39]

Hacking our food memories

There is also a really interesting line of research into the 'hacking' of food memories to bias people's food behaviours. For instance, researchers have shown that people's attitudes and behaviours around food can be subtly influenced simply by implanting false memories related to prior food-related experiences (e.g., telling someone that they once became ill after eating beetroot). Such misinformation, and the false memories that it can give rise to, can lead to significant behavioural change (e.g., lowered self-reported preference for and decreased consumption of beetroot). While the majority of the research conducted to date has been in the laboratory, there is growing interest in using such techniques to try to nudge people towards adopting healthier food behaviours. Could children, for instance, be encouraged to eat more vegetables by implanting false positive memories around previous pleasurable consumption episodes? And if they could, would it even be ethical to do so?[40]

Remember, remember . . .

Ultimately, all that stays with us, once the meal is over, is our memory of it; we have only our recollections of the great occasions, and of the terrible ones too. The stuff in the middle – well, that is mostly just forgotten. Chefs, at least those with an eye on the future, would like to create food experiences that are more memorable, that have more 'sticktion', in the words of the experience engineers. Their long-term success, after all, depends upon it.

It is our recollection of the taste and flavour of foods that determines which restaurants we go back to, which food and beverage brands we stay loyal to, and even how much we decide to eat. In fact, simply reminding someone of what they ate earlier in the day (for lunch, say) can be enough to lead to a significant reduction in their subsequent snack consumption, when compared to those who were encouraged to recollect what they ate for lunch the previous day instead.[41] Remembering what you have eaten recently can therefore be more important than you might have thought. Fortunately, the gastrophysics approach is increasingly assisting those who want you to recall more, as well as helping to improve the quality of life for those who can no longer remember.

And, although there isn't space to cover the topic in any depth here, consider too that the foods we eat can themselves also be used to evoke memories, as captured by the oft-cited case of Proust's madeleine. And just think of those so-called 'memory meals', like Thanksgiving in the US.[42]

Let me end, though, with Jean Anthelme Brillat-Savarin's quote on taste and flavour in ageing, from his classic book *The Physiology of Taste*, published in 1825. The famous French gastronome writes: 'The pleasures of the table, belong to all times and all ages, to every country and to every day; they go hand in hand with all our other pleasures, outlast them, and remain to console us for their loss.'[43] As the old polymath knew only too well, eating and drinking constitute some of life's most enjoyable experiences. When our memory of those pleasures go, one might ask: what else is left?

10. The Personalized Meal

You must have noticed how whenever you place an order in a branch of Starbucks the barista always asks for your name; then, when your beverage arrives, you find that it has been scrawled across the side of the cup. Now, this might seem to be necessary to avoid confusion at peak times, when there will be a large number of people standing expectantly at the counter, all waiting for their cappuccinos or skinny lattes. This is not merely a matter of operational convenience, though. Rather, this form of 'personalization' is company policy. Some even believe that it actually leads to a better experience for the customer. After all, one has the impression that the drink has been especially made for you.[1] The question that the gastrophysicist really wants an answer to here, though, is whether this (or any other) form of personalization can make whatever you consume taste better too.

Everyone loves personalization

That personalization sells was amply demonstrated by the phenomenal success of the 'Share a Coke' offer in 2013 and 2014, whereby consumers could buy a bottle of the fizzy black stuff with their name printed on the label (see Figure 10.1).[2] Make no mistake, this is superficial personalization (in the sense that the product itself hasn't changed). The drink is more or less the same the world over, and yet something about seeing your name emblazoned there on the front label changes the experience for you. So easy, so simple, and yet so incredibly effective: because of this campaign, sales increased for the first time in more than a decade.[3]

No wonder, then, that many other food and beverage companies have been scrambling to try and copy Coke's success with their own examples of personalization. Indeed, according to an article that

Figure 10.1. During the summers of 2013 and 2014 Coke's marketing strategy made headlines when it swapped the powerful equity of its own brand for the names of consumers across seventy countries. The campaign had originally kicked off in Australia in 2011.

appeared in *Forbes Magazine*, 'Personalization is not a trend. It is a marketing tsunami.' Late in 2015, for instance, Moët & Chandon set up a number of photo booths in branches of Selfridges across the UK where their customers could upload a snap of themselves on to the front of their Mini Moët champagne bottles. The perfect Christmas present, apparently. Vedett has also been encouraging people to customize its beer bottles with their own pictures, and Frito-Lay did something very similar, offering 10,000 bags of potato chips for people to personalize with photos of their 'favourite summer moments'.[4] In 2016, Kellogg's ran an offer whereby anyone who bought the requisite number of boxes of cereal could send off for their very own personalized spoon.*

---

* The writer Will Self managed to get a rather rude (or should that be puerile?) phrase printed on the front of his spoon; see W. Self, 'Finally my personalised spoon from Kellogg's® has turned up – and it's way better than I thought', *New Statesman*, 29

Have you heard of the 'self-prioritization effect'?

Why exactly should people respond differently to products that are associated with themselves? One possibility here relates to the 'self-prioritization effect'. Psychologists from Oxford recently discovered that arbitrary visual symbols (such as circles, squares and triangles), i.e., stimuli that have no intrinsic meaning, can nevertheless still take on a special significance just as soon as they become linked to us. In a typical study, one arbitrary stimulus (let's say a blue triangle) is associated with the self, while another is paired with a friend or someone else. The participants sitting in the laboratory are asked to press one button as rapidly as possible whenever they are shown the self-relevant object, and another whenever they see the stimulus belonging to the other person (perhaps a yellow square or a red circle). The results of a number of such studies have revealed that self-relevant objects are rapidly prioritized. That is, you see them sooner and respond to them faster than the other stimuli that have arbitrarily been classified as belonging to someone else. They have, in other words, become more salient because they are related or, in some sense, 'belong' to you.[5]

My suspicion is that a similar phenomenon may well be at play when consumers come across a Styrofoam cup of coffee or bottle of Coke with their name on it. And presumably the birthday cake that comes to the table will also taste better to the person whose birthday it is, for much the same reason.

Do you have a favourite mug? For me, it is the orange one with a cartoon pig on one side and a chicken on the other that I look for every morning when I make myself a cappuccino. I get annoyed if I find that it is still in the dishwasher. Of course, the coffee is the same whichever cup I drink from. But somehow the experience feels different; the drink just doesn't taste the same. It could be

September 2015 (http://www.newstatesman.com/culture/food-drink/2015/09/finally-my-personalised-spoon-kellogg-s-has-turned-and-it-s-way-better-i).

that the self-prioritization effect might, in some small way, help to explain why it is that beverages always seem to taste better drunk from a favourite mug. One might think of this as a kind of 'sensation transference' (sometimes referred to as 'affective ventriloquism'), a concept first introduced more than half a century ago by the legendary North American marketer Louis Cheskin. This is where the feelings we have about the cup, our very own cup or mug (i.e., all those warm feelings of ownership and familiarity), are transferred to our perception of the contents.[6] There is probably also a link here to the 'endowment effect': This favourite of the behavioural economists refers to the fact that we ascribe more value to things merely because we own them. This phenomenon is also known as the 'status quo bias'.[7]

As far as I am aware, the proper experiment has yet to be conducted,[8] but someone should really do it. All the budding gastrophysicist has to do is invite a group of people (thirty or forty is probably enough) to taste and evaluate coffee from their own cup and then coffee from someone else's mug (making sure, of course, to counterbalance the order of presentation). Maybe the coffee in the two cups is the same, or perhaps it's different – only the gastrophysicist knows for sure (or at least I hope they do). You almost don't have to do the study, though, in order to know which cup people would prefer to drink from. The funny thing is that when you quiz them people often feel almost embarrassed to admit they prefer their favourite mug, because at one level they believe it can't change the taste, that they are somehow just being silly. And yet, as a gastrophysicist, I firmly believe that this form of personalization really does make a difference to how much we enjoy the experience – a subtle one, perhaps, but significant nonetheless.

The 'cocktail party effect'

For many decades now, psychologists have been aware that your name seems to have a special meaning or significance: somehow it just 'pops out' of the background hubbub. You are probably most familiar with this happening when you suddenly become aware that someone else is talking about you while mingling at a noisy party, hence the name the 'cocktail party effect'.[9] But, at one level, maybe this prioritization of your own name is not so surprising given just how much experience you have had of hearing it all your life. What is so noteworthy about self-prioritization, though, by contrast, is just how rapidly it comes to influence our behaviour. It occurs almost as soon as an object becomes, in some sense, 'ours'. Even a silly blue triangle is treated differently, once it has been assigned to us. What is more, and supporting these behavioural changes, brain-imaging studies show that different neural circuits are activated by self- as compared to other-relevant stimuli.[10]

And that is not all: even the first letter of our last name has been shown to exert an influence over certain aspects of our behaviour. For instance, those whose surname starts with a letter that appears later in the alphabet are more likely to respond earlier in online auctions and limited-time offers, i.e., those with a surname beginning with the letter 'Z' are a little more impatient! This would appear to be tied to their names having been read out last at school, because married names don't exert anything like the same influence. There are a number of other intriguing phenomena, beyond this 'last-name effect'.[11] You might find it surprising, for example, that we all tend to prefer items that have a similar spelling as the first letters of our own name (the 'name-letter effect'). Moreover, marketers know that we like those products, brands and even potential partners more if their name happens to share at least a few letters with our own.[12]

Extending this line of reasoning to food might lead one to predict that we should all prefer, albeit ever so slightly, those dishes that share more letters with our own name as well. For me, *spicy* (*Spence*) food is the taste experience I always crave after a long trip. According to the research, the three shared letters with my surname may be doing at least some of the work here in terms of my liking for this sensation. Furthermore, whenever I order *chi*l*li* con carne, another of my favourite dishes, I cannot stop myself from thinking about all the letters it shares with my own first name (*Charles*). Why not try this for yourself? How many of your favourite dishes share letters with your

name? And, next time you meet a girl called Victoria, don't be surprised to find that she has something of a penchant for sponge cake.

Personalization in the restaurant

There are a number of ways in which the service offering can be personalized in the restaurant that are likely to enhance your meal
experience. One relatively simple example here comes from those restaurants like Ricard Camarena's Arrop in Valencia, Spain, where the
waiting staff take note of the type of bread that you choose when the
basket is first brought to the table. When they return, they will very
deliberately point to the bread that you selected last time and ask
whether you would like the same again, or whether you would prefer
something different. By this one act, they are letting you know, however subtly, that they are paying attention. Many other well-run
restaurants use similar techniques.

Another thoughtful example here comes from a restaurant that we
have heard a lot about in the previous chapters, namely the three-
Michelin-starred Fat Duck. The waiting staff there carefully observe
the diners at the start of the meal in order to figure out their handedness. A mental note is made of any lefties at the table and thereafter the
dinner service is aligned accordingly. The interesting thing, though, is
that no mention of this is made to those sitting at the table; in fact, all
that the less-observant diner may be aware of is that the experience
just seems to 'flow'. A few of the more attentive diners will probably
spot the personalization, and hopefully appreciate the effort and attention to detail that has gone into creating a dining experience especially
for them.[13]

'Where everybody knows your name'

Who doesn't like to be recognized when returning to a favourite
haunt? You know the sort of thing: 'Why, hello, Mr Spence, how nice
to see you again.' This has been christened the 'Cheers' effect, after

Figure 10.2. Charlie Trotter waiting with his staff for the arrival of a VIP (in this case, Chicago mayor Rahm Emanuel). A welcome guaranteed to make the diner feel special![14]

the Boston bar in the 1980s sitcom. While it is unlikely that the staff in your local branch of Pizza Hut will remember your name,* the ultra-high-end restaurants take things to a whole new level when it comes to making their guests feel special. The ultimate example has to be Charlie Trotter's famous refrain of 'Kerbside!', which would ring out through the kitchen of his namesake restaurant in Chicago. According to London-based chef Jesse Dunford Wood (who worked there for a while), that was the cue that a VIP was about to arrive. Everyone in the kitchen would then march outside and line up in front of the restaurant's entrance to welcome the guest (see Figure 10.2). Those of you who remember the hit TV series *Downton Abbey* will recognize this scene; staff at the house welcome the lord of the manor like this when he returns home after a long trip.

New York uber-restaurateur Danny Meyer caused something of a stir back in 2010 with the publication of *Setting the Table*, his memoir

---

* Though, as we saw in 'The Meal Remembered', even at the lower end of the food chain, the warmth of the welcome turns out to be the single most memorable aspect of a restaurant visit likely determining whether you will return.

of a life in the restaurant business. Meyer has been in charge of a string of famous restaurants, including Union Square Cafe, Gramercy Tavern and Eleven Madison Park. Time and again in his book, he stresses the importance of personalizing the service at his restaurants. For years, they have been storing information about the diners when the initial booking is taken, so as to ensure a familiar welcome when the guests arrive. In fact, they keep a file of regular guests and their gastronomic peccadilloes. You can imagine the sort of thing: does so-and-so prefer to be seated in the window or hidden away in an alcove? What is their first name and, more importantly, do they like to be recognized or do they prefer to remain anonymous? Do they have a fondness for Super Tuscans, perhaps, or, like Jay-Z, prefer white burgundies . . . ?[15]

While Meyer's New York restaurants are often cited as leading the way in terms of attentive and personalized service, many others claim to have adopted a similar policy. Famous Chicago venues such as Alinea, Next, Moto and iNG all try to find out something about those who are dining with them. According to Nick Kokonas, co-owner of Alinea, Next and The Aviary, they have kept a database of every single guest who has been on the premises since they opened. Initially, he says, the idea 'was simply to identify [guests] visually and thus greet them by name, like saying hello to an old friend you are greeting in your home'. Over time, though, this has morphed into the delivery of a more personalized experience for the diner. Even more surprising is the suggestion that the restaurateur might, on occasion, use such information to follow up on any of the regulars who haven't shown up in a while.[16]

How to make a first-time guest feel special

While it is easy to see how the restaurateur with a compendious Rolodex, or whatever the technological equivalent is these days, can make their regulars feel special, how can you give someone who has *never* been to one of your establishments the same experience?

Imagine for a moment how you would feel going to a restaurant in a new city where the doorman acknowledges you by name. Then you sit down, only to find that the waiter assigned to look after your table for the evening happens to come from your home town (somewhere far, far away). How freaky would that be? Don't worry, though, this is not ESP, rather just a sign that the restaurant has been googling you prior to your arrival. Justin Roller, for instance, the maître d' at Eleven Madison Park, is famous for googling every single one of the diners before they arrive, looking for anything that can help his staff to make diners feel both special and comfortable (i.e., almost as if they were at home). 'If, for example, Roller discovers it's a couple's anniversary, he'll then try to figure out which anniversary [. . .] All that googling pays off when the maître d' greets total strangers by name and wishes them a happy tenth anniversary before they've even taken off their coats. ("We want to evoke a sense of being welcomed home," [another staff member] says.)' It is hard not to notice how often commentators point to the outstanding customer service as an important part of the success of Meyer's restaurants. Now, at least you know the secret.[17]

Would it bother you if a restaurant googled you before you even walked through the door? Or would you welcome the practice because of the personalized service that you might receive as a result? Well, according to the results of a poll conducted back in 2010, nearly 40% of North Americans said that it was OK, their assumption being that it would lead to some kind of special treatment. A further 16% thought it was a little strange, but said that they could probably live with it. However, 15% of respondents thought it downright creepy. There is presumably a fine line to be drawn here between having a better experience as a result of personalization and feeling that one's privacy has somehow been violated. As one restaurant consultant interviewed by *The New York Times* put it: 'If you say, "I know you like a white Burgundy from the 1970s," that is creepy. Instead, you ask them what they like and point them in the direction of that white Burgundy.'[18]

When it slipped out in the press recently that The Fat Duck was googling its diners, several hundred reservations were immediately cancelled. Not a problem for a restaurant that is reputed to have 30,000 booking requests a day, but still a slight hiccup that one could have done without. The irony here is that the restaurant, like all those top North American venues, has actually been googling their guests for years. But that is beside the point: what is more interesting is the marked difference in reaction from North Americans. Perhaps the English are just that little bit more reserved.[19]

What's next in terms of personalization?

The service philosophy in many of these high-end restaurants is clearly fundamental to their offering. The aim, at least at the very top, is for people to want to come back to the restaurant because of the service. Remember that poor service is the number-one criticism of diners year after year. A professional attitude is important, obviously, and good food helps too, but *personalization* is key. It is, after all, one of the best ways to make the diner feel special. The more personalized the service, the more likely we are to enjoy the experience, the better we will remember the food as tasting and the larger the tip we will leave (though this is apparently another area where the British tend to be a little more reserved than their North American counterparts).[20]

The challenge, moving forward, at least as I see it, is how to take the idiosyncratic or one-off personalization of service at restaurants like Eleven Madison Park and industrialize it. For, ultimately, the canny restaurateur wants all of their diners to feel special, not just the lucky few who caught the maître d's eye when he googled them. Though, of course, when personalization becomes ubiquitous, it must surely lose some of its appeal. There is a very real danger of it coming across as artificially crafted as opposed to naturally friendly.

'Tell me when you were born and I will create a dish especially for you.' I can still remember a sentence to this effect appearing on the menu at The Fat Duck a decade or so ago, when the tasting menu was still optional. Nowadays, a more systematic approach to personalization has been adopted at the restaurant, one that revolves around nostalgia.* And rather than (or perhaps as well as) googling their diners, the restaurant staff ask about them directly. From the moment you manage to secure a booking (normally two months in advance), those working behind the scenes in Bray will be trying to find out some key information in order to personalize the experience tableside. The last time I went with my wife, we received a flurry of emails asking about our childhoods.

Part of the way in which such information is incorporated into the experience comes at the meal's end, when the miniature sweet shop is wheeled over to your table blowing cute little smoke rings from the chimney. This marvel of engineering looks like an ornate doll's house (and is rumoured to have cost more than a Rolls-Royce!). You are handed a coin and, when you insert it into the Sweet Shop (see Figure 10.3), drawers open and close in a seeming chaotic sequence. Eventually, though, the contraption† comes to what seems to be a haphazard stop, with one of the drawers left open. (It looks random, but, of course, it is not.) The waiter then hands the diner their very own bag of sweets from within the open drawer. The kinds of sweets that the diner finds in the bag should hopefully resonate with what they remember eating as a child, an update of 'the kid in a sweet shop' idea. Nostalgia is being used here to deliver a generic form of personalization (generic to those of a certain age, i.e., born in a specific decade). The hope is this interlude will help to trigger positive childhood memories and emotions that will come to colour the diner's recollection of the meal as a whole. The nostalgia/storytelling angle is still evolving.

While this level of personalization is currently restricted to

---

* In a lovely bit of publicity for the brand, it was the nostalgia element of a meal at The Duck that apparently saved one couple's marriage; see J. Tweedy, 'How dining at The Fat Duck saved a couple from divorce: Heston reveals warring lovers were reunited after eating "nostalgic" meal', *Daily Mail Online*, 16 December 2015 (http://www. dailymail.co.uk/femail/article-3362700/Heston-Blumenthal-says-dining-Fat-Duck-saved-couple-divorce.html).

† Though my favourite table automaton has to be the *Table Automaton featuring Diana and a Centaur*, created by Hans Jakob I. Backmann, in Augsburg (*c.*1602–6; https://artdone. wordpress.com/2016/05/10/celebration-125-years/hans-jakob-i-bachmann-table-automaton-featuring-diana-and-a-centaur-augsburg-ca-1602-06-khm-vienna/).

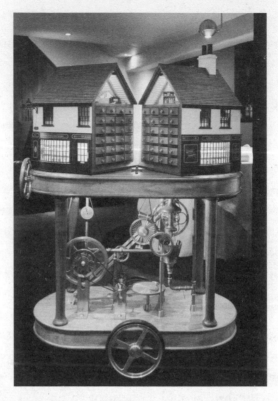

Figure 10.3. A personalized gift is waiting just for *you* inside one of the drawers of the Sweet Shop at meal's end at The Fat Duck restaurant.

high-end restaurants, it is not going to stay that way for long. There is already evidence that more mainstream venues are capitalizing on the various online tools, such as Venga (and OpenTable), that allow the restaurateur to pick up some useful 'diner-int.'. By integrating such guest management and loyalty programmes with a restaurant's point-of-sale systems, the staff can track their customers' average spend, their favourite items from the menu and even their preferred tipple. And – echoing what we saw earlier – some top restaurants use it to record the handedness of their diners. The Venga system is not cheap (currently coming in at somewhere around $149–$249 per month per location), but it is a price that a growing number of restaurateurs feel it worth paying to give their guests that VIP treatment. Here's a hint of what some are aspiring to: 'By the time a guest walks through the front doors at Ping Pong Dim Sum in Washington D.C., marketing manager Myca Ferrer can already be fairly certain what he or she will order.'[21] Perhaps such predictive software could also play a role in helping to reduce the phenomenal amount of food waste seen at most restaurants today.[22]

For those of you thinking about how to personalize your own dinner party, why not start by placing a name card where you want people to sit. You could use this crowd-control measure to keep all the bores together, as the Queen's party planner advises.[23] But who knows, it might make your guests enjoy their meal a little more too. And of course you could try googling anyone you don't know so well. It can't do any harm, after all – well, just as long as they either don't care or else don't find out!

At the chef's table

This concept ticks a number of the right boxes in terms of on-trend contemporary eating practices. A small number of diners get to sit around a central space behind which the chef prepares, or at the very least finishes, the dishes. It gives the solo diner something to look at, not to mention someone to talk to. It also enables diners to see their food being freshly prepared. And depending on the chef, there may well be an element of theatre and spectacle too. Crucially, there is also more scope for personalization. Typically, the format is a *prix fixe* ('fixed price') menu, as at the three-Michelin-starred chef's table at Brooklyn Fare, in New York, and 12 Chairs, in Shanghai.[24] It is hard not to be personal, after all, when the diner can look directly into the chef's eyes.

The ultimate in terms of personalized dining is the private chef. While this is something that you would normally expect to see only in the homes of the rich and famous, you can nevertheless find a few restaurants out there offering pretty much this level of service. Fuad's, in Houston, is one such venue. The head chef, Joseph Mashkoori, will come to your table and ask you what you'd like to eat. He might proffer a few suggestions, but he will insist that he is happy to cook whatever you want – from chateaubriand to a Philly steak sandwich. Meanwhile, in New York, Jehangir Mehta does something similar in his culinary venture called 'Me and You'. According to the website, the diner is promised: 'A unique private dining experience like none other. Every dish is custom crafted to suit your taste, play to your palate and inspire your juices.' The Solo Per Due restaurant in Vacone, Italy, also offers a very private meal experience, with only a single table for two.[25]

A matter of choice

At one level, this drive towards the personalization of the dining experience would seem to be at odds with the simultaneous rise of the tasting menu (where the diners are given virtually no choice about what they will eat).* The waiter may ask the diner about any allergies and dietary requirements they have, but that's pretty much it. In fact, most of the time the only decision that the diner has to worry about is whether or not to go for the wine pairing, assuming that there is one. But what, exactly, explains the growing popularity of the tasting menu? Isn't this, in some sense, the very antithesis of personalization?

Some commentators have wanted to link it to attempts by chefs and restaurateurs to increase the memorability of the meal. As we saw in 'The Meal Remembered', the more dishes the diner tries, the more opportunity there is for 'Sticktion'; tasting menus typically have more courses than one would choose if ordering from the normal à la carte menu. It is easy to imagine how, when everyone is enjoying the same food at the same time, it is going to enhance the feeling that one really is *sharing* a meal (see 'Social Dining'). And then, from the chef's perspective, perhaps the economies associated with serving only a limited range of seasonal produce, and being able to select one's best offerings, also helps make up for the lack of choice.† There are, of

---

* What is worse, at many of those restaurants where the tasting menu is still optional, they normally insist that either *everyone* at the table has it or else *no one* does.

† One could, perhaps, also frame it as the chef trying to wrest back some semblance of control, given the explosion of dietary requirements, allergies and intolerances that

course, those of a more cynical persuasion who see this as just another way to extract more cash from the diner, since tasting menus tend to command a higher price too. A more positive take, though, might simply be to say that diners don't like the process of choosing (and thus having to decide against all the things that they can't have). Perhaps there is also a link to the unwritten rule in the restaurant business that the better the venue, the smaller the number of options the diner is given.[26]

Some people are offended by the loss of choice. According to Tim Hayward, writing in the *Financial Times*: 'Menus without choice blaspheme against the doctrine of dining.'[27] However, it is not only the tasting menu where the diners' range of options is limited. From à la carte to those joints serving only a single dish, it can feel as if there is a steady reduction of choice across many styles of dining. In a sense, though, this is all just an extension of the *prix-fixe* menu or even the table d'hôte that used to be such a common feature of dining out in France and some other European countries.[28] Talking of which, one incredibly successful and long-running restaurant with limited choice is the L'Entrecôte chain in France (and London, New York, Bogotá and beyond). While there is a menu for drinks and desserts, there is only one option as far as the starter and main course is concerned: Salad, followed by steak (the diner is given the choice only of how they would like it cooked),[29] a delicious sauce (the recipe is kept secret) and French fries à *volonté*. Virtually no options, no personalization, and yet people queue, sometimes for more than an hour, to get a table (they do not take reservations). So, you have to ask yourself, how much choice do diners really want?

Well, on the one hand, it is certainly true that what the marketers used to tell us – namely, that more choice is *always* better – no longer seems to hold true (if it ever did). Give people too much choice and they can feel overloaded. If you *are* going to give diners choice, seven seems to be the magic number: seven starters, seven to ten mains, and seven desserts. Any less and there is a danger of there being too little

---

diners present with nowadays. When I see what friends who are chefs have to deal with, I sometimes wonder how they manage at all.

choice. Any more and the diner may find it difficult to decide. Of course, the trick for those restaurants wishing to offer more choices than such a fixed format allows is to break the menu up into a number of sections. How many? You guessed it: the recommendation is again seven.[30]

Rory Sutherland, Vice Chairman of the Ogilvy & Mather Group, UK, has an anecdote that fits nicely here about how airlines ended up selling far more cheap tickets as soon as they started to restrict the number of discount destinations they offered. Again, this would seem to run counter to sound economic principles. Surely, the more options there are, the more likely it is that a customer will find a destination they want. And yet the sales data emphatically shows the opposite to be true. Behavioural economists know only too well that we really can be paralysed by too much choice.[31] Presumably, that is why we have recently started to see the emergence of the 'condiment sommelier' in places like New York, whose job is to guide you through the options for, say, mustard or mayo when the range of alternatives becomes too formidable.[32]

The 'Ikea effect'

You know the situation: you are at home preparing a meal for your friends and you think that you have really excelled yourself this time, that the food tastes absolutely fabulous. Your guests, polite as always, tell you that the food is delicious. But what do they *really* think? My advice as a gastrophysicist is don't trust what they say; rather, watch what they do. The question still remains: are your guests just being polite, or do they perhaps taste the food differently because they didn't make it themselves?

Marketers have a name for the increased value that things seem to have if we make them ourselves: they call it the 'Ikea effect'. In other words, just because you assemble that wooden table yourself, it is worth more to you than if it had come pre-built. But while there is plenty of evidence that co-creation invests value in the outcome as far as flimsy furniture is concerned, what we want to know here is whether the same is true of a meal that you make for your friends. And does the answer depend on whether you are cooking from scratch versus cooking from a meal kit or partially prepared meal?[33]

Norwegian researchers have started to address these questions. In one series of experiments, they had various groups of individuals (not all students, you will be relieved to hear) make a meal from a kit in a kitchen laboratory. The researchers assessed what people said after they had prepared it themselves, or after being told that someone else had cooked it. Intriguingly, those who had prepared the meal themselves (or, to put it better, who thought that they had) rated it as tasting much better than those who believed that they were evaluating a dish made by someone else. This despite the fact that everyone was actually trying the same food (an Indian tikka masala dinner, if you must know). Furthermore, those who had to fry the meat and prepare the food as prescribed on the side of the package rated it as tasting better than those who merely had to stir and heat. In other words, the more involved the cook was in the act of creation, the better the end result tasted (at least to them).

So the chances are that if you make your friends dinner the food really will taste a little different to you, and this difference is likely to be more pronounced if you cooked from scratch (rather than 'cheating' with a pre-prepared meal). However, the bad news is that, if anything, the food probably tastes *better* to you than it does to anyone else (since they didn't make it). What all this means in practice is that you should get your friends involved in the kitchen, so that the food will taste better to them too.[34]

Making cake

There is an interesting link here to one of the marketers' favourite case studies, involving the Betty Crocker cake mix. According to the oft-recounted tale, this powdered cake mix failed on its launch in the marketplace back in the middle of the last century. The product's fortunes took off only once a certain marketing executive[35] had figured out that the product formula should be changed so that the home cook had to add an egg to the mix. This obviously increased the effort involved in baking the cake, something that any rational analysis would surely say was a bad idea. And yet sales steadily climbed. The suggestion was that by adding the egg the cook somehow became invested in the process – that is, they could feel more like they were actually cooking![36] And it is quite possible that the finished cake would actually taste better to whoever made it too, because of their greater involvement in its making – the Ikea effect again.

You find the Betty Crocker story recounted all over the place – even the top North American food writer Michael Pollan talks about it in one of his bestselling books.[37] It sounds almost too good a story to be true, right? Well, that is most probably because it is, at least according to a 2013 history of cake mixes that appeared in *Bon Appétit* magazine.[38] It turns out that a patent for 'a cake mix that required the home baker to add a fresh egg', rather than a powdered one, was granted back in 1935 to a company called P. Duff & Sons. It was in the 1950s that sales of cake mix (both with and without fresh eggs) stalled. The innovation that actually revived the fortunes of the cake mix was the introduction not of the egg but rather icing; that is, the explosion of interest in *customizing* one's cakes and buns with fancy designs was what really reinvigorated sales. So the Betty Crocker story is just a load of old baloney; nevertheless, the point about the importance of personalization still stands.

Before we leave this topic, I am going to have to pick a fight. And not just with anyone, but with Nobel Prize-winning experimental psychologist and now behavioural economist Daniel Kahneman. He is in print, in *The New York Times*, no less, trying to justify the claim that 'sandwiches taste better when someone else makes them'. This would certainly appear to be an idea that resonates with journalists around the world, as the story has been picked up by any number of news outlets. Nevertheless, if you trace the story back to its sources, it turns out that the assertion is based on nothing more than speculation. In other words, no one has, at least as far as I can tell, ever done the proper sandwich study. And, given what we have just learnt about the Ikea effect, I see little reason to believe that we would, in fact, prefer other people's sandwiches to our own. I don't know about you, but I certainly think my own sandwiches taste pretty darn good! And I am not alone on this one; from what people are saying on the online discussion forums it would seem that many others share my intuition.[39] Once again, then, another important piece of gastrophysics research waiting to be done.

I would like to end this chapter by asking: why is it OK to *customize* some dishes but not others? *Customization* can be thought of as a form of personalization, but one where control remains firmly in the hands of the consumer (or customer). It should feel empowering (rather than creepy). It was the opportunities for customization that were provided by the encouragement to ice one's cake that led to the revival in sales of cake mix. In the case of dining, customization occurs when the customer has some say over how their choice of dish is prepared, seasoned and/or served, everything from how spicy their curry through to how well done their burger. When the waiter offers to grate some Parmesan over your pasta at your local Italian, or you instinctively reach for the salt or pepper next time you go to the steakhouse – these are all examples of customization.[40] But when should we feel entitled to customize our food? Let me illustrate with an infamous tale from the annals of restaurant folklore.

'Would you be so kind as to pass the salt and pepper?'

Marco Pierre White, the first British celebrity chef to be awarded three Michelin stars, comes from my home town of Leeds, in the north of England. His cookbook *White Heat*[41] was one of the first I used when learning to cook, given to me by my sister on the occasion of my sixteenth birthday. It is still on (and often off) the shelf three decades later, containing, as it does, a truly sublime recipe for lemon tart. However, this chef first came to fame – or should that be public notoriety? – for throwing diners out of his restaurant should they be impertinent enough to ask for the salt and pepper. White said that it was an insult for any diner to want to season (or *customize*, we might say) their food in *his* restaurant. Seasoning a dish is, after all, the chef's remit, no? So, if a diner were to ask for the salt and pepper, then they could mean nothing but to insult the chef. Why so? Well, because they would be implying that the kitchens hadn't done their job properly. That, at least, is how the chef saw it.

In hindsight, one can view this incident as an early sign of the ascendance of the star (or *prima donna*) chef, no longer content to be hidden away in some dark and hot back room, never to be seen and rarely, if ever, to be acknowledged by the dining public. It presages, I would say, today's situation, whereby the celebrity chef has an open kitchen at the front of the restaurant for *everyone* to see.[42] The chef, in other words, is now very much the star of the show. They call the shots, and don't we all know it!

However, while it was Marco Pierre White's tantrum that first brought the question of customization to the table, as it were, it does seem to me that the salt and pepper have been slowly becoming a little more elusive in restaurants up and down the land. You certainly don't find them in the majority of high-end modernist establishments.* So, one might ask, is it not just choice that the chefs wish to restrict but also the diner's opportunity to customize their food? (At the same time, of course, the top restaurateurs are ramping up the level of personalization in other parts of the meal experience.)

So why do we customize some dishes but not others?

Come to think of it, I would never dream of asking for salt and pepper in The Fat Duck. But why not, I ask myself. (And why don't we have sugar and citric acid shakers to season our desserts too, I wonder.) It is partly about trusting the skills of the chef, or rather the culinary team beavering away in the kitchens. However, it partly relates to the nature of the food that is being served. Much of it is so unlike anything that I (or you, I presume) have ever had before that it can be hard to know quite what the chef had in mind, what they were

---

* My suspicion is that the pretensions of the chef/kitchen and the presence (or lack thereof) of the salt and pepper are most probably correlated.

trying to achieve. As such, I have no internal standard against which to judge the dishes. All I know for sure is that they are delicious. So, given that I do not really know what I am *supposed* to be tasting, I am really not sure what outcome I would be aiming for, were I to try seasoning it myself.

We can contrast this with the situation when you go out for a steak. Not just any steak, mind you; let's imagine you've just ordered the £140 Japanese wagyu 8oz rib-eye steak from star Austrian chef Wolfgang Puck's restaurant Cut, in The Dorchester Hotel on Park Lane, London.[43] In this case, I most definitely *do* expect to be asked how I would like my meat to be cooked. And there would be a riot if I didn't find the salt and pepper close at hand!* Notice that we are not so far off the price point of The Fat Duck (at least not once you've ordered your fries, sides and starter, etc.), and yet customization of the food is expected in one case and not even worth contemplating in the other. So price, or the skill of the kitchen, can't be the whole story here.

So what, then, *is* the difference? Well, I have had many steaks before, so I have some sort of internal standard that I am aiming for. My food memories may, of course, be inaccurate (see 'The Meal Remembered') but still I have an idea (or at least think I do) about the taste. Whenever a pepper grinder is available, though, I will habitually reach for it and add a liberal dose to my food, even before I have taken the first mouthful. How are we to explain such behaviour? Could it be that, by this simple act, I am somehow making the dish *mine* (customizing it, if you will), and that this somehow automatically makes the food taste better (in much the same way that it might when we rotate the plate just a little once the waiter has placed it down before us). Alternatively, however, I suppose it is also possible that we might all realize that our own taste preferences tend to reside at one end of the taste spectrum (for instance, I know that I like my food spicier than most other people), and hence that most food that

---

* Marco Pierre White now has a chain of steakhouses; we even have one of White's restaurants here in Oxford. I was tempted to go along and see what would happen if I asked for the salt and pepper myself, but it turns out that White actually has nothing to do with many of the restaurants that bear his name and image; they are just franchises.

has been prepared for the masses will taste better to us if we adapt it to our own personal profile. That said, perhaps the real distinction here relates to the degree of adulteration of the raw ingredients on their way to becoming a dish on the menu, and thereafter on my plate. So for relatively unprocessed foods, like steak, it is OK to season; it may even be expected. But as soon as that meat has been cooked in or presented with a sauce, say, then the justification for seasoning the dish ourselves is diminished somewhat, especially when we are in the hands of a top chef. And when it comes to the fabulous concoctions on the menu at The Fat Duck, they are so highly processed one may no longer have any real sense of quite what the raw ingredients were; that is, they have been transformed into something totally new. Under such conditions, then, there is simply less of an incentive for customization, as the goal of doing so is no longer so clear. Note that it is not that personalization is removed entirely in such venues, rather that it occurs in other aspects of the service or the meal.

My personal answer

So, in closing, let's return to the question of whether chef Marco Pierre White was right when he said that the diners had no right to customize his food. Is the seasoning of a dish really something that is best left to the chef (assuming that they have got the requisite number of Michelin stars)? Ultimately, isn't the diner always right?[44] And never forget, as we saw in the 'Taste' chapter, we all live in very different taste worlds. In fact, once you realize just how distinct and personal your taste, smell and even flavour perception really are, there can be no looking back. So, in the future, will the foods and drinks served in restaurants themselves start to match our own personal taste profiles? This was another of those far-reaching ideas that was prefigured by the Futurists. The movement's founder, F. T. Marinetti, wrote that '[w]e will create meals rich in different qualities, in which for each person dishes will be designed which take into account sex, character, profession, and sensibility'. Recently, consumers have been offered the opportunity to personalize the taste of everything from chocolate

(Maison Cailler) through to champagne (Duval-Leroy), while Illy has developed a new system to allow their customers to adjust the sensory profile of their coffee.[45]

So, given all of the above, I am sure you'll understand why I recommend that you leave the salt and pepper on the table the next time you throw a dinner party, no matter how good (you think) your cooking is, and no matter what you end up doing to the ingredients. As a gastrophysicist, I'd say that the fact that the diner chooses to season their dish shouldn't be thought of as an insult to the chef, but rather as a form of customization that recognizes the very different taste worlds in which we all live.

11.   The Experiential Meal

'Did you enjoy the experience?' This, the question that was asked time and again by the various service staff who ushered the guests through the different stages of Albert Adrià's 'About 50 days', held in London's Café Royal.[1] But since when did this become *the* question? Why not ask whether the diners enjoyed the food instead? Getting to the bottom of this matter is the topic of this chapter. It is the story of the ubiquitous rise of the 'experience industries', first predicted by Alvin Toffler in his bestselling 1970 book *Future Shock*.

B. Joseph Pine, II, and James H. Gilmore, building on Philip Kotler's early (1974) ideas about atmospherics deserve the credit for introducing the 'experience economy' to the marketplace.[2] Their point was simply that consumers are not really buying meals and drinks, or, for that matter, any other kind of product or service; rather, what people want is to enjoy, and increasingly to share, 'experiences'. These experiences are, by definition, multisensory. The realization, then, that dining out isn't really about fulfilling any kind of nutritional need helps make sense of why so many people now want to know the answer to Adrià's question.

Look around the world of fine dining and you find a growing number of chefs and restaurateurs promising to deliver multisensory dining experiences. For example, chef Andoni Aduriz (now chef at Mugaritz, in San Sebastián) had this to say of his time working at elBulli under the guidance of Ferran Adrià: 'For him, what mattered most was the experience, what one felt when eating at elBulli. Everything necessary would be done to create this experience.' Or take the following newspaper description of one of Marco Pierre White's franchise restaurants: 'The steakhouse, opened two years ago, is described on its website as being "all about the experience; the buzz, the atmosphere and enjoying the company of friends and family in a gorgeous comfortable surrounding."'[3]

On the rise of theatrical dining

As we will see in this chapter, restaurant cuisine is moving from its traditional function of providing nutrition – or restoration, as in the original meaning of the term 'restaurant' – towards becoming a medium for artistic expression.[4] Restaurants are becoming stages; the waiters and chefs at some of the world's top establishments are increasingly playing the role of actors and magicians. First there was atmospherics, then came the theatre, the storytelling and the magic at mealtimes: This really is the very heart and soul of 'off the plate' dining. It is all the rage amongst the chefs that you find jostling for position in the San Pellegrino list of the world's top fifty restaurants. Some have been tempted to suggest, though, that such lists may be exerting undue influence . . . 'At Eleven Madison Park [in New York . . .] Daniel Humm and Will Guidara devised an entire program of changes motivated in part by the perception that "the San Pellegrino voters reward restaurants with a strong sense of place, and of theatre." They included a three-card-monte dessert and – further belaboring the locavore trend – a cheese-and-beer course that emerged from an old-fashioned Central Park picnic basket.' Of course, not everyone is happy. As one commentator puts it: 'If the wine industry has become Parkerized, then the restaurant world might be said to have been Pellegrinoed.' Meanwhile, David Chang (of Momofuku fame) describes the archetypal '50 Best' restaurant thus: 'It's a Chinese restaurant by a guy who worked for Adrià, Redzepi, and Keller. He cooks over fire. Everything is a story of his terroir. He has his own farm and hand-dives for his own sea urchins.'[5] In fact, some have suggested that knowing just how important it is to sell 'the experience' may actually be what's driving many of the contemporary trends in the world of high-end dining and drinking.[6]

Sublimotion, in Ibiza, currently offers the world's most expensive restaurant meal. Should you be lucky enough to book a seat, you will need to find close to 1,500 euros a head. At that price, the twenty-course tasting menu can't just be about the food, now, can it?[7] The meal must be fabulous, that's a given. But there needs to be so much more. The underlying assumption is that diners are willing to pay a hefty premium for an *experience* rather than for a mere meal, no matter how tasty the food.

The rise of the open kitchen, not to mention the growing popularity of the 'at the chef's table' dining concept, can be framed in terms of their ability to transform the very preparation of food into a kind of theatre.[8] In fact, the tour of the kitchen is becoming an increasingly common component of a meal at many high-end restaurants. Juliet Kinsman, writing in the *Independent on Sunday* had the following to say: 'If you'd told proprietors then that one day diners would demand to eyeball the cooking team at work, they'd have blanched. Now butchery's brought to the fore and labour is part of the flavour – we want display as we dine. At ABaC Restaurant & Hotel in Barcelona, part of the thrill is being directed to your table (where you'll savour Jordi Cruz's two-Michelin-star 14- or 21-course menus) through the 200sq metre kitchen. And at the Typing Room in east London, I'm sure the fact that I could see Lee Wescott piping chestnut cream into the bowl made his dishes all the more delicious.'[9]

Even the removal of choice that we came across in 'The Personalized Meal' can be seen through the lens of experience design.[10] But this, let me tell you, is really only just the beginning. There is so much more that can, and increasingly is, being done. According to one textbook on restaurant design published in 2011, restaurants are more than 50% theatre (see Figure 11.1).[11] From where I am standing, that percentage would appear to be increasing year on year.

Have you come across theatrical plating?

There has been a growing tendency in recent years to turn the serving of food into a theatrical stunt too, this time by plating directly on to the table itself. At Alinea, in Chicago, for instance, a number of the desserts require a performance that can last for several minutes. For one dish, the waiters first lay a waterproof tablecloth over the table, before bringing all manner of sauces and ingredients. Next, one of the chefs emerges from the kitchen and starts to 'plate the table' in front of

Figure 11.1. Chef Jesse Dunford Wood sabre-ing the top off a bottle of sparkling wine. A most theatrical start to a meal at the chef's table at Parlour, in north London.

the amazed diners by breaking up the solid elements and painting with the sauces (both drop by drop and 'Jackson Pollocking' the table-top). Given all the practice that they have undoubtedly had, they manage to paint the dessert on to the table with great skill. Something similar happens at Sublimotion, in Shanghai, where the 'waiting staff appear with palettes of ingredients and "paint" an edible version of Gustav Klimt's "The Kiss" on the table'.[12] Once the chefs have finished their work, the diners tuck in, eating straight from the table.

Meanwhile, a little closer to home, one of the chefs I have worked with is Jesse Dunford Wood, who is famous for theatrically plating the dessert directly on to the chef's table (hidden away in a walled alcove between the kitchens and the rest of the restaurant) at Parlour, in Kensal Rise, north London. The chef wields a dangerous weapon: the blow torch! Headphones are handed out to every diner, playing tracks that have been specially chosen to be familiar and hopefully to trigger an emotional response. When I was last there, the iconic theme from *2001 A Space Odyssey* was up first, followed by Gene Wilder singing something from *Willy Wonka and the Chocolate Factory*. Scented smoke was pumped out from a hole in the wall – this was, in other words, a truly multisensory experience.[13] This example of performance around the plating of a dish is particularly interesting given that the chef himself listens to the same soundtracks as the diners while he works. Everyone at the table is connected in a private, but shared, sonic experience.[14]

There is, of course, always a danger that things can be taken too far: for example, Dive, the Los Angeles submarine-themed restaurant from Steven Spielberg. According to those who have been, the lighting was pretty extreme. There was a wall of monitors along one wall constantly flickering away, showing submarine-themed movie clips. One commentator described what would happen inside: 'Periodically, all lighting is extinguished except for intense red lights that whir and flash while a loudspeaker barks "Dive! Dive!"'[15] Sounds pretty intense, right? Perhaps a little too arousing? No wonder the restaurant closed down.

One way to deliver a memorable dining experience is by hosting it in a most dramatic or unusual location: venues such as the underwater restaurant in the Maldives or the Dinner in the Sky concept spring to mind (see Figure 11.2). A somewhat less extreme, though no less successful, version of the latter concept went by the name of The Electrolux Cube. For a while, this transparent structure was situated on top of the Royal Festival Hall, on London's South Bank, and a stream of Michelin-starred British chefs popped up there in order to serve the eighteen guests. When you add in a great view, the experience is undoubtedly elevated beyond your average pop-up dining event.[16] The cube has now appeared at a host of scenic locations around Europe: a rooftop overlooking the Piazza del Duomo in Milan, in Stockholm (atop the Swedish capital's Royal Opera House). It has been to Brussels too. Part of the success of this venture can presumably be put down to the fact that this is a limited offering (scarcity being highly valued in the restaurant sector at the moment).

'The everything else'

How would you like to go to a restaurant where the atmosphere changed from one course to the next? The top chefs with money to burn can do this using technology, others (on much smaller budgets)

Figure 11.2. *Top:* The underwater Ithaa restaurant opened in April 2005, seating up to fourteen diners and located five metres below sea level at the Conrad Maldives Rangali Island. *Bottom:* 'Dinner in the sky': these diners are having a unique experience suspended tens of metres above the ground. This is more about the experience than the food, one presumes.[17]

manage to achieve much the same effect by having their diners move from room to room as they switch between courses. As chef Grant Achatz put it (when contemplating overhauling the experience at his restaurant Alinea), 'perhaps diners will do a portion the tasting menu in one space, before moving to another environment that is completely different in its configuration, design elements, lighting, even aroma.'[18] Make no mistake about it: today, we are increasingly seeing a shift (often facilitated by technology) towards a much more dynamic and adventurous approach to the dining experience, one that involves storytelling, added theatre – oh, and possibly a dash of magic too. So, welcome to the all-new world of experiential dining.[19] And here we are not just talking about playing with the colour of the lighting, or synchronizing the music or soundscapes to match each and every one of the dishes. We have already come across a number of examples of ambient aroma being used to complement specific dishes (see 'Smell'). Some chefs, like Paco Roncero over in Ibiza's Hard Rock Hotel, have gone further: they are even playing with the atmosphere (i.e., the temperature and humidity) of the dining rooms they control.

The aim of the next generation of experience providers is to enhance, that is, to complement, what is hopefully already a fabulous product offering (not to distract you from a poor one), by optimizing 'the everything else'. It is the chefs who are really at the top of their game (those with two or three Michelin stars, and who appear in the San Pellegrino World's 50 Best Restaurants list year in, year out) who are innovating here; a number of them have realized that no matter how good the food they put on the plate, unless they are in control of 'the everything else', they really can't hope to optimize the experience for their diners. Of course, as was noted earlier, one could turn things around and suggest that these chefs are focusing on 'off the plate' dining precisely because they know that this is what the San Pellegrino judges are looking for. As one commentator put it: 'Chefs play to the list, mindful of its aesthetic preferences and its methodological weaknesses.'[20] The chefs are absolutely clear on this point: French chef Paul Pairet, of Ultraviolet in Shanghai, insists that the goal of changing the multisensory atmosphere on a course-by-course basis is to 'intensify the focus on the food, not distract from it'.[21]

Elsewhere, he has been quoted as saying: 'You can't escape from what I'm trying to convey. Everything will lead you to [develop] a strong focus on the dish.'[22] The technology-enabled atmospheric projections on the walls and tables of some of these futuristic dining rooms undoubtedly allow for more theatre and storytelling, elements that are key when it comes to trying to hold the diners' attention/interest over what may well be a 15–20+-course tasting menu.[23]

If you want to know what to expect, here's a description by one journalist who was lucky enough to dine at Pairet's Ultraviolet: 'Dinner starts dramatically with an apple wasabi sorbet, frozen and cut into wafers. A Gothic abbey appears on the walls, the air is filled with holy incense and AC/DC's "Hells Bells" assaults the ears.'[24] Meanwhile, an evening at Sublimotion has been described as: 'an emotional "theatre of the senses" . . . a night of gastronomy, mixology and technology'.[25] Ultraviolet bills itself as the first of its kind to bring together the latest in technology to create a fully immersive multisensory dining experience.[26] It opened in May 2012 to a maelstrom of interest from the world's media.★

Other no-holds-barred gastronomic events include the one-off Gelinaz dinners. The food at these events is whipped-up by some of the world's top chefs and the courses are interspersed with music, dance, magic and video. Just as well, I suppose, given that these events may last for anything up to eight hours.[27] The El Celler de Can Roca restaurant in Spain has regularly been voted top of the World's 50 Best Restaurants list in recent years. Back in May 2013, the chefs (the Roca brothers) worked with music director Zubin Mehta and visual artist Franc Aleu to create a fabulous twelve-course culinary opera called 'El Somni'. This one-off dinner was held for twelve carefully selected guests (talk about exclusive) in a specially designed rotunda in Barcelona. A once-in-a-lifetime experience – quite literally! An

---

★ The dates here are important in terms of assessing precedence: Ultraviolet opened in 2012, Sublimotion in 2014. And while a number of today's chefs may well fight over who deserves acknowledgement for precedence, really it was the Futurists who got there first (see 'Back to the Futurists'). One might wonder whether restaurants like these should be considered under the umbrella of 'eatertainment'; I suspect not, given the pejorative associations of the latter term.

amazing sound system was installed especially for the event and visually stunning projections surrounded the diners – no expense spared.
Indeed, I dread to think how much it must have cost to put this dining event on; there is no way that it could break even, even if the
diners were paying through the nose (which they weren't). My guess,
though, is that it was probably worth it for the many brands sponsoring this venture, given the huge amount of international publicity it
attracted.[28]

Performance at the table

Do you think that you'd enjoy your dessert more if it came out from
the kitchens at the same time as a cellist sat down next to you and played
a specially composed piece of music, or even just a sustained musical
chord? It would, at the very least, be a rather unique experience, would
it not?[29] Composing music especially for mealtimes certainly isn't new;
go back far enough (to the mid sixteenth century) and one finds *Tafelmusik* (literally 'table-music') being composed for feasts and other
special dining occasions.[30] Now, composers, artists and sonic designers
are once again taking up the challenge of designing music specifically
for the meal.[31] While once the music was, in some sense, composed for
the *occasion*, now it is designed to match the *food itself*.

Some of you may be wondering just what influence the atmospheric soundscapes and music being played in a growing number of
restaurants have on your experience? You might wonder about its
effect on the taste of the food, not to mention on how much you enjoy
the total experience. Note that we are not just talking about fancy
restaurant meals but also the local gastropub.[32] In an earlier chapter,
we saw how the sound of the sea could be used to enhance the taste of
oysters. Subsequent research by the Condiment Junkie multisensory
experiential design team has shown that playing the sounds of the
English summertime enhances the perceived fruitiness and freshness
of strawberries.[33] Put this evidence together with the literature on
sonic seasoning and it is clear that both sensory-discriminative (i.e.,
what it is) and hedonic ratings (how much you like it) of the food are

likely to be affected by what diners hear. All the more reason, then, to try and get it right.

One interesting challenge that crops up once you start to think about designing music, or soundscapes, to complement a particular dish (or even an entire meal) is that the structure and duration of each track probably has to be quite different from that of traditional music.[34] In fact, music that has been especially composed for the meal (or dish) probably has more in common with the kind of sonic backdrop that people design for video games, say, than a top-forty hit. Ideally, it should be a little bit repetitive, consistent over time, but with the potential to evolve seamlessly as the diner moves from one course (or level) to the next. This is exactly what the sound designer Ben Houge aims for in his innovative sonic installations. Back in 2012, for instance, Houge worked with chef Jason Bond on a series of dinners in Bond's restaurant Bondir in Cambridge, Massachusetts. Each table was outfitted with one loudspeaker for each diner, which resulted in a total of thirty channels of coordinated, real-time, algorithmic, spatially deployed sound that was designed to work even if different tables of diners arrived at different times.[35]

Storytelling at the table

According to an article that appeared in *The New York Times* back in 2012: 'Restaurants in the very top echelon these days – Noma in Copenhagen, Alinea in Chicago, Mugaritz and Arzak in Spain – sell cooking as a sort of abstract art or experimental storytelling.'[36] An excellent example is the Alice in Wonderland theme that runs through a number of the courses at The Fat Duck – the 'Mad Hatter's Tea Party' dish comes straight out of the pages of Lewis Carroll.[37] When the restaurant reopened late in 2015 (after refurbishment), Blumenthal turned to Lee Hall, the writer of the film *Billy Elliot*, to help weave the menu into a story, which means that 'the menu will now be a story. It will have an introduction and a number of chapters and the chapter headings that will give you an idea of what is coming.' The top chef hasn't stopped there, though. He has also been talking to the press about trying to redefine the very nature of the restaurant. The emphasis on narrative has been ramped up substantially: 'The fact is The Fat Duck is about storytelling. I wanted to think about the whole approach of what we do in those terms.'[38] All of this explains chef Jozef Youssef's decision to mount the details of the dishes that he served in his new 'Gastrophysics' dining concept in the books placed innocuously on the table. Meanwhile, at Alinea, chef Grant Achatz has been wondering what would change if dinner at his restaurant were to be like the set of a play.[39]

Magic is increasingly making an appearance at the dining table. At Eleven Madison Park in New York, for instance, they have been looking at introducing a card trick as part of the dessert service.[40] Blumenthal conferred with magicians while experimenting with a flaming sorbet that would ignite at the click of the waiter's fingers. According to one journalist: 'Blumenthal created it with a magician so, at the click of a waiter's fingers, the barley sorbet in a bowl of hidden compartments bursts alight, turning warm outside yet remaining ice-cold inside. As a fire crackles around the sorbet, a rolling vapour of whisky and leather transports you to some Scottish hunting lodge at Christmas.'[41] Unbelievably, the bowls are rumoured to cost £1,000 a piece.

Theatre at the table

Why does it always have to be dinner and a show? Why not simply combine the two and dine *while* watching the show, or where dinner *is*, in some sense, *the show*?[42] The dining events offered by Madeleine's Madteater in Copenhagen are often described as free-form experimental food theatre. As one journalist described it: 'It's art you experience with all five senses, the most satisfying performance in town. Madteater is precisely as its name translates: food theater. [. . .] We were transforming the act of eating into exactly that: an act. I felt equal parts diner, performer and audience member in a restaurant that channelled all at once the opera, an art gallery and a shrink's office. It was strange. It was delicious.'[43]

Figure 11.3. The Tickets Bar in Barcelona, a recent collaboration between the brothers Ferran and Albert Adrià.

What exactly would you think you were looking at were you to walk past the shopfront shown in Figure 11.3? It looks like some kind of theatre, right? But this is actually a tapas bar! According to one description: 'The atmosphere draws equal inspiration from the theater and the circus, with a nod to Willy Wonka and his overstimulating chocolate factory. It's a place where chefs toil away at various work-stations, waiters prance around like theater ushers and morsels of food arrive with the flair and mystery of a vaudeville act.'[44] So, as dining becomes more theatrical, more entertaining, it becomes natural to make the restaurant *look* like a theatre too.

Not only that but one might also consider the idea of selling tick-ets for the show. In fact, this is exactly what American chef Grant Achatz decided to do with his Chicago restaurant Next. Anyone want-ing to eat there can simply buy a ticket in advance from the website. And just like the theatre (and with airlines), cheaper seats are of-fered for off-peak shows/meals. So a seat at Monday lunchtime will cost you less than a prime-time seat on a Saturday evening. It's an intriguing concept. No surprise, then, that a number of other restaur-ants and pop-up dining events have subsequently adopted a similar model;[45] on the Ultraviolet website, for instance, you are encouraged to 'book your seats now.'[46]

In the years to come, we are going to see a continued blurring of the boundary between theatrical and culinary experience.[47] Take the highly innovative Punchdrunk Theatre company.[48] Still incredibly vivid in my memory, as I am sure in many others', is 'Sleep no more', the multi-story retelling of Shakespeare's *Macbeth* in an abandoned warehouse block in New York City. This was an immersive theatre experience like no other. So, as actors, singers and magicians increasingly make their way into the dining rooms, the question arises: what would you get if you mashed up something like Punchdrunk Theatre with multisensory dining/drinking? Well, funny you should mention that, because the people behind Punchdrunk opened a restaurant. According to founder Felix Barrett, they originally developed a distinct narrative for the restaurant involving a cast of twelve actors. However, when the concept was trialled, the feeling was that ' "people weren't ready to watch theater" while they ate. Expense, he suggested, was also a factor. So now there are fewer, less formal theatrical accompaniments to the catering.'[49]

Perhaps it is because I happen to be married to a Colombian that I think so, but there really is nothing quite like Andrés Carne de Res. This restaurant on the outskirts of Bogotá has actors, musicians, magicians and a host of other performers wandering haphazardly between the tables in an atmospheric higgledy-piggledy assortment of wooden shacks. Sorry, that is the best I can do to describe it.* You really need to experience it for yourself. Best go in the evening though, when, after the food has been served, the tables turn into an impromptu dance floor. Taking people through different spaces, rather than using technology to create different atmospheres within the same space, can be a low-tech and, more importantly, lower-cost solution to delivering experiential dining. After all, not everyone

---

* In 2016, this restaurant was rightly voted one of South America's top 50; see http://www.theworlds50best.com/latinamerica/en/The-List/41-50/Andres-Carne-de-Res.html. And for those of you who are worrying, the country is now much safer than all those scare-mongering TV shows like to make out.

has the budget or technical support offered to star chefs like Paul Pairet, Paco Roncero or the three brothers Roca! Reducing the expense associated with delivering the multisensory experience offers the opportunity to provide something that becomes a little more scaleable.

One example of the low-tech approach comes from Gingerline's 'Chambers of Flavour' experience in London. Small groups of diners enjoy a four- or five-course meal, each course being served in a different room with a contrasting theatrical experience. According to Suz Mountford, founder of this immersive dining enterprise: 'Guests book having no idea of what to expect and can journey through anything from enchanted forests to a spaceship to a sunset beach, meeting all sorts of crazy characters along the way . . . We've always wanted to firmly cast the dining experience as a creative space to stimulate not just the taste buds but all of the senses.' Actors, dancers, and performers are again all part of the experience.[50]

Amongst all this talk of spectacle at the table, I would be remiss not to mention the truly spectacular meal held in February 1783 in Paris. Alexandre Balthazar Laurent Grimod de la Reynière, son of a wealthy tax farmer and nephew of one of Louis XVI's ministers, hosted a dinner in which hundreds of spectators watched the proceedings from a gallery – converting hospitality into some kind of theatrical show. Invitations to the meal took the form of ornate burial announcements. Just take the following description:

Like a banquet of freemasons, to which contemporaries compared it, Grimod's supper made heavy use of arcane ritual and semi-democratic pretensions [. . .] Grimod's guests had to pass through an entrance hall and a series of chambers before reaching a darkened waiting room, and, finally, the inner sanctum of the dining room. In one room, heralds dressed in Roman robes examined the guests' invitations; in the next, an armed and helmeted 'strange and terrifying monk' subjected them to further scrutiny; before a note-taking fellow in lawyer's garb greeted and interrogated the twenty-two guests [. . .] in the final stage of initiation, two hired hands dressed as choir boys perfumed the guests with incense.[51]

This spectacular dinner was so far ahead of its time that it deserves to be remembered to this day. One could even frame it as a piece of performance art with food at the centre – this, from almost 250 years ago!

Performance art with food

Looking back over the last half-century or so, one finds many examples of performance artists incorporating food, the preparation of a dish and/or its consumption into their work. The idea goes right back to the usual suspects, the Futurists, who 'aimed to marry art and gastronomy, to transform dining into a type of performance art'.[52] But there are many more recent exponents in this space; just take, for instance, Alison Knowles, an American experimental artist who served up a salad for 300 people to eat while listening to Mozart at the Tate Modern in London (see Figure 11.4 for one such 'happening'). This participatory event, *Make a Salad*, grew out of the artists' collective Fluxus movement back in the 1960s.[53] This piece plays on the notion of consistency – a theme that we will come back to in the final chapter. As the artist herself puts it: 'The salad will be made again for

Figure 11.4. *Make a Salad* (1962–) by Alison Knowles, a participatory performance art piece with food (normally for several hundred people).

several hundred spectators [. . .] Beginning the event, a Mozart duo for violin and cello is followed by production of the salad by the artist and eating of the salad by the audience. The salad is always different as Mozart remains the same.'[54]

Nothing, though, could ever match the harrowing ordeal faced by the sixteen guests who attended Barbara Smith's six-course *Ritual Meal* (1969). This performative event started with the invitees waiting outside someone's home for an hour. They were told repeatedly by a voice over a Tannoy to: 'Please wait, please wait.' On being let in, the guests were immersed in a space that was filled with the loud, pulsing sound of a beating heart. Videos of open-heart surgery were projected on the walls and ceiling. If that sounds bad enough, wait till you hear what happened next:

> Eight waiters (four men wearing surgical scrubs and masks and four women wearing masks, black tights, and leotards) led them to a table. Prior to entering the house, the guests had to put on surgical scrubs [. . .] The guests were then served a meal like they had never seen before. In keeping with the surgical 'theme,' the eating utensils were surgical instruments. Meat had to be cut with scalpels. Wine, served in test tubes, resembled blood or urine. In this charged atmosphere, ordinary food took on extraordinary connotations, an effect that Smith enhanced by the preparation and presentation of the food. Pureed fruit was served in plasma bottles. Raw food, such as eggs and chicken livers, that had to be cooked at the table were included in the dinner along with plates of cottage cheese embedded with a small pepper resembling an organ. Although the food was actually quite good, the dining experience was intensely uncomfortable for the guests, who couldn't put down their wine/test tubes and were sometimes forced to eat with their hands.[55]

You can get an idea of what it must have been like from a close-up of the hands of one of the guests at the performance (see Figure 11.5).

Is food art? Traditionally the answer has been very definitely no, the key distinction being that the viewer/diner was not 'disinterested', in Wittgenstein's terminology.[56] And yet it is clear that many chefs are increasingly taking their inspiration from the world of art,

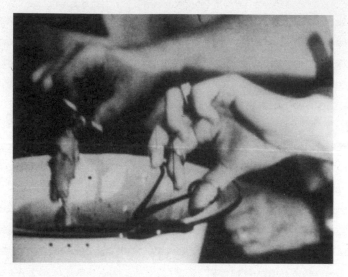

Figure 11.5. Barbara Smith's *Ritual Meal* (1969). Still feeling hungry?

some even referring to themselves as 'artists'.[57] Certainly, as we move away from the quaint notion that dining out has only to do with nutrition and sustenance,[58] we may even see the emergence of artistic dishes that do not necessarily taste that good. Something like this has already started to happen! Just take one of the latest menus at the Mugaritz restaurant, in San Sebastián. There's a dish (only one, mind) that the chef knows may be difficult for the diner to enjoy (a local delicacy of dried fish) and yet which plays an important part in the story of the meal that is served in this rural spot. The dish stays on the menu despite the negative comments of some diners online. As chef Andoni explains, in his book: 'There was a key moment in the evolution of Mugaritz when we realized that we were serving certain things that, objectively, weren't "good", but which had great emotional power. The dish "Roasted and Raw Vegetables, Wild and Cultivated Shoots and Leaves", for example, is eaten in a sort of altered state of consciousness [. . .] Plant bitterness is difficult to overcome, and this dish, out of context, could even be branded unpleasant. It is undoubtedly a proposition that creates mild discomfort.'[59]

Getting to the bottom of whether food can ever be considered as an art form, and whether chefs (at least the best of them) should really be thought of as artists is a debate that is likely to rumble on and on.[60]

It is certainly not something that I have any realistic hope of resolving here, and certainly not in a paragraph or two. My guess, for what it is worth, is that it will become increasingly natural to consider the top chefs as artists. And at some point in the not too distant future, we will all come to wonder why we ever didn't!

On the future of experiential dining

Any of you who find it difficult to imagine the experience of dining out being any different in the future from the way it has been in the past, just remember, the restaurant as we know it is actually a fairly modern invention, coming into existence in the latter half of the seventeenth and early eighteenth century in Paris. Perhaps it's time for the format to be updated.[61] A less radical way to think about this, though, is perhaps to consider the relative distribution (and pricing) of different kinds of restaurant changing. Researchers who have analysed the situation distinguish between the *fête spéciale*, 'where dining has been elevated to an event of extraordinary stature', the amusement restaurant and the convenience restaurant.[62] The rise of multisensory experiential dining is likely to enlarge the first two categories at the cost of the latter.

Rest assured – or be very afraid, depending on your perspective.[63] Before too long, many of our more mundane everyday dining and drinking experiences, be they in chain restaurants, hotels, food and wine shops, at home, or perhaps even in the air, will be accompanied by some sort of multisensory experiential cues. The hope (at least for me) is that these atmospheric stimuli will have been scientifically designed by one of the growing number of gastrophysicists out there to modulate, and preferably to enhance, some aspect (or aspects) of your tasting experience. Given what we know about the limits of the diner's brain, gastrophysicists interested in multisensory experience design are well placed to help to deliver experiences that are both stimulating and memorable without being overpowering.

There is also growing interest in moving from 'eatertainment' to 'edutainment', exemplified by Heston's Dinner in London. At this restaurant, the stories behind each of the dishes, relating to British food history, are retold. A similar dose of storytelling was also a part of the chef Jozef Youssef's 'Mexico' concept from Kitchen Theory. Take, for example, 'The Venison Dance' dish, which was preceded by a short video of the dance performed by the State Ballet of Mexico. The 'Memories of Oaxaca' dish served as part of the same concept also had a video to help set the scene for diners.[64]

In closing, it is important to stress the key challenge faced by theme restaurants. Given that diners know what to expect the second time around, those who wish to stage 'experiences' have to work constantly to refresh them. As the successful New York restaurateur Danny Meyer notes: 'Showmanship can be a tricky pursuit [. . . because] the more theatrical it gets, there does come a point when you just can't see that play another time.'[65] On the flipside (see 'The Meal Remembered'), though – there is also the comfort that comes with knowing exactly what we are going to eat.

12.   Digital Dining

Would your cocktail taste as nice or your dinner as delicious if you found out that it had been prepared by a robot? Would you trust a robot chef to season your food for you? Really? And how would you feel about being served by a robot waiter? Sounds like science fiction, right? Well, these things are already becoming reality, albeit in only a small number of venues so far. Like it or not, a whole host of digital technologies have become ever more closely intertwined with our everyday experience of food and drink. There are already digital menus out there that will send your order straight to the bar or kitchen. Pizza Hut even trialled a 'subconscious' menu that could, or so it was claimed, magically read your mind and tell you the three toppings that you wanted on your pizza, without you having to say a word. The menu tracked the customers' eye movements while they viewed the digitally displayed options. No worries, though, if you happen not to like what your subconscious chose, as you could always stare at the restart button and begin the process all over again! I have to say, though, that this example smacks of marketing gimmickry rather than a serious attempt to envision the restaurant of the future.[1] Elsewhere, though, one comes across headlines talking of a '*Star Trek* "replicator" that can recreate ANY meal in 30 seconds'.[2]

Soon there may no longer be any need to reach for the sugar. At least, not if you know about the latest findings from the field of sonic seasoning, which can be dispensed by the mobile devices that so many of us now carry around with us. In fact, digital artefacts will probably become a ubiquitous part of our multisensory food and drink experiences in the near future. It is more than likely that we will first come across these new technologies at the tables of some of the world's top modernist restaurants or cocktail bars. But from there, it is only a few short steps to their introduction in chain restaurants and the home environment.* Furthermore, many global food and beverage brands are eager to get a slice of the action here too. So let's jump straight in and see what tomorrow's digital dinners might look like, and who – or should that be *what*? – exactly might be making them.[3]

Would you like a 3D food printer?

Read the newspapers and you'd be hard pressed not to come away with the belief that the 3D food printer is going to be the all-new must-have gadget in the home kitchen. If you haven't heard about the Foodini, the Bocusini or the (rather more mundanely titled) 3D Systems ChefJet Pro 3D, then you are probably not an 'it' cook.[4] Chefs are now using 3D printers to impress diners by creating foods the likes of which they have never seen before (see Figure 12.1 for one beautiful example of 3D food printing). So will the 3D food printer really become the microwave of the future?[5] It is obviously in the best interests of the manufacturers to try and convince you that it is, but I beg to differ. Don't get me wrong, it's not that I see *no* use for them, it's just that my best guess is that they will end up in the kitchens of a few fancy modernist restaurants and in food innovation centres, not in the home kitchen (one of the few exceptions to my earlier predictions that what starts in the modernist kitchen and restaurant will trickle down and eventually find itself in the local restaurant or even the home.)[6]

One chef who has apparently become enamoured with the possibilities offered by 3D food printing is Paco Perez. He uses one at his restaurant La Enoteca, at the Hotel Arts in Barcelona, to create food forms that could not otherwise be made (intricate architectural renditions of famous buildings, for instance). I was also intrigued by the recent release of an inkjet printer that allows the barista to create amazing latte art (e.g., showing realistic portraits of famous people). And until his untimely death a few years ago, Homaru Cantu was

---

* See Buster Keaton's short comedy film *The Electric House* (1922) for an early take on what the digital dining room of the future might look like.

Figure 12.1. 'Caesar's Flower of Life': seasoned bread in 3D-printed sacred-geometric 'Flower of Life' design, with assorted flowers and vegetables. An eclectic eight-course dinner was 3D-printed and served using byFlow Focus 3D-Printers, prepared with fresh natural ingredients and borrowing innovative multisensory techniques from the world of molecular gastronomy.

famous for printing edible menus for his restaurant Moto, in Chicago. The inventive chef managed this trick by hacking a regular printer.[7] In May 2013, NASA offered him a six-month Phase 1 study contract to develop printing technologies capable of combining shelf-stable macronutrients, micronutrients and flavours to produce personalized food products for long-duration space missions. Cantu generated lots of excitement when he started talking to the press, especially when he mentioned the future of printing 3D pizzas in space. When the story broke, though, NASA was outraged. A number of those working in the area felt that all the press coverage largely demeaned (and distracted from) the serious science underpinning the development of adequate food supplies for long-distance space flight. No wonder, then, that funding for Cantu's project was not extended to the next stage.[8]

Currently, commercial food printers are still a little too expensive for widespread home uptake, coming in at somewhere around $1,000 apiece. In time, that figure will probably come down, as it always does with technology. But even if they were giving them away, I still wonder who would be daft enough to get one for their home kitchen. Should you be wondering why I have become so curmudgeonly all of a sudden, well, you need to ask yourself how long it actually takes to print each perfectly formed morsel of food. My guess is that you would probably have to start preparing your very own unique pasta shapes a good few days in advance if you were thinking of inviting all your friends over for dinner. So while you could make it yourself with your new kitchen toy, don't forget we still have those quaint things called shops. And if you do find yourself craving a printer, just ask yourself who, exactly, is going to clean the tubes out each time after you have used your glistening new kitchen appliance. You might also wonder how much your electricity bill would go up. So, I ask you once again, is it really worth it?

3D-printed foods have novelty value, yes. But, beyond that, what new food experiences could they possibly give you? Is there a unique selling point here? What is it, exactly, that this machine allows you to do that you couldn't do otherwise? And even if – for some unknown reason – sales of 3D food printers do eventually pick up, my guess is that you will find all these sorry devices stored away lonely and unloved, collecting dust at the back of the cupboard, within a couple of years of the hype reaching its apex. That is to say, they will end up in exactly the same position as so many of those home breadmakers and food processors – the *must-have* kitchen appliances of recent decades.*

However, I could imagine creating the perfectly shaped chocolate, one that snugly fit the contours of the average human tongue and could therefore deliver a more intense flavour hit than any one of those seemingly accidentally shaped confections currently out there in the marketplace.[9] Just imagine a chocolate that stimulated all of your taste buds simultaneously. But as soon as the perfect shape had

---

* That said, wait a few decades more, and they are likely to cycle back into fashion once again; K. Mansey, 'Gadgets of the 1970s get their fizz back', *The Sunday Times*, 15 December 2013, p. 23.

been identified (assuming, of course, that there is one), then the industrial production lines could be reconfigured to create it on a massive scale.

So, if you believe me when I say that we are not going to be eating 3D-printed dinners at home any time soon, then where else might we first experience the intrusion of digital technology into our food and beverage experiences? Well, many of you will have come across the next one already, namely the digital menu.

Do you like ordering from a digital menu?

No, me neither, though one does increasingly find digital menus in trendy high-end bars and restaurants. Now, this ought to make sense, right, at least on paper? No more worries that your waiter will forget something when they stubbornly refuse to write anything down when taking the orders. Digital menus should also allow for any changes to the vintages of the wines to be updated on the list in pretty much real time. This would at least deal with one of my pet peeves – when restaurants have one vintage on the wine list and then bring out a (generally inferior) younger bottle because they have run out of the good one, often without telling you (and expecting you to be happy to pay the same price for it). And, in theory, the digital menu should enable the restaurateur or bar owner to incorporate any seasonal dishes or drinks on to their menu, thus allowing them to bin the blackboard – you know, the one with the daily specials chalked all over it.

And yet, if you are anything like me, you will have realized that it just doesn't feel right. Maybe it is because I am not a millennial, but I have to say that when someone insists that I order from a digital menu, the dining/drinking experience is somehow diminished. Why so? Well, there are a couple of reasons: on the one hand, it is important, I think, to remember that dining is a fundamentally *social* activity (see 'Social Dining'). Part of the reason that we go to a restaurant or bar in the first place is for the interaction with the staff. Danny Meyer, one of the most successful restaurateurs of our time, nailed it, I think, when he said: 'Despite high-tech enhancements, restaurants will always remain a hands-on, high-touch, people-oriented business. Nothing will ever replace shaking people's hands, smiling, and looking them in the eye as a genuine means of welcoming them. And that is why hospitality – unlike widgets – is not something you can stamp out on an assembly line.'[10]

There are no two ways about this: digital menus take away from that social exchange and make for a much more transactional kind of experience. Some would say that placing an order from a digital menu is just cold. I, for one, am glad to see that restaurateurs and bar owners are slowly coming to their senses and getting rid of their digital menus. And not before time either, if you ask me. The only place where they make sense, I think, is in those venues where all that anyone wants is a rapid and efficient transaction, say, when grabbing a quick bite to eat at an airport.*

The other problem is that most digital menus look pretty much identical to the printed version. Why? Surely, going digital should be an opportunity to do something radically different. If a modernist chef were to present you with one, someone like Grant Achatz in Chicago, or Juan Maria Arzak in Spain, you *know* that it wouldn't simply be a digital replica of a paper menu. One of the few interesting examples is the digital menu that one sees on the table at Inamo, an Asian fusion restaurant in London. Not only can you order simply by touching the relevant item projected on to the table's surface but you can also see what the various dishes look like before ordering them.†
The digital nature of the interaction also means that presenting pictures of the dishes doesn't seem anything like as tacky as it would if exactly the same images were to be displayed on a regular printed menu.[11] There is, then, a tangible benefit here for the diner from 'going

---

* Intriguingly, though, the one place where you do find such impersonal service quite regularly is in Japan, where, on entering many restaurants/noodle bars, you find a picture menu from which you can order, and then wait for someone to place the food in front of you.
† In a way, this is again reminiscent of Japan, where many restaurants have a selection of hyper-realistic plastic meals on display at the entrance, showing what the food options look like. Indeed, I have just such a magnetic sushi set proudly displayed on the filing cabinets in my office.

digital'. You can even order a taxi home directly from the table-top! Of course, given the growing interest in where the food we eat comes from, one could also imagine being shown information about the history of the ingredients, which farm they came from, etc.

Digital menus also offer the possibility of delivering a more curated meal. Take Mother in Stockholm, where the menus are also projected directly on to the table-top, and the diners are quizzed about which foods they prefer. Then, a number of recommendations are made for dishes that the diner might like. (The only missed opportunity here is the fact that the system does not keep your details from previous visits – a trick of personalization that, as we have seen, is likely to make a meal both more enjoyable and more memorable.)

One other interesting use of digital interactive menus comes from The Weeny Weaning restaurant opened by Ella's Kitchen in 2014. This, the world's first sensory restaurant for babies, has been designed to encourage healthy eating from an early age, or so the blurb goes. According to one report: 'Little ones will be seated in highchairs at interactive tables, from which they will be able to choose from their very own digital menu, allowing them to order their own mains and desserts [. . .] Depending on the number of times they tap a particular food icon over a 30-second period, the digital menu responds accordingly and the waiters bring the children their selected choice of food.'[12] The next generation are likely to be much more open to this kind of digital interface with food, no matter whether or not they were exposed to it as a baby.

Tasting the tablet

Why would you use a real plate when you can use a tablet instead? (Or should I have said that the other way around?) This is, though, one of the ways in which technology has already started to change our visual experience in the modernist restaurant; there are chefs who serve some of their food from tablet computers rather than plates. (Who knows, perhaps this is the perfect way to make use of all those surplus tablets sitting around in all the restaurants and bars where they have figured out that digital menus are a waste of time!) Chef Andreas Caminada recently served one of the courses at his Swiss restaurant on top of a tablet displaying the image of a round white plate: an ironic take on digital plating.[13]

A few years ago, we were playing around with the idea of serving seafood from a tablet (see Figure 12.2).[14] The diner would catch a glimpse of the sun glinting off the waves and the sand on the seashore, so realistic that they could almost touch it. That was the hope anyway! Combine the sight of the seashore with the sounds of the sea (about which more below), and the seafood may well taste better. At the very least, plating from a tablet should offer the creative chef increased freedom as far as the storytelling around a dish is concerned. While so far this is happening only in a few cutting-edge restaurants, one could imagine, in the future, all of us repurposing our own tablet computers at mealtimes.

Figure 12.2. How long before high-end restaurants start serving food from a tablet rather than a round white plate? Top Spanish chef Elena Arzak says: 'At Arzak in San Sebastián, certain dishes are served over a digital tablet: grilled lemons with shrimp and patchouli sit atop a fired-up grill with the noise of crackling flames. [. . .] We experimented with serving the dish on and off the tablet and diners always said that having the image and the sound intensified the flavours of the dish and made it even more enjoyable. We're keen to use new technology to further augment the meal.'[15]

Some of you out there will be appalled: why on earth would I spend good money on a tablet computer only to eat off the damn thing? you may be muttering darkly to yourself. What, exactly, is so wrong with the good old-fashioned round white plate? Has the professor actually lost his marbles? Don't get me wrong, I am certainly not saying that the tablet is going to be the ideal option for every kind of food.[16] Even I can't imagine that it would be much fun trying to eat a big juicy steak, say, from a tablet. Probably best to keep serving this one on a wooden board.[17] Canapés and other finger foods might be more the thing, at least till you get the hang of this all-new digitally enhanced plateware.

In my defence, though, let me at least point out that some tablets are waterproof so you could, I presume, put them in the dishwasher straight after use, should the need arise. (Perhaps the professor has lost his marbles after all!) I can easily foresee how serving food off a tablet could also provide the ideal means of ensuring the perfect colour contrast between your food and the plateware (or, in this case, the tablet screen) against which it is viewed (see the 'Sight' chapter). Ultimately, though, I think that eating from a tablet will go mainstream only if the diner's experience of the dishes served in the modernist restaurant is radically altered, i.e., enhanced, by whatever is shown on the screen. And for those of you who might be thinking that the price would be too high, just remember that at some of the world's top restaurants individual plates designed for just one dish have been known to come in at more than £1,000 a pop. Plating off a tablet might seem cheap by comparison.

Would you like to eat cheesecake on Mars?

This is what the creators of Project Nourished are offering with their recent attempt to combine virtual reality with food. Developed by Kokiri Labs, in LA, Project Nourished is described as a ' "gastronomical virtual reality experience." This mashup of molecular gastronomy and virtual reality allows users to "experience fine dining without concern for caloric intake and other health-related issues." '[18] Their strapline is 'Wouldn't you like to eat cheesecake on Mars?'[19] Said almost rhetorically – for how could you resist? You must, at the very least, be a little curious. And, given what we have seen in previous chapters – how much impact context, atmosphere and environment have on dining and drinking – you can be sure that your cheesecake probably would taste a little different if you were wearing one of these headsets, assuming, that is, that the experience is suitably immersive,[20] and all that virtual space dust doesn't blow into your eyes! Looking a little further into the future, it is interesting to consider how the various new augmented and virtual reality (AR and VR respectively) technologies are going to enable the diner of tomorrow to eat one food while simultaneously viewing another.[21]

So what exactly do we all have to look forward to, in this mash-up of food and virtual reality? Well, here's a hint of what may be to come, at least if the techno-geeks have their way: 'As for Project Nourished, here's the deal: You put on your VR headset, [. . .] you lift your "food detection sensor," which, at this stage of development, looks like nothing more than a wooden fork with two prongs wrapped in tinfoil; you eat a hydrocolloid – a flexible substance that is "viscous, emulsifiable, and low caloric" – that has been shaped in a 3-D printed mold to "add physical characteristics to the 'faux' food." Then, with the help of a motion sensor, an aromatic diffuser, and a bone conduction transducer [. . .] you experience a gourmet meal with no downside in the way of calories, carbs, or allergy-inducing ingredients.'[22] Don't tell me you are still not convinced!

The more fundamental worry, though, is simply that the idea of eating cheesecake on Mars doesn't seem like an especially congruent combination, at least not to me. Perhaps we'd be better off matching the food to the environment that we are immersed in via the headset, maybe a strawberry-flavoured space cube (i.e., something that astronauts were given to eat on their space missions)? But then again, perhaps not, knowing how bad they were rumoured to taste.[23]

Some of the limitations associated with simulating a given environment when restricted to just visual VR are brought home if you imagine how realistic an experience you would get were you to try recreating the experience of dining in a plane, for example. You'd miss the background noise, the lack of air humidity, the lowered cabin pressure (see 'Airline Food'). You'd also fail to capture the pressure on your knees when the person in front suddenly decides to recline their seat during the meal service. So it really wouldn't be the same kind of experience at all now, would it? Vision is undoubtedly important, but without the other sensory cues, it is unlikely to be all that immersive – at least for those more *extreme* environments that we might want to simulate. I do wonder, though, whether such VR applications might not find a use amongst elderly patients. Perhaps it could be used to take them back to an earlier time, providing visual cues from the past. After all, playing the music of a bygone era has been shown to help enhance consumption.[24]

Augmented reality dining

Augmented reality involves superimposing artificial visual stimuli over the actual scene. So, for instance, the AR system utilized by my colleague Katsuo Okajima and his colleagues in Japan can update the visual appearance of food or drink in real time. Just imagine: you put on the commercially available headset, and at first you can see the sushi that you ordered on the plate in front of you. Then, simply by moving your hand over the plate (abracadabra-style), the fish suddenly changes from tuna to salmon, say. Move your hand over the plate again, and now it is eel. Not only that but you can pick up what looks like eel sushi and even take a bite without destroying the illusion.

So what's the point of it, some of you are no doubt wondering. Well, in some of our preliminary research, we have been able to demonstrate that changing how food looks can result in people saying different things about the taste, as well as the perceived texture, of cake, ketchup and sushi. Here, one could perhaps imagine consumers viewing what looks like highly desirable but unhealthy food while actually eating a healthy alternative. And, looking a few years into the future, I can well imagine how we might all end up craving some virtual sushi, when the seas have been fished to extinction and the real thing is nothing but a dim and distant memory (sorry to be so depressing).[25]

One other intriguing example of the intelligent use of AR headsets while dining comes from researchers who have been looking into whether they can trick you into feeling fuller sooner. They aim to do this simply by making the food that you see through the headset (e.g., a biscuit) look larger than it actually is.[26] However, much though I like the idea of AR and VR dining, my best guess is that we are still some years off seeing these headsets at the dining table, even at the world's most avant-garde restaurants. The limitation here is as much the cost as the fact that it may interfere with the social interaction around a meal.[27]

Have you heard of the 'Sound of the Sea'?

To date, the more immediate uptake of digital technology at the dining table has been linked to the personalized delivery of sound, not vision. Here I am thinking of soundscapes and musical compositions, for instance, that the diner or drinker listens to while enjoying a specific dish or drink. In an earlier chapter, we saw how Heston Blumenthal first became interested in the important, if neglected, role of this sense after we gave him a few sonic chips to nibble on in the Crossmodal Research Laboratory. The chef went away and, together with his talented crew in Bray, started to explore different ways of bringing sound to the table, digitally. The first iteration of 'sonic cutlery' they came up with was trialled 'secretively' for some of the restaurant's regular clientele, but, unbeknownst to the serving staff who were working the pass, a journalist was sitting incognito in the restaurant that day. When he caught wind of the fact that the diners on the *other* table had been served a dish that he himself hadn't had, he immediately summoned the waiting staff over, announced himself and demanded to know what was going on. There was nothing for it but to let him try out the sonic headphones too. The result: a few days later, who should appear staring intently out from the pages of *The Sunday Times* than that same reporter.[28] The all-new eating utensil for the techno-enhanced twenty-first century had just been 'outed'.

Putting the over-ear headphones on tended to mess with the expensive hairdos of a certain section of the clientele, leaving them feeling anything but alluring, and the headphones were unceremoniously

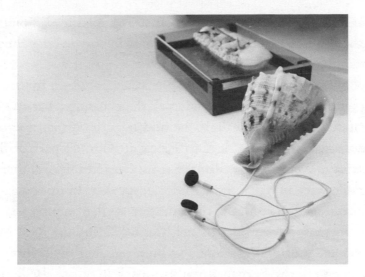

Figure 12.3. The 'Sound of the Sea' seafood dish (for a number of years, the signature dish served on the tasting menu at Heston Blumenthal's Fat Duck restaurant) provides an excellent example of how digital technology can be used to enhance the multisensory dining experience. In research conducted here in Oxford with the chef, we were able to demonstrate that seafood tastes significantly more pleasant (but no more salty) when people listen to waves crashing gently on the beach and seagulls flying overhead than while listening to restaurant cutlery noises or – surprise, surprise – modern jazz!

withdrawn from service almost as soon as they had been placed on the carefully ironed white tablecloths. It was, as they say, time to go back to the drawing board. Roll the clocks forward a couple of years, and what do those who are lucky enough to have booked themselves a seat at The Duck find? Well, for one of the courses, the waiter arrives at the table holding a plate of seafood in one hand: sashimi resting on a 'beach' made of tapioca and breadcrumbs with foam. With his other hand, the waiter passes the diner a seashell from which dangles a pair of MP3 earbuds (see Figure 12.3), and encourages the guest to insert the earbuds before they eat. What the diner hears, assuming that he or she does as told,* is the sound of the sea: waves

---

* According to the team at The Fat Duck, in the eight years or so that this dish has been on the menu, only one diner (a French chef) has not put the earbuds in, saying imperiously that he already knew what the sea sounded like!

crashing on the beach, a few seagulls flying overhead. Some have found the combination of sound and food to be so powerful they have been brought to tears.[29]

Since the 'Sound of the Sea' dish first appeared on the menu in Bray, a number of other chefs (and even the occasional barista)* have incorporated personalized digital sound into the dishes they serve. For instance, at El Celler de Can Roca, in Girona, part of Spain's *nueva cocina* movement, the culinary team created a dessert that came to the table with an MP3 player and loudspeaker. In this case, diners were encouraged to consume their dessert while listening to a football commentator describing Lionel Messi dodging the Real Madrid defenders and scoring Barcelona's winning goal at the teams' classic confrontation in the Bernabeu stadium back in 2012.[30] Brilliant! It's got both high emotion and storytelling, though I bet it tastes better if you are not a fan of the losing team. (And presumably the dish works better if you actually like football.) Meanwhile, the Michelin-starred chefs at Casamia in Bristol sometimes served a picnic basket with an MP3 player that, when opened in the restaurant, would play the sounds of the English summer.

Surprising spoons

There has also been growing interest in delivering sounds digitally from within the diner's mouth. Chefs, musicians, designers and culinary artists have all become interested in delivering personalized music/soundscapes to accompany specific tasting experiences. Just take, for example, the limited edition Bompas & Parr baked beans spoons, available for £57. Each spoon had an MP3 player hidden inside. If you bought one, you wouldn't hear anything until you put the spoon into your mouth. Then the sound waves would travel via your teeth and jawbone through to your inner ear. The flavour–music combinations in this case included Cheddar cheese with a rousing bit of Elgar, fiery

---

* Top barista Rasmus Helgebostad made a sonically enhanced coffee drink as part of his entry in the 2011 Norwegian barista championships.

chilli with a Latin samba, blues for the BBQ-flavoured beans and Indian sitar music for the curry-flavoured beans![31] While the diner could hear the music, the person sitting next to them would hear nothing. It remains to be seen just how congruent these musical selections were, and whether they really did enhance the flavour of the food.

Meanwhile, over in the Netherlands, Dutch pianist Karin van der Veen has been offering people the chance to taste digital bonbons, De Muziekbonbon. The idea is simple, really, if a little strange. You put the chocolate, which comes with a wire attached, into your mouth, and as you clench the piezoelectric strip (which vibrates when an electric current is passed through it) embedded in the bonbon between your teeth, you can faintly hear the bone-conduced sound of a piano resonating through your jawbone and carrying on all the way to your inner ear.* A pleasant if most unusual experience, as I am sure you can imagine. While I enjoyed my chance to try one of these multisensory treats, it is certainly not something that I can see going 'mass market' any time soon. I am just not sure that the increased enjoyment is really worth the effort, at least not once you have done it the first time. It is also rather antisocial, in the sense that all conversation is prevented while the 'musical bit' is clenched firmly between your teeth! On the plus side, though, I suppose that you could say that it aids concentration, and hence enhances the experience, nudging the taster towards a more mindful approach to consumption.

Digital flavour delivery

Researchers in Japan have been working to deliver food aromas to match whatever you might see through your AR headset. However, one look at this device (see Figure 12.4) tells you all you need to know about how soon you will be seeing this *beauty* in the modernist restaurant or gadget store. Never!† As is too often the case, the tech-

---

* The thing about bone-conduction is that there tends to be much more emphasis on the lower frequencies of sound.

† Though, it did make an appearance in Japan's entry to Expo '15 in Italy.

Figure 12.4. Hmmm, tasty! Sometimes I worry that human–computer interaction (HCI for short) researchers may spend a little too much of their time thinking about what is possible at the intersection between technology and food and perhaps not enough time considering what is actually likely to be applicable or even desirable out there in the real world. Even the most innovative modernist chef would, I presume, baulk at the idea of having their diners put one of these devices on. And there we were, thinking that over-ear headphones were intrusive!

nology developed to explore the potential connection, or digital interface, with food (or food aromas in this case) fails to consider the aesthetic appeal of their designs. Big mistake!★

The more plausible mainstream delivery of food scents will, I believe, come from plug-ins, like Scentee. This has already been used for one marketing-led intervention in the US, namely, the Oscar Meyer bacon-scented alarm clock app. You simply insert a small

★ That said, the all-new Nosulus Rift headset can deliver aromas via a very sleek black space-age headset; the only problem is it is exclusively designed to emit one un-pleasant smell to go with a *South Park* video game called *The Fractured but Whole* (see http://nosulusrift.ubisoft.com/?lang=en-US#!/introduction).

plug-in device to your mobile and set the time, and you will be woken up by the sound of sizzling bacon and the matching smell![32] Meanwhile, in Spain, top chef Andoni has been using digital scent delivery to extend the interaction with his diners; those who have booked a visit to Mugaritz can get to experience the actions, aromas and sounds that accompany one of the more multisensory of the dishes on the tasting menu in advance by downloading the appropriate app (see Figure 12.5).[33] On making a circular motion to virtually grind the spices that are visible on the screen of their mobile device, the user not only hears the sound of mortar on pestle but also gets a blast of spicy aroma up their nostrils (via the scent-enabled plug-in). These are exactly the same actions, sounds and aromas that they will subsequently experience on tasting the dish itself while they are sitting in the restaurant. One of the aims is to use multisensory stimulation in order to help build up anticipation in the mind of the diner prior to their arrival at the restaurant. Who knows, the expectant diner may even start to salivate.

However, while such digital smell delivery is certainly practical, the fundamental problem is whether anyone is going to buy the refill. This, in a way, contributed to the demise of DigiScents (a digital smell delivery company started during the internet boom years) a couple of

Figure 12.5. A scent-enabled app helping to build anticipation in the mind of the diner booked into the Mugaritz restaurant in Spain.

decades ago (and at no small expense to investors, it should be said).[34] My best guess is that, while the technology works, there isn't really any consumer desire, or need, for such digital innovation just yet. And without that, these dreams of digital smell are likely to fail, just like those previous attempts.

How would you fancy eating with a vibrating fork?

As we saw back in the 'Touch' chapter, the world of cutlery design is set for a revolution. Part of the change will come from the introduction of new forms, materials and textures for cutlery, but another strand of future development may well revolve around the emergence of digitized, or augmented, eating utensils. For, at least according to the human–computer interaction community, this may also radically transform the way in which we will interact with our food in the years to come. Just imagine a fork that vibrates to let you know that you are gobbling your food a little too fast! No, really! (See Figure 12.6.)

Perhaps the most interesting example of digitally augmented

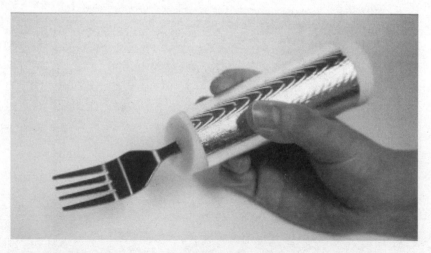

Figure 12.6. One of the ways in which digital technology could make its way on to your dining table in the future. This is an early prototype of the HAPIfork, a Japanese gadget designed to modify our eating behaviour.[35]

cutlery goes by the name of Gravitamine. This utensil cleverly creates the illusion of weight in the hands of its users.[36] Given what we saw in the 'Touch' chapter, I can well imagine how such a digital solution would enhance the diner's meal experience. Though you could be forgiven for wondering whether it wouldn't be more convenient to invest in some genuinely heavy cutlery instead of having to recharge it on a regular basis. One other market where digitally enhanced cutlery may have a promising future is for those patients who find it difficult to control their hand movements – sufferers of Parkinson's disease, for instance, whose tremors can lead to food spillage. In fact, one innovative company has already come out with some anti-shake cutlery to help combat this particular problem.[37]

Electric taste

Researchers can now deliver rudimentary taste sensations simply by electrocuting your tongue in the right way.[38] Any volunteers? Come on now, it isn't as unpleasant as it sounds. Sadly, though, it isn't anything like as pleasurable as many of the press reports would have you believe either! According to some journalists, this all-new approach holds the promise of a never-ending sequence of flavours delivered by your digital device. All you need is a power supply and a stimulator pressed against your tongue. Indeed, the world's press went wild when researchers recently launched a digital lollipop.[39] But hold on a minute, not so fast. The first question to ask before taking all the hype at face value is whether those who are writing about it have actually experienced electric taste for themselves. Too often, this would appear not to be the case! Instead, it seems that much of the time they are relying on second-hand reports by those who are promoting these devices.

I *have* tried some of these devices, and found the experience to be disappointing, to say the least. Now, perhaps I am just unlucky, since electric taste works better on some people's tongues than on others. However, even this approach's most ardent advocates admit that it is easier to get sour and metallic taste sensations than it is to get salty and umami . . . and sweetness – well, that is a real challenge. Thus the palate of digital tastes that one has to work with, even in the best-case scenario, is actually pretty limited, and that's for those in whom electrical stimulation of the tongue works well. I have little faith that things will improve much, experientially speaking, when the electrical stimulation device is embedded in the end of a spoon or in a piece of digital cutlery or glassware.[40]

But more importantly, even if all taste sensations could be rendered perfectly, you would still be left with a very *thin* dining experience. For anyone who has evaluated a weak pure tastant in solution will know only too well how unsatisfactory it is, even in the best of cases. And as we saw in the chapter on 'Smell', taste constitutes only a very small part of our multisensory flavour experiences anyway; all those fruity, floral, meaty, herbal notes that we enjoy while eating and drinking really come from the nose. In other words, you could never hope to evoke them by electrocuting anyone's taste buds. You should be electrocuting their nostrils instead (or as well) – likely to be an unpleasant, complicated and possibly even painful procedure.[41]

While the original aim of those developing digital taste was to remove the need to deliver an actual tastant, the approach that is now being developed involves augmented tasting. Some researchers, for instance, have been looking at what would happen if your tongue is zapped while you are looking at a tasty meal, say, or even while eating or drinking something real. It turns out that the hedonic response to electric taste was indeed changed when people were looking at gastroporn. Similarly, there is evidence to suggest that people's response to real food and beverage items can be modified by presenting electrical taste at the same time.[42] So, if diners were given salty sensations electrically while eating, would this mean that they wouldn't have to add so much salt? This was the conceit of the 'No Salt Restaurant', a two-day pop-up in Tokyo.[43] Diners got to eat with an Electro Fork, capable, apparently, of delivering some electrical seasoning. As a trial run, the restaurant offered a five-course saltless menu of salad, pork cutlets, fried rice, meatloaf and cake. My guess is that not many people would have wanted to repeat the experience.

Is this more about marketing than genuine health innovation? One important thing to bear in mind is that the role of salt isn't solely as a taste enhancer. It also plays a key role in determining food texture/ structure, and that is something that electric taste simply cannot help with. One further problem with this technology is that our brains seem exquisitely sensitive to how the sensation elicited by different compounds changes over time. This is part of the reason why, for instance, you can distinguish sugar from other artificial sweeteners like aspartame (because the taste sensation ramps up much sooner in one case than the other, and may well linger for longer too). So unless you can get the time-course of sensation associated with electrical taste right, the experience is never going to be as good as 'the real thing'.[44]

Digital technologies changing the food landscape

Go online nowadays and you can find all manner of apps that promise to provide assistance no matter what you want to know about or do to/with your food or drink. Switched-on celebrity chefs and lifestyle bloggers are only too willing to advise you what you should be eating, or else help you to prepare a new dish. This is, after all, big business these days.[45] There are also a growing number of smartphone apps that can interface with, and thus control, various kitchen devices. Just take the 'Bright Grill' as one representative example, an electric barbecue with an app that will alert you as soon as your sausages are cooked perfectly, even if you are away from the grill, so that hopefully you won't end up cremating them, as usual.[46] Don't such inventions make you wonder how we all managed in the good old pre-app days?

In fact, you can find pretty much anything you want at the app store nowadays.* Believe it or not, there is even one called the Egg-Calculator, brought out by ChefSteps. (This is one for all those of you

---

* Sadly, though, the one app that you won't find in the store was developed over in Japan by Kayac Inc. and was based on the results of our sonic chips study (see 'Sound'); Called Evercrisp, it could boost the crunchy, crackly and crispy sounds of the foods you were eating, thus enhancing the eating experience for all.

who are addicted to 'yolk porn' (see 'Sight'); it contains more 'protein in motion' shots than even the most ardent food junkie needs to get off on.) Using this app, there is no longer any excuse for your slow-cooked egg to turn out any other way than exactly how you wanted it. Meanwhile, there are now many price comparison apps that enable the savvy diner to scan their menu and compare how much the same item would cost at other restaurants. On occasion, in big cities like New York, you can find exactly the same bottle of wine being priced at four times what you would pay at another restaurant just down the block.[47] Wouldn't you want to know when you were paying through the nose?

The clever folk at Google have even come up with an app to help restaurant diners split the bill,[48] though the audience for this app is presumably declining somewhat given that more and more of us are dining out alone these days (see 'Social Dining'). And one does have to wonder whether a simple calculator wouldn't do just as well. Interestingly, aware that picking up the hefty tab at a fancy restaurant can put something of a downer on the meal experience of the person paying, and mindful of the fact that the end of an experience tends to be particularly memorable, some high-end restaurants now get their diners to part-pay for the meal in advance, so that the pain at the end isn't quite so bad. That is what I call intelligent design!

Another interesting development involves those sensory apps that allow you to access digital content by scanning the label of anything from a tub of Häagen-Dazs ice cream to a bottle of Krug champagne.[49] For instance, the Concerto app was designed to help the consumer pass the time after taking their ice cream out of the freezer before serving. Customers get their mobile device out and scan the QR code (that black and white patterned square on the lid of special packs), and the next thing you know musicians can be seen and heard 'magically' floating on top of the Häagen-Dazs. Each of the musical selections lasts for about two minutes – i.e., just long enough for the contents to soften slightly, according to the marketing blurb. Once the music draws to a close, the ice cream should be ready to serve.

There are other 'clever' apps that claim to be able to count the calories in a dish by analysing the picture you take of it, and many other new food-related technologies are in development.[50] One project, funded by Philips Research, investigated the feasibility of having people eat from plates that have digital scales embedded in them, in order to calculate the total amount of food that they had eaten.[51] Google's AI, called Im2Calories, is also training itself to count the calories in food photos. It is already accurate to within 20%.[52] But do you really want your technology to keep track of what you eat? What, every gram (or calorie) of it? Furthermore, it remains to be seen quite how accurate these devices really are. After all, the visual system, and the brain that supports it, have been fine-tuned over the course of human history to rapidly assess the energetic content of foods. This is precisely the sort of thing that our brain evolved for, and the evidence suggests that we manage to evaluate nutritive food sources in little more than the blink of the eye. However, even this fine bit of tackle sometimes gets it wrong, or at least our conscious minds do! So why expect the technology to do a better job?[53]

Do robot cooks make good chefs?

But to end, let's return to the question with which we started this chapter: what would you think if you found out that your dinner had been cooked by a robot chef? On the one hand, this should be a wonderful example of precision cooking. That is what we all want, isn't it? Exactly the same taste time after time? And yet, if the food or drink is made by machine, why bother going out to eat in the first place? Why not simply buy it from a packet straight from the production line, as sold in the supermarket?[54] But while the idea of a robot chef, cocktail maker, waiter or even dishwasher might sound completely futuristic, the truth is that the future is already here. For instance, at the Robot Restaurant in Harbin, China, twenty robots, each costing £20–30,000, work the kitchens and restaurant floor. They cook the dumplings and noodles, and wait tables too (see Figure 12.7), though they do need recharging every five hours or so.[55] And KFC recently introduced a robot-serviced restaurant in China.[56] Meanwhile, Royal Caribbean International teamed up with Makr Shakr to

Figure 12.7. Will robot chefs be making our dinner in the future?

create the world's first 'bionic bar' (i.e., a robot cocktail maker) aboard
the newest addition to its fleet, the *Quantum of the Seas*. Here's a
description of what those who have booked a berth can look forward
to while they are out at sea: 'Guests can place drinks orders via tablets
and then watch robotic bartenders mix their cocktails. Each robot can
produce one drink per minute and up to 1,000 drinks per day, according to Royal Caribbean.'[57]

I was recently approached by the start-up Momentum Machines,
who are about to roll out the first robot line/short order chef for
mainstream restaurant kitchens. They wanted to know what people
would think about food if they knew that it had been prepared by a
robot? Would they like it more, or would they perhaps be put off? Do
diners really care who makes their food, or are they instead just interested in the taste of the final product? My suspicion is that people will
rate food and drink differently (and if I had to guess, less positively) if
they are told that it has been created by a robot rather than a real person. As far as I can see, the real sticking point here is likely to be that
robots don't have very good tasting abilities.[58] Consequently, such
machines will do better when working with packaged (i.e., standardized) ingredients rather than with fresh produce whose quality/ripeness
may vary. I also suspect there is something about the predictability of
robot-cooking that makes it less appealing than when a real person
does all the work.

For better or for worse, then, there really can be little doubting that our future dinners will become increasingly intertwined with digital technologies. This is the case even at home. Indeed, for the home market, Moley Robotics' home-cooking chef should be available early in 2018 for around £50,000.[59] I can just see my wife's eyes lighting up at the prospect.

13.   Back to the Futurists

Did you hear about the spat between Paul Pairet and Paco Roncero, two of the world's top modernist chefs, whom we came across in 'The Experiential Meal'? The former accused the latter of stealing his ideas around multisensory experience design. Both chefs currently deliver single-sitting multicourse tasting menus in futuristic dining spaces in which the projections on the walls and table change on a course-by-course basis to complement each of the dishes. Not only that but the music and soundscapes, even the ambient scent and temperature are all designed to match the food. On the surface at least, the chefs' offerings are very similar. They both carefully control the atmosphere in order to deliver a truly multisensory dining *experience*.[1] It is atmospherics taken to the max, facilitated by the latest in technology.

The question remains, though: who deserves the credit? Well, actually, I would say, neither of them do. In this, the final chapter, I want to put the case that as far as modernist cuisine is concerned, the Italian Futurists got there first. They may not have had the knowledge then to bring their dreams to life, at least not in a tasty way, but this is exactly what the science of gastrophysics is increasingly enabling some of the world's top chefs – and eventually you at home – to do. It was the Futurists, after all, who were playing ambient soundscapes to complement the food they served, back in the 1930s; in their case, the sound of croaking frogs to accompany 'Total Rice', a dish of rice and beans, garnished with salami and frogs' legs. Just think about it: Heston Blumenthal only started serving the 'Sound of the Sea' seafood dish, the first multisensory attempt at the Michelin-starred level, back in 2007. And yet something very similar was taking place eighty years earlier, in Turin. No wonder, then, that some people have wanted to argue that the Futurists really were 'Heston's forerunner'.[2]

Similarly, the recent interest of modernist chefs in miscolouring their foods or playing on the assumptions/expectations of the diner – be it Joan and Jordi Roca's completely white dark chocolate sorbet or Blumenthal's own 'Beetroot and Orange Jelly' dish – are again pre-dated by the Futurists. These crazy Italians were deliberately miscolouring various familiar foods in order to discombobulate their dinner guests long before it became trendy for the modernist chef to do so. How would you feel about blue wine,* orange milk or red mineral water?[3] Exactly! They were a dab hand too when it came to brightly coloured cocktails.[4]

The Futurists were interested in touch too, creating what may well be the first painting (entitled *Sudan–Paris*, 1920) designed to be stroked.[5] They had their diners eat without the aid of cutlery, simply burrowing their faces in the plate at their restaurant, the Taverna del Santopalato (the Tavern of the Holy Palate), in Turin. And as we saw in the 'Touch' chapter, it was also in Italy, eighty years ago, that the guests were instructed to stroke their neighbours' pyjamas, made of different materials, while dining.

The interest in fragrance, and in the delivery of food aromas in new and unusual ways, was also something that the Futurists experimented with, waiters spraying atomized perfume directly into the faces of their diners, for example. You find echoes of their exploits in today's modernist food offerings. Chef Homaru Cantu described one of the dishes served at his restaurant Moto, in Chicago, thus: 'This is my favorite part of the meal. I get to pepper-spray our guests.'[6] This reminds me of: 'Aerofood: A signature Futurist dish, with a strong tactile element. Pieces of olive, fennel, and kumquat are eaten with the right hand while the left hand caresses various swatches of sandpaper, velvet, and silk. At the same time, the diner is blasted with a giant fan (preferably an aeroplane propeller) and nimble waiters spray him with the scent of carnation, all to the strains of a Wagner opera.'[7] How multisensory can you get? So, for anyone interested in disruptive multisensory design, a quick look back at the Futurists is perhaps the best place to start.

Just take the rise of 'off the plate' dining and the increasing

* Funnily enough, the Spanish company Gïk launched a blue wine in June 2016 (see 'Sight').

theatricalization of service that we came across in 'The Experiential Meal'. By now, you should be able to guess who was there first.[8] As Sophie Brickman notes in *The New Yorker*: 'The banquets and dinners that Marinetti lays out in *The Futurist Cookbook* [. . .] are as much little plays as they are feasts.'[9] Elsewhere, it has been suggested that the Futurists were interested in 'elevating the chef to the rank of sculptor, stage designer, and director of a performative event'.[10] Just take the following account of a dinner hosted in Bologna: 'the "culinary stratosphere" [. . .] was filled with "nutritious noises" of aeroplanes, complementing a mise-en-scène of food sculptures, inventive lighting effects, and amazing outfits for the waiters, designed by Depero.'[11] Remarkably, this event was held on the evening of 12 December 1931. All this, then, leads us to our next question . . .

Futurist cooking: was molecular gastronomy invented in the 1930s?[12]

It is the growing realization of just how many cutting-edge modernist food practices were actually first trialled by Marinetti and his colleagues that has led some to ask whether modernist cuisine really does have its roots back in the 1930s.[13] There are, in fact, a surprising number of similarities between what was happening in northern Italy back then and what one sees in restaurants around the world today. Just take a look at the tenets of the 'Futurist Manifesto' (see below), and then tell me that I am imagining things.[14]

According to the Futurists, then, the perfect meal requires:

1. Originality and harmony in the table setting (crystal, china, decor) extending to the flavours and colours of the foods.
2. Absolute originality in the food.
3. The invention of appetizing food sculptures, whose original harmony of form and colour feeds the eyes and excites the imagination before it tempts the lips.

4. The abolition of the knife and fork for eating food sculptures, which can give prelabial tactile pleasure.[15]

5. The use of the art of perfumes to enhance tasting. Every dish must be preceded by a perfume which will be driven from the table with the help of electric fans.

6. The use of music limited to the intervals between courses so as not to distract the sensitivity of the tongue and palate but to help annul the last taste enjoyed by re-establishing gustatory virginity.

7. The abolition of speech-making and politics at the table.

8. The use in prescribed doses of poetry and music as surprise ingredients to accentuate the flavours of a given dish with their sensual intensity.

9. The rapid presentation, between courses, under the eyes and nostrils of the guests, of some dishes they will eat and others they will not, to increase their curiosity, surprise and imagination.

10. The creation of simultaneous and changing canapés which contain ten, twenty flavours to be tasted in a few seconds. In Futurist cooking these canapés have by analogy the same amplifying function that images have in literature. A given taste of something can sum up an entire area of life, the history of amorous passion or an entire voyage to the Far East.

11. A battery of scientific instruments in the kitchens: ozonizers to give liquids and foods the perfume of ozone; ultra-violet ray lamps (since many foods when irradiated with ultra-violet rays acquire active properties, become more assimilable, preventing rickets in young children, etc.); electrolysers to decompose juices and extracts, etc. in such a way as to obtain from a known product a new product with new properties; colloidal mills to pulverize flours, dried fruits, drugs, etc.; atmospheric and vacuum stills; centrifugal autoclaves; dialysers. The use of these appliances will have to be scientific, avoiding the typical error of cooking foods under steam pressure, which provokes the destruction of active substances (vitamins, etc.) because of the high temperatures. Chemical indicators will take into account the acidity and alkalinity of the sauces and serve to correct possible errors: too little salt, too much vinegar, too much pepper or too much sugar.[16]

We have come across modernist chefs addressing pretty much every one of these issues in the preceding chapters. Indeed, the last point on the list sounds just like molecular gastronomy/modernist cuisine – call it what you will. While the names of the latest must-have kitchen gadgets have undoubtedly changed, the underlying idea is the same – science in the kitchen and the preservation of nutrients/ flavour (one of the major selling points of the sous vide technique).[17] I wonder what the Futurists would have made of sous vide, or, for that matter, the anti-griddle, popularized recently by chef Grant Achatz. This new modernist kitchen device can flash-freeze or semi-freeze any food that is placed on its chilled surface.

However, there are also some fundamental differences between what the Futurists were trying to achieve in Italy in the 1930s and what many of today's modernist chefs have in mind. The former were certainly not much interested in making their food taste good; rather, they wanted to provoke, to shock people out of their comfort zone, to stop them revelling in the past (in their ossified cultural and political institutions, as some described it). Today, by contrast, the world's most talented chefs are increasingly realizing that they need to control 'the everything else' in order to deliver the most stimulating, the most memorable and hopefully the most enjoyable dining experiences. The aim nowadays is to prepare the best-tasting food possible and to complement that with the most immersive and engaging multisensory stimulation 'off the plate' as well.

On reading about some of the Futurists' crazier ideas, I am reminded of Albert Einstein's quote: 'If at first the idea is not absurd, then there is no hope for it.'[18] Just take, for example, the incendiary suggestion that pasta should be banned. Marinetti argued that it interfered with critical thinking due to its heaviness in the stomach. He also objected to the fact that it is 'swallowed, not masticated'.[19] Could you imagine a more provocative suggestion in Italy? It is funny, though, to read contemporary descriptions of Futurist dinners, even when presented by sympathetic chefs or journalists; the end result rarely sounds all that palatable.[20] Just take the 'Libyan Aeroplane' dessert: glazed chestnuts marinated in eau de Cologne followed by milk (don't tell me, you're not hungry), served atop a pâté of apples, bananas, dates and sweet peas in the form of an aeroplane – the Futurists, after all, loved their machines (note the steam engine on the wall in Figure 13.1).[21]

There are other differences beyond the Futurists' lack of interest in how the food actually tasted. Marinetti had in mind a future where the calorific requirements of the populace would be met by pills and powders, giving the body 'the calories it needs as quickly as possible'.[22] His idea was that once the basic nutrition was taken care of, this would free up time for 'novel experiences for the mouth and tongue,

Figure 13.1. A Futurist banquet in Tunis, *c.*1931. I must say that the diners (including Filippo Tomaso Marinetti (1876–1944) himself, shown staring intently at the waiter) look much more conventionally attired than all that talk of textured jim-jams would have led one to believe.[23]

as well as for the fingers, nose and ears'.[24] The touch, sound and scent of the Futurists' dishes were actually meant as *substitutions* for the nutritive function of food. Marinetti himself was clear on this point, describing the latter as 'a dish I would not recommend for the hungry'.[25] By contrast, the modernist chef's food is designed to satisfy their guests' hunger, as well as to look good on the plate (i.e., feeding their mind). Though those who remember the heyday of nouvelle cuisine might disagree.

Of course, even Marinetti's ideas about the future of cuisine didn't arise out of nowhere; rather, one needs to look to Apollinaire (another of the Futurists), who hosted a dinner in Paris in September 1912 (see Figure 13.2). He called this new style of cooking '*gastro-astronomisme*', after the eighteenth-century astronomer Laval. One can already see this focus on feeding the mind: 'In the true style of nouvelle cuisine, these proto-Futurist culinary innovators did not cook in order to fill the stomach but to satisfy the cravings of the mind. Their intention was to create works of art; therefore "it is better not to have any hunger when tasting these new dishes".'[26] This comment clearly prefigures Marinetti's position, and yet it is the latter who remains the undisputed godfather of Futurism.

Fresh violettes without their stems seasoned with lemon juice
Monkfish cooked in eucalyptus
Rare sirloin steak seasoned with tobacco
Barded quail with licorice sauce
Salad seasoned with oil and marc (brandy)
Reblochon cheese seasoned with walnuts and nutmeg
Fruit

Figure 13.2. Menu from a proto-Futurist meal, Guillaume Apollinaire's '*Le Gastro-astronomisme ou la Cuisine Nouvelle*' (1912–13). Notice the similarities with nouvelle cuisine (e.g., in the unusual combinations of ingredients). In his book *Feast and Folly*, Allen Weiss explicitly draws parallels with the contemporary cooking practices of a number of top French chefs, including such luminaries as Michel Bras, Pierre Gagnaire and Alain Hacquard.[27]

Want to create your own Futurist party?

Here are my recommendations for those of you who want to go 'off
the beaten track' and create your very own Futurist party.[28] Given
that modernist chefs have taken so much of their inspiration from the
Futurists, there is really no reason why you at home shouldn't do the
same. So, my top tips include:

1. Cover your dining table (and if you have enough, the walls
   too) in aluminium foil. This was an exciting new material at
   the time the Futurists were doing their thing, both futuristic
   and technological. (Though it might be difficult to create quite
   the same sense of wonder in your guests today using foil.)[29]
2. The amuse bouche should probably consist of a dieting pill.
   Remember, Futurist food is all about feeding the mind, not
   the body.
3. Buy some atomizers and infuse small amounts of the herbs,
   spices or fruits that feature in your recipes in oil or water.
   Encourage your guests to spray before tasting the food, and
   encourage them to breathe deeply. Perhaps position a fan
   pointing towards the table and set it on full power. (I am
   assuming that you don't have a jet propeller to hand.)
4. Use variously textured place mats, or else give everyone a
   few swatches of materials like sandpaper, velvet and silk.
   Encourage your guests to rub the full range of materials
   while eating/drinking to see whether it brings out
   something different for them in the tasting experience. If
   any of your guests turn up in a velvet smoking jacket or a
   silk dress, so much the better! (But that may just be a north
   Oxford thing.)
5. Between courses, why not put some Wagner on the music
   system and turn the volume up loud?
6. Create a really fragrant sauce in a pan, then bring it round
   the table suggestively wafting it under your guests' nostrils
   before returning the pan to the kitchen untouched.

7. No need to bother with cutlery; simply encourage your guests to eat with their hands and/or burrow their faces into the dishes.

8. Get a selection of food colourings from the supermarket and mischievously add some to each and every one of the drinks you serve.

9. Make use of lashings of those crunchy silver balls used to decorate cakes. Symbolic of the machine age, these tasty cake toppings will be more palatable than the real ball-bearings that the Futurists used to stuff into their 'Chicken Fiat' dish.

10. Make sure you play the appropriate nature sounds to accompany your dishes: the sounds of the sea for seafood, croaking frogs for frogs' legs and mooing cows if you happen to be serving beef. Your guests may never look at a steak in quite the same way again . . .

11. And why not engage in a dash of sonic seasoning? Serve a bittersweet dish, something with dark chocolate or sweetened black coffee, perhaps. Then alternate between tinkling, high-pitched piano music and some low-pitched, brassy music and see whether the taste of the food you have prepared changes.

12. In terms of the food itself, I would go for straightforward nouvelle cuisine rather than full-on Futurist fayre. And if any of your know-it-all guests query why they aren't being offered 'Drum Roll of Colonial Fish', 'Excited Pig' or 'Clotted Blood Soup', or tell you that they were especially looking forward to 'Italian Breasts in the Sunshine', just remind them that Futurist cuisine actually started with Apollinaire's proto-Futurist banquet in Paris, back in 1912!

13. Whatever you do, don't serve PASTA!

You'll be sure to deliver a night to remember.

Looking forward to the future of food

As robot chefs become more popular (see 'Digital Dining'), the question of *who* makes our food, and *how*, will increasingly come to the fore. Similarly, as a growing number of the world's top chefs open multiple restaurants under their own brand name, the question we should all be asking ourselves is what, exactly, are we buying (buying into) when we go to these restaurants? In what sense have the dishes we order really been touched by the master's hand? Of course, we all want consistency, something that is uniformly good; none of us wants to be disappointed by a dish that just isn't up to scratch. And yet, if it were that simple, wouldn't a robot or production line do a better job of producing a consistent outcome? It could even be programmed to imitate the movements and mannerisms of the star chef – but is that really what we want? Certainly, when we discover that our favourite restaurants are buying their food in pre-prepared, we somehow feel disappointed, cheated even. Worryingly, this is something that a growing number of the larger restaurant chains appear to be guilty of these days.[30]

Tim Hayward, writing in the *Financial Times*, talks about the 'cult of inconsistency'. His suggestion is that we should be celebrating variability in the delivery of the foods we eat, not denigrating it (because it shows that there really was a fallible human hand at work in the process of creation). And that, after all, is what we actually all want when we eat out, isn't it? As one salmon smoker Hayward spoke to put it: 'Why would I want to be "consistent"? It's an artisanal product and variation is part of that. It's how people know it's not mass-produced.'[31]

So, as robot chefs and cocktail makers start to appear more frequently, what will we, as diners and drinkers, think? Our views concerning the preparation of food and drink might well change. Here, I am reminded of the Italian biscuit manufacturer who gives their bags of biscuits a natural, handmade feel (despite the fact that all of their products are created on a production line) by having one of the machines cut biscuits that are somewhat different in shape from all the rest. Put a couple of those unusually shaped biscuits in a bag and the consumer is likely to interpret this subtle cue to mean the product is 'made by hand', and perhaps enjoy the experience of eating them more as a result. That, at least, is the idea.

Will we still be so enthusiastic about the open kitchen concept in restaurants once the chef is robotic rather than human? At present, the new technology has novelty value for sure, but for how much longer? And there are a number of other emerging trends that may even presage the disappearance of the restaurant as we know it. Sounds unlikely? Crazy, even? And yet I believe that things may be slowly starting to move in that direction. There's a new presence on the streets of a growing number of cities – the green and black delivery boxes of Deliveroo whizzing around on motorbikes and cycles.* Other companies have gone one stage further; for example, if you live in central London (i.e., Zone 1), Supper will deliver high-end Michelin-starred cuisine direct to your front door.[32] So if, in the years to come, such home delivery services continue to grow at anything like the current pace (and the price comes down, as it always does), then the question on every restaurateur's lips really ought to be: will people continue to go out to eat in restaurants? Why, after all, should they bother, when they can enjoy the same dishes in the comfort of their own homes? Presumably this is exactly the same issue that cropped up when the possibility of watching the latest movies at home, rather than at the cinema, first became a reality.

What, exactly, is lost if you take the restaurant away from the food? Well, as we have seen in earlier chapters, the fact that no crockery, cutlery or napkins are provided by the home delivery services is likely to diminish the at-home experience (assuming that the cutlery

---

* In my local pizza restaurant, Mamma Mia's, in Jericho, north Oxford, for instance, it is becoming increasingly difficult to keep track of all the Deliveroo drivers queuing up to collect their deliveries. In fact, it is hard not to come away from a meal at the restaurant without having formed the impression that the majority of the trade for venues such as this is switching from *on-* to *off*-premises. Of course, you need the bricks-and-mortar establishment in order to distinguish yourself from the straight take-away (and charge a premium as a result).

used in high-end restaurants is likely to be of better quality and heavier than what most of us use at home). So, if you are tempted to try out these services, my tip is to choose your plateware and cutlery carefully. It really can make *all* the difference. Oh, and – as you'll know by now – be sure to get the music right too.

There has also been a recent push towards helping people to make their own food at home. A number of internet-based companies (think Blue Apron, HelloFresh and ChefSteps) now encourage people like you and me to make a chef-prepared recipe by sending you the ingredients in the right proportions and offering a step-by-step guide online on how to prepare the food. Should this trend continue, we might see more home chefs cooking healthier foods. These meals are likely to taste better too, given the Ikea effect we saw in 'The Personalized Meal'.

Another pressure on the traditional restaurant format comes from all those creative chefs who are busily transforming dining into something more akin to a show. Sure, there is still food involved, but it is no longer necessarily the central focus of the experience. Think of it as modernist 'eatertainment', if you will.[33] If you had been to one of chef Jozef Youssef's 'Gastrophysics' dinners in London in 2016, you would have heard the sound of a duck quacking, and then the duck being 'terminated' (imagine the sound of a meat cleaver on a heavy wooden chopping board, complete with the cracking of cartilage and bone), all before the duck course comes out from the kitchen. As the chef puts it: 'If [thinking about where their meal came from] makes the diner uncomfortable then they shouldn't be eating this animal in the first place.' The most important thing here is to keep the diner entertained, but, beyond that, there can be a more serious aim, to nudge the diner towards making healthier, more sustainable food choices.[34]

Already some of the language of dining is changing subtly; 'diners' are gradually being replaced by 'guests'. And what is more, you increasingly find yourself booking a ticket for the show, rather than a table for dinner (see 'The Experiential Meal'). As these trends continue to develop, the restaurant as we know it may soon disappear, or at least evolve into quite a different kind of experience (think of all those coffee shops selling books that morph into bookshops selling coffee).[35]

Big data and food

Looking forward, it is going to be interesting to see how big data and citizen science will change the design of the food experiences that we are all exposed to. We have already started to see linguists mining thousands of online menus to figure out how much we have to pay for each extra letter in a dish name on the menu – around six cents, apparently.[36] And then there are computer scientists out there busily comparing recipes from around the world in order to discover key flavour pairings that are intimately associated with the cuisine of a particular place (or region), ushering in a new area of science, known as 'computational gastronomy'.[37] The latest analysis of Indian recipes, for instance, shows that the chefs there tend to combine ingredients that are not harmonized – exactly the opposite pattern to what is seen in the rest of the world.[38]

What other genuine insights might emerge from this data-mining of large-scale food-relevant databases? Will a whole range of great-tasting but unusual new flavour combinations emerge from the likes of FoodPairing (who run a subscription website allowing chefs, cocktail makers and interested home chefs to figure out which combinations of ingredients share the same flavour compounds) and IBM's Chef Watson?[39] IBM's supercomputer Watson algorithmically analyses a database of thousands of recipes plus a database of flavour compounds found in thousands of ingredients, along with psychological findings about how people perceive different combinations of ingredients. This computer has no hands so merely comes up with unusual combinations that someone else makes: 'IBM is keen to stress that this is not about machines outdoing humans but rather working side by side with them [. . .] Heston Blumenthal better watch out'.[40] Will the diners of tomorrow be increasingly exposed to a whole range of new flavour combinations? The key point to remember here is that it is not about gastrophysicists competing against chefs, nor computers battling it out against humans. Rather, it is a question of how much more convincing a proposition we can make by bringing together the various disciplines. Meanwhile, researchers who surveyed over a million online restaurant reviews over a nine-year period from 2002 to 2011 across every state in the US found that good weather tends to result in our enjoying the experience of dining out that much more than when the weather is not so good.[41]

In the last few years, the Crossmodal Laboratory has been running a number of large-scale citizen-science experiments in museums and also online, to help provide information about the kind of design decisions that diners may appreciate: everything from assessing the orientation of the dish through to the wall colour and the background music. My guess is that the small-scale studies of the effect of the environment on the behaviour of diners (typically involving a few tens or hundreds of diners, as covered in 'The Atmospheric Meal') will soon be eclipsed by big-data studies (with the data perhaps acquired through the signals given off by diners' mobile devices). The number of participants that one can study will suddenly jump to the tens or hundreds of thousands.

This should hopefully allow for much better evidence-based decision-making in the design of food and beverage offerings. As a case in point, over the last year we have collected responses from more than 50,000 people who attended the 'Cravings' exhibition (either in person or online) while it was open to the public in London's Science Museum. The results have helped confirm some of our intuitions regarding how the presentation of food affects what we think about the flavour and how much we expect to like a dish. At the same time, however, the results have started to disprove some kitchen folklore – such as, for example, that odd numbers of items are preferred over even numbers on the plate.[42] We have recently extended this approach to look at the best orientation for a long straight element in a dish – think a seared spring onion, or a whole lobster. It turns out that people like linear elements to ascend from the bottom left to the top right. Another intriguing contrast that the latest research highlights is between the kinds of plating people will be willing to pay most for, and those they deem most creative.[43]

One other example of what is currently being done with big data analysis comes from Rupert Naylor, of Applied Predictive Technologies. He describes what the company he works for does for the restaurant chain as allowing them to: 'run control experiments, much in the same way that they test for the efficacy of new drugs [. . .] We do a control test on restaurants that show similar behaviour as a baseline, then we strip out all the noise – the data that may have affected sales anyway – to get to the truth.' The approach apparently helped Pizza Hut UK to increase average customer spend from £9 to £11. While this might not sound like much, rest assured it soon adds up.[44]

Synaesthetic experience design

Practitioners of multisensory experience design have, over many years, thoroughly explored those connections between the senses that are, in some sense, obvious. Just think of a dish of frogs' legs accompanied by the sound of croaking frogs, or a seafood dish accompanied by the sounds of the sea. Other restaurants, including Eleven Madison Park in New York, Casamia in Bristol and Ultraviolet in Shanghai, have tried various ways to recreate a picnic (or, more likely, the positive emotions that are associated in most people's minds with such events from their past): by using ceramic plates that imitate paper plates; by serving from a picnic basket, obviously; and probably by introducing the sounds, smells (scent of a field) and even associated visuals (think of a countryside scene). It is effective, most definitely, but it can seem a little clichéd, at least in hindsight.[45] Increasingly, though, chefs, culinary artists and experience designers are starting to move more into the world of synaesthetic design, where decisions about the experiences that are delivered to diners (and drinkers) are based on connections between our senses that are not so obvious, or literal.[46] Here, I am thinking of events like 'The Colour Lab', where ambient colours and music were used to alter the taste of a glass of wine.[47] This is synaesthesia-like, in that the connections between the senses are often surprising when first you hear about them (e.g., that sweet is high-pitched, pinky-red and round). And yet this form of design is fundamentally different from synaesthesia (the condition where people see letters, numbers and units of time in colour, say, or where sounds trigger smells). The key difference is that these newly discovered links between the senses tend to be shared between the majority of people.[48] It is these common yet often surprising associations – what are often referred to as 'crossmodal correspondences' – that allow for the creation of multisensory experiences that are both intriguing and truly meaningful. The growing body of gastrophysics research provides a number of insights for chefs and experience designers in this area.

Things begin to get really interesting, though, as soon as one starts playing with the correspondences as they relate to the chemical senses – that's taste, aroma and flavour.[49] This is not to say that adding more senses into the mix necessarily makes things easier as far as multisensory experience design is concerned. Just take Sean Rogg, talking about one of his recent events, part of the Waldorf Project, in which people were 'invited to taste colour'. The visitors, who had been asked to dress monochromatically for the event, got to drink fine wine and watch a host of dancers perform. According to the artist: 'Not only did the soundscapes have to sound like their respective colours, but I was asking the sound designer to pair his soundscapes to wine.'[50] That's a big ask. Nevertheless, despite the inherent challenges associated with working in this area, what is clear is that there has been an explosion of interest in synaesthetic design involving food and drink.[51]

There is no guarantee that different people will necessarily have the same reactions to such experiential events, but that is part of the fun. The rise in synaesthetic design, building on the surprising connections between the senses that we all share, goes hand in hand with the emergence of 'sensploration', the idea that consumers are increasingly curious to explore their own sensory world (or 'sensorium') and the hidden connections that can be found within each and every one of us.[52] And while sensory marketing seemed once to be all about making money, it now feels much more about the delivery of shared (and shareable) multisensory experiences (or at least it should).[53] For the culinary artist, it is as much about a journey of discovery. It is about exposing us all to the unusual, surprising, almost-synaesthetic connections between the senses. In fact, at least according to one recent report: '70 percent of United States-based Millennials now search for experiences that "promote their senses."' There are various explanations for this. One intriguing suggestion is that they are hungry for immersive engaging experiences, and as one commentator put it: 'As consumers grow evermore weary of constant digital bombardment, they seek out more authentic experiences to immerse themselves with a brand.'[54] So, while the experience economy continues to influence so many aspects of the marketplace, and marketing communications, it is time to get ready, I think, for the next 'sensory explosion' (as US marketing professor Aradna Krishna called it in an industry briefing back in 2013).

Have you heard of the Gesamtkunstwerk?

Ultimately, then, we are slowly making our way towards the *Gesamtkunstwerk*, food as a total work of art,[55] an installation, or experience, that engages all of the viewer's senses (if 'viewer' is still the right word?). The term is associated with the German composer Wagner, so no wonder that he was the composer of choice at the Futurists' dinner parties. In fact, it is hard to see how this objective of creating a work of art that stimulates all of the senses could ever be achieved without the involvement of food or drink.

The *Gesamtkunstwerk*, Futurism and various other artistic trends that were very much in vogue a century ago can all be linked, more or less directly, to the rise of the physiological aesthetic, in turn-of-the-century Europe. At that time, artists, including famous painters such as Seurat, were convening with scientists in order to design more pleasing experiences based on the emerging neuroscientific understanding of the mind of the viewer.[56] This interaction between the artists and the scientists undoubtedly led to a phenomenal wave of creativity, albeit one that ultimately fizzled out. The decline was probably due to the fact that it was the *wrong* science at the *wrong* time (brain science has undoubtedly come a long way in the intervening 120 years or so). And, as far as I can see, measuring brainwaves was never going to produce the information that the painters (and other artists) wanted to help them in the design of their compositions.[57]

However, roll the clocks forward and we are now increasingly seeing the merging of culinary artistry with the behavioural or psychological sciences. This is encompassed within the new science of gastrophysics. And when combined with the latest in design and technology, this new collaboration promises to deliver a food future that is utterly unlike anything we have seen before – unlike, even, the wildest dreams and ideas of the Italian Futurists.[58]

Looking forward to a healthy and sustainable food future

It is getting ever harder to think about the future of food without being conscious of the problem of climate change, the challenges associated with sustainability, and the rise of the mega-city. It is hard to say whether the solutions to sourcing our food in the future will come from vertical farming, lab-grown meat, increased entomophagy, or – heaven forbid – Soylent Green (a delicious green wafer advertised as containing high-energy plankton but actually secretly made of human remains). This dystopian prediction about the future of food came from Richard Fleischer's movie of the same name, set in the year 2022.* But I passionately believe that whatever the future holds, we will stand a far greater chance of achieving our goals by exploring the interface between modernist culinary artistry as it meets the latest in technology and design.[59] Ultimately, it is crucial to realize that changing behaviour is not simply a matter of informing people of what is good for them or what is sustainable for the planet. Other

---

* See C. Spence & B. Piqueras-Fiszman *The Perfect Meal: The Multisensory Science of Food and Dining* (Oxford: Wiley-Blackwell, 2014) for a discussion of Soylent Green and the new algal cuisine. The movie was based on Harry Harrison's 1966 novel *Make Room! Make Room!* Spotting a marketing opportunity, a California start-up recently started marketing a food product they call Soylent Green, which supposedly contains all the protein, carbohydrates, lipids and micronutrients that we need. Though what was most noticeable about the early formulations of this product was the extreme flatulence it induced in many of those who tried it.

strategies are needed to nudge people towards healthier and more sustainable food behaviours, approaches that are cognizant of the fact that our perception of food happens mostly in the mind, not in the mouth. My guess is that the notion of 'food hacking' is one that we are all going to become much more familiar with in the future.[60]

And from a more personal perspective, I believe that the future of gastrophysics holds fundamental challenges as well as many opportunities to really make a difference to the way many of us will interact with food in the years to come. I hope that some of the most exciting developments currently seen primarily in the high-end modernist restaurant will be scaled up for the masses; indeed, I am already seeing a huge increase in interest from a number of the world's largest food and beverage companies. As ever more of our food behaviours are taking place or being facilitated online, that will open up a whole new world in terms of the big-data analysis of food trends and behaviours. Chris Young, founder of ChefSteps, predicted that their website would be helping more than a million people to cook better meals at home by the end of 2016. This kind of interaction will generate huge amounts of data that can be used to figure out how best to personalize and enhance our food perception and our behaviour moving forward.

The scientific approach represented by the emerging field of gastrophysics will help quantify what really matters here, separating fact from fiction and intuition. Real progress will be made just as soon as more of us recognize that 'pleasure is not only found in the mouth', as chef Andoni Luis Aduriz put it, but also – mostly, in fact – in the mind. Actually, the chef is worth quoting at length here: 'What it comes down to is: *you don't have to like something for you to enjoy it* or, in other words, pleasure is not only found in the mouth. Predisposition, the ability to concentrate – the impulsive mechanisms of the brain – can completely modify the perception of something that, at first sight, would not even be considered food for humans. In the end, it isn't only about eating; it's also about discovering. We take advantage of the fact that we are always on the borderline between our conservative selves – the part that makes us creatures of habit, finding shelter and security in repetition – and our curious, and daring, selves, which

Figure 13.3. F. T. Marinetti gazing into the future.[61]

seek pleasure in the unknown, in the vertigo we feel when we try something for the first time, in risk and the unpredictability.'[62] And that, I think you will find, ladies and gentlemen, brings us right back to where we started! (See Figure 13.3.)

And to end: how about something healthy?

To close this book, I have summarized some of the key recommendation for anyone who wants to feel more satisfied while consuming less (i.e., to eat more healthily):[63]

1. Eat less – Obvious, I hear you say! But not everyone does it.
2. Hide food – you will be more tempted to snack if you can see the cookies in the jar than if they are in an opaque container. It really is a case of out of sight, out of mind. In fact, anything that you can do to make it more difficult to get your hands on food in the first place is likely to help too. This kind of nudge approach has often (not always, mind) been shown to be effective in helping reduce consumption.

3. Middle-aged and older adults should try drinking lots of water before every meal – half a litre thirty minutes before breakfast, lunch and dinner ought to do the trick. In one study, this led to a reduction in consumption at mealtimes of roughly forty calories. Plus all those extra bathroom breaks will no doubt help increase your physical activity![64]

4. If you happen to be fond of junk food, why not eat in front of a mirror, or else off a mirrored plate? Research suggests it can help reduce desire for, and consumption of, foods such as chocolate brownies. There is at least one famous actress who would apparently eat naked in front of a mirror. It would be interesting to see whether customers at those naked restaurants that have been popping up recently (see 'Social Dining') also eat less.[65] Try to eat slowly and mindfully. And yes, that does mean turning the TV off!

5. The more food sensations you can muster, the better. Stronger aroma, more texture – it all helps your brain to decide when it has had enough. In one of my favourite studies illustrating this point, people consumed far more calories when drinking apple juice as compared to apple puree, and when eating pureed apple as compared to apples. Exactly the same food in all three cases; all that differs are the textural cues the brain receives about how much it has consumed (and how much chewing is needed).[66] This is much the same reason why you should never use a straw to drink. It eliminates all the orthonasal olfactory cues that are normally such a large part of the enjoyment (see the 'Smell' chapter). Be sure to inhale the aroma of your food frequently; after all, this is where the majority of the pleasure resides. Whatever you do, don't drink iced water *with* your meals. It numbs the taste buds, plain and simple! Some researchers have even gone so far as to suggest that the North American preference for more highly sweetened foods may, in part, be linked to all the iced water they drink at mealtimes.[67]

6. Eat from smaller plates. This technique is especially effective

when people serve themselves. The numbers here are pretty staggering: if you eat from a plate that is twice as big, for example, you are likely to consume as much as 40% more food.[68]

7. Bowl food – eat from a heavy bowl without a rim, and hold the bowl in your hands while you eat; don't leave it sitting on the table. The weight in your hand is likely to trick your brain into thinking that you have consumed more, and you'll feel satiated sooner.

8. Eat from red plateware. In this context, red plates seem to trigger some kind of avoidance motivation.[69]

9. Eat with chopsticks rather than regular cutlery, or try eating with your non-dominant hand, or else with a smaller spoon or fork. Do anything, basically, that makes it more difficult for you to get the food into your mouth. Along much the same lines, artists and designers from thirty-five countries were recently tasked with making tableware that challenged eating norms and so encouraged slower, more mindful eating, at an Amsterdam supper club. Though you'd best watch your teeth if you ever try eating with the nail-filled spoon that one eager contributor created for the event.[70]

10. Oh, and one excellent tip from Yogi Berra: 'You better cut the pizza in four slices because I'm not hungry enough to eat six.'[71]

Notes

Amuse Bouche

1 See J. A. Cassiday, 'Sitophilia', in D. Goldstein (ed.), *The Oxford Companion to Sugar and Sweets* (Oxford: Oxford University Press, 2015), pp. 614–15.

2 J. S. Allen, *The Omnivorous Mind: Our Evolving Relationship with Food* (London: Harvard University Press, 2012). See also R. Tamal et al., 'Neuro cuisine: Exploring the science of flavour' – video, *Guardian Online*, 23 May 2016 (https://www.theguardian.com/lifeandstyle/video/2016/may/23/neuro-cuisine-exploring-the-science-of-flavour-video). As Heston Blumenthal put it: 'For me, the most important ingredient is the brain. We eat for two things: to survive and then for pleasure. The pleasure part needs the brain. The context of eating – the shape of the knife, the people you're with, the smell in the room – makes a difference. Whether you like it or not, it matters' (cited in G. Ulla, 'The future of food: Ten cutting-edge restaurant test kitchens around the world', *Eater*, 11 July 2017 (https://www.eater.com/archives/2012/07/11/ten-restaurant-test-kitchens.php)).

3 All of these factors together influence the final tasting experience, as does our mood, our genetic make-up and our prior history of exposure to different foods and flavour combinations. It all matters! See D. M. Masumoto, *Four Seasons in Five Senses: Things Worth Savouring* (New York: W. W. Norton & Co, 2003); D. Ackerman, *A Natural History of the Senses* (London: Phoenix, 2000).

4 D. Martin, *Evolution* (Lausanne: Éditions Favre, 2007).

5 C. Spence & B. Piqueras-Fiszman, *The Perfect Meal: The Multisensory Science of Food and Dining* (Oxford: Wiley-Blackwell, 2014). Of course, this is a two-way street; see P. Bongers et al., 'Happy eating: The underestimated role of overeating in a positive mood', *Appetite*, 67 (2013), 74–80. See also G. Sproesser, H. Schupp & B. Renner, 'The bright side

of stress-induced eating: Eating more when stressed but less when pleased', *Psychological Science*, 25 (2014), 58–65.

6 C. Humphries, 'Cooking: Delicious science', *Nature*, 486 (2012), S10–S11; N. Myhrvold & C. Young, *Modernist Cuisine: The Art and Science of Cooking* (La Vergne, TN: Ingram Publisher Services, 2011); J. Youssef, *Molecular Cooking at Home: Taking Culinary Physics out of the Lab and into Your Kitchen* (London: Quintet, 2013). See also P. Barnham et al., 'Molecular gastronomy: A new emerging scientific discipline', *Chemical Reviews*, 110 (2010), 2313–65.

7 Neurogastronomy has undoubtedly helped provide answers to some long-standing questions, like just what effect pricing, branding, labelling and product description have on perception (see 'Taste'), but it is important to remember that the isolated subject in the lab is a very long way removed from the way that most of us eat and drink.

The term 'neurogastronomy' was coined by G. M. Shepherd in 'Smell images and the flavour system in the human brain', *Nature*, 444 (2006), 316–21. See also G. M. Shepherd, *Neurogastronomy: How the Brain Creates Flavor and Why It Matters* (New York: Columbia University Press, 2012), and for a commentary, see my review of this book (*Flavour*, 1:21 (2012)).

8 O. G. Mouritsen, 'The emerging science of gastrophysics and its application to the algal cuisine', *Flavour*, 1:6 (2012); O. G. Mouritsen & J. Risbo, 'Gastrophysics—do we need it?', *Flavour*, 2:3 (2013); K. Parkers, 'Recipe for success: Teachers get inspiration from "gastrophysics"', *Physical Education*, 39 (2004), 19; T. A. Viglis, 'Texture, taste and aroma: Multi-scale materials and the gastrophysics of food', *Flavour*, 2:12 (2013).

9 It is not, note, gastronomy + astrophysics! Though this confusion has had some unintended consequences. See G. Weiss, 'Why is a soggy potato chip unappetizing?' *Science*, 293 (2001), 1753–4; J. Hinchliffe, 'Gastrophysics: Brisbane scientist Dr Stephen Hughes explores astrophysics in the kitchen', *ABC Brisbane*, 18 November 2015 (http://www.mobile.abc.net.au/news/2015-11-18/gastrophysics-brisbane-scientist-takes-astrophysics-to-kitchens/6948384).

And funnily enough, there is also a link to astronomy via the very first proto-Futurist meals held in Paris in 1912–13, called '*gastro-astronomisme*' after the then-famous French astronomer Laval; see G. Berghaus, 'The futurist banquet: Nouvelle Cuisine or performance art?', *New Theatre Quarterly*, 17(1) (2001), 3–17, esp. p. 9.

10 Using the term 'gastronomy' as a base helps to distinguish what the gastrophysicist does from the work of those sensory scientists trying to discover, for instance, just how much fish meal you can feed a chicken before you can taste it on the breast meat. You do have to feel sorry for all those tasting panels out there who may spend a whole year rating, say, frozen Brussels sprouts. Someone should do this kind of research – I'm just glad it's not me!

11 D. M. Green & J. A. Swets, *Signal Detection Theory and Psychophysics* (New York: Wiley, 1966). Most psychophysicists work on sight and sound, in large part because these are much simpler senses to control. That said, there may also be a certain snobbery here, with many philosophers and scientists considering sight and sound the *higher* rational senses. By contrast, taste, smell and touch are often described as the *lower* senses, with the unspoken inference that they are hardly worthy of study. The problem with studying taste and smell, two of the central components of flavour perception, is that they are messy to handle, and your 'guinea pigs' (the colloquial term for one's subjects, or participants) soon adapt, habituate or just become full.

12 M. Zampini & C. Spence, 'The role of auditory cues in modulating the perceived crispness and staleness of potato chips', *Journal of Sensory Science*, 19 (2004), 347–63.

13 H. Blumenthal, *Further Adventures in Search of Perfection: Reinventing Kitchen Classics* (London: Bloomsbury, 2007); C. Spence, M. U. Shankar & H. Blumenthal, '"Sound bites": Auditory contributions to the perception and consumption of food and drink', in F. Bacci & D. Melcher (eds.), *Art and the Senses* (Oxford: Oxford University Press, 2011), pp. 207–38.

14 A.-S. Crisinel et al., 'A bittersweet symphony: Systematically modulating the taste of food by changing the sonic properties of the soundtrack playing in the background', *Food Quality and Preference*, 24 (2012), 201–4; C. Spence, 'On crossmodal correspondences and the future of synaesthetic marketing: Matching music and soundscapes to tastes, flavours, and fragrance', in K. Bronner, R. Hirt & C. Ringe (eds.), *Audio Branding Academy Yearbook 2012/13* (Baden-Baden: Nomos, 2013), pp. 39–52.

15 'Louis Armstrong for starters, Debussy with roast chicken and James Blunt for dessert: British Airways pairs music to meals to make in-flight food taste better', *Daily Mail Online*, 15 October 2014 (https://www.dailymail.co.uk/travel/travel_news/article-2792286/british-airways-pairs-music-meals-make-flight-food-taste-better.html).

16 M. Zampini et al., 'Multisensory flavor perception: Assessing the influence of fruit acids and color cues on the perception of fruit-flavored beverages', *Food Quality & Preference*, 19 (2008), 335–43.

17 T. Turnbull, 'So who's the better chef: The one with Michelin stars or the geek?' *The Times*, 17 March 2016, pp. 6–7 (http://www.thetimes.co.uk/tto/life/food/article4714734.ece).

18 As Stefan Cosser, a former senior development chef in Bray, put it recently: 'Neurogastronomy is simply knowledge that can be added to a chef's repertoire like any other skill that applies to our job.' Quoted by N. Robinson in 'Which colours will arouse your customers?', *The Publican's Morning Advertiser*, 21 June 2016 (https://www.morningadvertiser.co.uk/Pub-Food/News/How-to-use-colour-in-pubs). Though some claims for the potential impact of applying neurogastronomy are undoubtedly exaggerated; see N. Robinson, 'Double pub sales with neurogastronomy?' *The Publican's Morning Advertiser*, 6 October 2015 (https://www.morningadvertiser.co.uk/Pub-Food/Food-trends/How-to-trick-diners-brains).

19 'Square plates are an "abomination", says *MasterChef* judge William Sitwell', *Daily Telegraph* (*Food & Drink*), 13 May 2014 (http://www.telegraph.co.uk/foodanddrink/10828052/Square-plates-are-an-abomination-says-MasterChef-judge-William-Sitwell.html).

20 C. Spence, 'Hospital food', *Flavour* (in press, 2017).

21 A. Gallace & C. Spence, *In Touch with the Future: The Sense of Touch from Cognitive Neuroscience to Virtual Reality* (Oxford: Oxford University Press, 2014).

22 The furry bowl idea was used by the Italian Futurists: 'tactile dinner parties organized by the Italian Futurists, in which courses were sometimes served from small bowls covered with different tactile materials' (F. T. Marinetti, *The Futurist Cookbook*, translated by S. Brill (1932; San Francisco: Bedford Arts, 1989), p. 125).

23 Why else, after all, do so many of us still continue to type with a QWERTY keyboard, despite all the evidence showing that other

keyboard configurations would allow us to type faster? See D. A. Norman, *The Design of Everyday Things* (London: MIT Press, 1998).

24 W. Welch, J. Youssef & C. Spence, 'Neuro-cutlery: The next frontier in cutlery design', *Supper Magazine*, 4 (2016), 128–9.

25 A. T. Woods, C. Michel & C. Spence, 'Odd versus even: A scientific study of the "rules" of plating', *PeerJ*, 4:e1526 (2016).

26 C. Spence & B. Piqueras-Fiszman, *The Perfect Meal: The Multisensory Science of Food and Dining* (Oxford: Wiley-Blackwell, 2014); A. Jordan, 'Sensing success: Melbourne store explores the five senses', *LS:N Global*, 18 December 2015 (https://www.lsnglobal.com/seed/article/16908/sensing-success-melbourne-store-explores-the-five-senses).

27 P. Kotler, 'Atmospherics as a marketing tool', *Journal of Retailing*, 49 (Winter 1974), 48–64.

28 F. T. Marinetti, *The Futurist Cookbook*, trans. S. Brill (1989) (1932; London: Penguin, 2014); 'Futurist cooking: Was molecular gastronomy invented in the 1930s?', *The Staff Canteen*, 25 April 2014 (https://www.thestaffcanteen.com/Editorials-and-Advertorials/futurist-cooking-was-molecular-gastronomy-invented-in-the-1930s); S. Brickman, 'The food of the future', *The New Yorker*, 1 September 2014 (https://www.newyorker.com/culture/culture-desk/food-future).

29 I had the 'pleasure' of being on a double bill with Michael on his home turf at a literary festival held in the middle of Dartmoor National Park ('The Perfect Meal', Professor Charles Spence & Michael Caines MBE in conversation, Chagford Literary Festival, 15 March 2015).

1. *Taste*

1 B. Wilson, *First Bite: How We Learn to Eat* (London: Fourth Estate, 2015); O. G. Mouritsen & K. Styrbaek, *Umami: Unlocking the Secrets of the Fifth Taste* (New York: Columbia University Press, 2014); B. Stuckey, *Taste What You're Missing: The Passionate Eater's Guide to Why Good Food Tastes Good* (London: Free Press, 2012).

There are those who want to argue that oleogustus, or fatty acid, should be considered as the sixth taste; see T. A. Gilbertson, 'Gustatory mechanisms for the detection of fat', *Current Opinion in Neurobiology*, 8

(1998), 447–52; R. S. J. Keast & A. Costanzo, 'Is fat the sixth taste primary? Evidence and implications', *Flavour*, 4:5 (2015); C. A. Running, B. A. Craig & R. D. Mattes, 'Oleogustus: The unique taste of fat', *Chemical Senses*, 40 (2015), 507–16.

Kokumi is, though, more of a mouth feeling than a taste, and is often associated with descriptors such as 'thickness', 'heartiness' and 'mouthfulness'. Garlic, onions and scallops all possess this quality. According to researchers, a group of chemicals known as gamma-glutamyl peptides give rise to such mouth sensations. However, a dedicated receptor has yet to be found, meaning that researchers are still undecided as to whether this is really the sixth, or should that be seventh, eighth or ninth taste . . . See O. G. Mouritsen, 'The science of taste', *Flavour*, 4:18 (2015). Intriguingly, kokumi is apparently capable of amplifying other tastes when presented in combination.

The latest 'new' taste was just reported: J. Hamzelou, 'There is now a sixth taste – and it explains why we love carbs', *New Scientist*, 2 September 2016 (https://www.newscientist.com/article/2104244-there-is-now-a-sixth-taste-and-it-explains-why-we-love-carbs/); T. J. Lapis, M. H. Penner & J. Lim, 'Humans can taste glucose oligomers independent of the hT1R2/hT1R3 sweet taste receptor', *Chemical Senses* 41 (2016), 755–62.

2 J. Delwiche, 'Are there "basic" tastes?', *Trends in Food Science & Technology*, 7 (1996), 411–15; R. P. Erikson, 'A study of the science of taste: On the origins and influence of the core ideas', *Behavioral and Brain Sciences*, 31 (2008), 59–105; B. Lindemann, 'Receptors and transduction in taste', *Nature*, 413 (2001), 219–25. One group of researchers recently summed it up as follows: 'Despite more than 2 millennia of reflection, consensus is lacking on what constitutes a basic taste quality, and whether taste is limited to a discrete set of taste "primaries"' (C. A. Running, B. A. Craig & R. D. Mattes, 'Oleogustus: The unique taste of fat', *Chemical Senses*, 40 (2015), 507–16, p. 507). Then there are a few eminent neuroscientists out there, like Charles Zucker, who are unwilling even to contemplate the existence of anything beyond the five well-known basic tastes unless a new class of receptor can be found in the taste buds on the tongue.

In mice, each of the five basic tastes is represented in its own cortical field (see X. Chen et al., 'A gustotopic map of taste qualities in the mammalian brain', *Science*, 333 (2011), 1262–6).

3 C. Spence, B. Smith & M. Auvray, 'Confusing tastes and flavours', in D. Stokes, M. Matthen & S. Biggs (eds.), *Perception and Its Modalities* (Oxford: Oxford University Press, 2015), pp. 247–74.

4 While some authors correct themselves later in the text, there is a real danger of ambiguity here. Anyone reading the cover of Jennifer Mc-Lagen's recent book *Bitter: A Taste of the World's Most Dangerous Flavour, with Recipes* (London: Jacqui Small, 2015) would be excused for feeling confused; it is only inside, tucked away in the main text, that the author acknowledges that she intends to use the terms 'taste' and 'flavour' interchangeably. Some would have us believe, at least in the titles of their papers, that umami is a flavour not a taste; C. McCabe & E. T. Rolls, 'Umami: A delicious flavour formed by convergence of taste and olfactory pathways in the human brain', *European Journal of Neuroscience*, 25 (2007), 1855–64. For the standard view, see, e.g., Y. Kawamura & M. R. Kare, *Umami: A Basic Taste: Physiology, Biochemistry, Nutrition, Food Science* (New York: Marcel Dekker, 1987).

5 P. Rozin, '"Taste-smell confusions" and the duality of the olfactory sense', *Perception & Psychophysics*, 31 (1982), 397–401. This fabulous little paper is packed full of goodness, even a third of a century after its publication!

6 H. Nagata et al., 'Psychophysical isolation of the modality responsible for detecting multimodal stimuli: A chemosensory example', *Journal of Experimental Psychology: Human Perception & Performance*, 31 (2005), 101–9. Carbon dioxide sensing is also interesting in this regard; see W. B. Frommer, 'Biochemistry: CO_2mmon sense', *Science*, 327 (2010), 275–6; J. Chandrashekar et al., 'The taste of carbonation', *Science*, 326 (2009), 443–5.

7 M. Skinner et al., 'Investigating the oronasal contributions to metallic perception' (submitted to *International Journal of Food Science & Technology*, 2016).

8 D. P. Hanig, '*Zur Psychophysik des Geschmackssinnes*' ['On the psychophysics of taste'], *Philosophische Studien*, 17 (1901), 576–623; E. G. Boring,

Sensation and Perception in the History of Experimental Psychology (New York: Appleton, 1942).

9 And as soon as you realize how few taste receptors there are in the centre of the tongue, you might wonder whether adding seasoning to a food that is likely to end up there is a bad idea (a waste of taste, as it were). Pringles are interesting in this regard. Here you have a food that people are likely to insert in their mouths in the same way every time. If one were to asymmetrically distribute the seasoning on the side that is likely to touch the tongue first, and perhaps focus on distributing it around the edges that will immediately touch the taste buds (rather than the centre that will not), then one could perhaps deliver a more intense taste experience.

10 J. Chandrashekar et al., 'The receptors and cells for mammalian taste', *Nature*, 444 (2006), 288–94; V. B. Collings, 'Human taste response as a function of locus of stimulation on the tongue and soft palate', *Perception & Psychophysics*, 16 (1974), 169–74; A. O'Connor, 'The claim: The tongue is mapped into four areas of taste', *The New York Times*, 10 November 2008 (https://www.nytimes.com/2008/11/11/health/11real. html?_r=1).

11 A. Dornenburg & K. Page, *Culinary Artistry* (New York: John Wiley & Sons, 1996).

12 One of the early influences on Adrià may well have been his friend Miguel Sánchez Romera, who worked as a neurologist by day and ran his restaurant by night, using the neuroscientific insights about the senses and perception he picked up in his day job as inspiration for the dishes he prepared in the evenings (see Sánchez Romera, *La cocina de los sentidos* [*The Kitchen of the Senses*] (Barcelona: Editorial Planeta, 2003)). However, when this early neurogastronomist closed his restaurant just outside Barcelona and tried to make his name and fortune with his namesake restaurant Romera in New York City, it all went pear-shaped very rapidly. The critic Frank Bruni had this to say: 'Romera [. . .] has coined a whole new genre for his cooking, which favors squishy textures, kaleidoscopic mosaics of vegetable powders, and a wedding's worth of edible flowers. He calls it neurogastronomy, which "embodies a holistic approach to food by means of a thoughtful study of the organoleptic properties of each ingredient", or so says the restaurant's

Web site. Organoleptic means "perceived by a sense organ." I looked it up' ('Dinner and derangement', *The New York Times*, 17 October 2011 (https://www.nytimes.com/2011/10/18/opinion/bruni-dinner-and-derangement.html?_r=0)). Perhaps unsurprisingly the restaurant soon closed!

13 A. V. Cardello, 'Consumer expectations and their role in food acceptance', in H. J. H. MacFie & D. M. H. Thomson (eds.), *Measurement of Food Preferences* (London: Blackie, 1994), pp. 253–97; B. Piqueras-Fiszman & C. Spence, 'Sensory expectations based on product-extrinsic food cues: An interdisciplinary review of the empirical evidence and theoretical accounts', *Food Quality & Preference*, 40 (2015), 165–79.

14 B. J. Koza et al., 'Color enhances orthonasal olfactory intensity and reduces retronasal olfactory intensity', *Chemical Senses*, 30 (2005), 643–9; H. N. J. Schifferstein, 'Effects of product beliefs on product perception and liking', in L. Frewer, E. Risvik & H. Schifferstein (eds.), *Food, People and Society: A European Perspective of Consumers' Food Choices* (Berlin: Springer Verlag, 2001), pp. 73–96.

15 Early recipes for savoury ices can be found in A. B. Marshall, *Mrs A. B. Marshall's Cookery Book* (London: Robert Hayes, 1888).

16 M. Yeomans et al., 'The role of expectancy in sensory and hedonic evaluation: The case of smoked salmon ice cream', *Food Quality and Preference*, 19 (2008), 565–73.

17 Of course, when it comes to the art and science of naming, some plump for brevity to the point of pretentiousness, while others opt for long and sensorially descriptive labels instead. Fashions in naming come and go, just like they do for anything else; see C. C. Carbon, 'The cycle of preference: Long-term dynamics of aesthetic appreciation', *Acta Psychologica*, 134 (2010), 233–44; C. Spence & B. Piqueras-Fiszman, *The Perfect Meal: The Multisensory Science of Food and Dining* (Oxford: Wiley-Blackwell, 2014), ch. 3; P. R. Klosse et al., 'The formulation and evaluation of culinary success factors (CSFs) that determine the palatability of food', *Food Service Technology*, 4 (2004), 107–15.

18 Just think about what you infer from the glossiness or otherwise of a fish's eyes; T. Murakoshi et al., 'Glossiness and perishable food quality: Visual freshness judgment of fish eyes based on luminance distribution', *PLoS ONE*, 8(3) (2013), e58994.

19 R. H. Thaler & C. R. Sunstein, *Nudge: Improving Decisions about Health, Wealth and Happiness* (London: Penguin, 2008). See also T. M. Marteau, G. J. Hollands & P. C. Fletcher, 'Changing human behaviour to prevent disease: The importance of targeting automatic processes', *Science*, 337 (2012), 1492–5.

20 This is precisely what happened when the question was raised on an episode of BBC Radio 4's *The Kitchen Cabinet* when I was a panellist early in 2016. See G. Secci et al., 'Stress during slaughter increases lipid metabolites and decreases oxidative stability of farmed rainbow trout (*Oncorhynchus mykiss*) during frozen storage', *Food Chemistry*, 190 (2016), 5–11.

21 R. Smithers, 'A colin and chips? Sainsbury's gives unfashionable pollack a makeover', *Guardian*, 6 April 2009 (https://www.guardian. co.uk/business/2009/apr/06/sainsburys-pollack-colin-fish-stocks); 'To prevent sniggering, pudding is renamed Spotted Richard', *Daily Telegraph*, 9 September 2009, p. 14.

22 See H. Blumenthal, *The Big Fat Duck Cookbook*. (London: Bloomsbury, 2008). Though if you go back to the 1950s, one finds that Fanny Cradock, the famous British TV chef, was already serving brightly coloured (blue) mashed potatoes; see C. Ellis, *Fabulous Fanny Cradock* (Stroud: Sutton Publishing Limited, 2007). I have even heard it said that as a treat at the end of the Second World War, children were served mashed potatoes coloured red, white and blue.

23 At least, that is what the marketing research suggests; see C. Irmak, B. Vallen & S. R. Robinson, 'The impact of product name on dieters' and non-dieters' food evaluations and consumption', *Journal of Consumer Research*, 38 (2011), 390–405; B. Wansink, K. van Ittersum & J. E. Painter, 'How descriptive food names bias sensory perceptions in restaurants', *Food Quality and Preference*, 16 (2005), 393–400. There are many other naming effects; see C. Spence & B. Piqueras-Fiszman, *The Perfect Meal: The Multisensory Science of Food and Dining* (Oxford: Wiley-Blackwell, 2014), ch. 3, on the art and science of naming. The Futurists were playing much the same game with some of their strikingly named dishes back in the 1930s, like 'Divorced Eggs' and 'Italian Breasts in the Sunshine' (see 'Back to the Futurists').

24 'Supermarkets attacked over phoney farms', *Daily Mail*, 25 March 2016, p. 10 (http://www.dailymail.co.uk/news/article-3508843/Supermarkets-attacked-phoney-farms-Tesco-created-seven-fictitious-names-including-Rosedene-Nightingale-replace-Everyday-Value-discount-brand-s-not-alone.html).

25 O. Thring, 'Head chefs: What do you get when you take two ex-Noma chefs, add some mismatched cutlery and a bizarre array of animal parts? Bror, Copenhagen's most exciting restaurant', Norwegian Airline magazine, 2014 (https://www.norwegian.com/magazine/features/2014/01/head-chefs).

26 Y. Liu, & S. Jang, 'The effects of dining atmospherics: An extended Mehrabian-Russell model', *International Journal of Hospitality Management*, 28 (2009), 494–503.

27 With some chefs taking the latter to ridiculous extremes, as when you are told that there are no free tables at Damon Baehrel's legendary restaurant Basement Bistro in New York State for the next decade. However, as an exposé suggests, all might not be quite what it seems: N. Paumgarten, 'The most exclusive restaurant in America', *The New Yorker* (*Annals of Dining*), 29 August 2016 (https://www.newyorker.com/magazine/2016/08/29/damon-baehrel-the-most-exclusive-restaurant-in-america).

28 A. L. Aduriz, *Mugaritz: A Natural Science of Cooking* (New York: Phaidon, 2014), p. 25.

29 J. A. Brillat-Savarin, *Physiologie du goût* [*The Philosopher in the Kitchen/The Physiology of Taste*] (Brussels: J. P. Meline, 1835); published as *A Handbook of Gastronomy*, trans. A. Lazaure (London: Nimmo & Bain, 1884).

30 That said, when we taste something that is completely unlike what we were expecting to taste, the discrepancy between expectation and experience tends to lead us to give more extreme sensory-discriminative and hedonic ratings than we would had we not been holding any particular expectations in the first place.

31 Cf. G. B. Northcraft & A. M. Neale, 'Experts, amateurs, and real estate: An anchoring-and-adjustment perspective on property pricing decisions', *Organizational Behavior and Human Decision Processes*, 39(1)

(1987), 84–97. Hence my advice to any of those restaurants that are crazy enough to allow their customers to pay what they want for the food they order: make sure you give people a high anchor price to begin with! (See D. Peters, 'Restaurant faces closure after customers take advantage of their "pay what you want" policy and fork out less than $3 a meal', *Daily Mail Online*, 23 August 2016 (http://www.dailymail.co.uk/news/article-3754538/Sydney-s-Lentil-vegan-restaurant-faces-closure-customers-advantage-pay-want-philosophy.html#ixzz4JPqyjOKn).)

32 D. Martin, 'The impact of branding and marketing on perception of sensory qualities', *Food Science & Technology Today: Proceedings*, 4(1) (1990), 44–9; S. M. McClure et al., 'Neural correlates of behavioral preference for culturally familiar drinks', *Neuron*, 44 (2004), 379–87; C. Spence, 'The price of everything – the value of nothing?', *The World of Fine Wine*, 30 (2010), 114–20.

33 S. M. McClure et al., 'Neural correlates of behavioral preference for culturally familiar drinks', *Neuron*, 44 (2004), 379–87. Intriguingly, subsequent research has shown that the effects of branding appear to be stronger in infrequent consumers, possibly hinting at a greater reliance on branding information in less experienced consumers; see S. Kühn & J. Gallinat, 'Does taste matter? How anticipation of cola brands influences gustatory processing in the brain', *PLoS ONE*, 8:4 (2013), e61569.

34 R. L. Allison & K. P. Uhl, 'Influence of beer brand identification on taste perception', *Journal of Marketing Research*, 1 (1964), 369; T. Davis (1987), 'Taste tests: Are the blind leading the blind?', *Beverage World*, 3 (April 1987), 42–4, 50, 85; D. Martin, 'The impact of branding and marketing on perception of sensory qualities', *Food Science & Technology Today: Proceedings*, 4(1) (1990), 44–9.

35 H. Plassmann et al., 'Marketing actions can modulate neural representations of experienced pleasantness', *Proceedings of the National Academy of Sciences of the USA*, 105 (2008), 1050–54; H. Plassmann & B. Weber, 'Individual differences in marketing placebo effects: Evidence from brain imaging and behavioural experiments', *Journal of Marketing Research*, 52 (2015), 493–510. Cf. R. Bagchi & L. Block, 'Chocolate cake please! Why do consumers indulge more when it feels more expensive?', *Journal of Public Policy & Marketing*, 30 (2011), 294–306.

36 Some critics have been tempted to query whether you would get the same results outside California! And the students may have been fooled, but surely the experts wouldn't be, would they? Well, while the neuroimaging research suggests that the pattern of brain activation seen in the wine expert is somewhat different from that seen in the social drinker, the wine experts certainly can be fooled, at least some of the time, into believing that what they are tasting is something very different from what it actually is; A. Castriota-Scanderbeg et al., 'The appreciation of wine by sommeliers: A functional magnetic resonance study of sensory integration', *NeuroImage*, 25 (2005), 570–78; L. Pazart et al., 'An fMRI study on the influence of sommeliers' expertise on the integration of flavor', *Frontiers in Behavioral Neuroscience*, 8:358 (2014).

37 Researchers have also demonstrated that expecting a very sweet drink while tasting a drink that is not as sweet as expected leads to increased reports of sweetness; what is more, it enhances activity in the primary taste cortex, relative to the same drink when no such expectation was present; J. B. Nitschke et al., 'Altering expectancy dampens neural response to aversive taste in primary taste cortex', *Nature Neuroscience*, 9 (2006), 435–42; A. T. Woods et al., 'Expected taste intensity affects response to sweet drinks in primary taste cortex', *Neuroreport*, 22 (2011), 365–9.

38 In another study, a savoury umami solution was described to people lying in a brain scanner. The descriptors included terms such as 'rich and delicious taste' versus 'monosodium glutamate'. Meanwhile, a solution to which a vegetable aroma had been added was described as 'rich and delicious flavour', 'boiled vegetable water' or 'monosodium glutamate'. Once again, the neural response to exactly the same solution changed as a function of the label that was given. Such results led Grabenhorst et al. to conclude that: 'top-down language-level cognitive effects reach far down into the earliest cortical areas that represent the appetite value of taste and flavour' ('How cognition modulates affective responses to taste and flavor: Top-down influences on the orbitofrontal and pregenul cortices', *Cerebral Cortex*, 18 (2008), 1549–59; p. 1549).

Elsewhere, researchers have demonstrated that labelling food as organic can give rise to increased activity in the part of the brain that is involved in controlling our motivation to acquire and eat food; N. S.

Linder et al., 'Organic labeling influences food valuation and choice', *NeuroImage*, 53 (2010), 215–20.

39　This, at least, would seem to be what is happening, judging from the research on the effects of beliefs regarding organic food, low-fat food, branding, etc., on the response of consumers. See, for example, E. C. Anderson & L. F. Barrett, 'Affective beliefs influence the experience of eating meat', *PLoS ONE*, 11(8) (2016), e0160424; W. J. Lee et al., 'You taste what you see: Do organic labels bias taste perceptions?', *Food Quality and Preference*, 29 (2013), 33–9; P. Sörqvist et al., 'Who needs cream and sugar when there is eco-labeling? Taste and willingness to pay for "eco-friendly" coffee', *PLoS ONE*, 8(12) (2013), 1–9.

　　　See B. Piqueras-Fiszman & C. Spence, 'Sensory expectations based on product-extrinsic food cues: An interdisciplinary review of the empirical evidence and theoretical accounts', *Food Quality & Preference*, 40 (2015), 165–79, for a review.

40　There is only so much spume, foam and espherification that the average person can stomach. And then, of course, there is always the potential backlash from those diners who may be concerned about all the artificial stuff they feel like they are ingesting; M. Campbell, 'World's top chef "poisons" diners with additives', *The Sunday Times*, 11 September 2009, p. 29; F. Govan, 'Famed El Bulli chef Ferran Adria accused of "poisoning" his diners', *Daily Telegraph*, 1 July 2008 (https://www.telegraph.co.uk/news/worldnews/europe/spain/1955806/Famed-El-Bulli-chef-Ferran-Adria-accused-of-poisoning-his-diners.html). And see J. Rayner, *The Man who Ate the World: In Search of the Perfect Dinner* (London: Headline, 2008), if you want to know what it feels like to eat too many fine-dining meals back to back.

41　G. B. Dijksterhuis, E. Le Berre & A. T. Woods, 'Food products with improved taste', Patent Identifier No. EP2451289 A1 (2010); G. Dijksterhuis, C. Boucon & E. Le Berre, 'Increasing saltiness perception through perceptual constancy created by expectation', *Food Quality & Preference*, 34 (2014), 24–8; C. Spence & B. Piqueras-Fiszman, *The Perfect Meal: The Multisensory Science of Food and Dining* (Oxford: Wiley-Blackwell, 2014); A. T. Woods et al., 'Flavor expectation: The effects of assuming homogeneity on drink perception', *Chemosensory Perception*, 3 (2010), 174–81.

One of the remarkable discoveries is that perception of mintiness in the nose is driven by sweetness in the mouth. Knowing that, the chewing gum manufacturers incorporate different sugars that will dissolve in the mouth at different rates. By so doing, they can prolong the minty sensation; J. M. Davidson et al., 'Effect of sucrose on the perceived flavor intensity of chewing gum', *Journal of Agriculture & Food Chemistry*, 47 (1999), 4336–40; R. Yatka et al., 'Chewing gum having an extended sweetness', US Patent 4,986,991 (1991).

42 J. Gerard, *The Herball or General Historie of Plants* (1597; Amsterdam: Theatrum Orbis Terrarum, 1974). See also H. Leach, 'Rehabilitating the "stinking herbe": A case study of culinary prejudice', *Gastronomica: The Journal of Food Culture*, 1(2) (2001), 10–15.

43 L. Mauer & A. El-Sohemy, 'Prevalence of cilantro (*Coriandrum sativum*) disliking among different ethnocultural groups', *Flavour*, 1:8 (2012); H. McGee, 'Cilantro haters, it's not your fault', *The New York Times*, 13 April 2010 (https://www.nytimes.com/2010/04/14/dining/14curious.html?_r=0).

44 That said, researchers have recently tested *c.* 25,000 people in order to determine the genetic variant that would appear to be responsible for this difference; N. Eriksson et al., 'A genetic variant near olfactory receptor genes influences cilantro preference', *Flavour*, 1:22 (2012). The soapy or pungent aroma is attributed to the presence of several aldehydes. As they note in the paper: 'It is suspected, although not proven, that cilantro dislike is largely driven by the odor rather than the taste' (p. 1).

45 J. E. Amoore & R. G. Buttery, 'Partition coefficients and comparative olfactometry', *Chemical Senses Flavour*, 3 (1978), 57–71; A. Keller et al., 'Genetic variation in a human odorant receptor alters odour perception', *Nature*, 449 (2007), 468–72; C. J. Wysocki & G. K. Beauchamp, 'Ability to smell androstenone is genetically determined', *Proceedings of the National Academy of Sciences USA*, 81 (1984), 4899–902; K. Lunde et al., 'Genetic variation of an odorant receptor OR7D4 and sensory perception of cooked meat containing androstenone', *PLoS ONE*, 7(5) (2012), e35259; P. Pelosi, *On the Scent: A Journey Through the Science of Smell* (Oxford: Oxford University Press, 2016); D. R. Reed & A. Knaapila, 'Genetics of taste and smell: Poisons and pleasures', *Progress in Molecular Biology and Translational Science*, 94 (2010), 213–40.

46 Coriander and androstenone are striking, though, because of the dras-
    tically different hedonic responses that they evoke in people. For most
    other chemicals, individual differences, though they may be pro-
    nounced, are seen more in terms of differing sensitivity to the chemical
    (evidencing itself as a change in the rated intensity) rather than in terms
    of the hedonic response. See D. R. Reed & A. Knaapila, 'Genetics of
    taste and smell: Poisons and pleasures', *Progress in Molecular Biology and
    Translational Science*, 94 (2010), 213–40, for further examples.

47 S. R. Jaeger et al., 'A Mendelian trait for olfactory sensitivity affects
    odor experience and food selection', *Current Biology*, 23 (2013), 1601–5; J.
    F. McRae et al., 'Identification of regions associated with variation in
    sensitivity to food-related odors in the human genome', *Current Biology*,
    23 (2013), 1596–1600; J. F. McRae et al., 'Genetic variation in the odor-
    ant receptor *OR2J3* is associated with the ability to detect the 'grassy'
    smelling odor, *cis*-3-hexan-1-ol', *Chemical Senses*, 37 (2012), 585–93.

    See also A. Abbott, 'Scientists have travelled the ancient Silk Road
    to understand how genes shape people's love for foods', *Nature*, 488
    (2012), 269–71; N. Pirastu et al., 'Genetics of food preferences: A first
    view from silk road populations', *Journal of Food Science*, 77(12) (2012),
    S413–18.

48 O. Styles, 'Parker and Robinson in war of words', *Decanter*, 14 April
    2004 (http://www.decanter.com/wine-news/parker-and-robinson-in-war-
    of-words-102172/). See also B. C. Smith, 'Same compounds: Different
    flavours?' in D. Chassagne (ed.), *Proceedings of Wine Active Compounds
    2008* (University of Bourgogne, Oenopluria Media, 2008), pp. 98–102.

    Here one is reminded of Cervantes' Don Quixote: 'It is with good
    reason, says Sancho to the squire with the great nose, that I pretend to
    have judgement in wine: this is a quality hereditary in our family. Two
    of my kinsmen were once called to give their opinion of a hogshead,
    which was supposed to be excellent, being old and of a good vintage.
    One of them tastes it, considers it; and after mature reflection pro-
    nounces the wine to be good, were it not for a small taste of leather
    which he perceived in it. The other, after using the same precautions,
    gives also his verdict in favour of the wine; but with the reserve of a
    taste of iron, which he could easily distinguish. You cannot imagine
    how much they were both ridiculed for their judgement. But who

laughed in the end? On emptying the hogshead, there was found at the bottom an old key with a leathern thong tied to it.' Cited in David Hume, 'Of the standard of taste', in *Of the Standard of Taste and Other Essays* (1757; Indianapolis: Bobbs-Merrill, 1965).

49 H.-R. Buser, C. Zanier & H. Tanner, 'Identification of 2,4,6-trichloro-anisole as a potent compound causing cork taint in wine', *Journal of Agricultural Food Chemistry*, 30 (1982), 359–62; V. Mazzoleni & L. Maggi, 'Effect of wine style on the perception of 2,4,6-trichloroanisole, a compound related to cork taint in wine', *Food Research International*, 40 (2007), 694–9; J. Prescott et al, 'Estimating a "consumer rejection threshold" for cork taint in white wine', *Food Quality and Preference*, 16 (2005), 345–9.

50 It is worth noting that these differences can make it difficult to obtain meaningful between-group ratings of taste. 'People use intensity descriptors to compare sensory differences: "This tastes strong to me; is it strong to you?" These comparisons are deceptive because they assume that intensity descriptors like strong denote the same absolute perceived intensities to everyone. This assumption is false' (L. M. Bartoshuk, K. Fast & D. J. Snyder, 'Difference in our sensory worlds: Invalid comparisons with labeled scales', *Current Directions in Psychological Science*, 14 (2005), 122–25; p. 122).

51 I. J. Miller & D. P. Reedy, 'Variations in human taste bud density and taste intensity perception', *Physiology and Behavior*, 47 (1990), 1213–19; see also J. A. Brillat-Savarin, *Physiologie du goût* [*The Philosopher in the Kitchen/The Physiology of Taste*] (Brussels: J. P. Meline, 1835); published as *A Handbook of Gastronomy*, trans. A. Lazaure (London: Nimmo & Bain, 1884); S. Eldeghaidy et al., 'The cortical response to the oral perception of fat emulsions and the effect of taster status', *Journal of Neurophysiology*, 105 (2011), 2572–81.

52 A. F. Blakeslee & A. L. Fox, 'Our different taste worlds: P. T. C. as a demonstration of genetic differences in taste', *Journal of Heredity*, 23 (1932), 97–107; A. F. Blakeslee & M. R. Salmon, 'Odor and taste blindness', *Eugenics News*, 16 (1931), 105–10; L. M. Bartoshuk, 'Separate worlds of taste', *Psychology Today*, 14 (1980), 48–9, 51, 54–6, 63; L. M. Bartoshuk, 'Comparing sensory experiences across individuals: Recent psychophysical advances illuminate genetic variation in taste perception', *Chemical Senses*, 25 (2000), 447–60.

53 D. M. Blank & R. D. Mattes, 'Sugar and spice: Similarities and sensory attributes', *Nursing Research*, 39 (1990), 290–93; J. Prescott, *Taste Matters: Why We Like the Foods We Do* (London: Reaktion Books, 2012).

54 Here, though, it is also worth noting that the bitter taste receptor T2R38 has been shown to regulate the mucosal innate defence of the human upper airway as well. This observation has led some to suggest that 'T2R38 is an upper airway sentinel in innate defense and that genetic variation contributes to individual differences in susceptibility to respiratory infection'; R. J. Lee et al., 'T2R38 taste receptor polymorphisms underlie susceptibility to upper respiratory infection', *Journal of Clinical Investigation*, 122 (2012), 4145–59; A. S. Shah et al., 'Motile cilia of human airway epithelia are chemosensory', *Science*, 325 (2009), 1131–4; B. Trivedi, 'Hardwired for taste: Research into human taste receptors extends beyond the tongue to some unexpected places', *Nature*, 486 (2012), S7.

55 C. Sagioglou & T. Greitemeyer, 'Individual differences in bitter taste preferences are associated with antisocial personality traits', *Appetite*, 96 (2016), 299–308; A. Sims, 'How you drink your coffee "could point to psychopathic tendencies"', *Independent*, 10 October 2015 (https://www.independent.co.uk/news/science/psychopathic-people-are-more-likely-to-prefer-bitter-foods-according-to-new-study-a6688971.html).

56 C. Sagioglou & T. Greitemeyer, 'Bitter taste causes hostility', *Personality and Social Psychology Bulletin*, 40 (2014), 1589–97 (study 3); E. Greimel et al., 'Facial and affective reactions to tastes and their modulation by sadness and joy', *Physiology & Behavior*, 89 (2006), 261–9; M. Zaraska, 'Sweet emotion: Accounting for different tastes', *i* (newspaper), 30 March 2015, pp. 26–7; R. S. Herz, 'PROP taste sensitivity is related to visceral but not moral disgust', *Chemosensory Perception*, 4 (2011), 72–9; F. Sheng, J. Xu & B. Shiv, 'Loss of sweet taste: The gustatory consequence of handling money', *EMAC2016*, Oslo, 24–7 May 2016, 53.

57 See C. Spence, 'Gastrodiplomacy: Assessing the role of food in decision-making', *Flavour*, 5:4 (2016).

58 C. Spence, 'The supertaster who researches supertasters', *BPS Research Digest*, 10 October 2013 (http://www.bps-research-digest.blogspot.co.uk/2013/10/day-4-of-digest-super-week-supertaster.html).

59 Perhaps the most appropriate way to think about taste is as a nutrient-poison detection system (the other role of the taste system being to

prepare the body to metabolize food and drink once it has been ingested); P. A. S. Breslin, 'An evolutionary perspective on food and human taste', *Current Biology*, 23 (2013), 409–18; R. S. J. Keast & A. Costanzo, 'Is fat the sixth taste primary? Evidence and implications', *Flavour*, 4:5 (2015); S. C. Woods, 'The eating paradox: How we tolerate food', *Psychological Review*, 98 (1991), 488–505.

It is interesting to note here that the two basic tastes that people confuse most frequently are sour and bitter, evolutionarily speaking the signals that a food should probably be rejected; A. F. Blakeslee & A. L. Fox, 'Our different taste worlds: P. T. C. as a demonstration of genetic differences in taste', *Journal of Heredity*, 23 (1932), 97–107; M. O'Mahony et al., 'Confusion in the use of the taste adjectives "sour" and "bitter"', *Chemical Senses and Flavour*, 4 (1979), 301–18.

And see D. Owen, 'Beyond taste buds: The science of delicious', *National Geographic Magazine*, 13 November 2015 (http://ngm.national-geographic.com/2015/12/food-science-of-taste-text), on the transition of taste from a means of finding food and avoiding poison to 'a route to extravagant adventure'.

60 See A. Gallace et al., 'Multisensory presence in virtual reality: Possibilities & limitations', in G. Ghinea, F. Andres & S. Gulliver (eds.), *Multiple Sensorial Media Advances and Applications: New Developments in MulSeMedia* (Hershey, PA: IGI Global, 2012), pp. 1–40.

61 There are, of course, exceptions to all of these generalizations. Just think of rhubarb: pinky-purply-green colour and yet intensely sour. Tomatoes and raspberries are a dark red, and yet taste more of umami and sourness respectively. However, on average, or so the scientists would have us believe, sweetness and redness are correlated in fruits (e.g., J. A. Maga, 'Influence of color on taste thresholds', *Chemical Senses and Flavor*, 1 (1974), 115–19; R. M. Pangborn, 'Influence of color on the discrimination of sweetness', *American Journal of Psychology*, 73 (1960), 229–38; L. Watson, *The Omnivorous Ape* (New York: Coward, McCann, & Geoghegan, 1971)). The luminance distribution is an important cue for judging the freshness of certain fruit and vegetables; see S. Péneau et al., 'A comprehensive approach to evaluate the freshness of strawberries and carrots', *Postharvest Biology Technology*, 45 (2007), 20–29. Perhaps, though, it is more a matter of greenness being associated with

unripeness and sourness; see F. Foroni, G. Pergola & R. I. Rumiati, 'Food color is in the eye of the beholder: The role of human tri-chromatic vision in food evaluation', *Scientific Reports*, 6:37034 (2016).

62 J. Chen, 'Tasting a flavor that doesn't exist', *The Atlantic*, 21 October 2015 (https://www.theatlantic.com/health/archive/2015/10/tasting-a-flavor-that-doesnt-exist/411454/).

63 Research shows that turning vegetables into superheroes can help; for instance, B. Wansink et al. reported that 8–11-year-old children ate twice as many 'X-ray Vision Carrots' as ordinary anonymous 'carrots'. This effect is apparently at least moderately long-lasting, given that the same results were observed in another month-long study ('Attractive names sustain increased vegetable intake in schools', *Preventive Medicine*, 55 (2012), 330–32).

64 R. Henry, 'Tasty – a tongue insured for £3m', *The Sunday Times*, 18 December 2011, p. 31; M. Costello, 'Costa insures the tongue that can tell sweet beans from sour for £10m', *The Times*, 9 March 2009, p. 41; 'Chocolate expert's £1m palate: Cadbury's worker has tastebuds insured – and must now avoid vindaloos and chilli peppers', *Daily Mail Online*, 6 September 2016 (http://www.dailymail.co.uk/news/article-3775150/Chocolate-expert-s-1m-palate-Cadbury-s-worker-tastebuds-insured-avoid-vindaloos-chilli-peppers.html).

2. Smell

1 Brits who are old enough to remember should just think of the Bisto Kids, engaging in a spot of orthonasal sniffing. See P. Rozin, ' "Taste-smell confusions" and the duality of the olfactory sense', *Perception & Psychophysics*, 31 (1982), 397–401; B. Piqueras-Fiszman & C. Spence, 'Sensory expectations based on product-extrinsic food cues: An inter-disciplinary review of the empirical evidence and theoretical accounts', *Food Quality & Preference*, 40 (2015), 165–79.

2 C. Spence, 'Just how much of what we taste derives from the sense of smell?', *Flavour*, 4:30 (2015); B. Stuckey, *Taste What You're Missing: The Passionate Eater's Guide to Why Good Food Tastes Good* (London: Free

Press, 2012); C. Spence, 'Multisensory flavour perception', *Cell*, 161 (2015), 24–35.

3 C. Spence, 'Oral referral: Mislocalizing odours to the mouth', *Food Quality & Preference*, 50 (2016), 117–28.

4 D. Owen, 'Beyond taste buds: the science of delicious', *National Geographic Magazine*, 13 November 2015 (http://ngm.nationalgeographic.com/2015/12/food-science-of-taste-text).

5 R. J. Stevenson & R. A. Boakes, 'Sweet and sour smells: Learned synaesthesia between the senses of taste and smell', in G. A. Calvert, C. Spence & B. E. Stein (eds.), *The Handbook of Multisensory Processing* (Cambridge, MA: MIT Press, 2004), pp. 69–83; R. J. Stevenson, R. A. Boakes & J. Prescott, 'Changes in odor sweetness resulting from implicit learning of a simultaneous odor-sweetness association: An example of learned synaesthesia', *Learning and Motivation*, 29 (1998), 113–32; R. J. Stevenson & C. Tomiczek, 'Olfactory-induced synesthesias: A review and model', *Psychological Bulletin*, 133 (2007), 294–309.

   The synaesthesia analogy seems misguided to me since the association between aroma and taste is ubiquitous rather than idiosyncratic, as synaesthesia is typically defined. Furthermore, what people typically experience is just a single unitary sweet flavour (not a separate inducer and a concurrent as in the case of synaesthesia proper). For these, and other reasons, I would argue that we should not confuse smelled sweetness with synaesthesia proper; both phenomena certainly exist, it's just that they are qualitatively different; M. Auvray & C. Spence, 'The multisensory perception of flavor', *Consciousness and Cognition*, 17 (2008), 1016–31.

6 N. Sakai et al., 'Enhancement of sweetness ratings of aspartame by a vanilla odor presented either by orthonasal or retronasal routes', *Perceptual and Motor Skills*, 92 (2001), 1002–8.

7 D. H. McBurney, V. B. Collings & L. Glanz, 'Temperature dependence of human taste responses', *Physiology & Behavior*, 11 (1973), 89–94; K. Talavera et al., 'Heat activation of TRPM5 underlies thermal sensitivity of sweet taste', *Nature*, 438 (2005), 1022–5; though see also H. R. Moskowitz, 'Effect of solution temperature on taste intensity in humans', *Physiology & Behavior*, 10 (1973), 289–92; A. Fleming, 'Hot or not? How serving temperature affects the way food tastes', *Guardian*

(*Word of Mouth* blog), 17 September 2013 (https://www.theguardian.com/lifeandstyle/wordofmouth/2013/sep/17/serving-temperature-affects-taste-food).

8 Note that chefs are by no means immune to such crossmodal influences; see, e.g., R. A. Boakes & H. Hemberger, 'Odour-modulation of taste ratings by chefs', *Food Quality and Preference*, 25 (2012), 81–6. That said, the gustatory (taste) effects of smell appears to be more response bias than anything else (meaning that it may change people's ratings under certain conditions but does not appear to actually affect their perceptual experience); see T. Linscott & J. Lim, 'Retronasal odor enhancement by salty and umami taste', *Food Quality & Preference*, 48 (2016), 1–10; though see also L. Levy et al., 'Taste memory induces brain activation as revealed by functional MRI', *Journal of Computer Assisted Tomography*, 23 (1999), 499–505. Interestingly, food companies have been pursuing the idea that they could deliver flavours that don't actually exist to consumers; see J. Chen, 'Tasting a flavor that doesn't exist', *The Atlantic*, 21 October 2015 (https://www.theatlantic.com/health/archive/2015/10/tasting-a-flavor-that-doesnt-exist/411454/); E. Le Berre et al., 'Reducing bitter taste through perceptual constancy created by an expectation', *Food Quality & Preference*, 28 (2013), 370–74.

9 P. Dalton et al., 'The merging of the senses: Integration of subthreshold taste and smell', *Nature Neuroscience*, 3 (2000), 431–2; see also P. A. Breslin, N. Doolittle & P. Dalton, 'Subthreshold integration of taste and smell: The role of experience in flavour integration', *Chemical Senses*, 26 (2001), 1035; C. Spence, 'Multisensory flavour perception', *Cell*, 161 (2015), 24–35.

10 Congruency is typically taken to mean that the aroma and taste should have co-occurred previously in the food and drink that the taster has been exposed to; see H. N. J. Schifferstein & P. W. J. Verlegh, 'The role of congruency and pleasantness in odor-induced taste enhancement', *Acta Psychologica*, 94 (1996), 87–105. It is an interesting question to consider whether there might also be a more fundamental sense of congruency in terms of some kind of innate perceptual similarity between the senses – which might hold even in the absence of any prior co-exposure to the component taste and smell stimuli; see T. Linscott & J. Lim, 'Retronasal odor enhancement by salty and umami taste',

Food Quality & Preference, 48 (2016), 1–10; L. E. Marks, 'Similarities and differences among the senses', *International Journal of Neuroscience*, 19 (1983), 1–12.

11 Look in the food science journals and you'll be amazed how many papers are being published by the large food companies on this theme; J. Delwiche & A. L. Heffelfinger, 'Cross-modal additivity of taste and smell', *Journal of Sensory Studies*, 20 (2005), 137–46; J. C. Pfeiffer et al., 'Temporal synchrony and integration of sub-threshold taste and smell signals', *Chemical Senses*, 30 (2005), 539–45; see also J. Chen, 'Tasting a flavor that doesn't exist', *The Atlantic*, 21 October 2015 (https://www.theatlantic.com/health/archive/2015/10/tasting-a-flavor-that-doesnt-exist/411454/).

12 R. J. Stevenson, J. Prescott & R. A. Boakes, 'The acquisition of taste properties by odors', *Learning and Motivation*, 26 (1995), 433–55. In fact, we start acquiring our taste preferences while we are in the womb; thus, a mother's diet during pregnancy really can influence the flavour preferences of their offspring after birth; see N. Bakalar, 'Partners in flavour: Our perception of food draws on a combination of taste, smell, feel, sight and sound', *Nature*, 486 (2012), S4–S5; A. Bremner, D. Lewkowicz & C. Spence (eds.), *Multisensory Development* (Oxford: Oxford University Press, 2012); J. Prescott, *Taste Matters: Why We Like the Foods We Do* (London: Reaktion Books, 2012).

Intriguingly, the latest research suggests that such effects of taste on aroma are seen only for nutritive tastes (e.g., sweet, umami and salt) and not for non-nutritive tastants (e.g., bitter and sour); see T. Linscott & J. Lim, 'Retronasal odor enhancement by salty and umami taste', *Food Quality & Preference*, 48 (2016), 1–10.

13 P. Rozin, '"Taste-smell confusions" and the duality of the olfactory sense', *Perception & Psychophysics*, 31 (1982), 397–401.

14 What could be going on in these rare cases where the impression differs so drastically between the orthonasal and retronasal routes? It has been suggested that the saliva in the mouth may be stripping off some of the volatile airborne molecules on their way to the back of the nose when we exhale after swallowing. According to an article that appeared in the *Financial Times*: 'saliva strips off about 300 of the 631 airborne chemicals that combine to form coffee's complex aroma, so you receive

only half of it retronasally' (L. Ge, 'Why coffee can be bittersweet', *FT Weekend Magazine*, 13 October 2012, p. 50). That said, I am not sure whether there is any data to back up such a claim, despite it being quoted in several places; A. Fleming, 'What makes eating so satisfying?', *Guardian*, 25 April 2013 (https://www.guardian.co.uk/lifeandstyle/ wordofmouth/2013/apr/23/what-makes-eating-so-satisfying). Individual variability in terms of salivation, then, may be one of the factors helping to explain why foods appear to taste different to different people; C. Spence, 'Mouth-watering: The influence of environmental and cognitive factors on salivation and gustatory/flavour perception', *Journal of Texture Studies*, 42 (2011), 157–71.

15 M. M. Mozell et al., 'Nasal chemoreception in flavor identification', *Archives of Otolaryngology*, 90 (1969), 367–73; J. Pierce & B. P. Halpern, 'Orthonasal and retronasal odorant identification based upon vapor phase input from common substances', *Chemical Senses*, 21 (1996), 529–43; B. C. Sun & B. P. Halpern, 'Identification of air phase retronasal and orthonasal odorant pairs', *Chemical Senses*, 30 (2005), 693–706; see also D. Small et al., 'Differential neural responses evoked by orthonasal versus retronasal odorant perception in humans', *Neuron*, 47 (2005), 593–605.

16 D. M. Zemke & S. Shoemaker, 'A sociable atmosphere: Ambient scent's effect on social interaction', *Cornell Hospitality Quarterly*, 49 (2008), 317–29; R. De Wijk & S. M. Zijlstra, 'Differential effects of exposure to ambient vanilla and citrus aromas on mood, arousal and food choice', *Flavour*, 1:24 (2012); See also 'Grant Achatz: The chef who couldn't taste', *NPR*, 29 August 2011 (https://www.npr.org/2011/08/29/ 139786504/grant-achatz-the-chef-who-couldnt-taste); C. Spence & J. Youssef, 'Olfactory dining: Designing for the dominant sense', *Flavour*, 4:32 (2015).

17 See C. Spence, 'Oral referral: Mislocalizing odours to the mouth', *Food Quality & Preference*, 50 (2016), 117–28.

18 Conigliaro's innovative design for 'The Rose' cocktail is, in some sense at least, analogous to Heston Blumenthal's use of sound to encourage the diners at The Fat Duck to reminisce about a pleasant childhood holiday with his 'Sound of the Sea' dish. See also L. Hyslop, 'Anyone for horseradish vodka? Meet cocktail wizard Tony

Conigliaro', *Daily Telegraph*, 16 September 2014 (http://www.telegraph. co.uk/foodanddrink/11099017/Anyone-for-horseradish-vodka-Meet-cocktail-wizard-Tony-Conigliaro.html).

19 S. Chu & J. J. Downes, 'Odour-evoked autobiographical memories: Psychological investigations of Proustian phenomena', *Chemical Senses*, 25 (2000), 111–16; S. Chu & J. J. Downes, 'Proust nose best: Odours are better cues of autobiographical memory', *Memory & Cognition*, 30 (2002), 511–18; R. S. Herz & J. W. Schooler, 'A naturalistic study of autobiographical memories evoked by olfactory and visual cues: Testing the Proustian hypothesis', *American Journal of Psychology*, 115 (2002), 21–32; S. T. Glass, E. Lingg & E. Heuberger, 'Do ambient urban odors evoke basic emotions?', *Frontiers in Psychology*, 5:340 (2014).

20 A. Gilbert, *What the Nose Knows: The Science of Scent in Everyday Life* (New York: Crown, 2008).

21 Coffee and chocolate are the two aromas that came out top in one study conducted in Germany and Japan; see S. Ayabe-Kanamura et al., 'Differences in perception of everyday odors: A Japanese–German cross-cultural study', *Chemical Senses*, 23 (1998), 31–8.

22 Or to put it better, when consumers were time and again shown to be unable to pick their favourite brand in a blind taste test; see e.g. R. I. Allison & K. P. Uhl, 'Influence of beer brand identification on taste perception', *Journal of Marketing Research*, 1 (1964), 36–9.

23 G. Mangan, 'Flipping your lid: New coffee cup aims at your nose', *CNBC*, 26 April 2014 (https://www.cnbc.com/2014/04/25/coffee-lid-gets-a-redesign-to-let-the-aroma-in.html); 'Craft beer brewers kick bottles, revamp cans to improve taste: Sly Fox Brewing Co. boasts "topless" cans', Associated Press, 2 July 2014 (http://www.nydailynews. com/life-style/eats/craft-beer-brewers-kick-bottles-revamp-cans-improve-taste-article-1.1388247).

24 K. G. Elzinga, C. H. Tremblay & V. J. Tremblay, 'Craft beer in the United States: History, numbers, and geography', *Journal of Wine Economics*, 10 (2015), 242–74.

25 C. Spence & I. Wan, 'Beverage perception & consumption: The influence of the container on the perception of the contents', *Food Quality & Preference*, 39 (2015), 131–40.

26 H. T. Fincks, 'The gastronomic value of odours', *Contemporary Review*, 50 (1886), 680–95.

27 N. Yeap, 'Airport superchef: Heston Blumenthal', *CNN Money*, 7 August 2015 (http://money.cnn.com/2015/08/07/luxury/heston-blumenthal-airport-chef-restaurant-london/). Though for an intriguing recent exception to the aroma-free airport claim, see S. Cable, 'What does Scotland smell like? Heathrow "Scent Globe" sprays fragrance that encapsulates heather-capped Highlands and crisp loch air', *Daily Mail Online*, 9 December 2014 (http://www.dailymail.co.uk/travel/travel_news/article-2866889/Heathrow-Scent-Globe-sprays-fragrance-encapsulates-Scotland-s-heather-capped-Highlands-crisp-loch-air.html).

28 C. Spence & J. Youssef, 'Olfactory dining: Designing for the dominant sense', *Flavour*, 4:32 (2015); 'Food hacking: Food perfume', *Munchies* – video, 13 April 2016 (https://munchies.vice.com/en/videos/food-hacking-food-perfume).

29 J. Youssef, *Molecular Cooking at Home: Taking Culinary Physics out of the Lab and into Your Kitchen* (London: Quintet, 2013).

30 C. Spence, 'Multisensory packaging design: Colour, shape, texture, sound, and smell', in M. Chen & P. Burgess (eds.), *Integrating the Packaging and Product Experience: A Route to Consumer Satisfaction* (Oxford: Elsevier, 2016), pp. 1–22.

31 C. Spence, 'On the psychological impact of food colour', *Flavour*, 4:21 (2015).

32 C. Morran, 'PepsiCo thinks its drinks aren't smelly enough, wants to add scent capsules', *Consumerist*, 17 September 2013 (https://consumerist.com/2013/09/17/pepsico-thinks-its-drinks-arent-smelly-enough-wants-to-add-scent-capsules/). Many food companies are becoming particularly interested in the use of specific food aromas to enhance perceived saltiness, sweetness, etc. See e.g. J. Busch et al., 'Salt reduction and the consumer perspective', *New Food*, 2 (2010), 36–9. And see the 'Taste' chapter for more on this theme.

33 C. Classen, D. Howes & A. Synnott, 'Artificial flavours', in C. Korsmeyer (ed.), *The Taste Culture Reader: Experiencing Food and Drink* (Oxford: Berg, 2005), pp. 337–42; C. Classen, D. Howes & A. Synnott, *Aroma: The Cultural History of Smell* (London: Routledge, 1994), pp. 197–200; R. Rosenbaum, 'Today the strawberry, tomorrow . . .', in N. Klein (ed.), *Culture, Curers and Contagion* (Novato, CA: Chandler &

Sharp, 1979), pp. 80–93; E. R. Shell, 'Chemists whip up a tasty mess of artificial flavors', *Smithsonion*, 17(1) (1979), 79.

34 Take the following quote from someone using a basil-scented Aroma-fork to augment their tomato and mozzarella salad and you will, I think, get the idea: 'This was reminiscent of a caprese salad, but like one of those caprese salads where the basil is a little weak. You know, just kind of anemic tasting . . . definitely not organic.' Quoted in C. Lower, 'I used an aromatherapy fork to confuse and disturb my body. My broccoli didn't magically taste like strawberry candy, but it did taste less like broccoli', *Xojane*, 9 October 2014 (https://www.xojane.com/fun/molecule-r-aromafork).

35 B. Wilson, *Swindled: From Poison Sweets to Counterfeit Coffee – The Dark History of the Food Cheats* (London: John Murray, 2009); Royal College of Physicians, *Every Breath We Take: The Lifelong Impact of Air Pollution* (report of a working party, February 2016).

36 F. T. Marinetti, *The Futurist Cookbook*, translated by S. Brill (1932; San Francisco: Bedford Arts, 1989), p. 43.

37 E. Waugh, *Vile Bodies* (London: Chapman & Hall, 1930), pp. 80–81.

38 Though the fact that you can absorb the alcohol makes this a poten-tially dangerous trend; see G. Heffer, 'Experts warn about "unsafe" new "Vaportini" drinking craze', *Daily Express*, 6 March 2014 (https://www.express.co.uk/news/uk/463441/Experts-warn-new-Vaportini-drinking-craze-is-unsafe).

39 S. Cuozzo, 'Bland cuisine and atmosphere don't boost Eat's silent din-ners', *New York Post*, 23 October 2013 (https:nypost.com/2013/10/23/bland-cuisine-and-atmosphere-dont-boost-eats-silent-dinners).

40 M. G. Ramaekers et al., 'Aroma exposure time and aroma concentra-tion in relation to satiation', *British Journal of Nutrition*, 111 (2014), 554–62; M. R. Yeomans & S. Boakes, 'That smells filling: Effects of pairings of odours with sweetness and thickness on odour perception and expected satiety', *Food Quality and Preference*, 54 (2016), 128–36. For evidence that increasing food aroma also reduces bite size for a semi-solid vanilla custard dessert, see R. A. de Wijk et al., 'Food aroma affects bite size', *Flavour*, 1:3 (2012).

41 See V. A. M. Demetros, 'The sweet smell of success', *The Crafts Report*, April 1997, recommending just this sort of approach.

42 H. F. A. Zoon, C. de Graaf & S. Boesveldt, 'Food odours direct specific appetite', *Foods*, 5:12 (2016); C. Spence, 'Leading the consumer by the nose: On the commercialization of olfactory-design for the food and beverage sector', *Flavour*, 4:31 (2015). It is, though, possible that such effects of pre-exposure to appetizing food aromas may have a more pronounced effect on the behaviour of restrained eaters; I. Fedoroff, J. Polivy & C. P. Herman, 'The effect of pre-exposure to food cues on the eating behavior of restrained and unrestrained eaters', *Appetite*, 28 (1997), 33–47.

43 S. Nassauer, 'Using scent as a marketing tool, stores hope it – and shoppers – will linger: How Cinnabon, Lush Cosmetics, Panera Bread regulate smells in stores to get you to spend more', *Wall Street Journal*, 20 May 2014 (http://www.wsj.com/articles/SB10001424052702303468704 579573953132979382). And for the latest, see M. A. A. M. Leenders, A. Smidts & A. El Haji, 'Ambient scent as a mood inducer in supermarkets: The role of scent intensity and time-pressure of shoppers', *Journal of Retailing and Consumer Services* (in press, 2017).

44 A. Robertson, 'Ghost Food: An art exhibit shows how we might eat after global warming. What would you do in a world without cod, chocolate, or peanut butter?', *The Verge*, 18 October 2013 (https://www. theverge.com/2013/10/18/4851966/ghost-food-shows-how-we-might-eat-after-global-warming).

3. Sight

1 G. L. Wenk, *Your Brain on Food: How Chemicals Control Your Thoughts and Feelings* (Oxford: Oxford University Press, 2015), p. 9; R. Wrangham, *Catching Fire: How Cooking Made Us Human* (London: Profile Books, 2010). See also J. S. Allen, *The Omnivorous Mind: Our Evolving Relationship with Food* (London: Harvard University Press, 2012).

2 J. Z. Young, 'Influence of the mouth on the evolution of the brain', in P. Person (ed.), *Biology of the Mouth: A Symposium, 29–30 December 1966* (Washington, DC: American Association for the Advancement of Science, 1968), pp. 21–35.

3 Such images, and aromas, led to a 24% increase in brain metabolism in one study when fasting participants could also taste the food and had to describe how they liked to eat it; see G.-J. Wang et al., 'Exposure to appetitive food stimuli markedly activates the human brain', *NeuroImage*, 212 (2004), 1790–97. See also M.-L. Bielser et al., 'Does my brain want what my eyes like? – How food liking and choice influence spatio-temporal brain dynamics of food viewing', *Brain & Cognition*, 110 (2016), 64–73; R. M. Piech, M. T. Pastorino & D. H. Zald, 'All I saw was the cake. Hunger effects on attentional capture by visual food cues', *Appetite*, 54 (2010), 579–82.

4 This gives me the opportunity to cite what has to be one of the best-titled papers ever: G. H. S. Razran, 'Salivating and thinking in different languages', *Journal of Psychology*, 36 (1936), 248–51.

5 W. K. Simmons, A. Martin & L. W. Barsalou, 'Pictures of appetizing foods activate gustatory cortices for taste and reward', *Cerebral Cortex*, 15 (2005), 1602–8. See also W. K. Simmons et al., 'Category-specific integration of homeostatic signals in caudal but not rostral human insula', *Nature Neuroscience*, 16 (2013), 1551–2.

6 L. Passamonti et al., 'Personality predicts the brain's response to viewing appetizing foods: The neural basis of a risk factor for overeating', *Journal of Neuroscience*, 29 (2009), 43–51, esp. p. 43. For an overview, see C. Spence et al., 'Eating with our eyes: From visual hunger to digital satiation', *Brain & Cognition*, 110 (2016), 53–63; and C. Spence, 'Mouth-watering: The influence of environmental and cognitive factors on salivation and gustatory/flavour perception', *Journal of Texture Studies*, 42 (2011), 157–71.

7 See Apicius, *Cooking and Dining in Imperial Rome*, trans. J. D. Vehling (Chicago: University of Chicago Press, 1936).

8 C. Cadwalladr, 'Jamie Oliver's FoodTube: Why he's taking the food revolution online', *Observer*, 22 June 2014 (https://www.theguardian.com/lifeandstyle/2014/jun/22/jamie-oliver-food-revolution-online-video).

9 E. Ochagavia, I. Anderson & P. Boyd, 'Crafty yoghurts: Can your taste-buds be tricked?' – video, *Guardian Online*, 27 June 2016 (https://www.theguardian.com/lifeandstyle/video/2016/jun/27/crafty-yoghurts-can-

your-tastebuds-be-tricked-video); E. Ochagavia, I. Anderson & P. Boyd, 'Jelly bean flavour: Is it all in the eyes?' – video, *Guardian Online*, 17 June 2016 (https://www.theguardian.com/lifeandstyle/video/2016/jun/17/jelly-bean-flavour-is-it-all-in-the-eyes-video).

10 In one classic study, French researchers investigated the effects of colour on people's perception of wine aroma. Participants (students enrolled on a degree course in wine in Bordeaux) were initially given a glass of white wine and instructed to describe its aroma. Next, they were asked to do the same with a glass of red wine. As one might have expected, the students used completely different descriptors for the aromas of the two wines: terms like citrus, lychee, straw and lemon for the white wine, and chocolate, berry, tobacco, etc. to describe the red. Finally, the students were given a third glass of wine to nose, and had to decide which of the aroma terms that they had chosen previously constituted the best match for the wine. This third glass looked exactly like the red wine but was, in fact, the same white wine, now coloured red. Surprisingly, they mostly chose the red wine odour descriptors. So they apparently no longer perceived the aromas in the coloured wine that they had previously reported when drinking the untainted white wine. This result powerfully demonstrates vision's dominance over orthonasal olfaction. Similar results have subsequently been reported in New Zealand wine experts (including professional wine tasters and wine makers) who actually tasted the wine. In particular, the experts' descriptions of the aroma of a barrique-fermented young Chardonnay that had been artificially coloured red were more accurate when the wine was served in an opaque glass than when it was served in a glass that was clear. G. Morrot, F. Brochet & D. Dubourdieu, 'The color of odors', *Brain and Language*, 79 (2001), 309–20; W. V. Parr, K. G. White & D. Heatherbell, 'The nose knows: Influence of colour on perception of wine aroma', *Journal of Wine Research*, 14 (2003), 79–101; C. Spence, 'The color of wine – Part 1', *The World of Fine Wine*, 28 (2010), 122–9; C. Spence, 'The color of wine – Part 2', *The World of Fine Wine*, 29 (2010), 112–19.

11 Just take C. S. Peirce, writing almost 150 years ago: 'Sight by itself informs us only of colors and forms. No one can pretend that the images of sight are determinate in reference to taste. They are, therefore, so far

general that they are neither sweet nor non-sweet, bitter nor non-bitter, having savor nor insipid' ('Some consequences of four incapacities', *Journal of Speculative Psychology*, 2 (1868), 140–57). Or take Helmholtz, who, a decade later, wrote: 'For example, one cannot ask whether sweet is more like red or more like blue' (*The Facts of Perception: Selected Writings of Hermann Helmholtz* (Middletown, CT: Wesleyan University Press, 1878)).

And for the opposite position, see e.g. B. Miller, 'Artist invites public to taste colour in ten-day event with dancers and wine at The Oval', *Culture24*, 3 February 2015 (http://www.culture24.org.uk/art/art516019-artist-invites-public-to-taste-colour-in-ten-day-event%20with-dancers-and-wine-at-the-oval).

12 X. Wan et al., 'Cross-cultural differences in crossmodal correspondences between tastes and visual features', *Frontiers in Psychology: Cognition*, 5:1365 (2014); C. Spence et al., 'On tasty colours and colourful tastes? Assessing, explaining, and utilizing crossmodal correspondences between colours and basic tastes', *Flavour*, 4:23 (2015); C. Velasco et al., 'Colour-taste correspondences: Designing food experiences to meet expectations or to surprise', *International Journal of Food Design*, 1 (2016), 83–102.

13 The 'correct', or, better-said, consensual, order is white (= salty), browny-black (= bitter), green (= sour), and red (= sweet). As to why so many of us order the colours in this way, it probably reflects the statistics of the environment. Just think about it: many fruits go from green and sour through to some reddish shade when ripe and sweet; C. Velasco et al., 'Colour-taste correspondences: Designing food experiences to meet expectations or to surprise', *International Journal of Food Design*, 1 (2016), 83–102.

14 The addition of colour (either congruent or incongruent) to a solution affects detection thresholds for the basic tastes; J. A. Maga, 'Influence of color on taste thresholds', *Chemical Senses and Flavor*, 1 (1974), 115–19. That said, the effect of colour on judgements of taste and flavour intensity appears to be more cognitive/decisional than perceptual in nature; see S. Hidaka & K. Shimoda, 'Investigation of the effects of color on judgments of sweetness using a taste adaptation method', *Multisensory Research*, 27 (2014), 189–205. Importantly, colour still influences

people's judgements even when they have been told to ignore it because it may be misleading.

For a review of the 200 or so studies published in this area, see C. Spence et al., 'Does food color influence taste and flavor perception in humans?', *Chemosensory Perception*, 3 (2010), 68–84; C. Spence, 'On the psychological impact of food colour', *Flavour*, 4:21 (2015).

15 J. Johnson & F. M. Clydesdale, 'Perceived sweetness and redness in colored sucrose solutions', *Journal of Food Science*, 47 (1982), 747–52.

16 For instance, Lyall Watson writing at the start of the 1970s: 'We have a deep-seated dislike of blue foods. Take a trip through a supermarket and see how many blue ones you can find. They are rare in nature and equally rare in our artificial hunting grounds. No sweet manufacturer ever successfully marketed a blue confection, and no blue soft drink or ice cream appeared on sale for very long' (*The Omnivorous Ape* (New York: Coward, McCann, & Geoghegan, 1971), pp. 66–7).

17 M. U. Shankar, C. Levitan & C. Spence, 'Grape expectations: The role of cognitive influences in color-flavor interactions', *Consciousness & Cognition*, 19 (2010), 380–90; see also G. Farrell, 'What's green. Easy to squirt? Ketchup!', *USA Today*, 10 July 2000, p. 2b; T. Triplett, 'Consumers show little taste for clear beverages', *Marketing News*, 28(11) (1994), 2, 11.

It is perhaps worth noting the importance of the glass in which such a drink is shown to the flavour meaning associated with it in the mind of the consumer. According to our research, the majority of Chinese and North American consumers, on seeing such a blue drink in a wine glass, will expect it to taste of blueberry; see X. Wan et al., 'Does the shape of the glass influence the crossmodal association between colour and flavour? A cross-cultural comparison', *Flavour*, 3:3 (2014).

18 J. Wheatley, 'Putting colour into marketing', *Marketing*, October 1973, 24–9, 67. Someone else who was fond of serving his guests blue food was Alfred Hitchcock: 'And all the food I had made up was blue! Even when you broke your roll. It looked like a brown roll but when you broke it open it was blue. Blue soup, thick blue soup. Blue trout. Blue chicken. Blue ice cream' (quoted in A. Hitchcock & S. Gottlieb (ed.), *Alfred Hitchcock: Interviews* (Jackson: University of Mississippi Press, 2003), p. 76).

19 A. Gallace, E. Boschin & C. Spence, 'On the taste of "Bouba" and "Kiki": An exploration of word-food associations in neurologically normal participants', *Cognitive Neuroscience*, 2 (2011), 34–46; M. Ngo, R. Misra & C. Spence, 'Assessing the shapes and speech sounds that people associate with chocolate samples varying in cocoa content', *Food Quality & Preference*, 22 (2011), 567–72.

   The Italian Futurists were interested in 'accentuating the taste of the dishes with novel combinations of shape, colour, and texture' (G. Berghaus, 'The futurist banquet: Nouvelle Cuisine or performance art?', *New Theatre Quarterly*, 17(1) (2001), 3–17; p. 15).

20 The only exception that we have come across so far are the Himba tribe of Kaokoland in rural Namibia; they do not associate either shape with carbonation, and for them, milk chocolate goes better with the more angular shape, while they associate dark chocolate with the rounder shape. However, like pretty much everyone else in the world, they too match the rounded shape with the sound of 'bouba' and the angular star-like shape with the sound of 'kiki'; see A. Bremner et al., ' "Bouba" and "Kiki" in Namibia? A remote culture make similar shape–sound matches, but different shape–taste matches to Westerners', *Cognition*, 126 (2013), 165–72. See also P. Liang et al., 'Invariant effect of vision on taste across two Asian cultures: India and China', *Journal of Sensory Studies*, 31 (2016), 416–22.

21 C. Spence & O. Deroy, 'Tasting shapes: A review of four hypotheses', *Theoria et Historia Scientiarum*, 10 (2013), 207–38; O. Deroy & C. Spence, '*Quand les goûts & les formes se répondent*', *Cerveau & Psycho*, 55 (2013), 74–9. People have even been shown to associate somewhat different shapes with different brands of beer; O. Deroy & D. Valentin, 'Tasting shapes: Investigating the sensory basis of cross-modal correspondences', *Chemosensory Perception*, 4 (2011), 80–90.

22 M. Bar & M. Neta, 'Humans prefer curved visual objects', *Psychological Science*, 17 (2006), 645–8; H. Leder, P. P. L. Tinio & M. Bar, 'Emotional valence modulates the preference for curved objects', *Perception*, 40 (2011), 649–55; L. Palumbo, N. Ruta & M. Bertamini, 'Comparing angular and curved shapes in terms of implicit associations and approach/avoidance responses', *PLoS ONE*, 10(10) (2015), e0140043; C. Spence, 'Managing sensory expectations concerning products and

brands: Capitalizing on the potential of sound and shape symbolism', *Journal of Consumer Psychology*, 22 (2012), 37–54; D. R. Reed & A. Knaapila, 'Genetics of taste and smell: Poisons and pleasures', *Progress in Molecular Biology and Translational Science*, 94 (2010), 213–40.

Note that this account can be extended to sonic properties too, with the latest evidence suggesting that there are good reasons to find certain sounds unpleasant; see F. Macrae, 'Why high-pitch tunes scare us: Jarring notes "brings out animals in us" by putting us on alert to predators', *Daily Mail Online*, 13 June 2012 (http://www.dailymail.co.uk/news/article-2158466/Why-high-pitch-tunes-scare-Jarring-notes-brings-animals-putting-alert-predators.html#ixzz23mdj1Znz). One problem with this, though, is that it would seem to predict that bitter will be associated with 'high-pitched tunes' – it is not!

23 C. Spence et al., 'Crossmodal correspondences: Assessing the shape symbolism of foods having a complex flavour profile', *Food Quality & Preference*, 28 (2013), 206–12. There is also research to be done here on pairing pasta shapes with the appropriate sauces – something that the Italians are passionate about; C. Spence & O. Deroy, 'On the shapes of tastes and flavours', *Petits Propos Culinaires*, 97 (2012), 75–108; C. Hildebrand & J. Kenedy, *The Geometry of Pasta* (London: Boxtree, 2010).

24 The first exemplar was Bass and Co.'s red triangle, trademarked in the UK back in 1822. See E. Dichter, 'The strategy of selling with packaging', *Package Engineering Magazine*, July 1971, 16a–16c; C. Velasco et al., 'Predictive packaging design: Tasting shapes, typographies, names, and sounds', *Food Quality & Preference*, 34 (2014), 88–95; S. J. Westerman et al., 'Product design: Preference for rounded versus angular design elements', *Psychology & Marketing*, 29 (2012), 595–605; S. J. Westerman et al., 'The design of consumer packaging: Effects of manipulations of shape, orientation, and alignment of graphical forms on consumers' assessments', *Food Quality and Preference*, 27 (2013), 8–17.

Though how much of what we see currently on the shelves is just a result of copycat marketing is hard to say; F. Van Horen & R. Pieters, 'When high-similarity copycats lose and moderate-similarity copycats gain: The impact of comparative evaluation', *Journal of Marketing Research*, 44 (2012), 83–91.

25 C. Spence, 'Assessing the influence of shape and sound symbolism on the consumer's response to chocolate', *New Food*, 17 (2) (2014), 59–62; C. Spence, 'Shape', in D. Goldstein (ed.), *The Oxford Companion to Sugar and Sweets* (Oxford: Oxford University Press, 2015), p. 607. Paul Young, an award-winning chocolatier from London, put it thus: 'the rectangular chunks of the old bars were integral to the taste of Dairy Milk' (quoted in A. Martin, 'Revolt over Cadbury's "rounder, sweeter" bars: Not only has the classic rectangle shape of a Dairy Milk changed, customers say they are more "sugary" too', *Daily Mail Online*, 16 September 2013 (http://www.dailymail.co.uk/news/article-2421568/Revolt-Cadburys-rounder-sweeter-bars-Not-classic-rectangle-shape-Dairy-Milk-changed-customers-also-sugary.html)). It is perhaps also worth thinking about how the change in shape affects the melt property in the mouth; F. Lenfant et al., 'Impact of the shape on sensory properties of individual dark chocolate pieces', *LWT – Food Science and Technology*, 51 (2013), 545–52.

26 M. Fairhurst et al., 'Bouba-Kiki in the plate: Combining crossmodal correspondences to change flavour experience', *Flavour*, 4:22 (2015).

27 D. Gal, S. C. Wheeler & B. Shiv, *Cross-modal influences on gustatory perception* (2007; unpublished manuscript (https://ssrn.com/abstract=1030197)). Though note that (plate) shape doesn't always influence taste perception, see B. Piqueras-Fiszman et al., 'Is it the plate or is it the food? Assessing the influence of the color (black or white) and shape of the plate on the perception of the food placed on it', *Food Quality & Preference*, 24 (2012), 205–8.

28 G. Van Doorn et al., 'Latté art influences both the expected and rated value of milk-based coffee drinks', *Journal of Sensory Studies*, 30 (2015), 305–15.

29 Q. (J.) Wang et al., 'Assessing the effect of shape on expected and actual chocolate flavour', *Flavour* (in press, 2017). Cf. Q. (J.) Wang et al., 'Sounds spicy: Enhancing the evaluation of piquancy by means of a customised crossmodally congruent soundtrack', *Food Quality & Preference*, 58 (2017), 1–9, for a similar argument regarding the influence of spicy music on spicy-tasting food.

30 See C. Michel, C. Velasco & C. Spence, 'Cutlery influences the perceived value of the food served in a realistic dining environment', *Flavour*, 4:27 (2015).

31 B. Piqueras-Fiszman et al., 'Is it the plate or is it the food? Assessing the influence of the color (black or white) and shape of the plate on the perception of the food placed on it', *Food Quality & Preference*, 24 (2012), 205–8; B. Piqueras-Fiszman, A. Giboreau & C. Spence, 'Assessing the influence of the colour/finish of the plate on the perception of the food in a test in a restaurant setting', *Flavour*, 2:24 (2013); P. C. Stewart & E. Goss, 'Plate shape and colour interact to influence taste and quality judgments', *Flavour*, 2:27 (2013). See also M. Fairhurst et al., 'Bouba-Kiki in the plate: Combining crossmodal correspondences to change flavour experience', *Flavour*, 4:22 (2015). There is also an interesting question here as to whether the effect of plate colour can be predicted on the basis of people's ratings of the taste of abstract pairs of colours; see A. T. Woods & C. Spence, 'Using single colours and colour pairs to communicate basic tastes', *i-Perception*, 7(4) (2016), DOI: 10.1177/2041669516658817; A. T. Woods et al., 'Using single colours and colour pairs to communicate basic tastes II; Foreground-background colour combinations', *i-Perception*, 7(5) (2016), DOI: 10.1177/2041669516663750.

32 B. Piqueras-Fiszman & C. Spence, 'Does the color of the cup influence the consumer's perception of a hot beverage?', *Journal of Sensory Studies*, 27(5) (2012), 324–31; G. Van Doorn, D. Wuillemin & C. Spence, 'Does the colour of the mug influence the taste of the coffee?', *Flavour*, 3:10 (2014).

 Intriguingly, a growing number of such results have now been documented across a range of sensory attributes and foods. As the results mount up, it becomes increasingly clear that what we taste is influenced not only by what a dish or drink looks like but also by the colour of the background against which it is seen – be that the cup, the plate or even the cutlery; see N. Guéguen & C. Jacob, 'Coffee cup color and evaluation of a beverage's "warmth quality"', *Color Research and Application*, 39 (2012), 79–81; V. Harrar & C. Spence, 'The taste of cutlery', *Flavour*, 2:21 (2013); P. Risso et al., 'The association between the colour of a container and the liquid inside: An experimental study on consumers' perception, expectations and choices regarding mineral water', *Food Quality and Preference*, 44 (2015), 17–25; Y. Tu, Z. Yang & C. Ma, 'The taste of plate: How the spiciness of food is affected by the color of the plate used to serve it', *Journal of Sensory Studies*, 31 (2016), 50–60.

33 Interestingly, the use of pale blue or red plates didn't change consumption relative to the white tableware baseline; T. E. Dunne et al., 'Visual contrast enhances food and liquid intake in advanced Alzheimer's disease', *Clinical Nutrition*, 23 (2004), 533–8. No wonder, given such striking results, that a number of start-ups have appeared in the last few years offering enhanced contrast for plateware for the home environment (e.g., R. Robbins, 'Can high-tech plates and silverware help patients manage disease?', *Stat News*, 28 December 2015 (https://www.statnews.com/2015/12/28/plates-silverware-health/); this dining set created by industrial designer Sha Yao, which uses bright primary colours; see http://www.eatwellset.com/).

34 This study was conducted at Salisbury District Hospital (UK); see S. Adams, 'How to rescue NHS food? Put it on a blue plate: Simple switch has helped elderly and weak patients eat nearly a third more', *Daily Mail Online*, 7 December 2013 (http://www.dailymail.co.uk/news/article-2520058/How-rescue-NHS-food-Put-blue-plate-Simple-switch-helped-elderly-weak-patients-eat-nearly-more.html). Remember, though, that the benefits of a given plate colour (on food consumption) are likely to depend on the visual properties (i.e., the hue) of the food that is served. So before you all go out and buy yourselves a set of blue crockery, be aware that this colour won't necessarily improve the taste of everything; see e.g. G. Van Doorn, D. Wuillemin & C. Spence, 'Does the colour of the mug influence the taste of the coffee?', *Flavour*, 3:10 (2014).

In Ystad, Sweden, older people have been shown to eat more white fish when it is served on a blue plate. Better contrast between the foreground food and the background colour is the probable explanation here; see B. Hultén, N. Broweus & M. van Dijk, *Sensory Marketing* (Basingstoke: Palgrave Macmillan, 2009), p. 119.

Meanwhile, in another study, yellow plateware was also found to increase dementia patients' consumption; S. Smyth, 'Yellow plates that help dementia patients to eat: Brightly coloured crockery said to make patients more likely to finish their food and put on weight', *Daily Mail Online*, 6 July 2016 (http://www.dailymail.co.uk/news/article-3676207/Yellow-plates-help-dementia-patients-eat.html).

35 B. Crumpacker, *The Sex Life of Food: When Body and Soul Meet to Eat* (New York: Thomas Dunne Books, 2006), p. 143.

36 N. Bruno et al., 'The effect of the color red on consuming food does not depend on achromatic (Michelson) contrast and extends to rubbing cream on the skin', *Appetite*, 71 (2013), 307–13; O. Genschow, L. Reutner & M. Wänke, 'The color red reduces snack food and soft drink intake', *Appetite*, 58 (2012), 699–702. In the latter study, people ate nearly twice as many pretzels when served off a white as opposed to a red plate, despite the colour of the plate not affecting tastiness ratings. Note that the effect wasn't about how appealing the food looked on the different plates either. The story becomes somewhat more complicated, though, when the colour is associated with a brand. Just think of Coca-Cola red.

37 *Still Hungry to Be Heard*, AgeUK report (2010; http://www.ageuk.org.uk/BrandPartnerGlobal/londonVPP/Documents/Still_Hungry_To_Be_Heard_Report.pdf); L. Bradley & C. Rees, 'Reducing nutritional risk in hospital: The red tray', *Nursing Standard*, 17 (2003), 33–7; R. Mehta & R. Zhu, 'Blue or red? Exploring the effect of color on cognitive task performances', *Science*, 323 (2009), 1226–9. Though see also K. M. Steele, 'Failure to replicate the Mehta and Zhu (2009) color-priming effect on anagram solution times', *Psychonomic Bulletin & Review*, 21 (2014), 771–6.

38 See G. J. Hollands et al., 'Portion, package or tableware size for changing selection and consumption of food, alcohol and tobacco (review)', *The Cochrane Library*, 9 (2015), for a recent review.

39 A. Barnett & C. Spence, 'When changing the label (of a bottled beer) modifies the taste', *Nutrition and Food Technology: Open Access*, 2:4 (2016); L. Cheskin, *How to Predict What People Will Buy* (New York: Liveright, 1957); A. Abad-Santos, 'What killed Coca-Cola's white can?', *The Wire*, 1 December 2011 (http://www.thewire.com/business/2011/12/what-killed-coca-colas-white-coke-can/45620/). Apparently the spare white cans were distributed to the airlines after being pulled from the shelves. See also L. Huang & J. Lu, 'Eat with your eyes: Package color influences the expectation of food taste and healthiness moderated by external eating', *The Marketing Management Journal*, 25(2) (2015), 71–87; Y. J. Tu & Z. Yang, 'Package colour and taste experiences: The

influence of package colour on liquid food taste experiences', *Journal of Marketing Science* (in press, 2017).

  For a review, see C. Spence, 'Multisensory packaging design: Colour, shape, texture, sound, and smell', in M. Chen & P. Burgess (eds.), *Integrating the Packaging and Product Experience: A Route to Consumer Satisfaction* (Oxford, UK: Elsevier, 2016), pp. 1–22.

40 See http://www.chefjacqueslamerde.com/home/; D. Galarza, 'Revealed: Instagram sensation Jacques La Merde is . . .', *Eater*, 28 January 2016 (https://www.eater.com/2016/1/28/10750642/revealed-instagram-sensation-jacques-la-merde-is).

41 C. Spence, Q. (J.) Wang, & J. Youssef, 'Pairing flavours and the temporal order of tasting', *Flavour* (in press, 2017).

42 O. Deroy et al., 'The plating manifesto (I): From decoration to creation', *Flavour*, 3:6 (2014); M. Y. Park, 'A history of how food is plated, from medieval bread bowls to Noma', *Bon Appétit*, 26 February 2013 (http://www.bonappetit.com/trends/article/a-history-of-how-food-is-plated-from-medieval-bread-bowls-to-noma).

43 K. Bendiner, *Food in Painting: From the Renaissance to the Present* (London: Reaction Books, 2004).

44 J. Yang, 'The art of food presentation', *Crave*, 2011; cited in C. Spence & B. Piqueras-Fiszman, *The Perfect Meal: The Multisensory Science of Food and Dining* (Oxford: Wiley-Blackwell, 2014), p. 113.

45 A. Cockburn, 'Gastro-porn', *New York Review of Books*, 8 December 1977 (https://www.nybooks.com/articles/1977/12/08/gastro-porn/). Subsequently, the term 'food porn' appeared in Rosalind Coward's 1985 book *Female Desires: How They Are Sought, Bought, and Packaged* (New York: Grove Press) to refer to food that has been carefully prepared so as to be irresistible. See also S. Poole, *You Aren't What You Eat: Fed Up with Gastroculture* (London: Union Books, 2012), p. 59.

46 A. McBride, 'Food porn', *Gastronomica*, 10 (2010), 38–46.

47 Quoted in E. Saner, 'Plate spinning: The smart chef's secret ingredient', *Guardian*, 12 May 2015 (https://www.theguardian.com/lifeandstyle/shortcuts/2015/may/12/plate-spinning-smart-chefs-secret-ingredient-food-on-plate); See also A. Buaya, 'From watermelon cake to rainbow lattes, gold leaf croissants and gelato roses: Australia's most Instagrammed foods of 2016 revealed', *Daily Mail Online*, 26 July

2016 (http://www.dailymail.co.uk/femail/food/article-3706552/Australia-s-Instragammed-foods-2016-revealed.html).

48 J. Prynn, 'Age of the Insta-diner: Restaurants drop ban on phones as foodie snaps become the norm', *Evening Standard*, 28 January 2016, p. 27.

49 A. F. Elliott, 'Lights, camera, broccoli! New restaurant concept built entirely around Instagram-worthy food serves meals on spinning plates with built-in phone stands', *Daily Mail Online*, 6 May 2015 (http://www.dailymail.co.uk/femail/article-3070928/Lights-camera-broccoli-New-restaurant-concept-built-entirely-Instagram-worthy-food-serves-meals-spinning-plates-built-phone-stands.html).

50 C. Michel et al., 'Rotating plates: Online study demonstrates the importance of orientation in the plating of food', *Food Quality and Preference*, 44 (2015), 194–202; see also E. Saner, 'Plate spinning: The smart chef's secret ingredient', *Guardian*, 12 May 2015 (https://www.theguardian.com/lifeandstyle/shortcuts/2015/may/12/plate-spinning-smart-chefs-secret-ingredient-food-on-plate).

51 C. Spence et al., 'Plating manifesto (II): The art and science of plating', *Flavour*, 3:4 (2014). Roland Barthes made an equivalent point about ornamental cookery in his *Mythologies*, when he was discussing the recipe photographs in *Elle*: 'This ornamental cookery is indeed supported by wholly mythical economics. This is an openly dream-like cookery, as proved by the photographs in *Elle*, which never show the dishes except from a high angle, as objects at once near and inaccessible, whose consumption can perfectly well be accomplished just by looking' (cited in A. Cockburn, 'Gastro-porn', *New York Review of Books*, 8 December 1977 (https://www.nybooks.com/articles/1977/12/08/gastro-porn/)).

52 A. Victor, 'Keep your background blurry, never use a flash and DON'T overuse filters: How to turn your dull food images into Instagram food porn in 12 simple steps', *Daily Mail Online*, 28 April 2015 (http://www.dailymail.co.uk/femail/food/article-3050116/12-tricks-help-beautiful-food-photos-Instagram.html).

53 C. Spence, Q. (J.) Wang & J. Youssef, 'Pairing flavours and the temporal order of tasting', *Flavour* (in press, 2017).

54 E. Carlo, '22 reasons you should definitely eat the yolk', *BuzzFeed*, 30 May 2015 (https://www.buzzfeed.com/emilycarlo/perfectly-captured-moments-of-yolk-porn#.ek1PyzD1J1).

55 On attentional capturing by energy-dense foods, see: U. Toepel et al., 'The brain tracks the energetic value in food images', *NeuroImage*, 44 (2009), 967–74; V. Harrar et al., 'Food's visually-perceived fat content affects discrimination speed in an orthogonal spatial task', *Experimental Brain Research*, 214 (2011), 351–6. On motion capturing attention, see S. L. Franconeri & D. J. Simons, 'Moving and looming stimuli capture attention', *Perception & Psychophysics*, 65 (2003), 999–1010; A. Açik, A. Bartel & P. König, 'Real and implied motion at the center of gaze', *Journal of Vision*, 14:2 (2014); J. Pratt et al., 'It's alive! Animate motion captures visual attention', *Psychological Science*, 21 (2010), 1724–30.

56 It was their 'Not just any food' campaign, launched in 2004, that led some to suggest that M&S be held responsible for introducing the unsuspecting British public to food porn; L. Ridley, 'Marks and Spencer is bringing back its food porn adverts', *Huffington Post*, 2 September 2014 (https://www.huffingtonpost.co.uk/2014/09/02/marks-and-spencer-food-pudding-advert-this-is-not-just-any_n_5751628.html).

57 Y. Gvili et al., 'Fresh from the tree: Implied motion improves food evaluation', *Food Quality and Preference*, 46 (2015), 160–65.

58 E. Hodgkin, 'How to make the chocolate ball dessert taking the web by storm: As hypnotic videos for melting domes go viral, we recreate the treat – with spectacular (and messy) results', *Daily Mail Online*, 18 February 2016 (http://www.dailymail.co.uk/femail/food/article-3451549/How-make-melting-chocolate-dome-dessert-gripping-web.html).

59 Y. Gvili et al., 'Fresh from the tree: Implied motion improves food evaluation', *Food Quality and Preference*, 46 (2015), 160–65.

60 C. Duboc, 'Munchies presents: Mukbang', *Munchies*, 17 February 2015 (https://munchies.vice.com/videos/munchies-presents-mukbang).

61 L. Braude & R. J. Stevenson, 'Watching television while eating increases energy intake. Examining the mechanisms in female participants', *Appetite*, 76 (2014), 9–16.

62 This perhaps helps to account for the success of *Buzzfeed Tasty* 'creators', one-minute meal preparation recipes filmed from just this first-person perspective (https://www.youtube.com/channel/UCJFp8uSYCjXOMnkUyb3CQ3Q/videos).

63 R. S. Elder & A. Krishna, 'The "visual depiction effect" in advertising: Facilitating embodied mental simulation through product orientation',

Journal of Consumer Research, 38 (2012), 988–1003. In fact, anything that can be done to enhance the ease with which our brains engage in a spot of embodied mental simulation is likely to increase the processing fluency of the situation. That, in turn, will most probably result in our having more positive feelings towards the food or drink, which is obviously just what the food companies, and the agencies they employ, want to hear! See L. W. Barsalou, 'Grounded cognition', *Annual Review of Psychology*, 59 (2008), 617–45; E. K. Papies, 'Tempting food words activate eating simulations', *Frontiers in Psychology*, 4:838 (2013); R. Reber, P. Winkielman & N. Schwartz, 'Effects of perceptual fluency on affective judgments', *Psychological Science*, 9 (1998), 45–8; O. Petit et al., 'Changing the influence of portion size on consumer behavior via mental imagery', *Journal of Business Research* (2016), DOI: 10.1016/j.jbusres.2016.07.021.

64 The increase in this case was in the region of 10–15%; C. P. Herman, J. M. Ostovich & J. Polivy, 'Effects of attentional focus on subjective hunger ratings', *Appetite*, 33 (2009), 181–93. This responsiveness to food images is described as external food sensitivity (or EFS for short).

65 L. Passamonti et al., 'Personality predicts the brain's response to viewing appetizing foods: The neural basis of a risk factor for overeating', *Journal of Neuroscience*, 29 (2009), 43–51; p. 43. See also R. Belfort-DeAguiar et al., 'Food image-induced brain activation is not diminished by insulin infusion', *International Journal of Obesity*, 40 (2016), 1679–86.

66 S. Howard, J. Adams & M. White, 'Nutritional content of supermarket ready meals and recipes by television chefs in the United Kingdom. Cross sectional study', *British Medical Journal* (2012), 345:e7607.

67 K. Ray, 'Domesticating cuisine: Food and aesthetics on American television', *Gastronomica*, 7 (2007), 50–63. And the unhealthy contents of their food may not be the worst of it, either; see 'Stop that! How poor on-screen hygiene of TV chefs – including not washing their hands – could be giving us food poisoning', *Daily Mail Online*, 23 April 2016 (http://www.dailymail.co.uk/news/article-3554788/How-poor-screen-hygiene-TV-chefs-including-not-washing-hands-giving-food-poisoning.html).

68 L. Pope, L. Latimer & B. Wansink, 'Viewers vs. doers. The relationship between watching food television and BMI', *Appetite*, 90 (2015), 131–5;

see Spence et al., 'Eating with our eyes: From visual hunger to digital satiation', *Brain & Cognition*, 110 (2016), 53–63 for a review.

69 C. Spence, 'Leading the consumer by the nose: On the commercialization of olfactory-design for the food and beverage sector', *Flavour*, 4:31 (2015); K. D. Vohs, 'Self-regulatory resources power the reflective system: Evidence from five domains', *Journal of Consumer Psychology*, 16 (2006), 217–23; K. D. Vohs & R. J. Faber, 'Spent resources: Self-regulatory resource availability affects impulse buying', *Journal of Consumer Research*, 33 (2007), 537–47.

70 F. M. Kroese, D. R. Marchiori & D. T. D. de Ridder, 'Nudging healthy food choices: A field experiment at the train station', *Journal of Public Health,* 38 (2016), e133–e137.

71 C. Michel et al., 'A taste of Kandinsky: Assessing the influence of the visual presentation of food on the diner's expectations and experiences', *Flavour*, 3:7 (2014); F. Macrae, 'If a meal looks good, we think it tastes better: Simply arranging food carefully on the plate can persuade diners to pay three times more for it', *Daily Mail Online*, 21 July 2015 (http://www.dailymail.co.uk/news/article-3168913/If-meal-looks-good-think-tastes-better-Simply-arranging-food-carefully-plate-persuade-diners-pay-three-times-it.html#ixzz42WZgXxHn).

72 K. Van Ittersum & B. Wansink, 'Plate size and color suggestibility: The Delboeuf Illusion's bias on serving and eating behavior', *Journal of Consumer Research*, 39 (2012), 215–28; A. McClain et al., 'Visual illusions and plate design. The effects of plate rim widths and rim coloring on perceived food portion size', *International Journal of Obesity and Related Metabolic Disorders*, 38 (2014), 657–62. In fact, in some of our latest work we have been looking to use this phenomenon, known as the Delboeuf Illusion, in order to help give the impression that the appropriate portion size is smaller than would otherwise be the case, all without changing the amount of food that is in the bowl.

73 T. M. Marteau et al., 'Downsizing: Policy options to reduce portion sizes to help tackle obesity', *British Medical Journal* (2015), 351:h5863.

74 C. K. Morewedge, Y. E. Huh & J. Vosgerau, 'Thought for food: Imagined consumption reduces actual consumption', *Science*, 330 (2010), 1530–33; see also J. Larson, J. P. Redden & R. Elder, 'Satiation from

sensory simulation: Evaluating foods decreases enjoyment of similar foods', *Journal of Consumer Psychology*, 24 (2014), 188–94.

75 A. Swerdloff, 'Eating the uncanny valley: Inside the virtual reality world of food', *Munchies*, 13 April 2015 (https://munchies.vice.com/en/articles/eating-the-uncanny-valley-inside-the-virtual-reality-world-of-food).

76 E. Royte, 'How "ugly" fruits and vegetables can help solve world hunger', *National Geographic*, 1 March 2016 (https://www.nationalgeographic.com/magazine/2016/03/global-food-waste-statistics/); R. Smithers, 'Asda puts UK's first supermarket wonky veg box on sale', *Guardian*, 5 February 2016 (https://www.theguardian.com/environment/2016/feb/05/asda-puts-uks-first-supermarket-wonky-veg-box-on-sale); 'Cutting food waste: reclaiming wonky veg', jamieoliver.com, 3 February 2016 (https://www.jamieoliver.com/news-and-features/features/reclaiming-wonky-veg/#IAx8K6186lm2xMAy.97).

77 Signalling is expensive in the world of flora and fauna, and hence most species will choose one route or the other – appeal to the eye or appeal to the nose, not both. There is actually research in the marketing journals addressing the question of who exactly chooses to buy oddly shaped fruits; see N. Loebnitz, G. Schuitema & K. G. Grunert, 'Who buys oddly shaped food and why? Impacts of food shape abnormality and organic labeling on purchase intentions', *Psychology and Marketing*, 32 (2015), 408–21; N. Loebnitz & K. G. Grunert, 'The effect of food shape abnormality on purchase intentions in China', *Food Quality and Preference*, 40 (2015), 24–30. See also P. Prokop & J. Fancovicová, 'Beautiful fruits taste good: the aesthetic influences of food preferences in humans', *Anthropologischer Anzeiger*, 69 (2012), 71–83.

78 C. Spence et al., 'Eating with our eyes: From visual hunger to digital satiation', *Brain & Cognition*, 110 (2016), 53–63; see also B. Crumpacker, *The Sex Life of Food: When Body and Soul Meet to Eat* (New York: Thomas Dunne Books, 2006). Alexander Cockburn is worth quoting at length here: 'Now it cannot escape attention that there are curious parallels between manuals on sexual techniques and manuals on the preparation of food; the same studious emphasis on leisurely technique, the same apostrophes to the ultimate, heavenly delights. True gastro-porn heightens the excitement and also the sense of the unattainable by

proffering colored photographs of various completed recipes. The gastro-pornhound can, in the Bocuse book for example, moisten his lips over a color plate of freshwater crayfish au gratin à la Fernand Point. True, you cannot get fresh crayfish in the United States or indeed black truffles, three tablespoons of which, cut into julienne, are recommended by Bocuse. No matter. The delights offered in sexual pornography are equally unattainable' ('Gastro-porn', *New York Review of Books*, 8 December 1977 (https://www.nybooks.com/articles/1977/12/08/gastro-porn/)).

79 It's interesting to note that posting videos of erotic banana-eating was recently banned in China; S. Malm, 'China bans erotic bananas: Live-video websites are told not to allow "suggestive" fruit-eating and hosts cannot wear stockings and suspenders', *Daily Mail Online*, 6 May 2016 (http://www.dailymail.co.uk/news/article-3577389/China-bans-erotic-bananas-Live-video-websites-told-not-allow-suggestive-fruit-eating-hosts-wear-stockings-suspenders.html).

80 Quoted from Max Ehrlich, *The Edict* (London: Severn House, 1972), p. 173.

4. Sound

1 C. Spence, 'Eating with our ears: Assessing the importance of the sounds of consumption to our perception and enjoyment of multisensory flavour experiences', *Flavour*, 4:3 (2015); R. S. Elder & G. S. Mohr, 'The crunch effect: Food sound salience as a consumption monitoring cue,' *Food Quality & Preference*, 51 (2016), 39–46. See also R. Tamal et al., 'Neuro cuisine: Exploring the science of flavour' – video, *Guardian Online*, 23 May 2016 (https://www.theguardian.com/lifeandstyle/video/2016/may/23/neuro-cuisine-exploring-the-science-of-flavour-video).

2 'How microwave meals are now on the menu at dinner parties', *Daily Mail Online*, 22 May 2016 (http://www.dailymail.co.uk/news/article-3603849/Third-guests-claim-not-bothered-served-ready-meal.html).

3 As the legendary North American salesman Elmer Wheeler put it, in his famous marketing mantra from the 1930s: 'Don't Sell the

Steak – Sell the SIZZLE!'; E. Wheeler, *Tested Sentences That Sell* (New York: Prentice Hall, Inc., 1938); http://www.elmerwheeler.net/.

4 I. P. Pavlov, *Conditioned Reflexes: An Investigation of the Physiological Activity of the Cerebral Cortex*, trans. and ed. G. V. Anrep (London: Oxford University Press, 1927). Similar associative learning effects have since been demonstrated in a variety of other species: See Y. Frolov, *Fish Who Answer the Telephone, and Other Studies in Experimental Biology*, trans. E. Graham (London: Kegan Paul, Trench, Tubner & Co., 1937), ch. 4.

5 S. Stummerer & M. Hablesreiter, *Food Design XL* (New York: Springer, 2010). I do think, though, that some rigorous gastrophysics testing may be needed here, before we put too much weight on this particular claim.

6 Specifically, the spectral contents of the sound associated with the operation of the coffee machine were either boosted or cut by 20 dB between the frequencies of 2.5 and 6.5 kHz; K. M. Knöferle, 'Using customer insights to improve product sound design', *Marketing Review St. Gallen*, 29(2) (2012), 47–53; K. M. Knöferle, 'It's the sizzle that sells: Crossmodal influences of acoustic product cues varying in auditory pleasantness on taste perception' (2011; unpublished ms).

7 C. Spence & M. Zampini, 'Auditory contributions to multisensory product perception', *Acta Acustica united with Acustica*, 92 (2006), 1009–25.

8 P. Samuelsson, 'Taste of sound – Composing for large scale dinners', keynote presentation given at the Sensibus Festival, Seinäjoki, Finland, 13–14 March 2014; see also C. Spence, 'Music from the kitchen', *Flavour*, 4:25 (2015).

9 See 'The sounds of Massimo Bottura by Yuri Ancarani & Mirco Mecacci' – video, *The New York Times Style Magazine*, 2016 (https://www.nytimes.com/video/t-magazine/100000004708074/massimo-bottura.html?smid=fb-share).

10 M. Zampini & C. Spence, 'The role of auditory cues in modulating the perceived crispness and staleness of potato chips', *Journal of Sensory Science*, 19 (2004), 347–63.

11 The latter are known affectionately as WEIRDos: Western, Educated, Industrialized, Rich and Democratic. See J. Henrich, S. J. Heine & A. Norenzayan, 'The weirdest people in the world?', *Behavioral and Brain Sciences*, 33 (2010), 61–135.

12 M. L. Demattè et al., 'Effects of the sound of the bite on apple perceived crispness and hardness', *Food Quality and Preference*, 38 (2014), 58–64; B. Wilson, 'The kitchen thinker: Are the crunchiest apples the most delicious?', *Daily Telegraph*, 29 September 2014 (http://www. telegraph.co.uk/foodanddrink/11110201/The-Kitchen-Thinker-Are-the-crunchiest-apples-the-most-delicious.html).

13 Why should the carbonation that once signalled overripe foods, e.g., fruit, that should be avoided now be one of the most desirable of attributes in drinks? J. Chandrashekar et al., 'The taste of carbonation', *Science*, 326 (2009), 443–5.

14 You can think of this as a kind of ventriloquism effect; see C. V. Jackson, 'Visual factors in auditory localization', *Quarterly Journal of Experimental Psychology*, 5 (1953), 52–65; D. Alais & D. Burr, 'The ventriloquist effect results from near-optimal bimodal integration', *Current Biology*, 14 (2004), 257–62.

15 M. Zampini & C. Spence, 'Modifying the multisensory perception of a carbonated beverage using auditory cues', *Food Quality and Preference*, 16 (2005), 632–41. See C. Spence, 'Eating with our ears: Assessing the importance of the sounds of consumption to our perception and enjoyment of multisensory flavour experiences', *Flavour*, 4:3 (2015) for a review. There is interesting research showing that sound can be used to modify mouthfeel and creaminess perception too; see D. Iijima & T. Koike, 'Change of mouthfeel by means of cross-modal effect using mastication sound and visual information of food', *IEICE Tech Report*, 113 (2013), 83–6 (in Japanese); G. van Aken, 'Listening to what the tongue feels', *Nizo*, 30 January 2013 (https://www.nizo.com/news/latest-news/67/listening-to-what-the-tongue-feels/); G. A. Van Aken, 'Acoustic emission measurement of rubbing and tapping contacts of skin and tongue surface in relation to tactile perception', *Food Hydrocolloids*, 31 (2013), 325–31.

16 M. Batali, *The Babbo Cookbook* (New York: Random House, 2002), cited in J. S. Allen, *The Omnivorous Mind: Our Evolving Relationship with Food* (London: Harvard University Press, 2012), p. 8; see also C. Dacremont, 'Spectral composition of eating sounds generated by crispy, crunchy and crackly foods', *Journal of Texture Studies*, 26 (1995), 27–43; G. Roudaut et al., 'Crispness: A critical review on sensory and material

science approaches', *Trends in Food Science & Technology*, 13 (2002), 217–27; P. Varela & S. Fiszman, 'Playing with sound', in C. Vega, J. Ubbink, & E. van der Linden (eds.), *The Kitchen as Laboratory: Reflections on the Science of Food and Cooking* (New York: Columbia University Press, 2012), pp. 155–65; P. Varela et al., 'Texture concepts for consumers: A better understanding of crispy-crunchy sensory perception', *European Food Research and Technology*, 226 (2007), 1081–90; Z. M. Vickers, 'Relationships of chewing sounds to judgments of crispness, crunchiness and hardness', *Journal of Food Science*, 47 (1981), 121–4. Perhaps it is time for a new vocabulary; see G. Dijksterhuis et al., 'A new sensory vocabulary for crisp and crunchy dry model foods', *Food Quality and Preference*, 18 (2007), 37–50; N. Whittle, 'Chef talk: Simon Hopkinson', *FT Magazine*, 25 May 2013, p. 45.

17 G. Weiss, 'Why is a soggy potato chip unappetizing?', *Science*, 293 (2001), 1753–4.

18 M. Batali, *The Babbo Cookbook* (New York: Random House, 2002), cited in J. S. Allen, *The Omnivorous Mind: Our Evolving Relationship with Food* (London: Harvard University Press, 2012), p. 8. There is simply too much that people want to label innate – more than can possibly be so.

19 J. S. Allen, *The Omnivorous Mind: Our Evolving Relationship with Food* (London: Harvard University Press, 2012), ch. 1. See also H. McGee, *On Food and Cooking: The Science and Lore of the Kitchen* (New York: Scribner, 1984), rev. edn published as *McGee on Food and Cooking: An Encyclopedia of Kitchen Science, History and Culture* (London: Hodder & Stoughton, 2004).

20 C. R. Luckett, J.-F. Meullenet & H.-S. Seo, 'Crispness level of potato chips affects temporal dynamics of flavor perception and mastication patterns in adults of different age groups', *Food Quality & Preference*, 51 (2016), 8–19.

21 M. Moss, *Salt, Sugar, Fat: How the Food Giants Hooked Us* (St Ives: W. H. Allen, 2013), p. 158; S. Eldeghaidy et al., 'The cortical response to the oral perception of fat emulsions and the effect of taster status', *Journal of Neurophysiology*, 105 (2011), 2572–81; I. E. De Araujo & E. T. Rolls, 'Representation in the human brain of food texture and oral fat', *Journal of Neuroscience*, 24 (2004), 3086–93; C. A. Running, B. A. Craig & R. D. Mattes, 'Oleogustus: The unique taste of fat', *Chemical Senses*, 40

(2015), 507–16; R. S. J. Keast & A. Costanzo, 'Is fat the sixth taste primary? Evidence and implications', *Flavour*, 4:5 (2015); A. Drewnowski & M. Schwartz, 'Invisible fats: Sensory assessment of sugar/fat mixtures', *Appetite*, 14 (1990), 203–17; J.-P. Montmayeur & J. Le Coutre, *Fat Detection: Taste, Texture, and Post-ingestive Effects* (Boca Raton: CRC Press/ Taylor & Francis, 2009).

22 O. Deroy, B. Reade & C. Spence, 'The insectivore's dilemma', *Food Quality & Preference*, 44 (2015), 44–55; J. House, 'Consumer acceptance of insect-based foods in the Netherlands: Academic and commercial implications', *Appetite*, 107 (2016), 47–58.

23 B. Stuckey, *Taste What You're Missing: The Passionate Eater's Guide to Why Good Food Tastes Good* (London: Free Press, 2012); see also A. Gmuer et al., 'Effects of the degree of processing of insect ingredients in snacks on expected emotional experiences and willingness to eat', *Food Quality & Preference*, 54 (2016), 117–27.

24 See O. Deroy, B. Reade & C. Spence, 'The insectivore's dilemma', *Food Quality & Preference*, 44 (2015), 44–55; J. S. Allen, *The Omnivorous Mind: Our Evolving Relationship with Food* (London: Harvard University Press, 2012), ch. 1.

25 P. Smith, 'Watch your mouth: The sounds of snacking', *Good*, 27 August 2011 (https://magazine.good.is/articles/watch-your-mouth-the-sounds-of-snacking).

26 We documented a 5% change in crunchiness perception following our manipulation of the sounds of crisp packets rattling, as compared to a *c.*15% change when the sound of the crunch of the crisp was modified; see C. Spence, M. U. Shankar & H. Blumenthal ' "Sound bites": Auditory contributions to the perception and consumption of food and drink', in F. Bacci & D. Melcher (eds.), *Art and the Senses* (Oxford: Oxford University Press, 2011), pp. 207–38.

27 B. Horovitz, 'Frito-Lay sends noisy, "green" SunChips bag to the dump', *USA Today*, 10 May 2010 (http://www.usatoday.com/money/industries/food/2010-10-05-sunchips05_ST_N.htm); S. Vranica, 'Snack attack: Chip eaters make noise about a crunchy bag; Green initiative has unintended fallout: A snack as loud as "the cockpit of my jet" ', *Wall Street Journal*, 10 August 2010 (http://online.wsj.com/news/articles/SB10001424052748703960004575427150103293906); S. Vranica, 'Sun

Chips bag to lose its crunch', *Wall Street Journal*, 6 October 2010 (http://online.wsj.com/article/SB10001424052748703843804575534182403878708).

28 For examples, see C. Spence, 'Multisensory packaging design: Colour, shape, texture, sound, and smell', in M. Chen & P. Burgess (eds.), *Integrating the Packaging and Product Experience: A Route to Consumer Satisfaction* (Oxford: Elsevier, 2016), pp. 1–22; C. Spence & Q. (J.) Wang, 'Sonic expectations: On the sounds of opening and pouring', *Flavour*, 4:35 (2015).

29 For reviews, see B. Piqueras-Fiszman & C. Spence, 'Sensory expectations based on product-extrinsic food cues: An interdisciplinary review of the empirical evidence and theoretical accounts', *Food Quality & Preference*, 40 (2015), 165–79; C. Spence & Q. (J.) Wang, 'Sonic expectations: On the sounds of opening and pouring', *Flavour*, 4:35 (2015).

30 M. Lindstrom, *Brand Sense: How to Build Brands through Touch, Taste, Smell, Sight and Sound* (London: Kogan Page, 2005), p. 12. See also M. A. Amerine, R. M. Pangborn & E. B. Roessler, *Principles of Sensory Evaluation of Food* (New York: Academic Press, 1965), p. 277.

31 E. Byron, 'The search for sweet sounds that sell: Household products' clicks and hums are no accident; Light piano music when the dishwasher is done?', *Wall Street Journal*, 23 October 2012 (http://www.wsj.com/articles/SB10001424052970203406404578074671598804116).

32 M. B. Sapherstein, 'The trademark registrability of the Harley-Davidson roar: A multimedia analysis', Boston College Intellectual Property & Technology Forum (1998; http://bciptf.org/wp-content/uploads/2011/07/48-THE-TRADEMARK-REGISTRABILITY-OF-THE-HARLEY.pdf). Though see A. D. Wilde, 'Harley hopes to add Hog's roar to its menagerie of trademarks', *Wall Street Journal* (Eastern edn), 23 June 1995, B1; O. El Akkad, 'Canadian court clears way to trademark sounds', *The Globe and Mail*, 28 March 2012 (http://www.theglobeandmail.com/globe-investor/canadian-court-clears-way-to-trademark-sounds/article4096387/).

33 I have, for the last few years, been head of sensory marketing at the JWT ad agency in London; this research, though, came from the Brazilian arm of the company; A. McMains, 'How JWT Brazil and Dolby captured the iconic sound of Coke being poured over ice', *Adweek*, 21

May 2015 (http://www.adweek.com/news/advertising-branding/how-jwt-brazil-and-dolby-captured-iconic-sound-experience-coke-being-poured-over-ice-164920).

34 J. Lee, 'How to make the perfect burger: Oxford food scientist claims to have answer: Oxford University chef says perfect burger is 7cm tall, should be eaten to music, given a name and should feel as good as it tastes', *Daily Telegraph*, 16 August 2015 (http://www.telegraph.co.uk/news/science/science-news/11823677/How-to-make-the-perfect-burger-Oxford-food-scientist-claims-to-have-answer.html); this from the chef-in-residence here in the Crossmodal Research Laboratory, Charles Michel, so it must be true! Though, that's not to say he didn't face some ribbing in the press; e.g., J. Sinnerton, 'Burger battle: Boffins can kiss my buns', *Brisbane Courier-Mail*, 5 August 2015 (http://www.couriermail.com.au/news/queensland/is-this-queenslands-best-burger-ohio-cafe-owner-yogesh-koshe-reckons-his-recipe-cant-be-beat/story-fnihsrf2-1227514375022).

35 H. Blumenthal, *Further Adventures in Search of Perfection: Reinventing Kitchen Classics* (London: Bloomsbury Publishing, 2007); H. Blumenthal, *The Big Fat Duck Cookbook* (London: Bloomsbury, 2008).

36 See J. S. Allen, *The Omnivorous Mind: Our Evolving Relationship with Food* (London: Harvard University Press, 2012), ch. 1.

37 A. T. Woods et al., 'Effect of background noise on food perception', *Food Quality & Preference*, 22 (2011), 42–7 (another intriguing piece of research funded by Unilever back in the day. And this is just the stuff they publish; imagine what else they know that is kept confidential). See C. Spence, 'Noise and its impact on the perception of food and drink', *Flavour*, 3:9 (2014) for a review. Perhaps under such noisy conditions the bone-conducted transmission of sound starts to become more influential.

38 L. D. Stafford, E. Agobiani & M. Fernandes, 'Perception of alcohol strength impaired by low and high volume distraction', *Food Quality and Preference*, 28 (2013), 470–74; L. D. Stafford, M. Fernandes & E. Agobiani, 'Effects of noise and distraction on alcohol perception', *Food Quality & Preference*, 24 (2012), 218–24; C. Spence, 'Music from the kitchen', *Flavour*, 4:25 (2015). Watch out, though: a recent study of family mealtimes suggested that loud background noise can lead to the

increased consumption of cookies; see B. H. Fiese, B. L. Jones & J. Jarick, 'The impact of distraction on food consumption and communication during family meals', *Couple and Family Psychology: Research and Practice*, 4 (2015), 199–211.

39 T. Sietsema, 'No appetite for noise', *Washington Post*, 6 April 2008 (https://www.washingtonpost.com/wp-dyn/content/article/2008/04/01/AR2008040102210_pf.html); T. Sietsema, 'Revealing raucous restaurants', *Washington Post*, 6 April 2008 (https://www.washingtonpost.com/wp-dyn/content/article/2008/04/04/AR2008040402735.html); A. Platt, 'Why restaurants are louder than ever', *Grub Street New York*, 13 July 2013 (http://www.grubstreet.com/2013/07/adam-platt-on-loud-restaurants.html); R. Cooke, 'Who wants a din with their dinner?', *Guardian*, 19 September 2016 (https://www.theguardian.com/lifeandstyle/2016/sep/19/rachel-cooke-who-wants-din-with-their-dinner?utm_source=esp&utm_medium=Email&utm_campaign=GU+Today+main+Docus+callout+200916&utm_term=191128&subid=16021322&CMP=EMCNEWEML6619I2), and for a possible solution, see T. Clynes, 'A restaurant with adjustable acoustics', *Popular Science*, 11 October 2012 (http://www.popsci.com/technology/article/2012-08/restaurant-adjustable-acoustics). And see C. Spence, 'Noise and its impact on the perception of food and drink', *Flavour*, 3:9 (2014) for a review. L. A. Pettit, 'The influence of test location and accompanying sound in flavor preference testing of tomato juice', *Food Technology*, 12 (1958), 55–7; C. Ferber & M. Cabanac, 'Influence of noise on gustatory affective ratings and preference for sweet or salt', *Appetite*, 8 (1987), 229–35; C. Buckley, 'Working or playing indoors, New Yorkers face an unabated roar', *The New York Times*, 19 July 2012 (https://www.nytimes.com/2012/07/20/nyregion/in-new-york-city-indoor-noise-goes-unabated.html?_r=0). As we will see in the 'Airline Food' chapter, when the background noise gets too loud, it can affect the taste of the food, and not for the better either! C. Spence, 'Noise and its impact on the perception of food and drink', *Flavour*, 3:9 (2014).

40 J. Gordinier, 'Who's rocking to the music? That's the chef', *The New York Times*, 23 April 2012 (https://www.nytimes.com/2012/04/25/dining/when-the-music-moves-the-chef-and-the-menu.html?pagewanted=all&_r=0).

41 Indeed, over and above what it does to the diners' experience, background music also plays an important, if often unacknowledged, role in helping to motivate the serving staff. As Colin Lynch, the executive chef of Barbara Lynch Gruppo, which comprises restaurants such as Menton and No. 9 Park, puts it: 'I don't think I've ever worked in a kitchen that didn't have some form of music in it. The whole energy of the kitchen changes. The speed at which people work changes depending what we listen to. During prep, you zone out. You're doing one thing for 45 minutes straight. It helps you keep that rhythm.' See D. First, 'Music to prep by: The tunes they name can lighten or quicken the mood before service', *Boston Globe*, 27 July 2011 (https://www.boston.com/ae/food/restaurants/articles/2011/07/27/food_and_music_are_complements_in_most_kitchens____before_its_time_to_focus_on_service/). See C. Spence, 'Music from the kitchen', *Flavour*, 4:25 (2015) for a review.

42 'Talking Scandinavian design with Space Copenhagen', *Design Curial*, 5 August 2013 (http://www.designcurial.com/news/talking-scandinavian-design-with-space-copenhagen/); C. Spence & B. Piqueras-Fiszman, *The Perfect Meal: The Multisensory Science of Food and Dining* (Oxford: Wiley-Blackwell, 2014); J. Moir, 'Why are restaurants so noisy? Can't hear a word your other half says when you dine out? Our test proves restaurants can be as loud as rock concerts', *Daily Mail Online*, 5 December 2015 (http://www.dailymail.co.uk/news/article-3346929/Why-restaurants-noisy-t-hear-word-half-says-dine-test-proves-restaurants-loud-rock-concerts.html).

43 J. Abrams, 'Mise en plate: The scenographic imagination and the contemporary restaurant', *Performance Research: A Journal of the Performing Arts*, 18(3) (2013), 7–14; p. 12. Indeed, as top French chef Alain Ducasse points out when talking about his new restaurant in Paris, at the Plaza Athénée, removing the table cloth from the table of a fine restaurant sends out a strong signal that the cuisine is also likely to be different (http://vimeo.com/108906437); picked up by P. P. Ferguson, 'Haute food: Modernity at table', in C. Korsmeyer (ed.), *The Taste Culture Reader: Experiencing Food and Drink* (2nd edn; Oxford: Bloomsbury, 2016), pp. 94–101.

44 P. Kogan, *Muzak-free London: A Guide to Eating and Drinking and Shopping in Peace* (London: Kogan Page, 1991).

45 G. Keeley, 'Spanish chefs want to take din out of dinner', *The Times*, 4 May 2016, p. 33 (http://www.thetimes.co.uk/article/spanish-chefs-want-to-take-the-din-out-of-dinner-cr3fpcg7p). See also R. Moulder & D. Lubman, 'Proposed guidelines for quiet areas in restaurants for hearing-impaired individuals', *Journal of the Acoustical Society of America*, 97 (1995), 3262.

46 Quote from hotel manager Edwin Kramer, of London's Edition Hotel, in L. Eriksen, 'Room with a cue', *The Journal*, Autumn 2014, 26–7; p. 27. See also D. Pardue, 'Familiarity, ambience and intentionality: An investigation into casual dining restaurants in central Illinois', in D. Beriss & D. Sutton (eds.), *The Restaurants Book: Ethnographies of Where We Eat* (Oxford: Berg, 2007), pp. 65–78; C. Spence, 'Music from the kitchen', *Flavour*, 4:25 (2015).

47 H. Hoby, 'Silence! The restaurant that wants you to eat without saying a word', *Guardian* (*Word of Mouth* blog), 10 October 2013 (https://www.theguardian.com/lifeandstyle/2013/oct/10/silence-restaurant-eat-without-saying-word); S. Kudel, 'Quiet at the table!', *FinnAir Blue Wings*, Summer 2016, p. 16; R. Lynch, 'Shhhh, New York City restaurant is serving up silent dinners', *LA Times*, 10 October 2013 (http://www.latimes.com/food/dailydish/la-dd-silent-dinner-service-brooklyn-20131010-story.html); N. Elali, 'What is a silent dinner party?', *Lebanon Now*, 30 January 2012 (https://now.mmedia.me/lb/en/reportsfeatures/what_is_a_silent_dinner_party); A. Majumdar, 'Eating their words', *Sydney Morning Herald*, 17 June 2011 (http://www.hedonics.com.au/SMH_SDPs.jpg).

48 C. Spence & B. Piqueras-Fiszman, 'Dining in the dark: Why, exactly, is the experience so popular?', *The Psychologist*, 25 (2012), 888–91; C. Spence & B. Piqueras-Fiszman, *The Perfect Meal: The Multisensory Science of Food and Dining* (Oxford: Wiley-Blackwell, 2014), ch. 8; B. Renner et al., 'Eating in the dark: A dissociation between perceived and actual food', *Food Quality & Preference*, 50 (2016), 145–51; B. Scheibehenne, P. M. Todd & B. Wansink, 'Dining in the dark. The importance of visual cues for food consumption and satiety', *Appetite*, 55 (2010), 710–13.

49 C. Spence, 'Noise and its impact on the perception of food and drink', *Flavour*, 3:9 (2014). See also C. Spence, 'Music from the kitchen', *Flavour*,

4:25 (2015); S. Cuozzo, 'Bland cuisine and atmosphere don't boost Eat's silent dinners', *New York Post*, 23 October 2013 (https://nypost.com/2013/10/23/bland-cuisine-and-atmosphere-dont-boost-eats-silent-dinners/).

50   The journalist Amy Fleming had this to say: 'I am sitting at my kitchen table eating chocolate in the name of science. (Turns out I'm pretty good at science.) I'm trying out some "sonic seasoning" whereby, if I listen to a low-pitched sound, my taste awareness somehow shrinks to the back of my tongue and focuses on the chocolate's bitter elements. When I switch to a high frequency, the floodgates to sweetness open up and my entire mouth kicks back in a warm, sugary bath. (Try it yourself at http://condimentjunkie.co.uk/blog/2015/4/27/bittersweet-symphony.) It is a curious sensation because it doesn't feel, to me at least, as if the chocolate tastes different. It is more that the sounds are twisting my grey matter, changing how it perceives the taste.' ('How sound affects the taste of our food', *Guardian*, 11 March 2014; https://www.theguardian.com/lifeandstyle/wordofmouth/2014/mar/11/sound-affects-taste-food-sweet-bitter). Note that since we conducted the original study, we have now compared the 'sweetness', 'sourness', 'saltiness' and 'bitterness' of 27 different tracks created by sound designers/composers with the aim of conveying a specific taste. See Q. (J.) Wang, A. Woods & C. Spence, ' "What's your taste in music?" A comparison of the effectiveness of various soundscapes in evoking specific tastes', *i-Perception*, 6(6) (2015), 1–23; Q. (J.) Wang & C. Spence, 'Unravelling the bittersweet symphony: Assessing the influence of crossmodally congruent soundtracks on bitter and sweet taste evaluation of taste solutions and complex foods' (submitted to *Food Quality & Preference*, 2016).

51   K. Kantono et al., 'Listening to music can influence hedonic and sensory perceptions of gelati', *Appetite*, 100 (2016), 244–55. See also K. Kantono et al., 'The effect of background music on food pleasantness ratings', *Psychology of Music*, 44(5) (2016), 1111–25; A. C. North, 'The effect of background music on the taste of wine', *British Journal of Psychology*, 103 (2012), 293–301; F. Reinoso Carvalho et al., 'Using sound-taste correspondences to enhance the subjective value of tasting experiences', *Frontiers in Psychology: Eating Behaviour*, 6:1309 (2015); F. Reinoso

Carvalho et al., 'Does music influence the multisensory tasting experience?', *Journal of Sensory Science*, 30 (2015), 404–12; F. Reinoso Carvalho et al., ' "Smooth operator": Music modulates the perceived creaminess, sweetness, and bitterness of chocolate', *Appetite*, 108 (2017), 383–90; A. Fiegel et al., 'Background music genre can modulate flavor pleasantness and overall impression of food stimuli', *Appetite*, 76 (2014), 144–52; 'The Sync Project: Taste in music, and tasting music', 1 June 2016 (https://syncproject.co/blog/2016/6/1/taste-in-music-and-tasting-music); T. M. Andrews, 'How the sounds you hear affect the taste of your beer', *Washington Post*, 23 June 2016 (https://www.washingtonpost.com/news/morning-mix/wp/2016/06/23/how-the-sounds-you-hear-affect-the-taste-of-your-beer/); L. Eplet, 'Pitch/fork: The relationship between sound and taste', *Scientific American*, 4 September 2011 (https://blogs.scientificamerican.com/food-matters/pitchfork-the-relationship-between-sound-and-taste/).

52 Take, for instance, Alex Hunt MW, writing about the influence of music on wine perception: 'This is genuinely curious, but I fail to find it especially interesting. For a start, there is scant practical value. Why spend time trying to unearth the music that will improve a particular wine when you could instead just try to find a wine you like in the first place? Secondly, there is no transcendent effect here. I see no claims, and have never experienced, a wine and music pairing that takes the wine to a completely different plane of enjoyment – it just tastes like better (or worse) wine. Lastly, the effect does not appear to be reciprocal: the right Chardonnay has no bearing on how the Beach Boys sound.' ('Can music improve wine?', jancisrobinson.com, 5 May 2015 (https://www.jancisrobinson.com/articles/can-music-improve-wine)).

53 See J. C. Stevens, L. M. Bartoshuk & W. S. Cain, 'Chemical senses and aging: Taste versus smell', *Chemical Senses*, 9 (1984), 167–79; J. C. Stevens et al., 'On the discrimination of missing ingredients: Aging and salt flavour', *Appetite*, 16 (1991), 129–40; J. M. Weiffenbach, 'Chemical senses in aging', in T. V. Getchell et al. (eds.), *Smell and Taste in Health and Disease* (New York: Raven Press, 1991), pp. 369–78; C. J. Wysocki & A. N. Gilbert, '*National Geographic* smell survey: Effects of age are heterogenous', *Annals of the New York Academy of Sciences*, 561 (1989), 12–28.

54 The title of Bonnell's 1966 article captures this idea: M. Bonnell, 'Add color, crunch, and flavor to meals with fresh produce. 2', *Hospitals*, 40(3) (1966), 126–30. And see C. R. Luckett, J.-F. Meullenet & H.-S. Seo, 'Crispness level of potato chips affects temporal dynamics of flavor perception and mastication patterns in adults of different age groups', *Food Quality & Preference*, 51 (2016), 8–19. Though it is also worth remembering that many older patients can have problems with chewing food (see H. Endo, S. Ino & W. Fujisaki, 'The effect of a crunchy pseudo-chewing sound on perceived texture of softened foods', *Physiology & Behavior*, 167 (2016), 324–31).

55 N. Koizumi et al., 'Chewing jockey: Augmented food texture by using sound based on the cross-modal effect' in *Proceedings of ACE '11, the 8th International Conference on Advances in Computer Entertainment Technology* (New York: ACM, 2011), article 21; see also M. Fellett, 'Smart headset gives food a voice', *New Scientist*, 5 December 2012 (https://www.newscientist.com/blogs/nstv/2011/12/smart-headset-gives-food-a-voice.html); Y. Hashimoto et al. 'Straw-like user interface: Virtual experience of the sensation of drinking using a straw', *Proceedings World Haptics 2007* (Los Alamitos, CA: IEEE Computer Society, 2007), pp. 557–8; Y. Hashimoto, M. Inami & H. Kajimoto, 'Straw-like user interface (II): A new method of presenting auditory sensations for a more natural experience', in M. Ferre (ed.), *Eurohaptics 2008, LNCS*, 5024 (Berlin: Springer-Verlag, 2008), pp. 484–93.

56 C. Antin, 'What does wine sound like?', *Punch*, 28 January 2014 (http://punchdrink.com/articles/what-does-wine-sound-like/).

5. *Touch*

1 A potential advantage of serving finger food that you might not have thought about is that it encourages diners to wash their hands first. The latest research suggests that, as a result, they will be a little more likely to make an indulgent food choice than those who have not washed them first; see C. M. Martins, L. G. Block & D. W. Dahl, 'Can hand washing influence hedonic food consumption?', *Psychology & Marketing*, 32 (2015), 742–50.

2 T. Field, *Touch* (Cambridge, MA: MIT Press, 2001); A. Montagu, *Touching: The Human Significance of the Skin* (New York: Columbia University Press, 1971).

3 O. G. Mouritsen & K. Styrbaek, *Mouthfeel: How Texture Makes Taste* (New York: Columbia University Press, 2017).

4 J. Scheide, 'Flavour and medium: Mutual effects and interrelationship', *Ice Cream & Frozen Confectionery*, January 1976, 228–30; C. Spence & B. Piqueras-Fiszman, 'Oral-somatosensory contributions to flavor perception and the appreciation of food and drink', in B. Piqueras-Fiszman & C. Spence (eds.), *Multisensory Flavor Perception: From Fundamental Neuroscience Through to the Marketplace* (Duxford, UK: Elsevier, 2016), pp. 59–79. See also C. M. Christensen, 'Effects of solution viscosity on perceived saltiness and sweetness', *Perception & Psychophysics*, 28 (1980), 347–53; T. A. Hollowood, R. S. T. Linforth & A. J. Taylor, 'The effect of viscosity on the perception of flavour', *Chemical Senses*, 27 (2002), 583–91.

5 J. H. F. Bult, R. A. de Wijk & T. Hummel, 'Investigations on multimodal sensory integration: Texture, taste, and ortho- and retronasal olfactory stimuli in concert', *Neuroscience Letters*, 411 (2007), 6–10; N. Roudnitzky et al., 'Investigation of interactions between texture and ortho- and retronasal olfactory stimuli using psychophysical and electrophysiological approaches', *Behavioural Brain Research*, 216 (2011), 109–15. See C. Spence & B. Piqueras-Fiszman, 'Oral-somatosensory contributions to flavor perception and the appreciation of food and drink', in B. Piqueras-Fiszman & C. Spence (eds.), *Multisensory Flavor Perception: From Fundamental Neuroscience Through to the Marketplace* (Duxford, UK: Elsevier, 2016), pp. 59–79 for a review.

6 A. M. Pederson et al., 'Saliva and gastrointestinal functions of taste, mastication, swallowing and digestion', *Oral Disease*, 8 (2002), 117–29. One of the surprising things is that the flavour (aroma) is pulsed when we swallow, and yet our perception of taste/flavour doesn't vanish between bursts of retronasal aroma. It would seem, then, that there is some sort of perceptual completion going on here.

7 All it takes to mislocalize taste out of the mouth is a Q-tip, a mirror and a realistic fake tongue: see C. Michel et al., 'The butcher's tongue illusion', *Perception*, 43 (2014), 818–24; N. Twilley, 'The fake-tongue

illusion', *The New Yorker*, 30 September 2014 (https://www.newyorker. com/tech/elements/butchers-tongue-illusion). This is a riff on the rubber hand illusion; see M. Botvinick & J. Cohen, 'Rubber hands "feel" touch that eyes see', *Nature*, 391 (1998), 756; H. H. Ehrsson, C. Spence & R. E. Passingham, 'That's my hand! Activity in premotor cortex reflects feeling of ownership of a limb', *Science*, 305 (2004), 875–7.

8 J. Todrank & L. M. Bartoshuk, 'A taste illusion: Taste sensation localized by touch', *Physiology & Behavior*, 50 (1991), 1027–31. For a review, see C. Spence, 'Oral referral: Mislocalizing odours to the mouth', *Food Quality & Preference,* 50 (2016), 117–28; though see also R. J. Stevenson, 'Flavor binding: Its nature and cause', *Psychological Bulletin*, 140 (2014), 487–510.

9 Quoted in G. Berghaus, 'The futurist banquet: Nouvelle Cuisine or performance art?', *New Theatre Quarterly*, 17(1) (2001), 3–17, p. 15.

10 Marinetti himself was a fascist, and a misogynist to boot, according to M. Halligan, *Eat My Words* (London: Angus & Robertson, 1990).

11 S. Brickman, 'The food of the future', *The New Yorker*, 1 September 2014 (https://www.newyorker.com/culture/culture-desk/food-future); F. T. Marinetti, *Il tattilismo* (Milan: Comoaedia, 1921); translated in English in L. Panzera & C. Blum (eds.), '*La futurista* : Benedetta Cappa Marinetti' (exhibition catalogue; Philadelphia: Goldie Paley Gallery, 1998), pp. 54–6. Though it should be noted that knives and forks were banned in the 'Futurist Manifesto on Cooking', see F. T. Marinetti & L. Colombo, *La cucina futurista: Un pranzo che evitò un suicidio* [*The Futurist Kitchen: A Meal That Prevented Suicide*] (1932; Milan: Christian Marinotti Edizioni, 1998), p. 42.

12 The chef also demonstrated a similar phenomenon in a whisky tasting for the press. As one commentator noted: 'After nosing and tasting the whisky with eyes closed and touching each surface it surprisingly tasted smokier and spicier when combined with the Velcro, while softer and rounder with velvet.' (M. Chambers, 'Think you know how to taste whisky? Think again . . .', *Huffington Post Lifestyle*, 24 September 2015 (https://www.huffingtonpost.co.uk/matt-chambers/think-you-know-how-to-taste-whisky_b_8183010.html)).

13 But see F. Dunlop, 'Spoon fed: How cutlery affects your food', *Financial Times*, 5 May 2012 (https://www.ft.com/cms/s/2/776ba1d4-93ee-11e1-bafo-00144feab49a.html#axzz2Ufqd6mmL).

14 I. Crawford, *Sensual Home: Liberate Your Senses and Change Your Life* (London: Quadrille, 1997).

15 G. M. Florio, 'How dirty is restaurant silverware? Here's the scoop on whether that fork is really clean', *Bustle*, 6 August 2015 (https://www.bustle.com/articles/102138-how-dirty-is-restaurant-silverware-heres-the-scoop-on-whether-that-fork-is-really-clean). Elvis Presley was so afraid of germs that he would take his own cutlery wherever he went so as not to have to use anyone else's; I. Ferris, 'How the King was all shook up over germs: Priscilla Presley reveals Elvis insisted on taking his own cutlery to other people's houses over fear of bugs', *Daily Mail Online*, 30 October 2015 (http://www.dailymail.co.uk/tvshowbiz/article-3296080/Priscilla-Presley-reveals-Elvis-insisted-taking-cutlery-people-s-houses-fear-germs.html).

16 W. Welch, J. Youssef & C. Spence, 'Neuro-cutlery: The next frontier in cutlery design', *Supper Magazine*, 4 (2016), 128–9.

17 Talking of which, see K. Gander, 'The supper club where diners are given spoons filled with nails to promote slow, mindful eating', *Independent*, 22 August 2016 (https://www.independent.co.uk/life-style/supper-club-mindful-eating-steinbeisser-slow-food-movement-amsterdam-california-a7196901.html).

18 Though many people think of the spork, the spoon-shaped eating utensil with short tines at the tip, as a relatively new addition to the cutlery family, the term first appeared in the English dictionary all the way back in 1909; see B. Wilson, *Consider the Fork* (London: Particular Books, 2012), p. 271.

19 B. Piqueras-Fiszman & C. Spence, 'Do the material properties of cutlery affect the perception of the food you eat? An exploratory study', *Journal of Sensory Studies*, 26 (2011), 358–62; C. Spence & B. Piqueras-Fiszman, 'Multisensory design: Weight and multisensory product perception', in G. Hollington (ed.), *Proceedings of RightWeight2* (London: Materials KTN, 2011), pp. 8–18; see also V. Harrar & C. Spence, 'The taste of cutlery', *Flavour*, 2:21 (2013). As important as the actual weight of the cutlery, I suspect, is whether it weighs *more* than expected or *less*, the latter rarely being a good thing.

20 See L. Berkowitz & E. Donnerstein, 'External validity is more than skin deep', *American Psychologist*, March 1982, 245–57.

21 C. Michel, C. Velasco & C. Spence, 'Cutlery influences the perceived value of the food served in a realistic dining environment', *Flavour*, 4:27 (2015).

22 The same distraction occurs in those restaurants where the waiter comes out and offers you a selection of cutlery to choose from. Whatever you do, best avoid the mistake made by the famous French chef Alain Ducasse when he opened his restaurant ADNY back in 2000. According to M. Steinberger (*Au Revoir to All That: The Rise and Fall of French Cuisine* (London: Bloomsbury, 2010), p. 168), New York diners were perturbed to be offered a dozen different ornate knives with which to cut their meat. As the critic Gael Greene put it: 'I'm not really amused being forced to choose my knife or my pen just so the house can show off how many it has assembled . . . It's vulgar' ('Gold-plate special', *New York Magazine*; https://nymag.com/nymetro/food/industry/features/3647/index2.html).

23 Z. Pelaccio, *Eat With Your Hands* (New York: Ecco, 2012). One childhood anecdote that resonates with many people is the joy associated with licking out the empty cake bowl with one's fingers after the cake had gone into the oven. Some of the illicit pleasure of this activity would surely be lost if you tried to accomplish the same task using a cold piece of cutlery instead.

24 C. Spence & B. Piqueras-Fiszman, *The Perfect Meal: The Multisensory Science of Food and Dining* (Oxford: Wiley-Blackwell, 2014).

25 Y. Martel, *Life of Pi* (New York: Harcourt, 2001), p. 7.

26 Actually, it looks like someone else has already done this study. They had a trained sensory panel in Arkansas and regular consumers in Brazil eat pizza either with their hands or with knife and fork. Surprisingly, ratings were pretty much identical in the two cases, except for an increase in aroma ratings for the cutlery-wielding Brazilians, a difference attributed to the knife releasing volatiles from tomato slices when cutting through the pizza; H.-S. Seo, 'Cross-cultural influences of eating behaviour and meal pattern on chemosensory perception of food', presentation given at the *17th International Symposium on Olfaction and Taste; JASTS 50th Annual Meeting*, 5–9 June 2016, Yokohama, Japan. The material properties of the chopsticks probably also matter; see e.g. T. Kariya & A. Hanasaki, *Oishinbo à la carte 20* (San Francisco: VIZ Media, 2006), pp. 150–61.

27 M. Barnett-Cowan, 'An illusion you can sink your teeth into: Haptic cues modulate the perceived freshness and crispness of pretzels', *Perception*, 39 (2010), 1684–6.

28 H. Furness, 'How to eat with one's fingers: The Debrett's guide to very modern etiquette, *Daily Telegraph*, 22 November 2012 (http://www.telegraph.co.uk/foodanddrink/foodanddrinknews/9696223/How-to-eat-with-ones-fingers-the-Debretts-guide-to-very-modern-etiquette.html).

29 E. McClelland, 'Recipe for a successful first date? Don't order a salad and pay the full bill (and make sure it is at least £50)', *Daily Mail Online*, 23 August 2015 (http://www.dailymail.co.uk/news/article-3208257/Recipe-successful-date-Don-t-order-salad-pay-bill-make-sure-50.html). Who knows whether this inspired the latest advertising from Carl's Jr; see H. Turnbull, 'Hayden Panettiere puts on a super-sexy display before sharing an intimate moment with a burger as she makes Carl's Jr. commercial debut', *Daily Mail Online*, 25 March 2016 (http://www.dailymail.co.uk/tvshowbiz/article-3510044/Hayden-Panettiere-makes-super-sexy-Carl-s-Jr-commerical-debut.html).

30 O. Deroy, B. Reade & C. Spence, 'The insectivore's dilemma', *Food Quality & Preference*, 44 (2015), 44–55.

31 See S. Poole, *You Aren't What You Eat: Fed Up with Gastroculture* (London: Union Books, 2012), pp. 44–5.

32 Quoted in J. Prescott, *Taste Matters: Why We Like the Foods We Do* (London: Reaktion Books, 2012), pp. 25–6.

33 B. Stuckey, *Taste What You're Missing: The Passionate Eater's Guide to Why Good Food Tastes Good* (London: Free Press, 2012), p. 93. See also C. Spence & B. Piqueras-Fiszman, 'Oral-somatosensory contributions to flavor perception and the appreciation of food and drink', in B. Piqueras-Fiszman & C. Spence (eds.), *Multisensory Flavor Perception: From Fundamental Neuroscience Through to the Marketplace* (Duxford, UK: Elsevier, 2016), pp. 59–79. And see A. Dornenburg & K. Page, *Culinary Artistry* (New York: John Wiley & Sons, 1996), p. 31, on the subject of comforting textures.

34 C. Baraniuk, 'Why "bowl food" might be tricking your brain', *BBC News Online*, 28 February 2016 (https://www.bbc.com/future/story/20160226-why-bowl-food-might-be-tricking-your-brain); M. Deacon,

'The amazing new food trend that's left Nigella bowled over', *Daily Telegraph*, 9 January 2016 (http://www.telegraph.co.uk/foodanddrink/12090015/The-amazing-new-food-trend-thats-left-Nigella-bowled-over.html).

35 Though I am not sure that a bowl full of baked beans would quite count as being *on trend* here. And you just better hope that you're not sitting opposite North American food critic Jeffrey Steingarten. He once described the bowl-food-eating lettuce munchers in the following derogatory terms: 'heads bowed, snouts brought close to their plastic wood-grained bowls, crunching and shovelling simultaneously' (*The Man Who Ate Everything: And Other Gastronomic Feats, Disputes, and Pleasurable Pursuits* (London: Headline, 1998), p. 177).

36 It is almost as if the diner, or consumer, cannot separate the food from the plateware or packaging in which it is presented; B. Piqueras-Fiszman & C. Spence, 'The weight of the container influences expected satiety, perceived density, and subsequent expected fullness', *Appetite*, 58 (2012), 559–62.

   Intriguingly, people prefer less salt in their soup when tasted from a beaker rather than when sampled with the aid of a spoon; see S.-Y. Jeon, E.-K. Lee & K.-O. Kim, 'The perceived saltiness of soup affected by tasting protocols', *Food Quality and Preference*, 35 (2014), 98–103.

37 K. Kampfer et al., 'Touch-taste-transference: Assessing the effect of the weight of product packaging on flavor perception and taste evaluation' (submitted to *PLoS ONE*, 2016).

38 L. Biggs, G. Juravle & C. Spence, 'Haptic exploration of plateware alters the perceived texture and taste of food', *Food Quality & Preference*, 50 (2016), 129–34. See also B. Piqueras-Fiszman & C. Spence, 'The influence of the feel of product packaging on the perception of the oral-somatosensory texture of food', *Food Quality & Preference*, 26 (2012), 67–73; Y. Tu, Z. Yang & C. Ma, 'Touching tastes: The haptic perception transfer of liquid food packaging materials', *Food Quality and Preference*, 39 (2015), 124–30; though see also B. G. Slocombe, D. A. Carmichael & J. Simner, 'Cross-modal tactile-taste interactions in food evaluations', *Neuropsychologia*, 88 (2016), 58–64.

39 L. E. Williams & J. A. Bargh, 'Experiencing physical warmth promotes interpersonal warmth', *Science*, 322 (2008), 606–7. (I have been trying to

convince someone to make me some bowls that wouldn't stand up unaided, i.e., bowls that would spill their contents unless held carefully in the diner's hands, for just this reason.)

See also A. McClain et al., 'Visual illusions and plate design. The effects of plate rim widths and rim coloring on perceived food portion size', *International Journal of Obesity and Related Metabolic Disorders*, 38 (2014), 657–62. This is all the more important when it is realized that such visual cues can modulate our consumption behaviour; see B. Wansink, J. Painter & J. North, 'Bottomless bowls: Why visual cues of portion size may influence intake', *Obesity Research*, 13 (2005), 93–100.

Furthermore, for all of those people out there for whom plating off slate is simply passé, and serving food from a brick or flowerpot is perhaps just a little too much, the bowl may well be the answer for you. Innovative (slightly), but not too 'out there'; see C. Spence & B. Piqueras-Fiszman, *The Perfect Meal: The Multisensory Science of Food and Dining* (Oxford: Wiley-Blackwell, 2014), ch. 4.

40 C. Spence & A. Gallace, 'Multisensory design: Reaching out to touch the consumer', *Psychology & Marketing*, 28 (2011), 267–308; A. Gallace & C. Spence, *In Touch with the Future: The Sense of Touch from Cognitive Neuroscience to Virtual Reality* (Oxford: Oxford University Press, 2014), ch. 11.

41 K. Hara, *Haptic: Awakening the Senses*, exhibition catalogue (Japan: Takeo Co., 2004); F. Lewis & R. Street, *Touch Graphics: The Power of Tactile Design* (Gloucester, MA: Rockport Publishers, 2003). The use of a distinctive, tactile feature may well increase the likelihood that the consumer will pick the product up from the shelf. This, if anything, will increase the likelihood of purchase; see 'Touch looms large as a sense that drives sales', *BrandPackaging*, 3(3) (1999), 39–41.

6. *The Atmospheric Meal*

1 Or as another restaurateur put it: 'Customers seek a dining experience totally different from home, and the atmosphere probably does more to attract them than the food itself.' ('More restaurants sell an exotic atmosphere as vigorously as food', *Wall Street Journal*, 4 August 1965,

p. 1; as cited in P. Kotler, 'Atmospherics as a marketing tool', *Journal of Retailing*, 49 (Winter 1974), 48–64; pp. 58–9.)

2 That said, there has been surprisingly little research into the restaurant environment; H. L. Meiselman et al., 'Demonstrations of the influence of the eating environment on food acceptance', *Appetite*, 35 (2000), 231–7. See also C. U. Lambert & K. M. Watson, 'Restaurant design: Researching the effects on customers', *Cornell Hotel and Restaurant Administration Quarterly*, 24(4) (1984), 68–76.

3 P. Kotler, 'Atmospherics as a marketing tool', *Journal of Retailing*, 49 (Winter 1974), 48–64, p. 48.

4 A. C. North, D. J. Hargreaves & J. McKendrick, 'In-store music affects product choice', *Nature*, 390 (1997), 132. See also A. C. North, D. J. Hargreaves & J. McKendrick, 'The influence of in-store music on wine selections', *Journal of Applied Psychology*, 84 (1999), 271–6.

5 R. E. Nisbett & T. D. Wilson, 'Telling more than we can know: Verbal reports on mental processes', *Psychological Review*, 84 (1977), 231–59.

6 R. Bell et al., 'Effects of adding an Italian theme to a restaurant on the perceived ethnicity, acceptability, and selection of foods', *Appetite*, 22 (1994), 11–24. Getting the ethnic feel of a restaurant right is big business for Italian restaurants, which, for many years now, have been amongst the most popular type of ethnic dining venue in North America; see K. Ray, 'Ethnic succession and the new American restaurant cuisine', in D. Beriss & D. Sutton (eds.), *The Restaurants Book: Ethnographies of Where We Eat* (Oxford: Berg, 2007), pp. 97–114.

7 One limitation of this study was that the names of the dishes also changed on the Italian theme days. In particular, ethnic food names were used, such that, for example, 'macaroni cheese' became '*macaroni gratinati*'. While this makes sense from the restaurateur's perspective, it does mean that it is difficult, if not impossible, to untangle the influence of the decor from that of the change in naming on the diners' responses.

8 In another, more recent study, people served Indian or Malay food were biased in their meal choices when music from one of the two countries was played in the background. These effects were especially pronounced in those who didn't have a strong preference for one or other type of cuisine in the first place; J. P. S. Yeoh & A. C. North, 'The effects of musical fit on choice between two competing foods', *Musicae*

Scientiae, 14 (2010), 127–38. See also D. Zellner et al., 'Ethnic congruence of music and food affects food selection but not liking', *Food Quality & Preference*, 56A (2017), 126–9.

9   M. Sheraton, *Eating My Words: An Appetite for Life* (New York: Harper, 2004), p. 172.

10 D. Sanderson, 'Chinese tastes better with Taylor Swift', *The Times*, 8 December 2015 (http://www.thetimes.co.uk/tto/science/article4635202.ece).

11 In the cafeteria study, classical music increased people's willingness to spend relative to an easy-listening pop-music control; A. C. North & D. J. Hargreaves, 'The effects of music on atmosphere and purchase intentions in a cafeteria', *Journal of Applied Social Psychology*, 28 (1998), 2254–73. For the restaurant study, see A. C. North, A. Shilcock & D. J. Hargreaves, 'The effect of musical style on restaurant customers' spending', *Environment and Behavior*, 35 (2003), 712–18; and see S. Wilson, 'The effect of music on perceived atmosphere and purchase intentions in a restaurant', *Psychology of Music*, 31 (2003), 93–112.

12 C. S. Areni & D. Kim, 'The influence of background music on shopping behavior: Classical versus top-forty music in a wine store', *Advances in Consumer Research*, 20 (1993), 336–40. See also J. Parkinson, 'What is shop music doing to your brain?', *BBC News*, 1 June 2016 (https://www.bbc.com/news/magazine-36424854).

13 A. Fiegel et al., 'Background music genre can modulate flavor pleasantness and overall impression of food stimuli', *Appetite*, 76 (2014), 144–52; J. Lanza, *Elevator Music: A Surreal History of Muzak, Easy-listening, and Other Moodsong* (Ann Arbor, MI: University of Michigan Press, 2004), p. 161; C. Spence, M. U. Shankar & H. Blumenthal, 'Sound bites': Auditory contributions to the perception and consumption of food and drink', in F. Bacci & D. Melcher (eds.), *Art and the Senses* (Oxford: Oxford University Press, 2011), pp. 207–38.

14 C. Jacob, 'Styles of background music and consumption in a bar: An empirical evaluation', *International Journal of Hospitality Management*, 25 (2006), 716–20; A. C. North & D. J. Hargreaves, 'The effects of music on responses to a dining area', *Journal of Environmental Psychology*, 16 (1996), 55–64; N. Stroebele & J. M. de Castro, 'Effects of ambience on food intake and food choice', *Nutrition*, 20 (2004), 821–38. N. Stroebele

& J. M. de Castro, 'Listening to music while eating is related to increases in people's food intake and meal duration', *Appetite*, 47 (2006), 285–9.

15 One can think of this as a kind of 'affective ventriloquism' or 'sensation transference'; see C. Spence & Q. (J.) Wang, 'Wine & music (II): Can you taste the music? Modulating the experience of wine through music and sound', *Flavour*, 4:33 (2015).

16 Cf. R. Pellegrino et al., 'Effects of background sound on consumers' sensory discriminatory ability among foods', *Food Quality & Preference*, 43 (2015), 71–8.

17 H. B. Lammers, 'An oceanside field experiment on background music effects on the restaurant tab', *Perceptual and Motor Skills*, 96 (2003), 1025–6; D. Pardue, 'Familiarity, ambience and intentionality: An investigation into casual dining restaurants in central Illinois', in D. Beriss & D. Sutton (eds.), *The Restaurants Book: Ethnographies of Where We Eat* (Oxford: Berg, 2007), pp. 65–78. See also B. Wansink & K. Van Ittersum, 'Fast food restaurant lighting and music can reduce calorie intake and increase satisfaction', *Psychological Reports: Human Resources & Marketing*, 111(1) (2012), 1–5.

18 R. E. Milliman, 'The influence of background music on the behavior of restaurant patrons', *Journal of Consumer Research*, 13 (1986), 286–9. In another study, diners in a cafeteria took significantly more bites per minute when fast-tempo instrumental, non-classical background music was played rather than slow-tempo music (122 vs 56 bpm resulted in 4.4 vs 3.2 bites per minute respectively); see T. Roballey et al., 'The effect of music on eating behavior', *Bulletin of the Psychonomic Society*, 23 (1985), 221–2. H. McElrea and L. Standing also reported that doubling the bpm of the background music increased the rate at which students drank soda in the lab; 'Fast music causes fast drinking', *Perceptual and Motor Skills*, 75 (1992), 362.

19 Quoted in C. Suddath, 'How Chipotle's DJ, Chris Golub, creates his playlists', *Businessweek*, 17 October 2013 (http://www.businessweek.com/articles/2013-10-17/chipotles-music-playlists-created-by-chris-golub-of-studio-orca).

20 There is a link to Muzak here. In fact, it is the same company (despite the fact that the means of transmission of the playlists has switched from satellite to internet), though the name has now changed (see

J. Lanza, *Elevator Music: A Surreal History of Muzak, Easy-listening, and Other Moodsong* (Ann Arbor, MI: University of Michigan Press, 2004)). The Applebees and Cheddar's restaurant chains also use such satellite radio services (D. Pardue, 'Familiarity, ambience and intentionality: An investigation into casual dining restaurants in central Illinois', in D. Beriss & D. Sutton (eds.), *The Restaurants Book: Ethnographies of Where We Eat* (Oxford: Berg, 2007), pp. 65–78).

21 C. Buckley, 'Working or playing indoors, New Yorkers face an unabated roar', *The New York Times*, 19 July 2012 (https://www.nytimes.com/2012/07/20/nyregion/in-new-york-city-indoor-noise-goes-unabated.html?_r=0).

22 Quoted in T. Clynes, 'A restaurant with adjustable acoustics', *Popular Science*, 11 October 2012 (http://www.popsci.com/technology/article/2012-08/restaurant-adjustable-acoustics).

23 Quite what your regular customers will make of it all is anyone's guess. Perhaps one should first narrow down to a specific genre, and then start experimenting with the tempo, loudness and even specific artists/songs.

   While it is easy to change the music, it is obviously much harder to change the colour scheme on the walls. So should you want to know what colour a new eating venue should be, one option is simply to crowd-source the question (i.e., ask a large number of people what they think). It turns out that you'll get fairly consistent answers about colour schemes: for instance, most people think that a steak restaurant should be painted a dark Bordeaux red (and that the music should be either jazz or classical), while the most appropriate colour scheme for a salad bar was judged to be lime green and the music playing in the background should be jazz, pop or soul; see M. Kontukoski et al., 'Imagined salad and steak restaurants: Consumers' colour, music and emotion associations with different dishes', *International Journal of Gastronomy and Food Science*, 4 (2016), 1–11.

24 So, for instance, researchers have shown that people tend to express more positive emotional associations when the eating occasion is appropriate (e.g., eating ice cream outdoors with friends on a sunny day as compared to eating ice cream with a small group of relative strangers indoors); B. Piqueras-Fiszman & S. R. Jaeger, 'The effect of

product-context appropriateness on emotion associations in evoked eating occasions', *Food Quality and Preference*, 40 (2014), 49–60. Likewise, another recent study found that people experience food intake more positively when it is consumed in an appropriate (and realistic) context; P. García-Segovia, R. J. Harrington & H.-S. Seo, 'Influence of table setting and eating location on food acceptance and intake', *Food Quality and Preference*, 39 (2015), 1–7. Taken together, then, these and other findings suggest that enhancing the match between the product offering and the environment heightens people's evaluation of a range of foods and beverages.

25 A. Shelton, 'A theatre for eating, looking and thinking: The restaurant as symbolic space', *Sociological Spectrum*, 10 (1990), 507–26; p. 522. That said, a few years ago McDonald's invested in softer chairs for one of their store rebranding exercises, the logic being that this change in the feel of their chairs would make their stores somehow appear friendlier; see P. Barden, *Decoded: The Science Behind Why We Buy* (Chichester: John Wiley & Sons, 2013); B. Hultén, N. Broweus & M. van Dijk, *Sensory Marketing* (Basingstoke: Palgrave Macmillan, 2009), pp. 144–5. Other venues that apparently use erect, hard seatbacks and hard finishes to discourage their diners from lingering include the Outback Steakhouse chain; S. K. A. Robson, 'Turning the tables: The psychology of design for high-volume restaurants', *Cornell Hotel and Restaurant Administration Quarterly*, 40(3) (1999), 56–63. The contrast case here is Starbucks, with their beat-up sofas that do encourage one to linger. Starbucks is, in fact, one of the poster-child examples of the success of the experience economy (see 'The Experiential Meal'). However, as everyone knows, you pay for the privilege.

26 Quoted in B. Stuckey, *Taste What You're Missing: The Passionate Eater's Guide to Why Good Food Tastes Good* (London: Free Press, 2012), pp. 85–6.

27 S. S. Dazkir & M. A. Read, 'Furniture forms and their influence on our emotional responses toward interior environments', *Environment & Behavior*, 44 (2012), 722–34; see also O. Vartanian et al., 'Impact of contour on aesthetic judgments and approach-avoidance decisions in architecture', *Proceedings of the National Academy of Sciences of the USA*, 110 (Supple. 2) (2013), 10446–53.

28 M. Visser, *The Rituals of Dinner: The Origins, Evolution, Eccentricities, and Meaning of Table Manners* (London: Penguin Books, 1991). See also R. Pellegrino et al., 'Effects of background sound on consumers' sensory discriminatory ability among foods', *Food Quality & Preference*, 43 (2015), 71–8.

29 There is a parallel here with the 'white cube', the term coined by Brian O'Doherty to describe the traditional ideology behind the design of gallery spaces. Interestingly, as we will see in 'The Experiential Meal', things are starting to change in both environments, as the experience becomes a whole lot more multisensory; B. O'Doherty, *Inside the White Cube: The Ideology of the Gallery Space* (San Francisco: Lapis Press, 1976); B. O'Doherty, *Beyond the Ideology of the White Cube* (Barcelona: MACBA, 2009).

30 A. Shelton, 'A theatre for eating, looking and thinking: The restaurant as symbolic space', *Sociological Spectrum*, 10 (1990), 507–26; p. 525.

31 It may also prime what Amy Trubek calls 'the taste of place', in *The Taste of Place: A Cultural Journey into Terroir* (London: University of California Press, 2009).

32 Nevertheless, such an authentic approach to the design of the environment can be very successful, commercially speaking. Just take the phenomenal rise of the Eataly chain of US restaurants and stores; anyone who shops there will know just what an immersive multisensory experience it is. See also C. Spence et al., 'Store atmospherics: A multisensory perspective', *Psychology & Marketing*, 31 (2014), 472–88.

33 J. McKinley, 'Order a Mai Tai and save paradise', *The New York Times*, 3 April 2009 (https://www.nytimes.com/2009/04/05/fashion/05tonga. html?_r=1&sq=Tonga%20Room&st=cse&adxnnl=1&scp=1&adxnnlx= 1239148998-kpKIJAoiS68aMaVKbRR3fg).

34 'Welcome to the experience economy', *Harvard Business Review*, 76(4) (1998), 97–105; p. 104.

35 In the best-case scenario, atmospherics – in both these early examples, a recognizable natural environment – will augment and enhance the customers' perception of the food and beverages. However, there is always the danger that all the smoke and mirrors are being used to distract the customer from what might be a fundamentally poor product offering. Indeed, in the closing decades of the twentieth century, the view amongst many commentators was that any restaurant or bar that

was investing so much in the atmosphere was most likely not paying enough attention to the food or drink; see D. Goldstein, 'The play's the thing: Dining out in the new Russia', in C. Korsmeyer (ed.), *The Taste Culture Reader: Experiencing Food and Drink* (Oxford: Berg, 2005), pp. 359–71. More sizzle than substance, as Elmer Wheeler might have said. And that was often the way it used to be; perhaps it still is at the 'lower end' of the food chain (at the higher end, it's about the multisensory atmosphere *as well as* the great-tasting food).

36 The rich opportunities afforded by this multisensory approach to the marketing of food and drink have not gone unnoticed by the large retailers either. The Safeway chain in California, for example, have started playing the sound of gathering thunderclouds in their stores just before they spritz their fruit and veg displays with a little water; B. Stuckey, *Taste What You're Missing: The Passionate Eater's Guide to Why Good Food Tastes Good* (London: Free Press, 2012), p. 127. Other retailers have been perfuming a whole host of retail spaces with food scents; see C. White, 'The smell of commerce: How companies use scents to sell their products', *Independent*, 16 August 2011 (https://www.independent.co.uk/news/media/advertising/the-smell-of-commerce-how-companies-use-scents-to-sell-their-products-2338142.html). See also C. Spence, 'Leading the consumer by the nose: On the commercialization of olfactory-design for the food and beverage sector', *Flavour*, 4:31 (2015); C. Spence et al., 'Store atmospherics: A multisensory perspective', *Psychology & Marketing*, 31 (2014), 472–88. Bloomingdale's of New York also created something of a stir when they engaged all five of the senses with their Christmas window displays in 2015. The taste window featured a red polar bear on a mountain of red and white peppermints, and a dispenser in front of the window offered passers-by a wrapped sweet treat.

37 Quoted in C. Rintoul, 'The next chef revolution', 'Food is the New Internet' blog (https://medium.com/food-is-the-new-internet/the-next-chef-revolution-dfe75f0820d2#.k62loo2e8). See also B. Palling, 'Juicy profits are off the menu', *The Sunday Times* (*News Review*), 20 September 2015, p. 4.

38 For an early mention of dining in the dark in literature, see J. L. Borges & A. Bioy-Casares, 'An abstract art', in L. Golden (ed.), *A Literary Feast*

(New York: The Atlantic Monthly Press, 1993), pp. 70–73; p. 73. Certainly, you shouldn't go there for the food; see C. Spence & B. Piqueras-Fiszman, 'Dining in the dark: Why, exactly, is the experience so popular?', *The Psychologist*, 25 (2012), 888–91; C. Spence & B. Piqueras-Fiszman, *The Perfect Meal: The Multisensory Science of Food and Dining* (Oxford: Wiley-Blackwell, 2014), ch. 8. As one commentator put it: 'The encompassing pleasure of a good meal depends on its setting' (A. T. Anderson, 'Table settings: The pleasures of well-situated eating', in J. Horwitz & P. Singley (eds.), *Eating Architecture* (Cambridge, MA: MIT Press, 2004), pp. 247–58; p. 247).

39 See also J. S. A. Edwards et al., 'The influence of eating location on the acceptability of identically prepared foods', *Food Quality and Preference*, 14 (2003), 647–52; H. L. Meiselman et al., 'Demonstrations of the influence of the eating environment on food acceptance', *Appetite*, 35 (2000), 231–7; C. Petit & J. M. Sieffermann, 'Testing consumer preferences for iced-coffee: Does the drinking environment have any influence?', *Food Quality and Preference*, 18 (2007), 161–72; K. Ryu & H. Han, 'New or repeat customers: How does physical environment influence their restaurant experience?', *International Journal of Hospitality Management*, 30 (2011), 599–611. One can only imagine the impact of the environment on diners in those restaurants and bars that have started opening up in public toilets . . . see e.g. A. Magrath, 'Would YOU eat here? Inside the London bars and cafes opening in abandoned public toilets (just remember to wash your hands!)', *Daily Mail Online*, 9 October 2014 (http://www.dailymail.co.uk/travel/travel_news/article-2786195/Would-YOU-eat-Inside-London-bars-cafes-opening-abandoned-public-toilets-just-remember-wash-hands.html).

40 S. K. A. Robson, 'Turning the tables: The psychology of design for high-volume restaurants', *Cornell Hotel and Restaurant Administration Quarterly*, 40(3) (1999), 56–63.

41 J. Bergman, 'Restaurant report: Ultraviolet in Shanghai', *The New York Times*, 10 October 2012 (https://www.nytimes.com/2012/10/07/travel/restaurant-report-ultraviolet-in-shanghai.html).

42 M. Steinberger, *Au Revoir to All That: The Rise and Fall of French Cuisine* (London: Bloomsbury, 2010), p. 78.

43 See C. Spence, M. U. Shankar & H. Blumenthal, '"Sound bites": Auditory contributions to the perception and consumption of food and drink', in F. Bacci & D. Melcher (eds.), *Art and the Senses* (Oxford: Oxford University Press, 2011), pp. 207–38.

44 As Blumenthal said at the time: 'Sound can really enhance the sense of taste.'

45 C. Spence et al., 'On tasty colours and colourful tastes? Assessing, explaining, and utilizing crossmodal correspondences between colours and basic tastes', *Flavour*, 4:23 (2015); C. Spence, 'Managing sensory expectations concerning products and brands: Capitalizing on the potential of sound and shape symbolism', *Journal of Consumer Psychology*, 22 (2012), 37–54; K. M. Knöferle & C. Spence, 'Crossmodal correspondences between sounds and tastes', *Psychonomic Bulletin & Review*, 19 (2012), 992–1006. See also K. M. Knöferle et al., 'That sounds sweet: Using crossmodal correspondences to communicate gustatory attributes', *Psychology & Marketing*, 32 (2015), 107–20.

46 See M. J. Intons-Peterson, 'Imagery paradigms: How vulnerable are they to experimenters' expectations?', *Journal of Experimental Psychology: Human Perception and Performance*, 9 (1983), 394–412; R. Rosenthal, *Experimenter Effects in Behavioral Research* (New York: Appleton-Century-Crofts, 1966); R. Rosenthal, 'Covert communication in the psychological experiment', *Psychological Bulletin*, 67 (1967), 356–67.

47 Originally, I had wanted everyone to receive a different glass of whisky in each room, so that they really wouldn't know whether they were tasting the same drink or not. Hence, or so I thought, any atmospheric effects might be larger (given people's greater uncertainty about what exactly they were tasting). I had also wanted to counterbalance the order in which people experienced the three environments. Practically speaking, though, none of this was possible given the constraints of the event, and ultimately, the fact that the glass didn't change turned out to be one of the most powerful aspects of the experience. For when people came to the end of 'The Sensorium', they could look back over their scorecard and see that they had said different things about the whisky in each of the rooms. Yet they *knew* that the whisky had not left their hands the whole time. What clearer evidence could you want for the

so-called 'Provencal Rosé Paradox' (the experience that many of us in northern Europe have had of foods and wines that tasted great when we were on holiday somewhere in the Mediterranean never quite living up to the memory when we bring them back home)? We did, though, conduct a more carefully controlled pre-test back in the gastrophysics lab, where the participants were given a series of different drinks to evaluate in different orders; see C. Velasco et al., 'Assessing the influence of the multisensory environment on the whisky drinking experience', *Flavour*, 2:23 (2013), experiment 1. See also A. T. Woods et al., 'Flavor expectation: The effects of assuming homogeneity on drink perception', *Chemosensory Perception*, 3 (2010), 174–81, on the unity assumption (roughly speaking, the assumption that food and drink products will not suddenly change their taste). And for more on the 'Provençal Rosé Paradox', see C. Spence & B. Piqueras-Fiszman, *The Perfect Meal: The Multisensory Science of Food and Dining* (Oxford: Wiley-Blackwell, 2014), ch. 9; B. C. Smith, *The Emotional Impact of a Wine and the Provençal Rose Paradox* (2009; unpublished manuscript).

48 S. Chadwick & H. Dudley, 'Can malt whiskey be discriminated from blended whisky? The proof. A modification of Sir Ronald Fisher's hypothetical tea tasting experiment', *British Medical Journal*, 287 (1983), 1912–15. For the experts, see K.-Y. Lee et al., 'Perception of whisky flavour reference compounds by Scottish distillers', *Journal of the Institute of Brewing*, 106 (2012), 203–8; C. Spence, 'The price of everything – the value of nothing?', *The World of Fine Wine*, 30 (2010), 114–20.

49 On this occasion, though, I had learnt my lesson, and we used a well-drilled team of budding young actors to guide the public through the experience over four days. We were also able to roughly counterbalance the order in which people were exposed to the environments on the first two versus the last two days (this being one of the limitations of the 'Singleton' event). Additionally, the wine was served in black tasting glasses so that we could be sure that the lighting manipulation didn't have any effect on the colour of the wine itself; C. Spence, C. Velasco & K. Knoeferle, 'A large sample study on the influence of the multisensory environment on the wine drinking experience', *Flavour*, 3:8 (2014). For similar results, see Q. (J.) Wang & C. Spence, 'Assessing the influence of the multisensory atmosphere on the taste

of vodka', *Beverages*, 1 (2015), 204–17; Q. (J.) Wang & C. Spence, 'Assessing the effect of musical congruency on wine tasting in a live performance setting', *i-Perception*, 6(3) (2015), 1–13. For similar work on the effect of coloured lighting, see D. Oberfeld et al., 'Ambient lighting modifies the flavor of wine', *Journal of Sensory Studies*, 24 (2009), 797–832.

50 A. C. North, 'The effect of background music on the taste of wine', *British Journal of Psychology*, 103 (2012), 293–301. For reviews, see C. Spence & Q. (J.) Wang, 'Wine & music (I): On the crossmodal matching of wine & music, *Flavour*, 4:34 (2015); C. Spence & Q. (J.) Wang, 'Wine & music (II): Can you taste the music? Modulating the experience of wine through music and sound', *Flavour*, 4:33 (2015); E. Schreuder et al., 'Emotional responses to multisensory environmental stimuli: A conceptual framework and literature review', *SAGE Open*, Jan–Mar 2016, 1–19. As B. E. Stein and M. A. Meredith put it in their now classic textbook on the neurophysiology of multisensory integration: 'Integrated sensory inputs produce far richer experiences than would be predicted from their simple coexistence or the linear sum of their individual products [. . .] The integration of inputs from different sensory modalities not only transforms some of their individual characteristics, but does so in ways that can enhance the quality of life' (*The Merging of the Senses* (Cambridge, MA: MIT Press, 1993), p. xi).

51 See C. Spence, C. Velasco & K. Knoeferle, 'A large sample study on the influence of the multisensory environment on the wine drinking experience', *Flavour*, 3:8 (2014), for further input from members of the public who attended the event.

This kind of result really should make all the food and drinks companies out there think a little more carefully about whether testing their new products in those sterile white desk cubicles (often under red lighting to obscure the colour of the food or drink) is really the best idea; see also C. T. Simons, 'Use of immersive technology in consumer sensory and acceptance testing', patent filed March 2014; D. Hathaway, 'The use of immersive technologies to improve consumer testing: The impact of multiple immersion levels on data quality and panelist engagement for the evaluation of cookies under a preparation-based scenario', thesis contributing to an MSc at Ohio State University, 2015.

52 The fact that we tested regular people who just happened to walk in off
    the street leaves unanswered the question of whether or not wine
    experts would be similarly affected. Perhaps not. They may have learnt,
    as a result of extensive training, to latch on to something in the tasting
    experience and filter out 'the everything else'. But even professional
    wine tasters often try to minimize any kind of environmental sensory
    distraction when tasting, suggesting that even they find it hard to com-
    pletely ignore such environmental effects; E. Peynaud, *The Taste of
    Wine: The Art and Science of Wine Appreciation*, trans. M. Schuster (Lon-
    don: Macdonald & Co., 1987).

53 It is sometimes suggested that while this kind of thing might work for
    a cheap wine (i.e., one that is unbalanced), it wouldn't have so much
    effect on a more expensive vintage, well balanced between sweetness
    and acidity. However, the key point to remember here (as we saw in
    'Taste') is that we all live in very different taste worlds; hence, no mat-
    ter how good the wine, what is balanced for one drinker may well not
    taste that good to the next person. Thus, I would argue that there is a
    role for this all-new form of sensory seasoning, regardless of the price
    point. Undoubtedly, not everyone will agree!

54 C. Spence & Q. (J.) Wang, 'Wine & music (I): On the crossmodal
    matching of wine & music', *Flavour*, 4:34 (2015). There are many more
    recommendations here; for example, in research we conducted with
    the London Symphony Orchestra and the Antique Wine Company, we
    found that people matched Pouilly-Fumé (a French white wine) and
    Mozart's Flute Quartet in D. By contrast, a Château Margaux (a French
    Bordeaux red) was matched with Tchaikovsky's String Quartet No. 1;
    see C. Spence et al., 'Looking for crossmodal correspondences between
    classical music & fine wine', *Flavour*, 2:29 (2013).

55 E. Lampi, 'Hotel and restaurant lighting', *Cornell Hotel and Restaurant
    Administration Quarterly*, 13 (1973), 58–64, p. 59.

56 H. J. Suk, G. L. Park & Y. Kim, '*Bon appétit!* An investigation about the
    best and worst color combinations of lighting and food', *Journal of Lit-
    erature and Art Studies*, 2 (2012), 559–66. See also S. K. A. Robson,
    'Turning the tables: The psychology of design for high-volume restau-
    rants', *Cornell Hotel and Restaurant Administration Quarterly*, 40(3) (1999),
    56–63.

57 S. Cho et al., 'Blue lighting decreases the amount of food consumed in men, but not in women', *Appetite*, 85 (2015), 111–17.

58 B. Wansink & K. Van Ittersum, 'Fast food restaurant lighting and music can reduce calorie intake and increase satisfaction', *Psychological Reports: Human Resources & Marketing*, 111(1) (2012), 1–5. Different colour schemes can be used to induce a differing mood or emotional response in diners too; see C. Jacquier & A. Giboreau, 'Perception and emotions of colored atmospheres at the restaurant', *Predicting Perceptions: Proceedings of the 3rd International Conference on Appearance* (Edinburgh, 2012), pp. 165–7; C. Jacquier & A. Giboreau, 'Customers' lighting needs and wants at the restaurant', paper presented at the 30th EuroCHRIE Annual Conference, Lausanne, 25–7 October 2012; B. Manav, R. G. Kutlu & M. S. Küçükdoğu, 'The effects of colour and light on space perception', *Colour and Light in Architecture First International Conference 2010 Proceedings*, pp. 173–7.

59 S. K. A. Robson, 'Turning the tables: The psychology of design for high-volume restaurants', *Cornell Hotel and Restaurant Administration Quarterly*, 40(3) (1999), 56–63. Who knows whether this links to Lyall Watson's early observation that '[s]uccessful restaurants seldom have large windows' (*The Omnivorous Ape* (New York: Coward, McCann, & Geoghegan, 1971), p. 151). However, it also prevents curious types from staring through the windows and putting the diners off their food; 'Mmm, that looks nice! Best restaurant in the world forced to deter voyeuristic "food tourists" by creating a rocky garden outside', *Daily Mail Online*, 27 June 2014 (http://www.dailymail.co.uk/travel/article-2670953/Noma-restaurant-forced-deter-food-tourists-rocky-garden.html).

60 David Ashen of D-Ash design, quoted in R. S. Baraban & J. F. Durocher, *Successful Restaurant Design* (Hoboken, NJ: John Wiley & Sons, 2010), p. 236. See also C. Sester et al., ' "Having a drink in a bar": An immersive approach to explore the effects of context on beverage choice', *Food Quality and Preference*, 28 (2013), 23–31; and see C. Spence & B. Piqueras-Fiszman, *The Perfect Meal: The Multisensory Science of Food and Dining* (Oxford: Wiley-Blackwell, 2014), ch. 9, for a review.

61 See the video at https://travel-brilliantly.marriott.com/our-innovations/goji-kitchen-and-bar. As Matthew von Ertfelda, VP of Innovation for

Marriott, says in the video: 'We will not only feed your appetite but we will ignite your senses.'

62   A few venues have been experimenting with allowing their customers to choose the music that is played via an app; see P. Ranscombe, 'Cafés to give customers a say on music', *Scotland on Sunday*, 30 January 2013 (https://www.scotsman.com/business/management/cafes-to-give-customers-a-say-on-music-1-2747830). Also interesting on this theme is T. Connelly, 'Pepsi Max and KFC team up for musical hidden camera ad', *The Drum*, 7 July 2015 (https://www.thedrum.com/news/2015/07/07/pepsi-max-and-kfc-team-musical-hidden-camera-ad).

63   See N. Lander, 'Blumenthal's magnificent flying machine', *Financial Times*, 27–8 February 2016, p. 33. There is a danger here that the way a new dish tastes in the test kitchens may be different from how it presents itself in the restaurant because of just such differences in lighting, though it is, as yet, unclear whether the gradual changes in lighting utilized at The Fat Duck will have the same impact as the more dramatic and sudden changes that were incorporated into 'The Colour Lab', say. Why so? Well, because the research suggests that we tend not to notice slow changes taking place in the environment. In the laboratory, there have been some wonderful demonstrations of this, as when people are asked to look at an image of a Paris street scene on a computer screen. Unbeknownst to them, the colour of one of the parked cars changes over the course of thirty seconds, and virtually no one notices; see D. J. Simons, S. L. Franconeri & R. L. Reimer, 'Change blindness in the absence of a visual disruption', *Perception*, 29 (2000), 1143–54.

64   And one can easily imagine that what is first trialled in the restaurant setting will soon be found in food stores and supermarkets too, designed to make us spend more. See C. Spence et al., 'Store atmospherics: A multisensory perspective', *Psychology & Marketing*, 31 (2014), 472–88; C. Spence, 'Leading the consumer by the nose: On the commercialization of olfactory-design for the food and beverage sector', *Flavour*, 4:31 (2015).

7. *Social Dining*

1 Especially having seen what happened to the documentary filmmaker Morgan Spurlock when he ate all his meals at the chain for a month (*Super Size Me*, 2004); see S. Cockcroft, 'That really IS a Happy Meal! McDonald's staff throw a surprise birthday party for a lonely 93-year-old widower who has gone to McDonald's almost every day since 2013', *Daily Mail Online*, 20 November 2015 (http://www.dailymail.co.uk/news/article-3327184/That-really-Happy-Meal-Lonely-93-year-old-gone-McDonald-s-day-death-wife-thrown-surprise-birthday-party-restaurant.html).

2 These figures are from 'Death of the family meal as one in four eat alone: Skipping dinner also increasingly common as our busy lifestyles take over', *Daily Mail Online*, 29 September 2015 (http://www.dailymail.co.uk/news/article-3252811/Death-family-meal-one-four-eat-alone.html). The precise figure is 44% meals eaten alone, up from around one in three meals eaten alone in the UK back in 1980; A. Ellson, 'Food wasted as dinner for one becomes the norm', *The Times*, 1 July 2015, p. 23 (http://www.thetimes.co.uk/tto/money/consumeraffairs/article4484213.ece). The situation has been deteriorating for years now. A Mintel survey in 2001, for example, found that three-quarters of British families had already abandoned regular meals, and 20% never sat down together to eat (quoted in Sue Palmer, *Toxic Childhood* (London: Orion Books, 2006), p. 34), while a *Grocery Retailing Report* in 2006 suggested that 51.1% of meals were eaten alone, as compared to 34.4% in 1994 (quoted in C. Steel, *Hungry City: How Food Shapes Our Lives* (London: Chatto & Windus, 2008), p. 339).

It is, though, worth noting that the rose-tinted view that family meals were commonplace in the past may be something of a myth, or at least only true at certain points in our history. A century ago families were not actually dining as a unit here in the UK (nor presumably in many other industrialized countries). Rather, the mother would eat with her children, and later, when the breadwinner came home from work, he would probably consume the meal that had been prepared by his wife alone and in silence; see J. P. Johnston, *A Hundred Years Eating:*

Food, Drink and the Daily Diet in Britain Since the Late Nineteenth Century (Dublin: Gill & Macmillan, 1977), p. 13.

3 'Dinner for one – now that's my kind of date', 14 April 2016 (https://www.theguardian.com/commentisfree/2016/apr/13/dinner-for-one-date-solo-dining-eat?utm_source=esp&utm_medium=Email&utm_campaign=GU+Today+main+NEW+H&utm_term=167009&subid=16021322&CMP=EMCNEWEML6619I2).

4 Y. Tani et al., 'Combined effects of eating alone and living alone on unhealthy dietary behaviors, obesity and underweight in older Japanese adults: Results of the JAGES', *Appetite*, 95 (2015), 1–8; A. I. Conklin et al., 'Social relationships and healthful dietary behaviour: Evidence from over-50s in the EPIC cohort, UK', *Social Science & Medicine*, 100 (2014), 167–75; J. A. Fulkerson et al., 'A review of associations between family or shared meal frequency and dietary and weight status outcomes across the lifespan', *Journal of Nutrition Education and Behavior*, 46(1) (2014), 2–19; S. Goldfarb, W. L. Tarver & B. Sen, 'Family structure and risk behaviors: The role of the family meal in assessing likelihood of adolescent risk behaviors', *Psychology Research and Behaviour Management*, 7 (2014), 53–66.

5 A. Hammons & B. H. Fiese, 'Is frequency of shared family meals related to the nutritional health of children and adolescents? A metaanalysis', *Pediatrics*, 127 (2011), e1565–e1574. Similarly, survey reports suggest that those families who tend to watch a lot of TV at mealtimes have offspring who consume more soda, pizza, salty snacks and fewer vegetables than those families where the TV typically remains off at mealtimes; see K. Coon et al., 'Relationships between use of television during meals and children's food consumption patterns', *Pediatrics*, 107 (2001), e7.

Quote from H. F. Harlow, 'Social facilitation of feeding in the albino rat', *Journal of Genetic Psychology*, 41 (1932), 211–20, p. 211.

6 To put the problem into perspective, one in seven adults currently lives alone in the US, and nearly 50% of older adults eat by themselves; while in the UK, the number of one-person households increased by more than a million between 1996 and 2013 (7.6 million of us were living alone by 2013, and today the figure is likely to be even higher); see K. K. Quigley, J. R. Hermann & W. D. Warde, 'Nutritional risk among

Oklahoma congregate meal participants', *Journal of Nutrition Education and Behavior*, 40(2) (2008), 89–93. Though note that the precise figure in terms of eating/living alone varies by gender, race, and age; see J. A. Marshall et al., 'Indicators of nutritional risk in a rural elderly Hispanic and non-Hispanic white population: San Luis Valley Health and Aging Study', *Journal of the American Dietetic Association*, 99(3) (1999), 315–22.

7 Perhaps indicative of this change, the traditional family roast is very much in decline here in the UK, marked by some truly awful newspaper headlines like: 'Roast in peace: Is this the end of a great tradition?' (*Guardian* editorial, 5 March 2016, p. 28). See also S. Poulter, 'Why cooking a Sunday roast is now too much of a chore: Number of meals eaten in past year drops by 55 million', *Daily Mail Online*, 24 August 2015 (http://www.dailymail.co.uk/news/article-3209505/Why-cooking-Sunday-roast-chore-Number-meals-eaten-past-year-drops-55-million.html).

What is also symptomatic, I think, of the changing culture is the rate at which British pubs, another traditional public space for socializing, are closing down across the land. Once again, many believe this change to be having a detrimental effect on the very fabric of society (at least amongst males); see R. Smithers, 'Number of pubs in UK falls to lowest level for a decade', *Guardian*, 4 February 2016 (https://www.theguardian.com/lifeandstyle/2016/feb/04/uk-pubs-lowest-number-for-decade-2015-camra-beer-tax-campaign).

8 Y. Tani et al., 'Combined effects of eating alone and living alone on unhealthy dietary behaviors, obesity and underweight in older Japanese adults: Results of the JAGES', *Appetite*, 95 (2015), 1–8.

9 And, according to recent research, living alone is more detrimental to our health than cancer, because it leads to suppression of the immune system; O. Moody, 'Lonely people more likely to die early', *The Times*, 24 November 2015, p. 17; see also K. Hafner, 'Researchers confront an epidemic of loneliness', *The New York Times*, 5 September 2016 (https://www.nytimes.com/2016/09/06/health/lonliness-aging-health-effects.html?_r=0).

10 L. Dubé et al., 'Nutritional implications of patient-provider interactions in hospital settings: Evidence from a within-subject assessment of mealtime exchanges and food intake in elderly patients', *European*

Journal of Clinical Nutrition, 61 (2007), 664–72; C. Paquet et al., 'More than just not being alone: The number, nature, and complementarity of meal-time social interactions influence food intake in hospitalized elderly patients', *The Gerontologist*, 48 (2008), 603–11. Meanwhile, other research has shown that elderly patients in a medical ward tend to consume more if they eat together in a supervised dining room rather than when eating individually at their own bedside. The increase in energy intake was not insubstantial either, coming in at a little over 30%; see L. Wright, M. Hickson & G. Frost, 'Eating together is important: Using a dining room in an acute elderly medical ward increases energy intake', *Journal of Human Nutrition and Dietetics*, 19 (2006), 23–6.

11 According to a study conducted by the Waste and Resource Action Program, those living alone throw away £290 of food and drink per year, £90 more than those living with others; see B. Webster, 'People living alone blamed for increase in wasted food', *The Times*, 7 November 2013 (http://www.thetimes.co.uk/tto/news/uk/article3915057.ece). See also A. Ellson, 'Food goes to waste as more of us dine alone', *The Times*, 1 July 2015, p. 23 (http://www.thetimes.co.uk/tto/money/consumeraffairs/article4484213.ece).

   Companies have sprung up to try to combat this, delivering exactly the ingredients needed to cook specific meals, e.g., Blue Apron in the US, or HelloTasty in the UK.

12 The actual figure is 43%, reported in A. Ellson, 'Food goes to waste as more of us dine alone', *The Times*, 1 July 2015, p. 23 (http://www.thetimes.co.uk/tto/money/consumeraffairs/article4484213.ece).

13 See https://www.youtube.com/watch?v=-c6DNB7zWBA.

14 E. Blass et al., 'On the road to obesity: Television viewing increases intake of high-density foods', *Physiology & Behavior*, 88 (2006), 597–604; L. Braude & R. J. Stevenson, 'Watching television while eating increases energy intake. Examining the mechanisms in female participants', *Appetite*, 76 (2014), 9–16; C. Chapman et al., 'Watching TV and food intake: The role of content', *PLoS ONE*, 9:e100602 (2014); M. Hetherington et al., 'Situational effects on meal intake: A comparison of eating alone and eating with others', *Physiology & Behavior*, 88 (2006), 498–505; U. Mathur & R. J. Stevenson, 'Television and eating: Repetition enhances food intake', *Frontiers in Psychology*, 6:1657 (2015).

See also K. Musick & A. Meier, 'Assessing causality and persistence in associations between family dinners and adolescent well-being', *Journal of Marriage and Family*, 74 (2012), 476–93; J. Brunstrom & G. Mitchell, 'Effects of distraction on the development of satiety', *British Journal of Nutrition*, 9 (2006), 761–9; J. Ogden et al., 'Distraction, the desire to eat and food intake. Towards an expanded model of mindless eating', *Appetite*, 62 (2013), 119–26; A. Tal, S. Zuckerman & B. Wansink, 'Watch what you eat: Action-related television content increases food intake', *JAMA Internal Medicine*, 174 (2014), 1842–3.

One study found that: 'While watching "dull" shows, the women taking part binged on 52 per cent more food than during the "entertaining" comedy programme. And the trend held up across different media, the researchers found. Participants ate 35 per cent less when watching the "engaging" episode on TV, compared with reading about insects' (L. Parry, 'How boring TV makes you fat: Dull programmes make us eat 50% more in front of the box', *Daily Mail Online*, 16 July 2014 (http://www.dailymail.co.uk/health/article-2694306/Bored-whats-box-Switch-Dull-TV-shows-make-women-binge-52-exciting-programmes-study-finds.html)). See also E. Robinson et al., 'Eating attentively: A systematic review and meta-analysis of the effect of food intake memory and awareness on eating', *The American Journal of Clinical Nutrition*, 97 (2013), 728–42.

15 See N. Burton & J. Flewellen, *The Concise Guide to Wine and Blind Tasting* (Exeter, Devon: Acheron Press, 2014), p. vii. As Montaigne once wrote: 'There is no preparation so sweet to me, no sauce so appetising, as that which is derived from society' (*The Complete Works of Montaigne*, trans. D. M. Freame (London: Hamish Hamilton, 1958), p. 846).

16 D. Chen & P. Dalton, 'The effect of emotion and personality on olfactory perception', *Chemical Senses*, 30 (2005), 345–51; N. K. Dess & D. Edelheit, 'The bitter with the sweet: The taste/stress/temperament nexus', *Biological Psychology*, 48 (1998), 103–19; T. P. Heath et al., 'Human taste thresholds are modulated by serotonin and noradrenaline', *Journal of Neuroscience*, 26 (2006), 12664–71; O. Pollatos et al., 'Emotional stimulation alters olfactory sensitivity and odor judgment', *Chemical Senses*, 32 (2007), 583–9; K. Smith, 'Mood makes food taste different', *Nature*, 6 December 2006 (http://www.nature.com/news/2006/061204/full/news061204-5.html).

More gastrophysics research is definitely needed here, but for this you need multiple subjects and, in some cases, what you really want to vary is the interpersonal dynamic between those who are dining together, which is tricky. This means that studying the social aspects of dining tends to be a more difficult area to investigate empirically than many of the others covered elsewhere in this book.

17 M. Jones, *Feast: Why Humans Share Food* (Oxford: Oxford University Press, 2008); N. D. Munro & L. Grosman, 'Early evidence (ca. 12,000 B. P.) for feasting at a burial cave in Israel', *Proceedings of the National Academy of Sciences of the USA*, 107 (2010), 15362–6; G. Simnel, 'Sociology of the Meal' (1910), trans. M. Symons, in *Food and Foodways*, 5(4) (1994), 345–50; M. Visser, *The Rituals of Dinner: The Origins, Evolution, Eccentricities, and Meaning of Table Manners* (London: Penguin Books, 1991).

18 See C. Steel, *Hungry City: How Food Shapes Our Lives* (London: Chatto & Windus, 2008), pp. 212–13; See also M. Hanefors & L. Mossberg, 'Searching for the extraordinary meal experience', *Journal of Business and Management*, 9 (2003), 249–70.

19 M. aan het Rot et al., 'Eating a meal is associated with elevations in agreeableness and reductions in dominance and submissiveness', *Physiology & Behavior*, 144 (2015), 103–9; and for a review, see C. Spence, 'Gastrodiplomacy: Assessing the role of food in decision-making', *Flavour* 5:4 (2016).

20 K. Davey, 'One in three people go a week without eating a meal with someone else, Oxford University professor finds', *Oxford Mail*, 13 April 2016 (http://www.oxfordmail.co.uk/news/14422266.One_in_three_people_go_a_week_without_eating_a_meal_with_someone_else__Oxford_University_professor_finds/); see also http://www.thebiglunchers.com/index.php/2016/04/table-for-one-the-importance-of-eating-together/.

One can only wonder what happened to the endorphin levels of those lucky enough to bag a table at the latest London naked dining pop-up Bunyadi; see J. Dunn, 'See you in the buff-et! A naked restaurant is opening in London . . . and there are already 10,000 people on the waiting list', *Daily Mail Online*, 21 April 2016 (http://www.daily-mail.co.uk/news/article-3551335/See-buff-et-naked-restaurant-opening-London-10-000-people-waiting-list.html).

21 H. Rumbelow, 'Tired of takeaways? Try supper in a stranger's home with the Airbnb of dining', *The Times* (*Times2*), 19 November 2015, pp. 6–7.

22 Camille Rumani, co-founder of the VizEat site.

23 V. I. Clendenen, C. P. Herman & J. Polivy, 'Social facilitation of eating among friends and strangers', *Appetite*, 23 (1994), 1–13; J. M. de Castro & E. M. Brewer, 'The amount eaten in meals by humans is a power function of the number of people present', *Physiology & Behavior*, 51 (1992), 121–5; J. M. de Castro et al., 'Social facilitation of the spontaneous meal size of humans occurs regardless of time, place, alcohol or snacks', *Appetite*, 15 (1990), 89–101. For a review, see C. P. Herman, D. A. Roth & J. Polivy, 'Effects of the presence of others on food intake: A normative interpretation', *Psychological Bulletin*, 129 (2003), 873–86; J. M. de Castro, 'Family and friends produce greater social facilitation of food intake than other companions', *Physiology & Behavior*, 56 (1994), 445–55.

24 G. I. Feunekes, C. de Graaf & W. A. van Staveren, 'Social facilitation of food intake is mediated by meal duration', *Physiology & Behavior*, 58 (1995), 551–8; R. C. Klesges et al., 'The effects of selected social variables on the eating behaviour of adults in the natural environments', *International Journal of Eating Disorders*, 3 (1984), 35–41; S. J. Goldman, C. P. Herman & J. Polivy, 'Is the effect of a social model on eating attenuated by hunger?', *Appetite*, 17 (1991), 129–40.

25 D. Ariely & J. Levav, 'Sequential choice in group settings: Taking the road less traveled and less enjoyed', *Journal of Consumer Research*, 27 (2000), 279–90.

26 C. Spence, 'Noise and its impact on the perception of food and drink', *Flavour*, 3:9 (2014).

27 R. Cornish, 'Din and dinner: Are our restaurants just too noisy?', *Good Food*, 13 August 2013 (http://www.goodfood.com.au/good-food/food-news/din-and-dinner-are-our-restaurants-just-too-noisy-20130805-2r92e.html).

28 The study, conducted by OpenTable, was quoted in A. Victor, 'Table for one, please! Number of solo diners DOUBLES in two years as eating alone is viewed as liberating rather than a lonely experience', *Daily Mail Online*, 13 July 2015 (http://www.dailymail.co.uk/femail/food/article-3156420/OpenTable-study-reveals-number-solo-diners-DOUBLES-two-years.html).

29 W. Smale, 'Your solo dining experiences', *BBC News (Business)*, 31 July 2014 (https://www.bbc.co.uk/news/business-28542359).

30 Quoted in Nell Frizzell, 'Dinner for one – now that's my kind of date', *Guardian*, 14 April 2016 (https://www.theguardian.com/commentisfree/ 2016/apr/13/dinner-for-one-date-solo-dining-eat?utm_source=esp& utm_medium=Email&utm_campaign=GU+Today+main+NEW+H& utm_term=167009&subid=16021322&CMP=EMCNEWEML6619I2). Frizzell agrees with him: 'Solo chomping is never a sign of social failure – it is evidence that you have sufficient sense of self to be able to enjoy life's most regular, essential pleasure entirely in your own company. It speaks of confidence, not commiseration.'

31 As one consultant put it: '[S]avvy restaurants around the world are trying to make themselves more welcoming to solo diners, for example by fitting more bar seating, or encouraging waiting staff to be more attentive to customers sitting on their own.' Quotes in this paragraph from B. Balfour, 'Tables for one – the rise of solo dining', *BBC News Online*, 24 July 2014 (https://www.bbc.co.uk/news/business-28292651).

32 A. S. Levine, 'New York today: Where to eat alone', *The New York Times*, 11 February 2016 (https://www.nytimes.com/2016/02/11/nyregion/ new-york-today-where-to-eat-alone.html?_r=0).

33 Van Goor also says that 'eating alone is the most extreme form of feeling disconnected in our culture'. Note that dining at Eenmaal does not seem to be about stopping by for a bite to eat, but *rather* actually making a statement by deliberately booking to eat alone. Both quotes from B. Balfour, 'Tables for one – the rise of solo dining', *BBC News Online*, 24 July 2014 (https://www.bbc.co.uk/news/business-28292651).

    See also S. Muston, 'The blissful silence of a peaceful meal for one', *Independent*, 16 January 2015 (https://www.independent.co.uk/ life-style/food-and-drink/features/the-blissful-silence-of-a-peaceful-meal-for-one-9981463.html).

34 D. Jurafsky, *The Language of Food: A Linguist Reads the Menu* (New York: Norton, 2014); S. Muston, 'On the menu: Sharing plates and family-style dining are in, courses are out', *Independent*, 14 November 2013 (https://www.independent.co.uk/life-style/food-and-drink/features/ on-the-menu-sharing-plates-and-family-style-dining-are-in-courses-are-out-8940492.html).

35 Research shows that people are less likely to argue, or get into a conflict situation, when sitting at a round table than when sitting at an angular one: 'Circular-shaped seating arrangements prime a need to belong while angular shaped seating arrangements prime a need to be unique'; R. (J.) Zhu & J. J. Argo, 'Exploring the impact of various shaped seating arrangements on persuasion', *Journal of Consumer Research*, 40 (2013), 336–49; R. Reilly, 'How to avoid those dinner party squabbles: Sit your guests around a circular table and NOT a square one', *Daily Mail Online*, 19 June 2013 (http://www.dailymail.co.uk/sciencetech/article-2344388/How-avoid-dinner-party-squabbles-Sit-guests-circular-table-NOT-square-one.html).

Now that I think about it, I wonder if we like our meal more if the shape of the plate matches the shape of the table? Round plates for round tables, and square plates for angular tables?

36 H. Armstrong, 'Sharing tables with strangers: Do we British have a problem with sharing?', *Guardian*, 23 September 2009 (https://www.theguardian.com/lifeandstyle/wordofmouth/2009/sep/23/sharing-table-restaurants). The London-based trend-forecasting consultancy The Future Laboratory were ahead of the curve when they homed in on the growth of more conviviality and sharing at the table as one of the future food trends back in 2008; see C. Sanderson et al., *CrEATe: Eating, Design and Future Food* (Berlin: Gestalten, 2008), pp. 190–91.

37 A. J. N. Rosny, *Le Péruvian à Paris* (1801), quoted in R. L. Spang, *The Invention of the Restaurant* (Cambridge, MA: Harvard University Press, 2000), p. 64.

38 See http://mellajaarsma.com/installations-and-costumes/i-eat-you-eat-me/; S. Smith, *Feast: Radical Hospitality in Contemporary Art* (Chicago: IL: Smart Museum of Art, 2013), pp. 212–19.

39 https://blogs.uchicago.edu/feast/2012/02/i_eat_you_eat_me.html.

40 http://www.marijevogelzang.nl/studio/eating_experiences/Pages/sharing_dinner.html.

41 P. Barden et al., 'Telematic dinner party: Designing for togetherness through play and performance', in *Proceedings of the ACM Conference on Designing Interactive Systems 2012* (New York: ACM, 2012), pp. 38–47; R. Comber et al., 'Not sharing sushi: Exploring social presence and connectedness at the telematic dinner party', in J. H.-J. Choi, M. Foth

& G. Hearn (eds.), *Eat, Cook, Grow: Mixing Human–Computer Inter-actions with Human–Food Interactions* (Cambridge, MA: MIT Press, 2014), pp. 65–79. See also J. Wei et al., 'CoDine: An interactive multi-sensory system for remote dining', in *Proceedings of the 13th International Confer-ence on Ubiquitous Computing* (New York: ACM, 2011), pp. 21–30.

42 Quoted in R. Comber et al., 'Not sharing sushi: Exploring social pres-ence and connectedness at the telematic dinner party', in J. H.-J. Choi, M. Foth & G. Hearn (eds.), *Eat, Cook, Grow: Mixing Human–Computer Interactions with Human–Food Interactions* (Cambridge, MA: MIT Press, 2014), pp. 65–79; p. 71.

43 Another solution is *mukbang* (see 'Sight'). One interesting open ques-tion is whether the consumption behaviour of those who dine with such 'virtual' partners is influenced by what the person seen onscreen consumes; cf. P. Pliner & N. Mann, 'Influence of social norms and palatability on amount consumed and food choice', *Appetite*, 42 (2004), 227–37. See also S. Zhou, M. A. Shapiro & B. Wansink, 'The audience eats more if a movie character keeps eating: An unconscious mechanism for media influence on eating behaviors', *Appetite*, 108 (2017), 407–15.

44 H. Rumbelow, 'Tired of takeaways? Try supper in a stranger's home with the Airbnb of dining', *The Times* (*Times2*), 19 November 2015, pp. 6–7; K. Forster & R. David, 'Future of food: How we share it', *Guardian*, 13 September 2015 (https://www.theguardian.com/technology/2015/sep/13/future-of-food-how-we-share). Camille Rumani gave the latest figures at a FrenchConnectLondon Foodtech event (19 April 2016), at which we were both speaking.

45 The explosion in food-delivering services is huge business. The Ele.me start-up in China (the name roughly translates from Mandarin as 'Are you hungry now?'), for example, raised well in excess of $2 billion dol-lars; see B. Solomon, 'Chinese food delivery start-up Ele.me snags $1.25 billion from Alibaba', *Forbes*, 13 April 2016 (http://www.forbes.com/sites/briansolomon/2016/04/13/chinese-food-delivery-startup-ele-me-snags-1-25-billion-from-alibaba/#2a7bbed52ccb). Late in 2016, Face-book also announced that it would soon be enabling people to order restaurant meals via their Facebook page; see N. I. Pesce, 'Facebook launches new food delivery service, continues to colonize your life', *New York Daily News*, 21 October 2016 (http://www.nydailynews.

com/life-style/facebook-launches-new-food-delivery-service-article-
1.2839653). Amazon also started a restaurant home-delivery service in
central London in September 2016.

46 See, e.g., A. Garretón, 'Top 5 *puertas cerradas*', *The Argentina Independent*, 30 May 2012 (http://www.argentinaindependent.com/life-style/
food-drink/top-5-puertas-cerradas/). Note that there has been a long
tradition of dining in people's apartments in Cuba; J. Cooke, 'The new
way to eat in Cuba', *Saveur*, 27 October 2015 (http://www.saveur.com/
best-paladares-restaurants-in-havana-cuba).

47 It will be interesting to see how such technology impacts on the proportion of meals we eat outside the home. According to C. Steel
(*Hungry City: How Food Shapes Our Lives* (London: Chatto & Windus,
2008), p. 207), more than a third of the food we eat in the UK is consumed outside the home, and that figure is expected by some to rise to
close to 50%, thus mirroring the US.

8. Airline Food

1 Based on just such anecdotal observations, my colleagues and I hypothe-
sized that there had to be something special about umami; see C.
Spence, C. Michel & B. Smith, 'Airplane noise and the taste of umami',
Flavour, 3:2 (2014).

2 Tomato juice tastes better in the air, scientists reveal. Thank you, scientists . . . see *ShortList Magazine* (https://www.shortlist.com/home/
tomato-juice-tastes-better-in-the-air); A. Burdack-Freitag et al., 'Odor
and taste perception at normal and low atmospheric pressure in a simulated
aircraft cabin', *Journal für Verbraucherschutz und Lebensmittelsicherheit* [*Journal
of Consumer Protection and Food Safety*], 6 (2011), 95–109. See also S. Jackson,
'Why do we drink tomato juice on planes?', *The Pulse*, 2 October 2014.

3 See R. Foss, *Food in the Air and Space: The Surprising History of Food and
Drink in the Skies* (Lanham, MA: Rowman & Littlefield, 2014); Q. Xie,
'When plane food WAS first class: Vintage photos show passengers
being served lobster, caviar and cream cakes during the golden age of
flying', *Daily Mail Online*, 6 May 2016 (http://www.dailymail.co.uk/
travel/travel_news/article-3572879/Forget-beige-unappetising-plane-

food-Vintage-photos-passengers-served-lobster-caviar-cream-cakes-golden-age-flying.html). See also K. Kovalchik, '11 things we no longer see on airplanes' (http://mentalfloss.com/article/51270/11-things-we-no-longer-see-airplanes).

4 M. Mannion, 'BA2012 pops up in London', *Business Traveller*, 10 April 2012 (https://www.businesstraveller.com/news/ba2012-pops-up-in-london); E. Berry, 'Food: History of airline meals', *Nowhere Mag*, February 2013 (http://nowheremag.com/2013/02/food-history-of-airline-meals/).

5 See C. McGuire, 'Please sir, can I have some . . . thing different? Social media craze charts the worst aeroplane food served on flights (and the full English is a top offender)', *Daily Mail Online*, 29 September 2015 (http://www.dailymail.co.uk/travel/travel_news/article-3253262/Please-sir-thing-different-Social-media-craze-charts-worst-aeroplane-food-served-flights-English-offender.html) for a number of unsavoury examples.

6 G. de Syon, 'Is it really better to travel than arrive? Airline food as a reflection of consumer anxiety', in L. C. Rubin (ed.), *Food for Thought: Essays on Eating and Culture* (Jefferson, NC: McFarland & Co., 2008), pp. 199–209.

7 D. Michaels, 'Test flight: Lufthansa searches for savor in the sky', *Wall Street Journal*, 27 July 2010 (http://www.wsj.com/articles/SB10001424052748703294904575384954227906006). I wonder, though, whether it might not also make sense for the chefs to try creating their dishes in this rarefied, not to mention exceedingly noisy, multisensory atmosphere too. There is, after all, evidence to suggest that background noise doesn't just influence how we taste but also affects the way in which we prepare and season foods and drinks; see C. Spence, 'Music from the kitchen', *Flavour*, 4:25 (2015).

8 Calorie estimate from a survey conducted by Jetcost.co.uk; see *The Sunday Times (Travel)*, 15 March 2015, p. 3. According to Dr Charles Platkin, who does an annual calorie count of the food offered by the big airlines, the average number of calories per item in the air was 360 in 2012; T. Thornhill, 'The best and worst airline food of 2015 revealed by diet expert (who discovers one inflight meal with more calories than TWO Big Macs)', *Daily Mail Online*, 26 December 2015

(http://www.dailymail.co.uk/travel/travel_news/article-3374770/The-best-worst-airline-food-2015-revealed-diet-expert-discovers-one-meal-calories-TWO-Big-Macs.html). See also J. Busch et al., 'Salt reduction and the consumer perspective', *New Food*, 2 (2010), 36–9.

9 G. I. Howe, *Dinner in the Clouds: Great International Airline Recipes* (Corona del Mar, CA: Zeta, 1985).

10 K. Severson, 'What's cooking in First Class? Eating and indulging, you can bet it's not peanuts', *The New York Times*, 16 April 2007 (https://www.nytimes.com/2007/04/16/business/businessspecial3/16eats.html?_r=0). Since 2011, Air France has entrusted one of their business-class menu dishes to renowned French chefs such as Joël Robuchon, Guy Martin, Michel Roth, Thibaut Ruggeri, Régis Marcon and Anne-Sophie Pic. According to a press release (1 October 2015), François Adamski was to work with the Servair culinary studio (see 'François Adamski signs new dishes in Air France Business cabin'; http://corporate.airfrance.com/fileadmin/dossiers/documents/press_releases/CP_Adamski_2015.09-en.pdf).

11 D. M. Green & J. S. Butts, 'Factors affecting acceptability of meals served in the air', *Journal of the American Dietetic Association*, 21 (1945), 415–19; P. García-Segovia, R. J. Harrington & H. Seo, 'Influences of table setting and eating location on food acceptance and intake', *Food Quality and Preference*, 39 (2015), 1–7.

12 H. K. Ozcan & S. Nemlioglu, 'In-cabin noise levels during commercial aircraft flights', *Canadian Acoustics*, 34 (2006), 31–5; C. Spence, 'Noise and its impact on the perception of food and drink', *Flavour*, 3:9 (2014).

13 See Y. Kawamura & M. R. Kare, *Umami: A Basic Taste: Physiology, Biochemistry, Nutrition, Food Science* (New York: Marcel Dekker, 1987); O. G. Mouritsen & K. Styrbaek, *Umami: Unlocking the Secrets of the Fifth Taste* (New York: Columbia University Press, 2014); M.-J. Oruna-Concha et al., 'Differences in glutamic acid and 5'-ribonucleotide contents between flesh and pulp of tomatoes and the relationship with umami taste', *Journal of Agricultural and Food Chemistry*, 55 (2007), 5776–80.

14 K. S. Yan & R. Dando, 'A crossmodal role for audition in taste perception', *Journal of Experimental Psychology: Human Perception & Performance*, 41 (2015), 590–96. Ironically, the first study to assess the impact of sound

on taste used tomato juice as the test stimulus and the researcher obtained a null result, thus effectively putting the brakes on further research in this area. Had she chosen pretty much *anything* else for her participants to taste, or at least something that wasn't quite so rich in umami, the results may well have looked rather different, and the last seventy years of research on sound and taste might have taken a very different course; see L. A. Pettit, 'The influence of test location and accompanying sound in flavor preference testing of tomato juice', *Food Technology*, 12 (1958), 55–7.

See S. McCartney, 'The secret to making airline food taste better' – video, *Wall Street Journal*, 13 November 2013 (http://live.wsj.com/video/the-secret-to-making-airline-food-taste-better/8367EF44-52DD-41C4-AC4A-FFA6659F3422.html#!8367EF44-52DD-41C4-AC4A-FFA6659 F3422) on BA's umami-enhanced menu.

Given such differences between taste perception in the air versus on the ground, one might question the wisdom of Virgin Airline's gimmick of encouraging their passengers to choose from one of five taste cubes (bitter, sweet, salty, sour and umami) in the clubhouse so that a perfect cocktail can be prepared in the air. Taste perception will undoubtedly change from one environment to the other; see S. Lawrence, 'In-flight cocktails: Airlines' latest innovation', *Daily Telegraph* (*Luxury*), 29 March 2013 (http://www.telegraph.co.uk/luxury/travel/1248/in-flight-cocktails-airlines-latest-innovation.html).

15 C. Ferber & M. Cabanac, 'Influence of noise on gustatory affective ratings and preference for sweet or salt', *Appetite*, 8 (1987), 229–35.

16 A. T. Woods et al., 'Effect of background noise on food perception', *Food Quality & Preference*, 22 (2011), 42–7.

17 M. G. Skift, 'British Airways has a playlist that it hopes will make its food taste better', *Skift*, 17 October 2014 (https://skift.com/2014/10/17/british-airways-has-a-playlist-that-it-will-make-its-food-taste-better/); A. Victor, 'Louis Armstrong for starters, Debussy with roast chicken and James Blunt for dessert: British Airways pairs music to meals to make in-flight food taste better', *Daily Mail Online*, 15 October 2014 (http://www.dailymail.co.uk/travel/travel_news/article-2792286/british-airways-pairs-music-meals-make-flight-food-taste-better.html).

18 A.-S. Crisinel et al., 'A bittersweet symphony: Systematically modulating the taste of food by changing the sonic properties of the soundtrack playing in the background', *Food Quality and Preference*, 24 (2012), 201–4.

19 J. A. Maga & K. Lorenz, 'Effect of altitude on taste thresholds', *Perceptual and Motor Skills*, 34 (1972), 667–70. As to why this should be so, no one is quite sure.

20 B. Raudenbush & B. Meyer, 'Effect of nasal dilation on pleasantness, intensity and sampling behaviors of foods in the oral cavity', *Rhinology*, 39 (2001), 80–83. For a review, see C. Spence & J. Youssef, 'Olfactory dining: Designing for the dominant sense', *Flavour*, 4:32 (2015).

21 B. Smith, 'Drinking at 30,000 feet', *Prospect Magazine*, July 2014, p. 84. See also B. Tyrer, 'Why plonk tastes posh at 35,000ft', *The Sunday Times* (*Travel*), 20 May 2014, p. 8. (http://www.thesundaytimes.co.uk/sto/travel/Your_Travel/article1410455.ece).

22 See J. Beck, 'Why airplane food is so bad', *The Atlantic*, 19 May 2014 (https://www.theatlantic.com/health/archive/2014/05/the-evolution-of-airplane-food/371076/).

23 For a review of the literature on glassware, see C. Spence, 'Crystal clear or gobbletigook?', *The World of Fine Wine*, 33 (2011), 96–101; C. Spence & I. Wan, 'Beverage perception & consumption: The influence of the container on the perception of the contents', *Food Quality & Preference*, 39 (2015), 131–40.

24 B. Piqueras-Fiszman et al., 'Tasting spoons: Assessing how the material of a spoon affects the taste of the food', *Food Quality and Preference*, 24 (2012), 24–9. The challenge here, if the airlines do come up with some particularly tasty cutlery, will be to keep one's hands on the spoons given how many seem to go missing, even in the most respectable of institutions; see D. A. Fahrenthold & F. Sonmez, 'Stick a fork in Hill's "green" cutlery', *Washington Post*, 5 March 2011, A1, A4; M. S. C. Lim, M. E. Hellard & C. K. Aitken, 'The case of the disappearing teaspoons: Longitudinal cohort study of the displacement of teaspoons in an Australian research institute', *British Medical Journal*, 331 (2005), 1498–1500. Qantas actually has an online store to market the ceramics that they use for their in-flight service; see M. C. O'Flaherty, 'Flying at the height of luxury', *The Times* (*Raconteur*), 17 March 2015, pp. 10–11.

25 K. Kovalchik, '11 things we no longer see on airplanes' (http://mentalfloss.com/article/51270/11-things-we-no-longer-see-airplanes); A. Toffler, *Future Shock* (New York: Random House, 1970), pp. 206–11.

9. The Meal Remembered

1 See F. E. Reichheld & W. E. Sasser, Jr., 'Zero defections: Quality comes to services', *Harvard Business Review*, 68(5) (1990), 105–11.

2 W. James, *Principles of Psychology* (New York: Henry Holt, 1890).

3 B. L. Fredrickson, 'Extracting meaning from past affective experiences: The importance of peaks, ends, and specific emotions', *Cognition and Emotion*, 14 (2000), 577–606; E. Robinson, 'Relationships between expected, online and remembered enjoyment for food products', *Appetite*, 74 (2014), 55–60; H. L. Roediger, Y. Dudai & S. M. Fitzpatrick (eds.), *Science of Memory: Concepts* (Oxford: Oxford University Press, 2007); Y. Weinstein & H. L. Roediger, 'Retrospective bias in test performance. Providing easy items at the beginning of a test makes students believe they did better on it', *Memory & Cognition*, 38 (2010), 366–76.

4 E. N. Garbinsky, C. K. Morewedge & B. Shiv, 'Interference of the end: Why recency bias in memory determines when a food is consumed again', *Psychological Science*, 25 (2014), 1466–74; E. Robinson, J. Blissett & S. Higgs, 'Peak and end effects on remembered enjoyment of eating in low and high restrained eaters', *Appetite*, 57 (2011), 207–12; E. Robinson, J. Blissett & S. Higgs, 'Changing memory of food enjoyment to increase food liking, choice and intake', *British Journal of Nutrition*, 108 (2012), 1505–10; E. Rode, P. Rozin & P. Durlach, 'Experienced and remembered pleasure for meals: Duration neglect but minimal peak, end (recency) or primacy effects', *Appetite*, 49 (2007), 18–29; E. H. Zandstra, B. J. Hauer & M. F. Weegels, *Our Changing Memory for Food* (2008; unpublished research report to the British Feeding and Drinking Group, 2005).

5 L. P. Carbone, *Clued in: How to Keep Customers Coming Back Again and Again* (Upper Saddle River: Prentice Hall, 2004); L. P. Carbone & S. H. Haeckel, 'Engineering customer experiences', *Marketing Management*, 3(3) (1994), 8–19.

6 See A. Fulton, 'Sensorium: A feast for the senses and memories', *NPR*, 31 May 2011 (https://www.npr.org/2011/05/31/136811291/sensorium-a-feast-for-the-senses-and-memories).

7 B. Piqueras-Fiszman & C. Spence, 'Sensory expectations based on product-extrinsic food cues: An interdisciplinary review of the empirical evidence and theoretical accounts', *Food Quality & Preference*, 40 (2015), 165–79; L. Sela & N. Sobel, 'Human olfaction: A constant state of change-blindness', *Experimental Brain Research*, 205 (2010), 13–29.

8 P. Dalton & C. Spence, 'Selective attention in vision, audition, and touch', in R. Menzel (ed.), *Cognitive Psychology of Memory* (vol. 2 of *Learning and Memory: A Comprehensive Reference*; Oxford: Elsevier, 2008), pp. 243–57.

9 See also J. Rayner, 'What makes a meal really memorable?', *Guardian*, 18 February 2016 (https://www.theguardian.com/lifeandstyle/2016/feb/18/what-makes-a-meal-really-memorable).

10 J. C. Pfeiffer et al., 'Taste-aroma interactions in a ternary system: A model of fruitiness perception in sucrose/acid solutions', *Perception & Psychophysics*, 68 (2006), 216–27.

11 L. Sela & N. Sobel, 'Human olfaction: A constant state of change-blindness', *Experimental Brain Research*, 205 (2010), 13–29.

12 A. T. Woods et al., 'Flavor expectation: The effects of assuming homogeneity on drink perception', *Chemosensory Perception*, 3 (2010), 174–81; E. Le Berre et al., 'Reducing bitter taste through perceptual constancy created by an expectation', *Food Quality & Preference*, 28 (2013), 370–74; J. Busch et al., 'Salt reduction and the consumer perspective', *New Food*, 2 (2010), 36–9. Alternatively, one can contrast high and low salt elements, or layers, say within a food product. Of course, this strategy will not work for all foods: Oreos spring to mind here as a particularly challenging case, given that some people like to start with the dark biscuit, whereas others always start with the cream filling.

13 L. Hall et al., 'Magic at the marketplace: Choice blindness for the taste of jam and the smell of tea', *Cognition*, 117 (2010), 54–61; though see also C. Petitmengin et al., 'A gap in Nisbett and Wilson's findings? A first-person access to our cognitive processes', *Consciousness and Cognition*, 22 (2013), 654–69.

14 Check out the video for yourself at https://www.youtube.com/watch?v=_VPclo4Adh8.

15 T. Davis, 'Taste tests: Are the blind leading the blind?', *Beverage World*, 3 (1987), 42–4, 50, 85; D. Martin, 'The impact of branding and marketing on perception of sensory qualities', *Food Science & Technology Today: Proceedings*, 4(1) (1990), 44–9. See also J. Robinson, 'Lager? It all tastes the same: Drinkers struggle to distinguish between big name brands in blind taste tests', *Daily Mail Online*, 13 August 2014 (http://www.dailymail.co.uk/news/article-2723583/Lager-It-tastes-Drinkers-struggle-distinguish-big-brands-blind-taste-tests.html).

16 C. Spence, 'The price of everything – the value of nothing?', *The World of Fine Wine*, 30 (2010), 114–20; V. Harrar et al., 'Grape expectations: How the proportion of white grape in Champagne affects the ratings of experts and social drinkers in a blind tasting', *Flavour*, 2:25 (2013).

17 N. Burton & J. Flewellen, *The Concise Guide to Wine and Blind Tasting* (Exeter, Devon: Acheron Press, 2014); B. Smith, 'Wine: Secrets of blind tasting', *Prospect Magazine*, November 2016 (http://www.prospectmagazine.co.uk/magazine/wine-secrets-of-blind-tasting).

18 L. P. Carbone & S. H. Haeckel, 'Engineering customer experiences', *Marketing Management*, 3(3) (1994), 8–19; p. 8.

19 Ever since the arrival of nouvelle cuisine back in the 1970s, the amuse gueule, or amuse bouche as it is now more popularly known, has become a regular feature in many restaurants; see W. Grimes, 'First, a little something from the chef . . . Very, very little', *The New York Times*, 22 July 1998 (https://www.nytimes.com/1998/07/22/dining/first-a-little-something-from-the-chef-very-very-little.html?pagewanted=all&src=pm). And what was once nothing more than a complimentary 'palate cleanser', has now evolved into something far more exotic, at least at the high end. This 'unexpected' gift can lift the diner's mood, it may tantalize their appetite and, crucially, it provides a very simple means of making an impression; B. Wansink, *Mindless Eating: Why We Eat More Than We Think* (London: Hay House, 2006). As one journalist writing for *The New York Times* put it, recommending that we should *all* be surprising our guests the next time we host a dinner party: 'As it often is in restaurants, the amuse-bouche can be one of the most memorable moments of the meal' (M. Clark, 'Tiny

come-ons, plain and fancy', *The New York Times*, 30 August 2006 (https://www.nytimes.com/2006/08/30/dining/30amus.html)).

20 B. L. Fredrickson & D. Kahneman, 'Duration neglect in retrospective evaluations of affective episodes', *Journal of Personality & Social Psychology*, 65 (1993), 45–55.

21 B. B. Murdock, 'The serial position effect of free recall', *Journal of Experimental Psychology*, 64 (1962), 482–8; N. H. Anderson & A. Norman, 'Order effects in impression formation in four classes of stimuli', *Journal of Abnormal and Social Psychology*, 69 (1964), 467–71. The latter research reports that imagined meals starting with three highly liked dishes (followed by three less-liked dishes) were rated more highly by 140 students than when the imagined meal started with the less-liked dishes instead.

22 B. Wansink & L. W. Linder, 'Interactions between forms of fat consumption and restaurant bread consumption', *International Journal of Obesity*, 27 (2003), 866–8. It is, though, perhaps worth bearing in mind here that not all studies have demonstrated enhanced memory for peak, end (recency) or primacy effects when it comes to our memories of meals: see E. Rode, P. Rozin & P. Durlach, 'Experienced and remembered pleasure for meals: Duration neglect but minimal peak, end (recency) or primacy effects', *Appetite*, 49 (2007), 18–29. However, in the latter study, statistical power was pretty low (only 20 participants were tested in the critical Chinese buffet meal study, Experiment 3), as the authors themselves readily acknowledge; furthermore, the peak dish wasn't as outstanding as one might expect to find in a fancy restaurant, which may also help to explain the lack of a peak effect. In other words, more gastrophysics research needed here!

23 In this case, though, don't ask me to vouch for the quality of the food! K. A. LaTour & L. P. Carbone, 'Sticktion: Assessing memory for the customer experience', *Cornell Hospitality Quarterly*, 55 (2014), 342–53.

24 Take, for example, the iconic Meat Fruit dish at Dinner in the Mandarin Oriental hotel in London (http://www.four-magazine.com/articles/291/chicken-liver-meat-fruit). This famous chicken liver dish, a beautiful example of trompe-l'oeil, looks exactly like a mandarin (just be sure not to eat the stalk). I bet that anyone who has been to the

restaurant will remember that dish, no matter what else they forget. Whenever I go back, I do want to order it again . . . and again. That I know this dish will be on the menu when I return next is most certainly comforting in some way.

25 F. Gobet, M. Lloyd-Kelly & P. C. R. Lane, 'What's in a name? The multiple meanings of "chunk" and "chunking"', *Frontiers in Psychology*, 7:102 (2016).

26 A menu, or a bag of sweets, basically anything to help the diner remember their meal. As one restaurant critic's comments clearly communicate: 'I came away from my third meal at The Fat Duck in Bray, Berkshire, somewhat the poorer financially but with strong memories to compensate [. . .] And I can still savour the four hours we spent at the table, not least thanks to the bag of retro-styled sweets that one chooses from a trolley at the end of the meal' (N. Lander, 'Blumenthal's magnificent flying machine', *Financial Times*, 27 February 2016, p. 33).

27 C. Spence & B. Piqueras-Fiszman, *The Perfect Meal: The Multisensory Science of Food and Dining* (Oxford: Wiley-Blackwell, 2014).

28 B. Piqueras-Fiszman & C. Spence, 'Sensory incongruity in the food and beverage sector: Art, science, and commercialization', *Petits Propos Culinaires*, 95 (2012), 74–118; see also M. Hanefors & L. Mossberg, 'Searching for the extraordinary meal experience', *Journal of Business and Management*, 9 (2003), 249–70.

29 R. B. Chase & S. Dasu, 'Want to perfect your company's service? Use behavioral science', *Harvard Business Review*, 79(6) (2001), 79–84.

30 E. L. Robinson & S. Higgs, 'Memory and food: Leaving the best till last', *Appetite*, 57 (2011), 538. See also E. N. Garbinsky, C. K. Morewedge & B. Shiv, 'Interference of the end: Why recency bias in memory determines when a food is consumed again', *Psychological Science*, 25 (2014), 1466–74.

31 E. Robinson et al., 'Eating attentively: A systematic review and meta-analysis of the effect of food intake memory and awareness on eating', *American Journal of Clinical Nutrition*, 97 (2013), 728–42; see also Y. Cornil & P. Chandon, 'Pleasure as a substitute for size: How multisensory imagery can make people happier with smaller food portions', *Journal of Marketing Research*, 53(5) (2016), 847–64.

32 See S. Higgs & H. McVittie, 'Engagement with a computer game affects lunch memory and later food intake', *Appetite*, 59 (2012), 628.

33 J. K. O'Regan, 'Solving the "real" mysteries of visual perception: The world as an outside memory', *Canadian Journal of Psychology*, 46 (1992), 461–88.

34 B. Stuckey, *Taste What You're Missing: The Passionate Eater's Guide to Why Good Food Tastes Good* (London: Free Press, 2012).

35 A. Barnett & C. Spence, 'Assessing the effect of changing a bottled beer label on taste ratings', *Nutrition and Food Technology*, 2 (2016), 4.

36 R. Raghunathan, R. W. Naylor & W. D. Hoyer, 'The unhealthy = tasty intuition and its effects on taste inferences, enjoyment, and choice of food products', *Journal of Marketing*, 70(4) (2006), 170–84; B. Wansink, S. T. Sonka & C. M. Hasler, 'Front-label health claims: When less is more', *Food Policy*, 29 (2004), 659–67; B. Wansink, K. van Ittersum & J. E. Painter, 'How diet and health labels influence taste and satiation', *Journal of Food Science*, 69 (2004), S340–S346. Though intuitions about taste certainly vary. North Americans intuitively expect unhealthy foods to be tasty, but that intuition certainly isn't shared by the French; see C. O. C. Werle, O. Trendel & G. Ardito, 'Unhealthy food is not tastier for everybody: The "healthy = tasty" French intuition', *Food Quality and Preference*, 28 (2013), 116–21. On the flip side, even simply reading a word such as 'salt' or 'cinnamon' can activate many of the same brain areas as when a salty taste is actually experienced in the mouth; J. González et al., 'Reading *cinnamon* activates olfactory brain regions', *NeuroImage*, 32 (2006), 906–12. Indeed, more of our brain lights up when we merely think about (or anticipate) food than when we actually get to taste it; J. O'Doherty et al., 'Neural responses during anticipation of a primary taste reward', *Neuron*, 33 (2002), 815–26; M. Khan, 'PepsiCo R&D: A catalyst for change in the food and beverage industry', *New Food*, 16(4) (2015), 10–13.

37 E. P. Köster, 'Memory for food and food expectations: A special case?', *Food Quality and Preference*, 17 (2006), 3–5; J. Mojet & E. P. Köster, 'Flavor memory', in B. Piqueras-Fiszman & C. Spence (eds.), *Multisensory Flavour Perception: From Fundamental Neuroscience through to the Marketplace* (Amsterdam: Elsevier, 2016), pp. 169–84.

38 P. Rozin et al., 'What causes humans to begin and end a meal? A role for memory for what has been eaten as evidenced by a study of multiple meal eating in amnesic patients', *Psychological Science*, 9 (1998), 392–6.

Although, intriguingly, sensory-specific satiety is intact in these individuals; S. Higgs et al., 'Sensory-specific satiety is intact in amnesiacs who eat multiple meals', *Psychological Science*, 19 (2008), 623–8.

39 O. Franklin-Wallis, 'Lizzie Ostrom wants to transform people's lives through their noses', *Wired*, 3 October 2015 (http://www.wired.co.uk/magazine/archive/2015/11/play/lizzie-ostrom-smell); J. Morton, *MedTech Engine*, 6 January 2016 (https://medtechengine.com/article/appetite-stimulation-in-dementia-patients/).

40 D. M. Bernstein & E. F. Loftus, 'The consequences of false memories for food preferences and choices', *Perspectives in Psychological Science*, 4 (2009), 135–9; D. M. Bernstein et al., 'False beliefs about fattening foods can have healthy consequences', *Proceedings of the National Academy of Sciences of the USA*, 102 (2005), 13724–31; D. M. Bernstein et al., 'False memories about food can lead to food avoidance', *Social Cognition*, 23 (2005), 11–14; C. Laney et al., 'Asparagus, a love story: Healthier eating could be just a false memory away', *Experimental Psychology*, 55 (2008), 291–300; A. Scoboria, G. Mazzoni & J. Jarry, 'Suggesting childhood food illness results in reduced eating behavior', *Acta Psychologica*, 128 (2008), 304–9.

41 S. Higgs, 'Memory for recent eating and its influence on subsequent food intake', *Appetite*, 39 (2002), 159–66; S. Higgs, A. C. Williamson & A. S. Attwood, 'Recall of recent lunch and its effect on subsequent snack intake', *Physiology & Behavior*, 94 (2008), 454–62. One of the more interesting questions here is what exactly it is that provokes us to eat in the first place. Obviously, in part, and under certain conditions, it is the physiological need state of hunger, but on a day-to-day basis, the evidence suggests that our memory of when we last ate is often just as important.

42 As one commentator noted: '[E]ating can serve as a medium for the act of remembering' (quoted in D. E. Sutton, *Remembrance of Repasts: An Anthropology of Food and Memory* (Oxford: Berg, 2001)). See also J. Mojet & E. P. Köster, 'Flavor memory', in B. Piqueras-Fiszman & C. Spence (eds.), *Multisensory Flavour Perception: From Fundamental Neuroscience through to the Marketplace* (Amsterdam: Elsevier, 2016), pp. 169–84; J. S. Allen, *The Omnivorous Mind: Our Evolving Relationship with Food* (London: Harvard University Press, 2012).

43 J. A. Brillat-Savarin, *Physiologie du goût* [*The Philosopher in the Kitchen/ The Physiology of Taste*] (Brussels: J. P. Meline, 1835); published as *A Handbook of Gastronomy*, trans. A. Lazaure (London: Nimmo & Bain, 1884), p. 14.

10. *The Personalized Meal*

1 V. Barford, 'Will you tell Starbucks your name?', *BBC News Magazine*, 14 March 2012 (https://www.bbc.co.uk/news/magazine-17356957). They do, of course, occasionally get it wrong, but I am sure that the beneficial effect on the tasting experience when they get it right far outweighs any annoyance or humour caused on the few occasions when there is a cock-up; see A. Konstantinides, 'Denial, Christ and Bratt walk into a Starbucks . . . Customers of the beloved coffee chain share the baristas most hilarious name fails', *Daily Mail Online*, 24 October 2016 (http://www.dailymail.co.uk/news/article-3858194/Starbucks-customers-share-hilarious-fails.html). I'm far more worried about being given tea in a paper cup that clearly states 'Starbucks coffee' on the front.

2 There were more than 1,000 names to choose from on the shelf, and 500,000 through their online store. The campaign resulted in over 998 million impressions on Twitter and more than 150 million personalized bottles were sold; see M. Hepburn, 'The share a Coke story' (https://www.coca-cola.co.uk/stories/history/advertising/share-a-coke/). Note that this campaign isn't only about *personalization* but also *sharing* (see 'Social Dining'), thus hitting two important touch points with consumers. A pedant could, I suppose, question whether this is really about *personalization* given that bottles with thousands of popular names were printed and placed in store (perhaps it is better thought of as mass customization).

3 Note, though, that the instantly recognizable colour scheme and typeface of Coke were kept, maintaining ownership even when the brand name itself no longer appeared on the front of the bottle or can; see C. Velasco et al., 'The taste of typeface', *i-Perception*, 6(4), 1–10.

4 See A. Dan, '11 marketing trends to watch for in 2015', *Forbes*, 9 November 2015 (http://www.forbes.com/sites/avidan/2014/11/09/11-marketing-trends-to-watch-for-in-2015/#63a810e67e83).

5 J. Sui, X. He & G. W. Humphreys, 'Perceptual effects of social salience: Evidence from self-prioritization effects on perceptual matching', *Journal of Experimental Psychology: Human Perception & Performance*, 38 (2012), 1105–17. While this self-prioritization effect was first demonstrated with arbitrary visual objects, here at the Crossmodal Research Laboratory, we have recently shown that it extends to sounds and even tactile stimuli too. So there would seem to be no good reason why this phenomenon wouldn't also influence our perception of smells and tastes. In other words, it may colour our response to flavourful food and beverages, like coffee and cola; S. Schäfer et al., 'Self-prioritization in vision, audition, and touch', *Experimental Brain Research*, 234 (2016), 2141–50.

6 See C. Spence & A. Gallace, 'Multisensory design: Reaching out to touch the consumer', *Psychology & Marketing*, 28 (2011), 267–308; C. Spence & B. Piqueras-Fiszman, 'The multisensory packaging of beverages', in M. G. Kontominas (ed.), *Food Packaging: Procedures, Management and Trends* (Hauppauge, NY: Nova Publishers, 2012), pp. 187–233.

7 R. Thaler, 'Toward a positive theory of consumer choice', *Journal of Economic Behavior and Organization*, 1 (1980), 39–60; see also L. Cramer & G. Antonides, 'Endowment effects for hedonic and utilitarian food products', *Food Quality and Preference*, 22 (2011), 3–10; W. Samuelson & R. Zeckhauser, 'Status quo bias in decision making', *Journal of Risk and Uncertainty*, 1 (1988), 7–59; C. K. Morewedge & C. E. Giblin, 'Explanations of the endowment effect: An integrative review', *Trends in Cognitive Sciences*, 19 (2015), 339–48; E. Van Dijk & D. Van Knippenberg, 'Trading wine: On the endowment effect, loss aversion and comparability of consumer goods', *Journal of Economic Psychology*, 19 (1998), 485–95.

8 Although a quick scan of the newspapers reveals a couple of academics claiming to have done something similar. In one study, for instance, Professor David Figgins from Huddersfield University had people come into the lab with their own favourite mug. They were blindfolded and given two cups of tea to taste, one in their own mug and one in a plain white mug. The participants simply had to say which tea tasted better. Over 70% picked the tea from their mug, 21% thought all the tea tasted the same and just 9% thought the tea in the plain mug was better. The tea was the same; all that differed was the mug. However,

while the evidence looks promising, no properly controlled peer-reviewed experiment has yet been published on the topic.

9 A. Bronkhorst, 'The cocktail party phenomenon: A review of research on speech intelligibility in multiple-talker conditions', *Acustica*, 86 (2000), 117–28.

10 G. W. Humphreys & J. Sui, 'Attentional control and the self: The Self-Attention Network (SAN)', *Cognitive Neuroscience*, 21 (2015), 1–13. Though see also A. M. García et al., 'Commentary: Attentional control and the self: The Self-Attention Network (SAN)', *Frontiers in Psychology*, 6:1726 (2015).

11 K. A. Carlson & J. M. Conard, 'The last name effect: How last name influences acquisition timing', *Journal of Consumer Research*, 38 (2011), 300–307.

12 C. M. Brendl et al., 'Name letter branding: Valence transfers when product-specific needs are active', *Journal of Consumer Research*, 32 (2005), 405–15; J. T. Jones et al., 'How do I love thee? Let me count the Js: Implicit egotism and interpersonal attraction', *Journal of Personality and Social Psychology*, 87 (2004), 665–83; cf. J. M. Nuttin, Jr, 'Narcissism beyond Gestalt and awareness: The name letter effect', *European Journal of Social Psychology*, 15 (1985), 353–61.

13 The same thing happens in the world's best sushi restaurant; see the documentary film *Jiro Dreams of Sushi*.

14 Trotter would have his staff pick up the discarded cigarette butts from off the sidewalk outside his namesake Chicago restaurant – nothing should detract from the experience! ('Charlie Trotter: He changed Chicago dining – and America's', *Time*, 18 November 2013, p. 12). In a similar vein, on a recent trip to The Fat Duck Research Kitchens in Bray, I spotted one of the junior kitchen staff vigorously scrubbing the pavement outside the front door with a bucket of hot soapy water, long before the day's guests would start to arrive at the restaurant. But don't think that such a practice is restricted only to high-end restaurants: believe it or not, in the early days the legendary Ray Kroc would spend his weekends scraping the gum off the forecourts of his McDonald's restaurants in the US (C. Steel, *Hungry City: How Food Shapes Our Lives* (London: Chatto & Windus, 2008), pp. 235–6).

15 D. Meyer, *Setting the Table: Lessons and Inspirations from One of the World's Leading Entrepreneurs* (London: Marshall Cavendish International,

2010); cf. F. E. Reichheld & W. E. Sasser, Jr, 'Zero defections: Quality comes to services', *Harvard Business Review*, 68(5) (1990), 105–11. Restaurants use acronyms such as FOM (Friend Of the Manager) or WW (Wine Whale); S. Craig, 'What restaurants know (about you)', *The New York Times*, 4 September 2012 (https://www.nytimes.com/2012/09/05/dining/what-restaurants-know-about-you.html?pagewanted=all&_r=0).

16 J. A. Heidemann, 'You've been Googled — bon appetit!', *Chicago Business*, 29 June 2013 (https://www.chicagobusiness.com/article/20130629/ISSUE03/306299997/youve-been-googled-bon-appetit); S. Craig, 'What restaurants know (about you)', *The New York Times*, 4 September 2012 (https://www.nytimes.com/2012/09/05/dining/what-restaurants-know-about-you.html?pagewanted=all&_r=0).

17 Quotes from A. Sytsma, 'Hardcore coddling: How Eleven Madison Park modernized elite, old-school service', *Grub Street*, 9 April 2014 (http://www.grubstreet.com/2014/04/eleven-madison-park-foh-staff-detailed-look.html?mid=huffpost_lifestyle). See also A. Spiegel, 'Visiting a restaurant? There's a good chance it's Googling you', *Huffington Post*, 14 April 2014 (http://www.huffingtonpost.com/2014/04/14/restaurants-googling-patrons_n_5132535.html); A. Waiter, *Waiter Rant: Behind the Scenes of Eating Out* (London: John Murray, 2009), p. 33. There is even a consulting business that has been built up on the back of this; see D. Cardwell, 'Spreading his gospel of warm and fuzzy', *The New York Times*, 23 April 2010 (https://www.nytimes.com/2010/04/25/nyregion/25meyer-ready.html?pagewanted=all).

One does have to wonder, though, what happens to the anonymous restaurant reviewer. Do they simply not get the same experience as the guests who may be reading their recommendations? Mimi Sheraton, who would famously wear a variety of wigs and outfits to disguise her identity when she was out reviewing restaurants in NYC, wrote: 'The longer I reviewed restaurants, the more I became convinced that the unknown customer has a completely different experience from either a valued patron or a recognized food critic; for all practical purposes, they might as well be in different restaurants' (*Eating My Words: An Appetite for Life* (New York: Harper, 2004), pp. 104–5); see also I. Parker, 'Pete Wells has his knives out: New York Times critic writes the reviews that make and break restaurants', *The New Yorker*, 12

September 2016 (https://www.newyorker.com/magazine/2016/09/12/
pete-wells-the-new-york-times-restaurant-critic).

18 'Lunchtime poll – investigating patrons', *CNN*, 10 August 2010 (https://
cnneatocracy.wordpress.com/2010/10/28/lunchtime-poll-investigating-
patrons/); quote from S. Craig, 'What restaurants know (about you)',
The New York Times, 4 September 2012 (https://www.nytimes.com/
2012/09/05/dining/what-restaurants-know-about-you.html?pagewanted=
all&_r=0).

This is reminiscent of those targeted ads that pop up on your
favourite websites; see M. Köster et al., 'Effects of personalized banner
ads on visual attention and recognition memory', *Applied Cognitive
Psychology*, 29 (2015), 181–92; S. Martin, 'Let's get personal: Use individual touches if you want your customers to remember you', *Business
Life*, April 2015, 45.

19 The chef has apparently been consulting with top UK magician Der-
ren Brown on how to find out about the diners at his restaurant without
them necessarily being aware of divulging any information. Though,
of course, the diner, on experiencing some surprising act of personalization, might assume that the restaurant must have been Googling,
thus defeating the object; J. Rayner, 'It's goodbye to snail porridge as
Heston Blumenthal bids to reinvent the restaurant again', *Guardian*, 23
August 2015 (https://www.theguardian.com/lifeandstyle/2015/aug/23/
goodbye-snail-porridge-heston-blumenthal-fat-duck-new-chapter-in-
food). Another Michelin-starred British chef, Jason Atherton, had this
to say: 'We certainly don't go Googling into people's private affairs . . .
There are going to be people who absolutely love it and there are going
to be people saying, "Don't go snooping into my affairs." ', J. Malvern,
'Heston's next trick is mindblowing', *The Times*, 24 August 2015, p. 3
(http://www.thetimes.co.uk/tto/life/food/article4536244.ece).

20 Research clearly shows that those waiting staff who use first names,
who casually touch the diner, and so on, end up with bigger tips, presumably because it results in a more personalized style of service. And
while the average tip size isn't necessarily the ideal measure of customer
satisfaction, it is currently the best metric that researchers have got; see
J. Surowiecki, 'The financial page: Check, please', *The New Yorker*, 5
September 2005, p. 58; B. Rind & P. Bordia, 'Effect of server's "thank

you" and personalization on restaurant tipping', *Journal of Applied Social Psychology*, 25 (1995), 745–51.

21 S. Miles, '6 tools restaurants can use for better guest intelligence', *Streetfight*, 22 July 2013 (http://streetfightmag.com/2013/07/22/6-tools-restaurants-can-use-for-better-guest-intelligence/).

22 See M. Orlando, 'Worms, sustainability, and innovation', The Conference, Malmö, Sweden, 16–17 August 2016 (http://videos.theconference.se/matt-orlando-worms-sustainability-and-innovation).

23 D. Bates, 'Top tip from Queen's party planner: Sit bores together, conversation is more important than the food and organize everything by phone', *Daily Mail Online*, 25 April 2016 (http://www.dailymail.co.uk/news/article-3556827/Top-tip-Queen-s-party-planner-Sit-bores-conversation-important-food-organise-phone.html).

24 'The rise of the no-choice restaurant', *Forbes Travel Guide*, 26 September 2012 (http://www.forbes.com/sites/forbestravelguide/2012/09/26/the-rise-of-the-no-choice-restaurant/#2739835d5f0a).

25 In fact, back in the very earliest days of the restaurant in Paris in the latter half of the eighteenth century, one would find both regular restaurant tables but also 'small separate apartments, where one is served just as one is at home [*comme chez soi*]' (R. L. Spang, *The Invention of the Restaurant: Paris and Modern Gastronomic Culture* (Cambridge, MA: Harvard University Press, 2000), p. 79).

26 Based on an analysis of a huge number of restaurant menus posted online, linguist Dan Jurafsky notes that 'expensive restaurants ($$$$) have half as many dishes as cheap ($) restaurants' (D. Jurafsky, *The Language of Food: A Linguist Reads the Menu* (New York: Norton, 2014), p. 12).

27 'Menus without choice blaspheme against the doctrine of dining', *FT Weekend Magazine*, 23 January 2016, p. 12.

28 See 'The rise of the no-choice restaurant', *Forbes Travel Guide*, 26 September 2012 (http://www.forbes.com/sites/forbestravelguide/2012/09/26/the-rise-of-the-no-choice-restaurant/#2739835d5f0a); R. L. Spang, *The Invention of the Restaurant: Paris and Modern Gastronomic Culture* (Cambridge, MA: Harvard University Press, 2000).

29 Though even here the choice is restricted to rare, medium or well done. If you want your meat medium-rare, for example, best go somewhere

else. Another idiosyncratic feature of this restaurant is that it does not accept bookings. All that waiting outside also helps add to the seeming desirability of the venue, though one might want to frame it as the very antithesis of hospitality; see A. A. Gill, 'Table Talk: BAO Fitzrovia', *The Sunday Times* (*Magazine*), 14 August 2016, p. 48.

30 C. Spence & B. Piqueras-Fiszman, *The Perfect Meal: The Multisensory Science of Food and Dining* (Oxford: Wiley-Blackwell, 2014).

31 Ogilvy's talk is summarized at https://www.warc.com/Content/News/ N34910_Behavioural_economics_is_effective__.content?PUB=Warc% 20News&CID=N34910&ID=00be1349-4c3d-4b81-81e3-31f01402d325&q= sutherland&qr. See also S. S. Iyengar, *The Art of Choosing: The Decisions We Make Everyday – What They Say About Us and How We Can Improve Them* (USA: Little, Brown, 2010); S. S. Iyengar & M. R. Lepper, 'When choice is demotivating: Can one desire too much of a good thing?', *Journal of Personality and Social Psychology*, 79 (2000), 995–1006; E. Osnos, 'Too many choices? Firms cut back on new products', *Philadelphia Inquirer*, 27 September 1997, D1, D7; B. Schwartz, *The Paradox of Choice: Why More Is Less* (London: HarperCollins, 2005). Note also the more choices that we are given at mealtimes, the more we eat: D. A. Levitsky, S. Iyer & C. R. Pacanowski, 'Number of foods available at a meal determines the amount consumed', *Eating Behaviors*, 13 (2012), 183–7.

32 See R. O. Bellomo, 'Condiment sommeliers are taking over New York City', *Time Out* (*New York*), 26 February 2015 (https://www.timeout. com/newyork/blog/condiment-sommeliers-are-taking-over-new-york-city); E. Kis, 'Maille mustard boutique's sommelier classes up our condiment game', *Metro* (*US*), 17 December 2014 (http://www.metro. us/lifestyle/maille-mustard-boutique-s-sommelier-classes-up-our-condiment-game/zsJnlr---IBboIYw4kUTgY/).

33 M. I. Norton, D. Mochon & D. Ariely, 'The IKEA Effect: When labor leads to love', *Journal of Consumer Psychology*, 22 (2012), 453–60. Kit meals are ideal for this kind of research, inasmuch as their taste/flavour is more or less standardized, while still allowing the maker to feel involved in the creation process. As one commentator succinctly put it, they allow 'self-expression for the time-deprived' (quote from J. Rossant, 'Somewhat individual', *Forbes*, 157 (1996), 152). See also S. V. Troye & M. Supphellen, 'Consumer participation in coproduction: "I made it

myself" effects on consumers' sensory perceptions and evaluations of outcome and input product', *Journal of Marketing*, 76 (2012), 33–46.

34 Ikea's new pop-up dining concept in Shoreditch, London, would seem to be playing to this; see I. Blake, 'Fancy a flat-pack dinner? FEMAIL visits Ikea's new restaurant where you cook your own gourmet meals – and there's not a meatball in sight', *Daily Mail Online*, 9 September 2016 (http://www.dailymail.co.uk/femail/food/article-3781301/Fancy-flat-pack-dinner-FEMAIL-visits-Ikea-s-new-restaurant.html).

35 Not just any old marketing executive either: it was Ernst Dichter, one of Louis Cheskin's long-term collaborators. Both were emigrés who fled the chaos and persecution in central Europe in the middle of the last century. For a history of the period, see: L. R. Samuel, *Freud on Madison Avenue: Motivation Research and Subliminal Advertising in America* (Oxford: University of Pennsylvania Press, 2010).

36 See M. Y. Park, 'A history of the cake mix, the invention that redefined "baking"', *Bon Appétit*, 26 September 2013 (https://www.bonappetit.com/entertaining-style/pop-culture/article/cake-mix-history); S. Shapiro, *Something from the Oven: Reinventing Dinner in 1950s America* (London: Penguin, 2005).

37 M. Pollan, *Cooked: A Natural History of Transformation* (London: Penguin Books, 2013).

38 M. Y. Park, 'A history of the cake mix, the invention that redefined "baking"', *Bon Appétit*, 26 September 2013 (https://www.bonappetit.com/entertaining-style/pop-culture/article/cake-mix-history).

39 D. Kahneman, 'Why do sandwiches taste better when someone else makes them?', *The New York Times*, 2 October 2011 (http://query.nytimes.com/gst/fullpage.html?res=9E0DE2DE123EF931A3575 3C1A9679D8B63); see also https://ask.metafilter.com/21673/Why-does-a-sandwich-taste-better-if-someone-else-makes-it.

40 S. S. Sundar & S. S. Marathe, 'Personalization versus customization: The importance of agency, privacy, and power usage', *Human Communication Research*, 36 (2010), 298–322.

41 This book had a profound effect on many chefs. Just take Anthony Bourdain: 'I don't know if I can adequately convey to you the impact that *White Heat* had on me, on the chefs and cooks around me, on

subsequent generations' (quoted in C. Rintoul, 'The next chef revolution', *Food is the New Internet* (https://medium.com/food-is-the-new-internet/the-next-chef-revolution-dfe75f0820d2#.k62loo2e8)).

42 C. Rintoul, 'The next chef revolution', *Food is the New Internet* (https://medium.com/food-is-the-new-internet/the-next-chef-revolution-dfe75f0820d2#.k62loo2e8). But the very first show kitchen was, in fact, installed in the Brighton Pavilion by the Prince Regent, later George IV, for French chef Marie Antoin (Antonin) Carême. It was completed by John Nash in 1818 and was more than 7 metres tall. As noted in one contemporary description: 'A kitchen so close to the Banqueting Room was unusual for the day. It gave George IV the opportunity to impress his guests with his new facilities and he often escorted his guests around the Great Kitchen as part of his tour of the state apartments. It was one of the first "show" kitchens.'

43 J. Lanchester, 'Restaurant: Cut, London W1 – review: John Lanchester spends £240 on Planet Rich, so you don't have to', *Guardian*, 7 October 2011 (https://www.theguardian.com/lifeandstyle/2011/oct/07/cut-wolfgang-puck-london-review); M. Norman, 'Restaurant review: Cut, London: There is nice food at Cut in London, but is head chef Mr Puck a robbin' goodfellow?', *Daily Telegraph*, 24 October 2011 (http://www.telegraph.co.uk/foodanddrink/restaurants/8838203/Restaurant-review-Cut-London.html).

44 As the author of *Waiter Rant* notes: 'When people go out to eat, they don't want to hear the word no' (London: John Murray, 2009), p. 39. See also H. G. Parsa, 'Why restaurants fail', *Cornell Hotel & Restaurant Administration Quarterly*, 46 (2005), 304–22; B. D. Wolf, 'Cameron: Now, most likely to succeed', *Columbus Dispatch*, 23 October 2003.

45 F. T. Marinetti, 'Nourishment by Radio', in F. T. Marinetti, *The Futurist Cookbook*, translated by S. Brill (1932; San Francisco: Bedford Arts, 1989), p. 67.

For customization examples see M. Kim, 'Nestlé unveils personalized luxury chocolates', *Candy Industry*, 1 March 2013 (https://www.candyindustry.com/articles/85018-nestl--unveils-personalized-luxury-chocolates); E. B. York, 'Chocolate gets personal: Nestle to launch customization pilot program', *Chicago Tribune*, 20 October 2011 (http://articles.chicagotribune.com/2011-10-20/business/ct-biz-1020-

chocolate-20111020_1_chocolate-business-premium-chocolate-chocolate-preferences); P. Kakaviatos, 'Crafting personalized champagne at Duval-Leroy', *MyInforms.com*, 19 June 2016 (http://myinforms.com/en-us/a/36869357-crafting-personalized-champagne-at-duval-leroy/); and the video at R. Mattu, 'Future of food', *Financial Times*, 22 June 2016 (http://video.ft.com/4972852572001/Future-of-food/Companies).

As Richard Newcombe of the New Zealand Institute for Plant and Food Research, Auckland, puts it: '[k]nowing how genes determine olfaction could lead to the creation of recipes for food or other products that are tailored to an individual's molecular odour profile' (J. Howgego, 'Sense for scents traced down to genes: Genetic factors for ability to smell violets and certain foods point to a future of "personalized" scents and flavours', *Nature Neuroscience*, 1 August 2013 (http://www.nature.com/news/sense-for-scents-traced-down-to-genes-1.13493)). D. R. Reed & A. Knaapila note that 'perhaps no single human trait has as many person-to-person differences as the ability to taste and smell' ('Genetics of taste and smell: Poisons and pleasures', *Progress in Molecular Biology Translational Science*, 94 (2010), 213–40; p. 215).

11. *The Experiential Meal*

1 '50 Days By Albert Adrià: Who is this Spanish chef and why is his London residency worth getting excited about?', *Evening Standard*, 1 December 2015 (http://www.standard.co.uk/goingout/restaurants/50-days-by-albert-adri-who-is-this-spanish-chef-and-why-is-his-london-residency-is-worth-getting-a3127366.html).

2 The experience economy can be seen as the natural progression of Kotler's concept of selling the *total* product and not just the *tangible* (or should that be edible?) product; see P. Kotler, 'Atmospherics as a marketing tool', *Journal of Retailing*, 49 (Winter 1974), 48–64; B. J. Pine, II, & J. H. Gilmore, 'Welcome to the experience economy', *Harvard Business Review*, 76(4) (1998), 97–105; B. J. Pine, II, & J. H. Gilmore, *The Experience Economy: Work Is Theatre & Every Business is a Stage* (Boston, MA: Harvard Business Review Press, 1999). Or as one recent article put it: 'Experiences have overtaken products as the must-have purchase' (L.

Xiong, T. Loo & R. Chen, 'The adorkable Chinese Post-oo teens', *Admap*, February 2016, 44–5). No matter what you think about the approach, the experience economy continues to exert a profound influence over many aspects of our daily lives as consumers (at least in the West).

3 Quotes from A. L. Aduriz, *Mugaritz: A Natural Science of Cooking* (New York: Phaidon, 2014), p. 18; J. Simpson & J. Mattson, 'TV chef's grubby steakhouse mixed raw and cooked meat', *The Times*, 26 May 2014, p. 18 (http://www.thetimes.co.uk/tto/news/uk/article4100051.ece).

4 R. L Spang, *The Invention of the Restaurant: Paris and Modern Gastronomic Culture* (Cambridge, MA: Harvard University Press, 2000).

5 Quotes from L. Collins, 'Who's to judge? How the World's 50 Best Restaurants are chosen', *The New Yorker* (*Annals of Gastronomy*), 2 November 2015 (https://www.newyorker.com/magazine/2015/11/02/whos-to-judge).

6 Indeed, the idea has spread far and wide; from the high street to holidays, everyone is, or so it often seems, talking about 'the experience'; see e.g. J. Bacon, 'Consumers value stores' appearance and atmosphere', *Marketing Week*, 16 April 2014 (http://www.marketingweek.co.uk/trends/trending-topics/shopper-behaviour/consumers-value-stores-appearance-and-atmosphere/4010022.article); K. Hilton, 'Psychology: The science of sensory marketing', *Harvard Business Review*, March 2015, 28–31; L. Spinney, 'Selling sensation: The new marketing territory', *New Scientist*, 2934, 18 September 2013 (https://www.newscientist.com/article/mg21929340-400-selling-sensation-the-new-marketing-territory/).

7 This compared to just £400 at Ultraviolet the last time I checked; see B. Palling, 'Fork it over: Are the world's priciest restaurants worth the expense?', *Newsweek*, 4 December 2015 (http://www.pressreader.com/usa/newsweek/20151204/282089160685916). Talking of phenomenally expensive one-off dinners, one should not forget to mention the £15,000-a-head Epicurean Masters of the World at the Dome Restaurant in Bangkok, in 2007; see 'Blow out! History's 10 greatest banquets', *Independent*, 10 February 2007 (https://www.independent.co.uk/life-style/food-and-drink/features/blow-out-historys-10-greatest-banquets-435763.html).

8 As journalist Juliet Kinsman says: 'It's official: dinner is no longer just about tasting your meals, it's about seeing them made and taking part' ('Give us a butcher's . . . for diners, seeing is believing', *Independent on Sunday*, 7 June 2015, p. 59). Note that it wasn't always so, though; see H. Levenstein, *The Paradox of Plenty: A Social History of Eating in Modern America* (Oxford: Oxford University Press, 1993). See also H. Armstrong, 'Tried and tasted: Chef's tables', *Daily Telegraph* (*Luxury*), 26 October 2012 (http://www.telegraph.co.uk/luxury/drinking_and_dining/49459/tried-and-tasted-chefs-tables.html); C. Hargreaves, 'The hottest restaurant tables now put diners right in the thick of the action – the kitchen', *Independent*, 22 January 2015 (https://www.independent.co.uk/life-style/food-and-drink/features/the-hottest-restaurant-tables-now-put-diners-right-in-the-thick-of-the-action-the-kitchen-9996892.html). Hargreaves notes: 'The "chef's table", whereby punters sit at a table near or in the kitchen and eat a menu set by the chef, is also becoming more down-to-earth. [. . . At Gordon Ramsay's] Maze, for instance, for £125 per head you get a balcony seat in the pulsating heart of the busy Michelin-starred kitchen from where, as the blurb puts it, you "absorb the kitchen theatrics" (hopefully not of the *Kitchen Nightmares* variety) as you eat a seven-course meal.' Note that this trend also fits with our growing desire to see food freshly prepared in front of us – think of the phenomenal popularity of food trucks these days.

9 J. Kinsman, 'Give us a butcher's . . . for diners, seeing is believing', *Independent on Sunday*, 7 June 2015, p. 59. Though some would rather the chefs get back into the kitchen and leave the front-of-house service to the professionals; see M. O'Loughlin, 'The Frog, London E1: "I can't shake the feeling that this is a rather silly restaurant"', *Guardian* (*Life & Style*), 5 August 2016 (https://www.theguardian.com/lifeandstyle/2016/aug/05/the-frog-london-e1-restaurant-review-marina-oloughlin).

10 As Tim Hayward puts it: 'By removing choice, chefs can reposition themselves; they are no longer part of a team meeting customers' desires, making them happy, but instead artist-performers whose work you pay to experience' ('Menus without choice blaspheme against the doctrine of dining', *FT Weekend Magazine*, 23 January 2016, p. 12).

11 J. R. Walker, *The Restaurant: From Concept to Operation* (6th edn; Hoboken, NJ: Wiley, 2011), p. 53. Indeed, the link to the theatre goes back to the earliest days of the restaurant in Paris: 'Descriptions of restaurants often emphasized their specular – and spectacular – functions. Like a theatre, a restaurant was a stable frame around an ever-changing performance, a stage where fantasies might be brought to life.' (R. L Spang, *The Invention of the Restaurant: Paris and Modern Gastronomic Culture* (Cambridge, MA: Harvard University Press, 2000), p. 236).

12 M. Moore, 'Taste the difference: Sublimotion vs. Ultraviolet', *Financial Times*, 28 August 2015 (https://www.ft.com/cms/s/2/0a4f62f0-4ca2-11e5-9b5d-89a026fda5c9.html#slide0).

13 Something very similar also going on at the chef's table just down the road at the Mall Tavern in Notting Hill: 'Take it from us, the theatrical, *Pulp Fiction* meets *Charlie and the Chocolate Factory*, part theatre part dessert is worth the booking alone' (A. Carter, 'The best chef's tables in London', *Stylenest*, 14 August 2014 (http://www.stylenest.co.uk/food/best-of-food/the-best-chefs-tables-in-london/)).

14 C. Spence, 'Music from the kitchen', *Flavour*, 4:25 (2015); R. Kitson, 'Food for thought: Chef teams with Heston's professor to serve up mind-bending dining fun', *Evening Standard*, 18 November 2015 (https://www.standard.co.uk/goingout/restaurants/food-for-thought-chef-teams-with-heston-s-professor-to-serve-up-mindbending-dining-fun-a3117396.html).

15 S. K. A. Robson, 'Turning the tables: The psychology of design for high-volume restaurants', *Cornell Hotel and Restaurant Administration Quarterly*, 40(3) (1999), 56–63, p. 60.

The acerbic UK restaurant critic A. A. Gill was spot on when he wrote – after a traumatic trip to The Rainforest Cafe with OPK (other people's kids) – that '[e]ver since the first public dining-room, restaurateurs have been asking, "What shall we make it look like?" The difference recently has been that, whereas the set used to complement the food, now the food is an adjunct to the marketing – or "the experience", as it is invariably called' (*Table Talk: Sweet and Sour, Salt and Bitter* (London: Weidenfeld & Nicolson, 2007), p 102). Here, it would be interesting to get some reactions from kids, for whom the atmosphere is designed, rather than from the parents and guardians who inevitably get dragged along to pick up the tab.

16 J. O'Ceallaigh, 'Restaurant review: The Cube by Electrolux, Royal Festival Hall', *Daily Telegraph* (*Luxury*), 17 July 2012 (http://www.telegraph.co.uk/luxury/drinking_and_dining/3690/restaurant-review-the-cube-by-electrolux-royal-festival-hall.html); as the journalist put it: 'Two knockout courses, the exceptional view and one-off experience meant we were sailing into best-meal-in-recent-memory territory . . .'

17 I suspect that dining while suspended high up in the air increases arousal in much the same way as in the study where people were found to be more likely to agree to a date if propositioned on a flimsy bridge (the idea being that people misattribute the arousal they feel because of being suspended over a canyon, say, to the person who just asked them out on a date); D. G. Dutton & A. P. Aron, 'Some evidence for heightened sexual attraction under conditions of high anxiety', *Journal of Personality and Social Psychology*, 30 (1974), 510–17. See also 'The world's strangest restaurants: Where to eat dinner in the sky – or an alien's stomach', *Daily Mail Online*, 14 March 2014 (http://www.dailymail.co.uk/travel/article-2586920/The-worlds-strangest-restaurants-Where-eat-dinner-sky-aliens-stomach.html#ixzz4DqQfAgOU).

18 Quoted in G. Ulla, 'Grant Achatz plans to "overhaul the experience" at Alinea', Eater.com, 23 November 2011 (https://eater.com/archives/2011/11/23/grant-achatz-planning-major-changes-at-alinea.php#more).

19 Edmund Weil, owner of London's Nightjar, says of the theatrical approach taken to service in his cocktail bar: 'Nobody takes their drinks more seriously than the barmen [. . .] but their aim is to give their guests an experience they won't have elsewhere. I don't want anything extraneous – something interesting conceptually but that doesn't enhance the drink in any way. It has to be something that matches and heightens the customer's experience. A lot of it is about bringing more of the senses into enjoying the cocktail – and offering a visual treat' (quoted in J. Sissons, 'Theatrical cocktails', *Financial Times* (*How to Spend It*), 20 March 2014 (https://howtospendit.ft.com/drink/50303-theatrical-cocktails)).

20 Quoted in L. Collins, 'Who's to judge? How the World's 50 Best Restaurants are chosen', *The New Yorker* (*Annals of Gastronomy*), 2 November 2015 (https://www.newyorker.com/magazine/2015/11/02/whos-to-judge).

21 J. Bergman, 'Restaurant report: Ultraviolet in Shanghai', *The New York Times*, 3 October 2012 (https://www.nytimes.com/2012/10/07/travel/restaurant-report-ultraviolet-in-shanghai.html).

22 Quoted in M. Joe, 'Dishing it out: Chefs are offering diners a multisensory experience', *South China Morning Post*, 10 January 2014 (https://www.scmp.com/magazines/style/article/1393915/dishing-it-out-chefs-are-offering-diners-multisensory-experience).

23 Perhaps, then, one can see the increasing drive towards the more theatrical, the more experiential, as one of the ways in which the wily restaurateur can keep the punters coming to their bricks-and-mortar establishments; J. Abrams, 'Mise en plate: The scenographic imagination and the contemporary restaurant', *Performance Research: A Journal of the Performing Arts*, 18(3) (2013), 7–14.

24 S. Pigott, 'Appetite for invention', *Robb Report*, May 2015, 98–101, p. 99. Or take this awe-inspiring description: 'The scene resembles a religious procession, not the opening salvo of a 22-course degustation menu. As images of candles flicker on the walls, a bell begins to toll and an angelic hymn fills the room. Ten servers wearing ball caps and blue aprons slowly enter from a side door, heads bowed reverently over the silver bowls in their hands' (J. Bergman, 'Restaurant report: Ultraviolet in Shanghai', *The New York Times*, 3 October 2012 (https://www.nytimes.com/2012/10/07/travel/restaurant-report-ultraviolet-in-shanghai.html)).

25 Roncero boasts that his is 'the first gastronomic show in the world' (quoted in B. Palling, 'Fork it over: Are the world's priciest restaurants worth the expense?', *Newsweek*, 4 December 2015 (http://www.pressreader.com/usa/newsweek/20151204/282089160685916)). See also A. Jakubik, 'The workshop of Paco Roncero', *Trendland: Fashion Blog & Trend Magazine*, 23 July 2012 (https://trendland.com/the-workshop-of-paco-roncero/).

26 For a review on the topic of multisensory immersion, see A. Gallace et al., 'Multisensory presence in virtual reality: Possibilities & limitations', in G. Ghinea, F. Andres & S. Gulliver (eds.), *Multiple Sensorial*

Media Advances and Applications: New Developments in MulSeMedia (Hershey, PA: IGI Global, 2012), pp. 1–40.

27 According to the organizer Andrea Petrini, the aim is to deliver 'a multisensory food performance' (quoted in M. Joe, 'Dishing it out: Chefs are offering diners a multisensory experience', *South China Morning Post*, 10 January 2014 (https://www.scmp.com/magazines/style/article/1393915/dishing-it-out-chefs-are-offering-diners-multisensory-experience)).

28 See http://www.elsomni.cat/en/el---somni/dinner/. Jean Roca, head chef of El Celler de Can Roca, recently took a 'gastronomic opera' on tour around Spain.

29 This was one of Grant Achatz's ideas when he talked about overhauling the experience at Alinea in 2011 (G. Ulla, 'Grant Achatz plans to "overhaul the experience" at Alinea', Eater.com, 23 November 2011 (https://eater.com/archives/2011/11/23/grant-achatz-planning-major-changes-at-alinea.php#more)). It would also, hopefully, avoid the terrible music one is sometimes accosted by while eating out; see J. Mariani, 'How restaurant music got so bad: A brief history', *Esquire*, 13 August 2013 (https://www.esquire.com/food-drink/restaurants/a24222/restaurant-music-history/).

30 According to P. Jalkanen, *Salonkimusiikki* (Helsinki: WSOY, 2003), restaurateurs started to hire permanent ensembles to play music in their restaurants in the mid nineteenth century too. This soon became a status symbol for the venue concerned. See also J. T. Lang, 'Sound and the city: Noise in restaurant critics' reviews', *Food, Culture and Society*, 17(4) (2014), 571–89; W. Littler, 'Tafelmusik takes off', *Music Magazine*, xii (1989), 14–17.

31 E. Peralta, 'The sounds of asparagus, as explored through opera', *The Salt*, 1 June 2012 (https://www.npr.org/blogs/thesalt/2012/05/29/153950254/the-sounds-of-asparagus-as-explored-through-opera); C. Spence, 'Music from the kitchen', *Flavour*, 4:25 (2015).

32 For example, 'Eating sound at The Truscott Arms' in Maida Vale, London, which aimed to use food to: 'deepen the understanding of sound. Connections are made between timbre, texture and taste that in turn help to refresh the audience's perceptions of flavour and sound.' The event was a collaboration between chef Barry Snook and composer/

improviser Sam Bailey (see https://www.studiotheolin.com/eating-sound/).

33 Condiment Junkie, 'Strawberries say summer' (press release, 2014; http://condimentjunkie.co.uk/case-study/strawberries-say-summer).

34 A mismatch in timing was the problem with Marina Abramovic's 'Volcano Flambé' dessert served at the restaurant Park Avenue Winter (F. Fabricant, 'Marina Abramovic's art doubles as dessert', *The New York Times*, 11 January 2011 (https://www.nytimes.com/2011/01/12/dining/12art.html?_r=0)). This dish was accompanied by a talk delivered by an MP3 player; however, the performance artist's talk lasted only three minutes or so, while the act of eating the dessert lasted longer, so there was a disparity between the duration of the two sensory experiences (according to one of my colleagues who tasted this delight).

35 See B. Houge, 'Food opera: Merging taste and sound in real time', *New Music Box*, 11 September 2013 (you can hear an excerpt from the score: http://www.newmusicbox.org/articles/food-opera-merging-taste-and-sound-in-real-time/?utm_source=rss&utm_medium=rss&utm_campaign=food-opera-merging-taste-and-sound-in-real-time); see also J. Irwin, 'Listen up for Food Opera – a live, sonic interactive dinner', *Kill Screen*, 20 June 2013 (https://killscreendaily.com/articles/articles/food-opera-your-new-favorite-interactive-eating-experience/).

36 J. Gordinier, 'A restaurant of many stars raises the ante', *The New York Times*, 27 July 2012 (https://www.nytimes.com/2012/07/28/dining/eleven-madison-park-is-changing-things-up.html).

37 See J. Gerard, 'Heston Blumenthal: My new Alice in Wonderland menu', *Daily Telegraph*, 1 July 2009 (http://www.telegraph.co.uk/foodanddrink/restaurants/5700481/Heston-Blumenthal-my-new-Alice-in-Wonderland-menu.html).

38 Quotes from J. Rayner, 'Blue sky thinking', *Observer Food Monthly*, 23 August 2015, pp. 18–22, pp. 21–22; see also the *Heston's Feasts* and *Fantastical Food* TV shows. Happily, a typical play, movie and multicourse tasting menu are all around the same length (2–3 hours), which should mean, at least in principle, that the timescale of experience design co-incides across these various different media. See also C. Spence, 'Multisensory experience design', invited presentation given at the

Digital Biscuit conference, Dublin, Ireland, 29–30 January 2015 (http://www.digitalbiscuit.com/archive/2015/).

39 G. Ulla, 'Grant Achatz plans to "overhaul the experience" at Alinea', Eater.com, 23 November 2011 (https://eater.com/archives/2011/11/23/grant-achatz-planning-major-changes-at-alinea.php#more).

40 As Will Guidara, business partner of executive chef Daniel Humm, put it in one interview: 'We're not looking to impress people. We want to entertain them.' The magic was designed by Jonathan Bayme and Dan White, illusionists from a company called Theory11; see J. Gordinier, 'A restaurant of many stars raises the ante', *The New York Times*, 27 July 2012 (https://www.nytimes.com/2012/07/28/dining/eleven-madison-park-is-changing-things-up.html). No matter whether or not the restaurant staff are literally magicians, they should nevertheless still be aiming to generate surprise and to play with their diners' expectations; M. Hanefors & L. Mossberg, 'Searching for the extraordinary meal experience', *Journal of Business and Management*, 9 (2003), 249–70.

41 J. Gerard, 'Heston Blumenthal: My new Alice in Wonderland menu', *Daily Telegraph*, 1 July 2009 (http://www.telegraph.co.uk/foodanddrink/restaurants/5700481/Heston-Blumenthal-my-new-Alice-in-Wonderland-menu.html). Notice here how ambient aroma is used to set the scene (see 'Smell').

42 S. Mountfort, 'Like Heston meets Crystal Maze', *Metro*, 9 December 2015, p. 49. Elizabeth Carter, editor of the *Good Food Guide*, described Heston Blumenthal's Fat Duck restaurant thus: 'It is food as theatre, the way the waiters interact with the table, it is like a performance' (quoted in 'Fat Duck wins award despite scare', *BBC News*, 18 August 2009 (http://news.bbc.co.uk/1/hi/england/berkshire/8208103.stm)).

43 K. Sekules, 'Food for thought. Copenhagen's coolest dinner theatre', *The New York Times*, 19 January 2010 (http://tmagazine.blogs.nytimes.com/2010/01/19/food-for-thought-copenhagens-coolest-dinner-theater/).

44 M. Johanson, 'Not just food: These restaurants hit all five senses', *CNN Travel*, 12 August 2015 (http://edition.cnn.com/2015/08/11/travel/best-multisensory-restaurants/).

45 Pushing the analogy with other forms of entertainment, trailers (short movies) are released whenever a new menu is launched at Next too; P. Wells, 'In Chicago, the chef Grant Achatz is selling tickets to his new restaurant', *The New York Times*, 10 May 2010 (https://www.nytimes.com/2010/05/05/dining/05achatz.html).

46 And according to the website: 'The experience unfolds as a play. Food leads. Each course is enhanced with its own taste-tailored atmosphere: lights, sounds, music, scents, projection, images and imagination . . . and food.'

47 'The idea of fusing food with the arts to enhance one or the other isn't exactly new, but it takes a certain amount of creativity and a whole lot of sweat. Last year, we brought you the story of chef and entrepreneur Bryon Brown, who was frustrated that his memories of food are less about the food than the atmosphere. So he created a theatrical project called *Sensorium* that employed actors to harness diners' sight, smell, texture and sound to enhance their memories' (E. Peralta, 'The sounds of asparagus, as explored through opera', *The Salt*, 1 June 2012 (https://www.npr.org/blogs/thesalt/2012/05/29/153950254/the-sounds-of-asparagus-as-explored-through-opera)). Or take the following description of one multi-course, multisensory dining experience going by the name of 'Inspiracle' (from chef Rob Sidor in Philadelphia) to get an idea of the sort of thing that is going on: 'We're talking visual effects, auditory effects, scents and vapors and all their palatal effects. Plus things that engage the attendees, table-side: siphon infused sauces, edible "perfume", and smoking cinnamon at the table . . . you'll be completely immersed in six-courses (plus a few surprises) of body-shaking, mind-baffling food' (quoted in A. Tewfik, 'Multi-sensory pop-up with Inspiracle', *Philadelphia* (*Phillymag*), 14 January 2014 (http://www.phillymag.com/foobooz/2014/01/17/inspiracle-pop-will-byob/)).

   See also J. Rayner, ' "Molecular gastronomy is dead." Heston speaks out', *Observer Food Monthly*, 17 December 2006 (http://observer.guardian.co.uk/foodmonthly/futureoffood/story/0,,1969722,00.html); G. Achatz, 'Food tasting or art installation?', *The Atlantic*, 23 April 2009 (http://food.theatlantic.com/back-of-the-house/food-tasting-or-art-installation.php).

48 F. Rose, 'How Punchdrunk Theatre reels 'em in with immersive story-telling', *Wired*, 13 March 2012 (https://www.wired.com/2012/03/punchdrunk-theatre-immersive/); see also https://www.theguardian.com/stage/punchdrunk.

49 A. Soloski, 'Sleep No More: From avant garde theatre to commercial blockbuster', *Guardian*, 31 March 2015 (https://www.theguardian.com/stage/2015/mar/31/sleep-no-more-avant-garde-theatre-new-york). Felix Barrett is quoted in the article thus: ' "What we're doing with the bar and the restaurant are experiments, research," he said. "How do you tell a story through food? How do you have a three-course meal that has a narrative?", See also 'Sleep No More adds high-end restaurant to its New York roster', *Guardian*, 26 November 2013.

50 S. Mountfort, 'Like Heston meets Crystal Maze', *Metro*, 9 December 2015, p. 49.

'In Cristoforo Messisbugo's classic 1549 text "Banquets, Compositions of Courses and General Table Design" ["*Banchetti, compozizioni de vivande et apparecchio generale*"], a detailed manual for court banquets, "each course had its own particular music or form of spectacle, all perfectly integrated into the serving of the food in a way that we would categorise in modern parlance as a happening" [. . .] These banquets might, for example, be preceded by the performance of a farce and a concert' (quoted in P. McCouat, 'The Futurists declare war on pasta', *Journal of Art in Society* (2014; http://www.artinsociety.com/the-futurists-declare-war-on-pasta.html)).

51 A. S. Weiss, *Feast and Folly: Cuisine, Intoxication and the Poetics of the Sublime* (Albany, NY: State University of New York Press, 2002, pp. 103–4).

52 P. McCouat, 'The Futurists declare war on pasta', *Journal of Art in Society*, 2014 (http://www.artinsociety.com/the-futurists-declare-war-on-pasta.html).

The Italian Futurists provided the inspiration for much of what is going on in contemporary theatrical dining: Just take *The Banquet for Ultra Bankruptcy*, developed for Art Laboratory Berlin in 2013. During each of the five performances (each for six guests) of the six-course meal, selected foods were combined with images, sounds and scents. Each course had been designed as an aesthetic experience, to recreate some of the Futurists' ideas. The underpinning idea here was to allow the audience to participate in a host of simultaneous multisensory experiences (http://artlaboratory-berlin.org/assets/pdf/A_Banquet_for_Ultra_Bankruptcy-Statement.pdf).

53 See R. Brooks, 'Tate tosses up super-salad as art', *The Sunday Times* (*News*), 16 March 2008, p. 9. See also R.-L. Goldberg, *Performance Art: From Futurism to the Present* (London: Thames & Hudson, 1979); S. Smith, *Feast: Radical Hospitality in Contemporary Art* (Chicago: IL: Smart Museum of Art, 2013), pp. 66–75; B. Morais, 'Salad as performance art', *The New Yorker*, 26 April 2012 (https://www.newyorker.com/culture/culture-desk/salad-as-performance-art); B. Kirshenblatt-Gimblett, 'Playing to the senses: Food as a performance medium', *Performance Research*, 4 (1999), 1–30.

54 C. A. Jones (ed.), *Sensorium: Embodied Experience, Technology, and Contemporary Art* (Cambridge, MA: MIT Press, 2006), p. 19.

55 J. Klein, 'Feeding the body: The work of Barbara Smith', *PAJ: A Journal of Performance and Art*, 21(1) (1999), 24–35, p. 25.

56 C. Spence, 'The art and science of plating', in N. Levent & I. D. Mihalache (eds.), *Food and Museums* (London: Bloomsbury Academic, 2017), pp. 237–53. See also J. Carey, *What Good Are the Arts?* (London: Faber & Faber, 2005); A. Dornenburg & K. Page, *Culinary Artistry* (New York: John Wiley & Sons, 1996), p. 2.

57 E. D. Barba, 'My cuisine is tradition in evolution', *Swide*, 23 April 2013; V. Félix, 'Pierre Gagnaire', *Télérama*, 15 December 2011 (http://www.telerama.fr/monde/pierre-gagnaire-la-cuisine-ce-n-est-pas-toujours-de-l-art-heureusement,76142.php).

58 'No one sits down to eat a plate of nutrients. Rather, when we sit down for a meal, we are seeking physical as well as emotional and psychological nourishment'; quoted from L. G. Block et al., 'From nutrients to nurturance: A conceptual introduction to food well-being', *Journal of Public Policy Marketing*, 30 (2011), 5–13.

59 A. L. Aduriz, *Mugaritz: A Natural Science of Cooking* (New York: Phaidon, 2014), p. 42.

60 See, for example, S. Pollard, 'Food as art, or pie in the sky?', *The Times* (*times2*), 17 May 2007, p. 6 (http://www.thetimes.co.uk/tto/life/food/article1778612.ece); E. Tefler, 'Food as art', in A. Neill & A. Ridley

(eds.), *Arguing About Art: Contemporary Philosophical Debates* (2nd edn; London: Routledge, 2002), pp. 9–27.

61 R. L. Spang, *The Invention of the Restaurant: Paris and Modern Gastronomic Culture* (Cambridge, MA: Harvard University Press, 2000). That certainly seemed to be what Blumenthal was hinting at when quoted earlier in the chapter: 'As he reveals in an exclusive interview in *Observer Food Monthly*, when The Fat Duck reopens at the end of September after a six-month relocation to Australia while the original was being refurbished, it won't simply be a place for dinner; it will be a "story". "In the sense that we cook food and it's served to people, we're a restaurant. But that's not much, is it?" says Blumenthal' (quoted in J. Rayner, 'Blue sky thinking', *Observer Food Monthly*, 23 August 2015, pp. 18–22, p. 20).

62 J. Finkelstein, *Dining Out: A Sociology of Manners* (New York: New York University Press, 1989), p. 68.

63 It is interesting how often people talk about becoming emotional and being brought to tears at such multisensory encounters; see M. Joe, 'Dishing it out: Chefs are offering diners a multisensory experience', *South China Morning Post*, 10 January 2014 (https://www.scmp.com/magazines/style/article/1393915/dishing-it-out-chefs-are-offering-diners-multisensory-experience); M. Pelowski, 'Tears and transformation: Feeling like crying as an indicator of insightful or "aesthetic" experience with art', *Frontiers in Psychology*, 6:1006 (2015); C. Spence & Q. (J.) Wang, 'Wine & music (III): So what if music influences taste?', *Flavour*, 4:35 (2015); C. De Lange, 'Feast for the senses: Cook up a master dish', *New Scientist*, 2896, 18 December 2012 (https://www.newscientist.com/article/mg21628962.200-feast-for-the-senses-cook-up-a-master-dish.html). See also S. Knapton, 'Why sparkling wine sounds like beans falling on a plastic tray', *Daily Telegraph*, 2 May 2015, p. 7; she describes people experiencing wine with a matching soundscape created by synaesthetic musician Nick Ryan as follows: 'The first volunteers to try listening to the scores while drinking said they felt physically transported to a different place. Others wept. Ryan said: "It seems to be a profoundly moving and engaging experience." '

64 As B. J. Pine and J. H. Gilmore note: 'The more senses an experience engages, the more effective and memorable it can be' ('Welcome to the experience economy', *Harvard Business Review*, 76(4) (1998), 97–105, p. 104).

65 Quoted in J. Gordinier, 'A restaurant of many stars raises the ante', *The New York Times*, 27 July 2012 (https://www.nytimes.com/2012/07/28/dining/eleven-madison-park-is-changing-things-up.html).

12. Digital Dining

1 A. Victor, 'Pizza Hut launches digital menu that reads your mind by tracking eye movement . . . and tells you what to order in 2.5 seconds', *Daily Mail Online*, 28 November 2014 (http://www.dailymail.co.uk/femail/food/article-2852847/Pizza-Hut-launches-digital-menu-reads-mind.html). Note also that while we do tend to gaze for longer on those things we like more, the relationship certainly isn't perfect; see S. Shimojo et al., 'Gaze bias both reflects and influences preference', *Nature Neuroscience*, 6 (2003), 1317–20; C. Simion & S. Shimojo, 'Early interactions between orienting, visual sampling and decision making in facial preferences', *Vision Research*, 46 (2006), 3331–5; see also L. N. van der Laan et al., 'Do you like what you see? The role of first fixation and total fixation duration in consumer choice', *Food Quality & Preference*, 39 (2015), 46–55.

2 M. Prigg, 'The Star Trek "replicator" that can recreate ANY meal in 30 seconds: App-controlled cooker mixes pods of ingredients', *Daily Mail Online*, 5 May 2015 (http://www.dailymail.co.uk/sciencetech/article-3069073/The-star-trek-replicator-recreate-meal-30-seconds-App-controlled-cooker-mixes-pods-ingredients.html).

3 The question on every food futurist's lips is how digital technology will change the way we eat and drink in the years to come. The stereotypical view, at least until very recently, was that it was best to keep the technology away from the food and drink, for fear of spilling something over the keyboard. However, things are set to change, and in some cases already have; see C. Spence & B. Piqueras-Fiszman, *The Perfect Meal: The Multisensory Science of Food and Dining* (Oxford: Wiley-Blackwell, 2014), ch. 10; E. Zolfagharifard, 'Could 2015 be the year of domestic robots and 3D printed food? Futurologist claims technology has reached a "tipping point"', *Daily Mail Online*, 24 December 2014 (http://www.dailymail.co.uk/sciencetech/article-2886424/Could-2015-year-domestic-robots-3D-printed-food.html). It is just as valid to

consider how food can influence the experiences of digital technology; that, though, is definitely a topic for another day.

4 K. Moskvitch, 'Printer produces personalised 3D chocolate', *BBC News*, 5 July 2011 (https://www.bbc.co.uk/go/em/fr/-/news/technology-14030720); H. Ledford, 'Foodies embrace 3D-printed cuisine: Printers unleash creative cookery, but will consumers bite?', *Nature News*, 20 April 2015; J. Prisco, ' "Foodini" machine lets you print edible burgers, pizza, chocolate', *CNN News*, 31 December 2015 (http://edition.cnn.com/2014/11/06/tech/innovation/foodini-machine-print-food/); L. Gibbons, '3D printing could be as valuable as internet to food industry: 3D printing could one day be as valuable to food manufacturers as the internet and help them reduce costs and energy and save production time, according to an expert in the field', foodmanufacture.co.uk, 12 November 2015 (https://www.foodmanufacture.co.uk/Supply-Chain/Food-industry-should-adopt-3D-printing); D. Periard et al., 'Printing food', in *Proceedings of the 18th Solid Freeform Fabrication Symposium* (Austin, TX: 2007), pp. 564–74; Italian pasta-maker Barilla has also teamed up with the Dutch research centre TNO to develop pasta-printing machines aimed at the mainstream restaurant sector; see 'Are you ready to eat food that looks like you?', 3D Food Printing Conference, 12 February 2015 (https://3dfoodprintingconference.com/food/ready-eat-pasta-looks-like/).

5 T. Do, 'Is the Foodini 3-D printer the microwave of the future?', *Food Republic*, 11 April 2016 (http://www.foodrepublic.com/2016/04/11/meet-the-foodini-the-freshest-3-d-food-printer-yet/).

6 Others have expressed similar reservations: 'Media technologies theorist Henry Jenkins (2006) would be sceptical of the notion that a new technology such as the PFP [Personal Food Printer] would displace current technologies, collapsing all kitchen appliances into a single all-mighty black box. Jenkins refers to this as the black-box fallacy' (quoted in G. Hearn & D. L. Wright, 'Food futures: Three provocations to challenge HCI', in J. H.-J. Choi, M. Foth & G. Hearn (eds.), *Eat, Cook, Grow: Mixing Human–Computer Interactions with Human–Food Interactions* (Cambridge, MA: MIT Press, 2014), pp. 265–78, pp. 273–4).

7 N. Koenig, 'How 3D printing is shaking up high end dining', *BBC News*, 2 March 2016 (https://www.bbc.co.uk/news/business-35631265);

E. Zolfaghifard, 'Now you can print your FACE on a latte: Coffee machine lets anyone create intricate foam art in just 10 seconds', *Daily Mail Online*, 26 June 2015 (http://www.dailymail.co.uk/sciencetech/article-3141099/Forget-hearts-print-FACE-latte-Coffee-machine-lets-create-intricate-foam-art-just-10-seconds.html); L. Goodall & S. Ceurstemont, 'Futuristic food: Chefs at a Chicago restaurant are using technology to change the way people perceive and eat food', *First Science*, 2014 (http://www.firstscience.com/SITE/ARTICLES/food.asp).

8 A. Jayakumar, 'Home-baked idea? NASA mulls 3D printers for food replication', *Guardian*, 4 June 2013 (https://www.guardian.co.uk/technology/2013/jun/04/nasa-3d-printer-space-food).

9 F. Lenfant et al., 'Impact of the shape on sensory properties of individual dark chocolate pieces', *LWT – Food Science and Technology*, 51 (2013), 545–52. There may also be a market for the precision dosing of medicines; see J. Wakefield, 'First 3D-printed pill approved by US authorities', *BBC News Online*, 4 August 2015 (https://www.bbc.com/news/technology-33772692).

10 D. Meyer, *Setting the Table: Lessons and Inspirations from One of the World's Leading Entrepreneurs* (London: Marshall Cavendish International, 2010), p. 93.

11 R. Verma, M. E. Pullman & J. C. Goodale, 'Designing and positioning food services for multicultural markets', *Cornell Hotel and Restaurant Administration Quarterly*, 40 (1999), 76–87. Picture menus are normally best avoided except in those cases where those one is serving may struggle with language or, like many older hospital patients, may have sight problems.

12 B. London, 'World's first sensory restaurant for BABIES complete with digital menus and interactive menus opens doors', *Daily Mail Online*, 5 June 2014 (http://www.dailymail.co.uk/femail/article-2649367/Worlds-sensory-restaurant-BABIES-complete-digital-menus-interactive-menus-opens-doors.html).

13 M. Johanson, 'Not just food: These restaurants hit all five senses', *CNN Travel*, 12 August 2015 (http://edition.cnn.com/2015/08/11/travel/best-multisensory-restaurants/). R. King, 'Eating light with plates of the future', *Fine Dining Lovers*, 4 April 2014 (https://www.finedininglovers.com/stories/multi-sensory-dining-splendur-andreas-caminada/).

14 C. Spence, 'Multisensory marketing' presentation, Zeitgeist Curator, Berlin, 30 August 2012.

15 Quoted in S. Pigott, 'Appetite for invention', *Robb Report*, May 2015, 98–101.

16 It is all too easy at around this point to start sounding like William Sitwell, one of the judges on TV's *MasterChef*, who said: 'No food should be served on a plate with a right angle' ('Square plates are an "abomination", says *MasterChef* judge William Sitwell. The BBC cookery judge says he is holding an amnesty on square plates at a food festival near his home and will dispose of square plates on behalf of owners', *Daily Telegraph*, 13 May 2014 (http://www.telegraph. co.uk/foodanddrink/10828052/Square-plates-are-an-abomination-says-MasterChef-judge-William-Sitwell.html)).

17 Though, for all of you who are nodding your head at this point, just ask yourself how hygienic that wooden board really is. My suspicion is that eating steak from a tablet would actually be a darn sight more hygienic than from one of those boards, which may not have been given enough time to dry in many commercial situations; E. Smallman, 'Fancy a slice of lemon in your water? You won't after reading this: From germ encrusted menus to what's under the waiter's nails, a top biologist on the hidden horrors in your restaurant', *Daily Mail Online*, 13 August 2014 (http://www.dailymail.co. uk/femail/article-2724281/Fancy-slice-lemon-water-You-won-t-reading-From-germ-encrusted-menus-s-waiter-s-nails-biologist-hidden-horrors-restaurant.html).

18 A. Swerdloff, 'Eating the uncanny valley: Inside the virtual reality world of food', *Munchies*, 13 April 2015 (https://munchies.vice.com/en/articles/eating-the-uncanny-valley-inside-the-virtual-reality-world-of-food).

19 A. Victor, 'Is this the future of food? Virtual reality experiment lets you eat anything you want without worrying about calories or allergies', *Daily Mail Online*, 8 January 2015 (http://www.dailymail. co.uk/femail/food/article-2901755/Virtual-reality-gastronomic-Project-Nourished-Kokiri-Lab-uses-Oculus-Rift-headsets-create-unique-dining-experiences.html). See also C. Chevalier, 'Chef Blumenthal and

Marshmallow laser feast are cooking up something weird', *VR Scout*, 16 April 2016 (https://vrscout.com/news/chef-blumenthal-marshmallow-laser-feast-cooking-something-weird/).

20 This certainly isn't something that should be taken as a given; see A. Gallace et al., 'Multisensory presence in virtual reality: Possibilities & limitations', in G. Ghinea, F. Andres & S. Gulliver (eds.), *Multiple Sensorial Media Advances and Applications: New Developments in MulSeMedia* (Hershey, PA: IGI Global, 2012), pp. 1–40.

21 J. Schöning, Y. Rogers & A. Krüger, 'Digitally enhanced food', *Pervasive Computing*, 11 (2012), 4–6; J. H.-J. Choi, M. Foth & G. Hearn (eds.), *Eat, Cook, Grow: Mixing Human–Computer Interactions with Human–Food Interactions* (Cambridge, MA: MIT Press, 2014).

22 A. Swerdloff, 'Eating the uncanny valley: Inside the virtual reality world of food', *Munchies*, 13 April 2015 (https://munchies.vice.com/en/articles/eating-the-uncanny-valley-inside-the-virtual-reality-world-of-food).

23 This is presumably the reason why some of these space cubes were still in circulation, untouched, several decades later; B. Crumpacker, *The Sex Life of Food: When Body and Soul Meet to Eat* (New York: Thomas Dunne Books, 2006).

24 For a review, see C. Spence, 'Hospital food', *Flavour* (in press, 2017).

25 Note that sushi is a good food stimulus for this kind of research, given the limited smell (or olfactory) component, hence allowing vision to play a bigger role; see K. Okajima & C. Spence, 'Effects of visual food texture on taste perception', *i-Perception*, 2(8) (2011); K. Okajima, J. Ueda & C. Spence, 'Effects of visual texture on food perception', *Journal of Vision*, 13 (2015), 1078; and see http://www.okajima-lab.ynu.ac.jp/demos.html for a video. Importantly, this updating of the visual image can be done without the need for any marker to be placed on the food itself, one of the limitations of other AR applications in this space. What is more, it doesn't even need a meta-cookie to help identify the food.

26 For example, in one experiment researchers were able to show that the participants in a laboratory study consumed less when the food that they had been given to eat (a large biscuit in this case) was made to look

larger than it actually was. Longer-term follow-up studies would be needed, though, before coming to any meaningful conclusions about the impact of such new technologies on people's eating behaviours in more naturalistic environments and over the longer term; see T. Narumi et al., 'Augmented perception of satiety: Controlling food consumption by changing apparent size of food with augmented reality', *Proceedings 2012 ACM Annual Conference Human Factors in Computing Systems* (Austin, TX: ACM, 2012).

27 C. Spence & B. Piqueras-Fiszman, 'Technology at the dining table', *Flavour*, 2:16 (2013). In some of the promotional videos for Paul Pairet's Ultraviolet restaurant, diners can be seen wearing these headsets; that said, my colleagues who have eaten there do not remember them being part of the service offering.

28 B. Dowell, 'Listen, this food is music to your ears', *The Sunday Times*, 29 August 2004 (http://www.thesundaytimes.co.uk/sto/news/uk_news/article236417.ece). Japanese researchers have developed a similar concept with their 'Mouth Jockey' headset, which detects the user's jaw movements and then plays back a specific pre-recorded sound; see 'Sound'.

29 C. Spence, M. U. Shankar & H. Blumenthal, ' "Sound bites": Auditory contributions to the perception and consumption of food and drink', in F. Bacci & D. Melcher (eds.), *Art and the Senses* (Oxford: Oxford University Press, 2011), pp. 207–38; C. De Lange, 'Feast for the senses: Cook up a master dish', *New Scientist*, 2896, 18 December 2012 (https://www.newscientist.com/article/mg21628962.200-feast-for-the-senses-cook-up-a-master-dish.html).

30 C. Spence, 'Multisensory meals and digital dining', *The Wired World in 2014*, special issue, 31 October 2013, p. 71.

31 'Musical spoons to go with your Heinz beans', *Advertising Age*, 28 March 2013 (http://adage.com/article/creativity-pick-of-theday/bompas-parr-design-musical-spoons-heinz-beans/240605/); see also M. F. Mickiewicz, 'Beatballs', *Protein*, 10 September 2014 (https://www.prote.in/feed/beatballs).

32 D. Griner, ' "Wake up and smell the bacon" with free alarm gadget from Oscar Meyer', *AdWeek*, 6 March 2014 (http://www.adweek.com/adfreak/wake-and-smell-bacon-free-alarm-gadget-oscar-mayer-156123).

The O-phone is operating in much the same space; see D. Bates, 'The mobile that sends SMELLS', *Daily Mail Online*, 13 June 2014 (http://www.dailymail.co.uk/sciencetech/article-2657288/The-mobile-sends-SMELLS-oPhone-receives-creates-scents-make-messages-memorable.html). See also J. T. Quigley, 'TV set that emits smells? This startup has what it takes to make it happen', *Tech in Asia*, 25 November 2014 (https://www.techinasia.com/tv-that-emits-smell-japan-startup-aromajoin).

33 'How to turn your smartphone into a "smell phone"', *BBC News Online*, 12 March 2014 (https://www.bbc.co.uk/news/technology-26526916).

34 C. Platt, 'You've got smell', *Wired*, 1 November 1999 (https://www.wired.com/1999/11/digiscent/); A. Dusi, 'What does $20 million burning smell like? Just ask DigiScents!', *StartupOver*, 19 January 2014 (http://www.startupover.com/en/20-million-burning-smell-like-just-ask-digiscents/).

35 A. Kadomura, 'EaTheremin', in *SIGGRAPH Asia 2011 Emerging Technologies* (New York: ACM, 2011), p. 7; K. Winter, 'The fork that talks! New Japanese gadget makes bizarre sounds while you eat', *Daily Mail Online*, 28 December 2012 (http://www.dailymail.co.uk/femail/article-2254192/The-fork-talks-New-Japanese-gadget-makes-bizarre-sounds-eat.html). There is also an augmented spoon (https://www.hapi.com).

36 M. Hirose et al., 'Gravitamine spice: A system that changes the perception of eating through virtual weight sensation', in *Proceedings of the 6th Augmented Human International Conference* (ACM, 2015), pp. 33–40.

37 Liftware to counteract hand tremors, weighing in at $295, has been on the market for nearly three years now. R. Robbins, 'Can high-tech plates and silverware help patients manage disease?', *Stat News*, 28 December 2015 (https://www.statnews.com/2015/12/28/plates-silverware-health/). See also C. Spence & B. Piqueras-Fiszman, 'Technology at the dining table', *Flavour*, 2:16 (2013); C. Spence & B. Piqueras-Fiszman, *The Perfect Meal: The Multisensory Science of Food and Dining* (Oxford: Wiley-Blackwell, 2014). And the latest research suggests that wearable technology which helps people monitor how many mouthfuls of food they have consumed can help reduce food intake;

P. W. Jasper et al., 'Effects of bite count feedback from a wearable device and goal setting on consumption in young adults', *Journal of the Academy of Nutrition and Dietetics* 16(11) (2016), 1785–93.

38　S. Liberatore, 'Who needs seasoning? Prototype electric fork SHOCKS your tongue to stimulate the taste of salt', *Daily Mail Online*, 29 March 2016 (http://www.dailymail.co.uk/sciencetech/article-3514391/Who-needs-seasoning-Prototype-electric-fork-SHOCKS-tongue-stimulate-taste-salt.html); N. Ranasinghe, K.-Y. Lee & E. Y.-L. Do, 'Fun-Rasa: An interactive drinking platform', in *Proceedings of the 8th International Conference on Tangible, Embedded and Embodied Interaction* (New York: ACM, 2014), pp. 133–6; N. Ranasinghe et al., 'Digital taste: Electronic stimulation of taste sensations', *Ambient Intelligence: Lecture Notes in Computer Science,* 7040 (2011), 3459; N. Ranasinghe et al., 'Digital flavor interface', in *Proceedings of the Adjunct Publication of the 27th Annual ACM Symposium on User Interface Software and Technology* (New York: ACM, 2014), pp. 47–8. But it should be noted that knowledge about electric taste has been around for more than forty years now; K.-H. Plattig & J. Innitzer, 'Taste qualities elicited by electric stimulation of single human tongue papillae', *Pflügers Archiv/European Journal of Physiology,* 361 (1976), 115–20 (these researchers managed to get sour, salty and bitter); G. von Békésy, 'Sweetness produced electrically on the tongue and its relations to taste theories', *Journal of Applied Physiology,* 19 (1964), 1105–13; G. von Békésy, 'Temperature coefficients of the electrical thresholds of taste sensations', *Journal of General Physiology,* 49 (1965), 27–35.

39　The digital lollipop was featured everywhere from the *New Scientist* to *Discovery News, Time, Gizmodo* and the *Daily Telegraph*. It was even selected as one of the 10 Best Innovations in 2014 by the Netexplo forum at UNESCO; see e.g. P. Marks, 'That tasty tingle: With an electrode on the tongue you can sample virtual food', *New Scientist,* 23 November 2013, p. 22 (https://www.newscientist.com/article/mg22029444-500-electrode-recreates-all-four-tastes-on-your-tongue/).

40　See A. Cuthbertson, 'Taste+ smart spoon and cup virtually enhance food flavours and restore taste to the elderly', *International Business Times,* 23 April 2015 (https://www.ibtimes.co.uk/taste-smart-spoon-cup-virtually-enhance-food-flavours-restore-taste-elderly-1497816).

41 The problem here is that smell involves pattern recognition rather than the stimulation of a single receptor. Electrically stimulating taste is a much easier proposition in so far as individual taste buds on the tongue code for different tastes such as bitter, sweet, salty, etc.; see G. M. Shepherd, *Neurogastronomy: How the Brain Creates Flavor and Why It Matters* (New York: Columbia University Press, 2012) for a pointillist account of smell. See also T. Weiss et al., 'From nose to brain: Un-sensed electrical currents applied in the nose alter activity in deep brain structures', *Cerebral Cortex*, 26 (2016), 4180–91.

42 K. Ohla et al., 'Visual-gustatory interaction: Orbitofrontal and insular cortices mediate the effect of high-calorie visual food cues on taste pleasantness', *PLoS ONE*, 7(3): e32434 (2012); here, though, one should consider whether the salivation that is induced by viewing delectable images of food might simply have improved the contact between electrode and tongue. See also C. Spence, 'Mouth-watering: The influence of environmental and cognitive factors on salivation and gustatory/ flavour perception', *Journal of Texture Studies*, 42 (2011), 157–71; N. Ranasinghe et al., 'Virtual ingredients for food and beverages to create immersive taste experiences', *Multimedia Tools and Applications*, 75(20) (2016), 12291–309.

43 A. Bolton, 'No salt, no problem! Japanese Electro Fork zaps flavour into your mouth', *CNet*, 31 March 2016 (https://www.cnet.com/uk/ news/no-salt-no-problem-japanese-electro-fork-zaps-salt-flavour-into-your-mouth/).

44 B. Stuckey, *Taste What You're Missing: The Passionate Eater's Guide to Why Good Food Tastes Good* (London: Free Press, 2012).

45 C. Cadwalladr, 'Jamie Oliver's FoodTube: Why he's taking the food revolution online', *Guardian*, 22 June 2014 (https://www.theguardian. com/lifeandstyle/2014/jun/22/jamie-oliver-food-revolution-online-video). There is even a cooking simulation computer game, *Cooking Mama*.

46 S. Griffiths, 'No more burnt burgers on the barbeque! $199 smartphone-controlled grill tells you when your food is perfectly cooked', *Daily Mail Online*, 31 July 2015 (http://www.dailymail.co.uk/ sciencetech/article-3181211/No-burnt-burgers-barbeque-199-smartphone-controlled-grill-tells-food-perfectly-cooked.html).

47 J. Chung, 'Cracking the code of restaurant wine pricing', *Wall Street Journal*, 15 August 2008 (http://online.wsj.com/article/SB1218756 95594642607.html); G. Roberts, 'The lowdown on restaurant markups. The inside story on why and how restaurants price their wines', *Wine Enthusiast*, 7 May 2010 (http://www.winemag.com/Wine-Enthusiast-Magazine/May-2010/The-Lowdown-on-Restaurant-Markups).

48 J. Bradley, 'New Google app splits restaurant bills', *Scotsman on Sunday*, 17 October 2013 (https://www.scotsman.com/lifestyle/food-drink/features/new-google-app-splits-restaurant-bills-1-3147140).

49 S. Baral, 'Krug app will tell you what music to pair with your champagne', *iDigitalTimes*, 7 October 2015 (http://www.idigitaltimes.com/krug-app-will-tell-you-what-music-pair-your-champagne-456934).

50 Philadelphia start-up Fitly is producing the $99 Smartplate equipped with cameras to photograph your food and sensors to weigh it too. It is supposed to calculate how many calories there are in your food. See also L. Stinson, 'One way to get your kid to eat veggies: Make it into a video game', *Wired*, 9 August 2015 (https://www.wired.com/2015/09/an-interactive-plate-that-turns-your-kids-meal-into-a-game/); S. Manton et al., 'The "Smart Dining Table": Automatic behavioural tracking of a meal with a multi-touch-computer', *Frontiers in Psychology*, 7:142 (2016).

51 E. Toet, B. Meerbeek & J. Hoonhout, 'Supporting mindful eating with the InBalance chopping board', in J. H.-J. Choi, M. Foth & G. Hearn (eds.), *Eat, Cook, Grow: Mixing Human–Computer Interactions with Human–Food Interactions* (Cambridge, MA: MIT Press, 2014), pp. 99–116.

52 E. Sofge, 'Google's A. I. is training itself to count calories in food photos', *Popular Science*, 29 May 2015 (https://www.popsci.com/google-using-ai-count-calories-food-photos).

53 J. Z. Young, 'Influence of the mouth on the evolution of the brain', in P. Person (ed.), *Biology of the mouth: A symposium presented at the Washington meeting of the American Association for the Advancement of Science, 29–30 December 1966* (Washington, DC: American Association for the Advancement of Science, 1968), pp. 21–35.

On the mis-estimation of calorie count, see B. Wansink & P. Chandon, 'Meal size, not body size, explains errors in estimating the calorie content of meals', *Annals of Internal Medicine*, 145(5) (2006), 326–32; B. J.

Rolls, E. L. Morris & L. S. Roe, 'Portion size of food affects energy intake in normal-weight and overweight men and women', *American Journal of Clinical Nutrition*, 76 (2002), 1207–13. It seems there is an important distinction here. For while our subconscious mind appears to be very good at identifying sources of nutrition in the environment, our conscious mind lags behind in terms of judging the calorie content of foods; see D. W. Tang, L. K. Fellows & A. Dagher, 'Behavioral and neural valuation of foods is driven by implicit knowledge of caloric content', *Psychological Science*, 25 (2014), 2168–76.

 The final question is meant to be rhetorical. One reason why an app might do a better job than the human mind is that the former is not subject to such phenomena as the Delboeuf Illusion; see K. Van Ittersum & B. Wansink, 'Plate size and color suggestibility: The Delboeuf Illusion's bias on serving and eating behavior', *Journal of Consumer Research*, 39 (2012), 215–28.

54 See T. Hayward, 'The cult of inconsistency', *FT Weekend Magazine*, 10 October 2014 (http://www.ft.com/intl/cms/s/0/41cb3e4c-4e66-11e4-bfda-00144feab7de.html).

55 A. Ward, 'Mechanic masterchef: Robots cook dumplings, noodles and wait tables at restaurant in China', *Daily Mail Online*, 13 January 2013 (http://www.dailymail.co.uk/news/article-2261767/Robot-Restaurant-Robots-cook-food-wait-tables-Harbin.html#ixzz2SWtkRc9S).

56 S. Liberatore, 'The future of fast food: KFC opens restaurant run by AI ROBOTS in Shanghai', *Daily Mail*, 6 May 2016 (http://www.dailymail.co.uk/sciencetech/article-3577192/The-future-fast-food-KFC-opens-restaurant-run-AI-ROBOTS-Shanghai.html#ixzz4DLZHz3mc); see also S. Liberatore, 'The robochef is here! Watch Kawasaki's new two-armed robot rustle up everything from pizza to sushi (and it'll even draw your portrait as well)', *Daily Mail*, 17 August 2016 (http://www.dailymail.co.uk/sciencetech/article-3745932/The-robochef-Watch-Kawasaki-s-new-two-armed-robot-rustle-pizza-sushi-ll-draw-portrait-well.html).

57 S. Curtis, 'Robotic bartender serves up drinks on world's first "smart ship": Royal Caribbean's *Quantum of the Seas* is the most technologically advanced cruise ship in the world', *Daily Telegraph*, 1 November 2014 (http://www.telegraph.co.uk/technology/news/11198509/Robotic-bartender-serves-up-drinks-on-worlds-first-smart-ship.html).

58 T. Fuller, 'You call this Thai food? The robotic taster will be the judge', *The New York Times*, 29 September 2014, A1 (https://www.nytimes.com/2014/09/29/world/asia/bad-thai-food-enter-a-robot-taster.html?_r=0).

59 R. Burn-Callender, 'The robot chef coming to a kitchen near you', *Daily Telegraph*, 6 October 2015 (http://www.telegraph.co.uk/finance/businessclub/11912085/The-robot-chef-coming-to-a-kitchen-near-you.html).

13. *Back to the Futurists*

1 Sublimotion describes itself as a 'merger of haute cuisine, gastronomy and . . . technology'; quoted in M. Moore, 'Taste the difference: Sublimotion vs. Ultraviolet', *Financial Times*, 28 August 2015 (https://www.ft.com/cms/s/2/0a4f62f0-4ca2-11e5-9b5d-89a026fda5c9.html#slide0).

2 See B. McFarlane and T. Sandham, 'Back to the Futurism', *The House of Peroni*, 2016 (https://thehouseofperoni.com/ie-en/lifestyle/back-futurism/).

3 The Roca brothers created the dish using distilled cocoa bean essence; although completely white, it tastes of dark chocolate; see 'Delicious science', *Harvard Magazine*, 14 September 2012 (https://harvardmagazine.com/2012/09/delicious-science). For Blumenthal's jelly, see H. Blumenthal, *The Big Fat Duck Cookbook* (London: Bloomsbury, 2008), pp. 138–43 (the dish is currently off the menu at The Fat Duck). See also F. T. Marinetti & L. Colombo, *La cucina futurista: Un pranzo che evitò un suicidio* [*The Futurist Kitchen: A Meal That Prevented Suicide*] (1932; Milan: Christian Marinotti Edizioni, 1998). Note that this early deliberate attempt to miscolour food also pre-dates what is, I think, the first scientific study to have been published on the topic, by the chemist H. C. Moir ('Some observations on the appreciation of flavour in foodstuffs', *Journal of the Society of Chemical Industry: Chemistry & Industry Review*, 14 (1936), 145–8).

4 'The *polibibita*, or cocktail, felt the force of Futurism too. Wine was drunk from petrol tanks, absinthe – officially banished from the country – was created in secret and Italian concoctions hitherto

enjoyed neat would be used as mixers. Futurists would drink in Milanese bars such as the Caffè Del Centro, and the Caffè Savini, which is still open today, and such elbow-bending anarchy led to the creation of an unusual array of counter-culture cocktails. The Italian cocktail scene was traditionally based on vermouth, bitters and classic decorations such as a lemon or orange zest but, "we want no part of it" said Marinetti, "the past"' (B. McFarlane & T. Sandham, 'Back to the Futurism', *The House of Peroni*, 2016 (https://thehouseofperoni.com/ie-en/lifestyle/back-futurism/)). For a review, see C. Spence, 'Futurist cocktails', *The Cocktail Lovers* (in press, 2017).

5　Marinetti published his manifesto on touch in Milan in 1921: *Il tattilismo* (Milan: Comoaedia, 1921), translated in English in Lisa Panzera & C. Blum (eds.), '*La futurista*: Benedetta Cappa Marinetti' (exhibition catalogue; Philadelphia: Goldie Paley Gallery, 1998), pp. 54–6. See also A. Gallace & C. Spence, *In Touch with the Future: The Sense of Touch from Cognitive Neuroscience to Virtual Reality* (Oxford: Oxford University Press, 2014).

6　The dish in question was a chilled citrus soup, finished at the table by the waiter spraying a little togarashi mist over the bowl; see P. Vettel, *Good Eating's Fine Dining in Chicago* (Chicago: Agate Digital, 2013).

7　F. T. Marinetti, *The Futurist Cookbook*, translated by S. Brill (1932; San Francisco: Bedford Arts, 1989).

8　In 'The Atmospheric Meal', we saw how the atmospherics influence the dining experience. While chefs today typically want to trigger certain positive emotions or else enhance the taste of the food, the Futurists had other ideas in mind: 'The surroundings and the dishes served at the Holy Palate were designed to put diners in unsettling situations, so as to totally engage their senses and shock them into breaking free from their normal everyday habits and expectations' (P. McCouat, 'The Futurists declare war on pasta', *Journal of Art in Society*, 2014 (http://www.artinsociety.com/the-futurists-declare-war-on-pasta.html)). See also Q. (J.) Wang, *Music, Mind, and Mouth: Exploring the Interaction between Music and Flavour Perception* (manuscript contributing towards an MSc at the Massachusetts Institute of Technology, Cambridge, MA, 2013).

9　S. Brickman, 'The food of the future', *The New Yorker*, 1 September 2014 (https://www.newyorker.com/culture/culture-desk/food-future).

10 G. Berghaus, 'The futurist banquet: Nouvelle Cuisine or performance art?', *New Theatre Quarterly*, 17(1) (2001), 3–17, p. 15.

11 G. Berghaus, 'The futurist banquet: Nouvelle Cuisine or performance art?', *New Theatre Quarterly*, 17(1) (2001), 3–17, p. 10. Depero was also responsible for designing the iconic triangular Campari Soda bottle, still in use nearly a century later.

12 This is the title of a recent article; see 'Futurist cooking: Was molecular gastronomy invented in the 1930s?', *The Staff Canteen*, 25 April 2014 (https://www.thestaffcanteen.com/Editorials-and-Advertorials/futurist-cooking-was-molecular-gastronomy-invented-in-the-1930s).

13 Marinetti published his infamous 'Manifesto of Futurist Cooking' in the *Gazzetta del Popolo* in Turin on 28 December 1930 (reprinted in F. T. Marinetti, *The Futurist Cookbook*, translated by S. Brill (1932; San Francisco: Bedford Arts, 1989), pp. 33–40).

14 There's a marked failure to name-check the Futurists throughout the genre of cookery writing (who knows, perhaps because of their fascist tendencies?). The one mention you find is in Elizabeth David's *Italian Food* (London: Barrie & Jenkins, 1987). The glaring absence of the Futurists is emphasized by the fact that modernist chefs are not themselves averse to putting out their own manifestos, e.g., F. Adrià et al., 'Statement on the "new cookery"', *Observer*, 10 October 2006 (https://www.theguardian.com/uk/2006/dec/10/foodanddrink.obsfoodmonthly). What is more, these contemporary chefs have shown themselves elsewhere to be intrigued by historical recipes, e.g., H. Blumenthal, *Historic Heston Blumenthal* (London: Bloomsbury, 2013).

15 It is interesting to note that they explicitly state that this will 'make food much more tasty' (F. T. Marinetti, *The Futurist Cookbook*, translated by S. Brill (1932; San Francisco: Bedford Arts, 1989), p. 84); cited in G. Berghaus, 'The futurist banquet: Nouvelle Cuisine or performance art?', *New Theatre Quarterly*, 17(1) (2001), 3–17.

16 See S. Smith (ed.), *Feast: Radical Hospitality in Contemporary Art* (Chicago: IL: Smart Museum of Art, 2013), p. 35 for this reproduction. Interestingly, many of the same ideas were first voiced in their manifesto on *culinaria futurista* published in 1920; Irba, '*Culinaria futurista*: Manifesto', *Roma futurista*, 9 May 1920 (cited in G. Berghaus, 'The

futurist banquet: Nouvelle Cuisine or performance art?', *New Theatre Quarterly*, 17(1) (2001), 3–17, p. 10).

17 Sous vide describes the technique of cooking in a vacuum, placing sealed meat or vegetables into a water bath at an exact (normally relatively low) temperature for much longer than one would normally cook food (up to 72 hours in some cases); see J. Rayner, 'The man who mistook his kitchen for a lab', *Observer*, 15 February 2004 (https://www.guardian.co.uk/lifeandstyle/2004/feb/15/foodanddrink.restaurants).

18 D. MacHale, *Wisdom* (London: Prion, 2002); from https://www-history.mcs.st-andrews.ac.uk/Quotations/Einstein.html.

19 D. Darrah, 'Futurist's idea on food finds Italy contrary', *Chicago Daily Tribune*, 11 December 1931; H. B. Higgins, 'Schlurrrp!: The case for and against spaghetti', in S. Smith (ed.), *Feast: Radical Hospitality in Contemporary Art* (Chicago: IL: Smart Museum of Art, 2013), pp. 40–47; P. McCouat, 'The Futurists declare war on pasta', *Journal of Art in Society*, 2014 (http://www.artinsociety.com/the-futurists-declare-war-on-pasta.html); R. Golan, 'Ingestion/Anti-pasta', *Cabinet*, 10 (2003), 1–5.

20 See 'Futurism for foodies', *Artnet*, 12 August 2011 (https://www.artnet.com/magazineus/news/artnetnews/the-futurist-cookbook.asp); S. Brickman, 'The food of the future', *The New Yorker*, 1 September 2014 (https://www.newyorker.com/culture/culture-desk/food-future); B. Hoyle, 'Recipe for revolution takes diners back to the Futurists', *The Times*, 28 January 2008, p. 23.

21 F. T. Marinetti, *The Futurist Cookbook*, translated by S. Brill (1932; San Francisco: Bedford Arts, 1989), p. 153.

22 F. T. Marinetti, *The Futurist Cookbook*, translated by S. Brill (1932; San Francisco: Bedford Arts, 1989), p. 65.

23 Talking of the fancy attire of the dinner guests, it is worth noting that 'the Futurist diet favoured the aristocrat. The resplendent table would shimmer with exotic liquor (Muscat wine, Marsala, Whiskey, Alkermes, and Vermouth), costly and hard to find ingredients (rose petals, caviar, cockscombs, and pineapple), and labor-intensive food (sculpted meat skyscrapers, Chicken Fiat, dried lettuce leaves, and hollow banana peels)' (H. B. Higgins, 'Schlurrrp!: The case for and against spaghetti',

in S. Smith (ed.), *Feast: Radical Hospitality in Contemporary Art* (Chicago: IL: Smart Museum of Art, 2013), pp. 40–47).

24 H. B. Higgins, 'Schlurrrp!: The case for and against spaghetti', in S. Smith (ed.), *Feast: Radical Hospitality in Contemporary Art* (Chicago: IL: Smart Museum of Art, 2013), pp. 40–47, p. 43.

25 F. T. Marinetti, *The Futurist Cookbook*, translated by S. Brill (1932; San Francisco: Bedford Arts, 1989), p. 84.

26 G. Berghaus, 'The futurist banquet: Nouvelle Cuisine or performance art?', *New Theatre Quarterly*, 17(1) (2001), 3–17, pp. 8–9; Berghaus also notes that '[i]n 1913, the journal *Fantasio* published an essay on "La cuisine futuriste", which contained a "Manifeste de la cuisine futuriste" by the chef Jules Manicave.' This latter including the line: 'We want a cuisine in tune with modern life and the latest scientific inventions.' And before even the proto-Futurists, there might be a link to the much earlier Satiricon of Petronius Arbitrius; see R. Golan, 'Ingestion/Anti-pasta', *Cabinet*, 10 (2003), 1–5.

27 In *Le Poète assassiné* (1916; Paris: Gallimard, 1992), pp. 258–9, reprinted and translated in A. S. Weiss, *Feast and Folly: Cuisine, Intoxication and the Poetics of the Sublime* (Albany, NY: State University of New York Press, 2002), pp. 114–15, pp. 145–6. See also Apollinaire, *Oeuvres en prose*, vol. 1 (Paris: Gallimard, 1912), pp. 402–3, translated in G. Berghaus, 'The futurist banquet: Nouvelle Cuisine or performance art?', *New Theatre Quarterly*, 17(1) (2001), 3–17, p. 9.

28 A. Fleming, 'Neuro cuisine: Exploring the science of flavour', *Guardian*, 23 May 2016 (https://www.theguardian.com/lifeandstyle/ng-interactive/2016/may/23/neuro-cuisine-exploring-the-science-of-flavour-tamal-ray); S. Pigott, 'Appetite for invention', *Robb Report*, May 2015, 98–101; https://kitchen-theory.com/tag/gastrophysics/.

29 G. Berghaus, 'The futurist banquet: Nouvelle Cuisine or performance art?', *New Theatre Quarterly*, 17(1) (2001), 3–17, p. 10.

30 See A. Waterlow, 'Is your restaurant dish really a ready meal? High street chains including Pizza Express and Frankie & Benny's "serve pre-prepared mains, desserts and even scrambled eggs" '*Daily Mail Online*, 19 January 2016 (http://www.dailymail.co.uk/femail/article-3404895/Is-restaurant-dish-really-ready-meal-High-street-chains-

including-Pizza-Express-Frankie-Benny-s-serve-pre-prepared-mains-desserts-scrambled-eggs.html). I must say that I have much the same issue with those restaurants, even Michelin-three-starred ones I know, that serve pod coffee. Maybe this is a space issue, but really, if you have chefs preparing the food freshly, why would you think that I didn't want the same attention to detail for my coffee afterwards?

31 T. Hayward, 'The cult of inconsistency', *FT Weekend Magazine*, 10 October 2014 (https://www.ft.com/content/41cb3e4c-4e66-11e4-bfda-00144feab7de). The top chefs are also increasingly starting to consider how exactly diners interact with their dishes. After all, even if the food on the plate is identical, the order in which the diner samples the various elements may fundamentally change the ensuing experience. According to chef Andoni: 'The dish "Roasted and Raw Vegetables, Wild and Cultivated Shoots and Leaves" [. . .] consists of hundreds of vegetables, leaves and herbs – something nobody would think of trying at home. One hundred ingredients make it impossible for the person preparing the dish to plate any two in the same way. It also makes it almost impossible for two diners to eat it in exactly the same way. This is one of the evocative powers of this recipe' (A. L. Aduriz, *Mugaritz: A Natural Science of Cooking* (New York: Phaidon, 2014), p. 42).

32 At the time of writing, you have to pay a £5.95 delivery cost and a 9% service change for this at-home experience. So with average prices ranging from £80–100 for a meal, one might well ask whether it's really worth it. The only obvious benefit, as far as I can see, is that you don't have to ask for a doggy bag should you find that you ordered more than you can manage. See S. Barns, 'Takeaways go gourmet: Busy professionals can now order Michelin-starred food straight from top restaurants to their home . . . but is it worth the hefty price tag?', *Daily Mail Online*, 8 September 2015 (http://www.dailymail.co.uk/femail/food/article-3225211/Supper-delivers-Michelin-starred-food-straight-London-s-restaurants-door.html#ixzz47OP6HN5a).

33 An early usage of this term comes from B. J. Pine, II, and J. H. Gilmore, the self-styled architects of the experience economy, who highlighted the fact that '[a]t theme restaurants such as the Hard Rock

Café, Planet Hollywood, or the House of Blues, the food is just a prop for what is known as "eatertainment." ' (*The Experience Economy: Work Is Theatre & Every Business Is a Stage* (Boston, MA: Harvard Business Review Press, 1999), p. 99).

34 C. Spence & J. Youssef, 'Constructing flavour perception: From destruction to creation and back again', *Flavour*, 5:3 (2016). The meal in question was coordinated by Kitchen Theory.

35 Again, perhaps this shouldn't come as a surprise given that the restaurant, as we know it, has been around only since the early 1800s; see R. L. Spang, *The Invention of the Restaurant: Paris and Modern Gastronomic Culture* (Cambridge, MA: Harvard University Press, 2000).

36 D. Jurafsky, *The Language of Food: A Linguist Reads the Menu* (New York: Norton, 2014).

37 Y.-Y. Ahn et al., 'Flavor network and the principles of food pairing', *Scientific Reports*, 1:196 (2011), 1–6; S. E. Ahnert, 'Network analysis and data mining in food science: The emergence of computational gastronomy', *Flavour*, 2:4 (2013). For an interactive graphic, see 'The Flavor Connection', *Scientific American*, 1 September 2013; R. B. Ness, *The Creativity Crisis: Reinventing Science to Unleash Possibility* (Oxford: Oxford University Press, 2015), p. 142.

38 A. Jain, N. K. Rakhi & G. Baglerb, 'Spices form the basis of food pairing in Indian cuisine', Cornell University Library, 2015 (http://arxiv.org/ftp/arxiv/papers/1502/1502.03815.pdf). See also R. A. Ferdman, 'Scientists have figured out what makes Indian food so delicious', *Washington Post*, 3 March 2015 (https://www.washingtonpost.com/news/wonk/wp/2015/03/03/a-scientific-explanation-of-what-makes-indian-food-so-delicious/).

39 On food pairing, see W. L. P. Bredie et al., 'Flavour pairing of foods: A physical-chemical and multisensory challenge for health promotion', *European Sensory Network*, 2015 (http://www.esn-network.com/index.php?id=1034); https://www.foodpairing.com/; and see C. Spence, Q. (J.) Wang & J. Youssef, 'Flavour pairing: A review', *Flavour* (in press, 2017), for a critical review of the flavour-pairing approach.

40 Quote from J. Wakefield, 'What would a computer cook for dinner?', *BBC News Online*, 7 March 2014 (https://www.bbc.co.uk/news/technology-26352743). See also *Cognitive Cooking with Chef Watson:*

Recipes for Innovation from IBM & the Institute of Culinary Education (Chicago: Sourcebooks, 2015). Those who have tried Watson's bartending skills haven't necessarily always been that impressed; see C. Trout, 'I trusted my gut to IBM's Watson and it gave me a fowl old-fashioned', *Engadget*, 15 May 2015 (https://www.engadget.com/2015/05/15/drinking-with-watson/).

41 The reviews were correlated with the weather reports at the time that they had been written. One of the report's authors summarized their findings as follows: 'the best reviews are written on sunny days between 70 and 100 degrees [. . .] Science has shown that weather impacts our mood, so a nice day can lead to a nice review. A rainy day can mean a miserable one.' Or take the following quote from the research paper itself: 'Reviews written on rainy or snowy days tend to have lower ratings than those written on days without rain or snow' (S. Bakhshi, P. Kanuparthy & E. Gilbert, 'Demographics, weather and online reviews: A study of restaurant recommendations by WWW 2014', *Proceedings of the 23rd International Conference on World Wide Web* (New York: ACM, 2014)); see also J. Maderer, 'A rainy day can ruin an online restaurant review: Weather helps determine whether a review will be positive or negative', *Georgia Tech News*, 2 April 2014 (https://www.news.gatech.edu/2014/04/02/rainy-day-can-ruin-online-restaurant-review); 'Fair-weather reviews? Study finds online restaurant reviews written on sunny days get top marks', *Daily Mail Online*, 5 April 2014 (http://www.dailymail.co.uk/news/article-2597835/Fair-weather-reviews-Study-finds-online-restaurant-reviews-written-sunny-days-marks.html).

42 For instance, we now have more than 5,000 people's responses to the question of which is the sweetest drink colour from studies that were simultaneously run online and as part of the 2015–16 'Cravings' exhibition; C. Velasco et al., 'Colour-taste correspondences: Designing food experiences to meet expectations or to surprise', *International Journal of Food Design*, 1 (2016), 83–102.

Several thousand have also given us their opinion on the importance of balance vs asymmetry on the plate. See C. Velasco et al., 'On the importance of balance to aesthetic plating', *International Journal of Gastronomy and Food Science* (5–6) (2016), 10–16. See also N. L. Garneau et al., 'Crowdsourcing taste research: Genetic and phenotypic predictors of bitter taste perception as a model', *Frontiers of Integrative Neuroscience*, 8:33 (2014).

43 J. Youssef et al., 'On the art and science of naming and plating food', *Flavour*, 4:27 (2015); J. Roque et al., 'Plating influences diners' perception of culinary creativity' (manuscript submitted for publication 2016).

44 Quoted in M. Wall, 'From pizzas to cocktails the data crunching way', *BBC News*, 18 August 2015 (https://www.bbc.co.uk/news/business-33892409). Applied Predictive Technologies is one of the companies currently offering this data-crunching service for the food industry. The late British writer and restaurant critic A. A. Gill commented on the fact that mom-and-pop places, the one-off restaurants, can't compete any more. It is all chains and economies of scale, at least in London. He thought this reduces choice as far as the restaurant-goer is concerned ('Table Talk: BAO Fitzrovia', *The Sunday Times Magazine*, 14 August 2016, p. 48).

45 See M. Moore, 'Taste the difference: Sublimotion vs. Ultraviolet', *Financial Times*, 28 August 2015 (https://www.ft.com/cms/s/2/0a4f62f0-4ca2-11e5-9b5d-89a026fda5c9.html#slide0); S. Pigott, 'Appetite for invention', *Robb Report*, May 2015, 98–101.

46 M. Haverkamp, *Synesthetic Design: Handbook for a Multisensory Approach* (Basel: Birkhäuser, 2014); C. Spence, 'Synaesthetic marketing: Cross sensory selling that exploits unusual neural cues is finally coming of age', *The Wired World in 2013*, special issue, 1 November 2012, 104–7; N. Scharping, 'Synesthesia mask lets you wake up and smell the colors', *Discover Magazine*, 23 February 2014 (https://blogs.discovermagazine.com/d-brief/2016/02/23/synesthesia-mask-helps-you-wake-up-and-smell-the-colors/#.V2UEU7srLIU).

47 C. Spence, C. Velasco & K. Knoeferle, 'A large sample study on the influence of the multisensory environment on the wine drinking experience', *Flavour*, 3:8 (2014).

48 See O. Deroy & C. Spence, 'Learning "arbitrary" crossmodal correspondences: Staying away from neonatal synaesthesia', *Neuroscience & Biobehavioral Reviews*, 37 (2013), 1240–53; O. Deroy & C. Spence, 'Weakening the case for "weak synaesthesia": Why crossmodal correspondences are not synaesthetic', *Psychonomic Bulletin & Review*, 20 (2013), 643–64;

O. Deroy & C. Spence, 'Training, drugs, and hypnosis: Artificial syn-aesthesia, or artificial paradises?', *Frontiers in Perception Science*, 4:660 (2013).

49 C. Spence, 'Crossmodal correspondences: A tutorial review', *Attention, Perception, & Psychophysics*, 73 (2011), 971–95; C. Spence, 'Managing sensory expectations concerning products and brands: Capitalizing on the potential of sound and shape symbolism', *Journal of Consumer Psych-ology*, 22 (2012), 37–54; O. Deroy, A.-S. Crisinel & C. Spence, 'Crossmodal correspondences between odors and contingent features: Odors, musical notes, and geometrical shapes', *Psychonomic Bulletin & Review*, 20 (2013), 878–96; C. Velasco et al., 'Assessing the role of the pleasantness of music (vs. white noise) and odours in olfactory perception', *Frontiers in Psychology*, 5:1352 (2014). What is so striking about traditional attempts to utilize synaesthetic design is how little of it has ever gone beyond the domain of the audiovisual. For instance, only a handful out of the 460 pages in Haverkamp's otherwise masterful 2014 volume have any-thing to do with the chemical senses; see C. Spence, 'Book review: Synaesthetic Design', *Multisensory Research*, 28 (2015), 245–8. Though it is in-teresting to note how few synaesthetic chefs there are: C. Spence, J. Youssef & O. Deroy, 'Where are all the synaesthetic chefs?', *Flavour*, 4:29 (2015).

50 B. Miller, 'Artist invites public to taste colour in ten-day event with dancers and wine at The Oval', *Culture 24*, 3 February 2015 (http://www.culture24.org.uk/art/art516019-artist-invites-public-to-taste-colour-in-ten-day-event%20with-dancers-and-wine-at-the-oval).

51 L. Barnett, 'Synaesthesia: When two senses become one', *Guardian*, 5 December 2011 (https://www.theguardian.com/lifeandstyle/2011/dec/05/synaesthesia-hearing-colours-mixing-senses); A. Fleming, 'How sound affects the taste of our food', *Guardian*, 11 March 2014 (https://www.theguardian.com/lifeandstyle/wordofmouth/2014/mar/11/sound-affects-taste-food-sweet-bitter); Q. (J.) Wang, *Music, Mind, and Mouth: Exploring the Interaction between Music and Flavour Perception* (manuscript contributing towards an MSc at the Massachusetts Institute of Tech-nology, Cambridge, MA, 2013).

52 D. Arroche, 'Never heard of Sensploration? Time to study up on epi-cure's biggest luxury trend', *LuxeEpicure*, 22 December 2015 (https://

www.justluxe.com/lifestyle/dining/feature-1962122.php); see also D. J. Pangburn, 'This soup is alive with the sound of music', *The Creators Project*, 30 September 2015 (https://thecreatorsproject.vice.com/blog/this-soup-is-alive-with-the-sound-of-music).

53 In one early event, in October 2013, a quartet from the London Symphony Orchestra played a selection of tracks that had been demonstrated to match each of four wines especially well to about 120 guests; see C. Spence et al., 'Looking for crossmodal correspondences between classical music & fine wine', *Flavour*, 2:29 (2013).

On 23 February 2016, the London Contemporary Orchestra held a live multisensory event called 'Sounds of Flavourites', in which music was paired with nine different flavours of Cadbury Dairy Milk; see K. Deighton, 'Cadbury collaborates with orchestra for multi-sensory live event', *Event Magazine*, 17 February 2016 (http://www.eventmagazine.co.uk/cadbury-collaborates-orchestra-multi-sensory-live-event/brands/article/1383795). There was also an intriguing opportunity to create music based on your favourite pair of chocolate flavours (http://cadburyflavourites.co.uk/), and even an album release (see 'Cadbury Dairy Milk releases music album The Sound of Flavourites', *Burton Mail*, 1 March 2016 (http://www.burtonmail.co.uk/Cadbury-Dairy-Milk-releases-music-album-Sound/story-28836333-detail/story.html)).

See also K. Monks, 'Magical organ gives "musical taste" a new meaning', *CNN*, 6 January 2015 (http://edition.cnn.com/2014/09/18/tech/innovation/multi-sensory-organ/); and see C. Spence & Q. (J.) Wang, 'Wine & music (III): So what if music influences taste?', *Flavour*, 4:35 (2015), for a review of musical wine-tasting events.

54 Quotes from Y. Arrigo, 'Welcome to the booming experience economy', *Raconteur (Future of Events & Hospitality)*, 362 (2016), 2–3.

55 'Through reinventing the overall experience as *Gesamtkunstwerk*, high-end chefs truly claim their place, as Carême articulated, in the pantheon of great artists' (quoted in J. Abrams, 'Mise en plate: The scenographic imagination and the contemporary restaurant', *Performance Research: A Journal of the Performing Arts*, 18(3) (2013), 7–14, p. 14).

56 The excitement of those heady times is captured in Robert Brain's wonderful book *The Pulse of Modernism: Physiological Aesthetics in Fin-du-siècle Europe* (London: University of Washington Press, 2015).

One of the key scientists at the interface with the arts was Charles Henry; see J. A. Argüelles, *Charles Henry and the Formation of a Psychophysical Aesthetic* (Chicago, IL: University of Chicago Press, 1972).

57 Others, though, might, disagree: see M. Boccia et al., 'Where does brain neural activation in aesthetic responses to visual art occur? Meta-analytic evidence from neuroimaging studies', *Neuroscience and Biobehavioral Reviews*, 60 (2016), 65–71; M. T. Pearce et al., 'Neuroaesthetics: The cognitive neuroscience of aesthetic experience', *Perspectives in Psychological Science*, 11(2) (2016), 265–79.

58 S. Pigott, 'Appetite for invention', *Robb Report*, May 2015, 98–101; S. Stummerer & M. Hablesreiter, *Food Design XL* (New York: Springer, 2010); S. Stummerer & M. Hablesreiter, *Eat Design* (Vienna: Metroverlag, 2013). And see Brainy Tongue for one example of such collaboration; this event brought together cognitive neuroscientists with some of the world's most creative chefs late in 2016 (http://brainytongue.com/).

59 Moving forward, I suspect that many of the key innovations will be at the interface between technology and sustainability; see, for instance, '*Das Gasthaus zum Übermorgen: Essen in fünf Dimensionen*', W.I.R.E., May 2015 (https://www.thewire.ch/de/events/serie/gasthaus-zum-uebermorgen). And the recently launched Too Good To Go app allows people to order food from restaurants that would otherwise be thrown away; see S. Fitzmaurice, 'New app lets you order food from top restaurants that would otherwise be thrown away – for just £2 PER DISH', *Daily Mail Online*, 8 August 2016 (http://www.dailymail.co.uk/femail/food/article-3729581/Too-Good-lets-order-food-restaurants-thrown-away.html).

60 J. Wapner, 'The flavor factory: Hijacking our senses to tailor tastes', *New Scientist*, 3 February 2016 (https://www.newscientist.com/article/2075674-the-flavour-factory-hijacking-our-senses-to-tailor-tastes/).

61 M. Ouellet et al., 'Is the future the right time?', *Experimental Psychology*, 57 (2010), 308–14; 'Marketing lessons from the Johnnie Walker brand', *Incitrio*, 9 May 2014 (http://incitrio.com/marketing-lessons-from-the-johnnie-walker-brand/).

62 A. L. Aduriz, *Mugaritz: A Natural Science of Cooking* (New York: Phaidon, 2014), pp. 42–3. Though recent discussion of the gut–brain

biome suggests that, in some sense, the mind is also to be found in the stomach. This another intriguing topic for another day.

63 You'll find references to much of the underpinning research scattered throughout the book; however, there are a few claims that we haven't come across yet, so here is some of the relevant research. You will find plenty more excellent suggestions in B. Wansink, *Mindless Eating: Why We Eat More Than We Think* (London: Hay House, 2006). For other sensible suggestions, see T. M. Marteau, G. J. Hollands & P. C. Fletcher, 'Changing human behaviour to prevent disease: The importance of targeting automatic processes', *Science*, 337 (2012), 1492–5; H. L. Meiselman, 'The role of context in food choice, food acceptance and food consumption', in R. Shepherd & M. Raats (eds.), *The Psychology of Food Choice* (Wallingford, Oxon: CABI, 2006), pp. 179–201.

64 E. A. Dennis et al., 'Water consumption increases weight loss during a hypocaloric diet intervention in middle-aged and older adults', *Obesity*, 18 (2010), 300–307.

65 A. Jami, 'Healthy reflections: The influence of mirror induced self-awareness on taste perceptions', *Journal of the Association for Consumer Research*, 1 (2016), 57–70.

66 C. de Graaf, 'Why liquid energy results in overconsumption', *Proceedings of the Nutrition Society*, 70 (2011), 162–70.

67 P. Mony et al., 'Temperature of served water can modulate sensory perception and acceptance of food', *Food Quality and Preference*, 28 (2013), 449–55.

68 S. S. Holden, N. Zlatevska & C. Dubelaar, 'Whether smaller plates and bowls reduce consumption depends on who's serving and who's looking: A meta-analysis', *Journal of the Association for Consumer Research*, 1 (2016), 134–46; http://www.smallplatemovement.org/; G. J. Hollands et al., 'Portion, package or tableware size for changing selection and consumption of food, alcohol and tobacco (Review)', *The Cochrane Library*, 9 (2015).

69 N. Bruno et al., 'The effect of the color red on consuming food does not depend on achromatic (Michelson) contrast and extends to rubbing cream on the skin', *Appetite*, 71 (2013), 307–13; O. Genschow, L. Reutner & M. Wänke, 'The color red reduces snack food and soft drink intake', *Appetite*, 58 (2012), 699–702.

70 J. Dunn, 'The easiest way to lose weight? Eat with a FORK – not a spoon – to control your portion sizes', *Daily Mail Online*, 29 December 2015 (http://www.dailymail.co.uk/health/article-3377504/Eating-fork-instead-spoon-make-lose-weight-researchers-say.html); X. Clay, 'Eating with chopsticks "helps lose weight"', *Daily Telegraph*, 16 January 2009 (http://www.telegraph.co.uk/news/health/news/4271430/Eating-with-chopsticks-helps-lose-weight.html); K. Gander, 'The supper club where diners are given spoons filled with nails to promote slow, mindful eating', *Independent*, 22 August 2016 (https://www.independent.co.uk/life-style/supper-club-mindful-eating-steinbeisser-slow-food-movement-amsterdam-california-a7196901.html).

71 Retrieved June 2016, from https://www.brainyquote.com/quotes/authors/y/yogi_berra.html.

Illustration Credits

Figure o.1: Reproduced with kind permission of Restaurant Denis Martin

Figure o.2: Courtesy of the author

Figure o.3: Courtesy of the Science Museum, London

Figure o.4: © Andy T. Woods, Charles Michel & Charles Spence, 2016

Figure 1.1: © National Academy of Sciences of the USA, 2008

Figure 1.2: © Oxford University Press

Figure 2.1: 'Jelly of Quail' © Ashley Palmer-Watts, reproduced with kind permission of Lotus PR and The Fat Duck

Figure 2.2: The Viora lid reproduced with kind permission of Barry Goffe; Crown's 360End™can reproduced with kind permission of Cormac Neeson

Figure 2.3: © A. Dagli Orti/DEA/Getty Images

Figure 2.4: © PARS International Corp, 2017

Figure 2.5: Courtesy of the author

Figure 3.1: Courtesy of the author

Figure 3.3: © Luesma & Vega SL

Figure 3.4: Foodography campaign created by BBR Saatchi & Saatchi on behalf of the Carmel Winery

Figure 3.5: C. Michel et al., 'Rotating plates: Online study demonstrates the importance of orientation in the plating of food', *Food Quality and Preference*, 44 (2015), 194– 202

Figure 3.6: © Roger Stowell/Getty Images

Figure 3.7: Reproduced with kind permission of KEEMI

Figure 3.8: Michel et al., 'Rotating Plates', in *Food Quality and Preference*

Figure 4.1: Reproduced with kind permission of Massimiliano Zampini

Figure 4.2: © HOANG DINH NAM/AFP/Getty Images

Figure 4.3: © Frito-Lay North America, Inc., 2017

Figure 4.4: Reproduced by kind permission of Naoya Koizumi

Figure 4.5: The Krug Shell, reproduced by kind permission of Krug Maison de Champagne

Figure 5.1: 'Tableware as Sensorial Stimuli, Rear Bump Spoon for Enhancing Colour & Tactility', Ceramic, 2012, courtesy of Jinhyun Jeon

Figure 5.2: Mulberry Textured Sensory Spoons, courtesy of Studio William

Figure 5.3: Courtesy of the author

Figure 5.4: Meret Oppenheim, *Object* (1936) © Artists Rights Society (ARS), New York / Pro Litteris, Zurich, 2017; rabbit spoon reproduced with kind permission of Charles Michel

Figure 5.5: 'Counting Sheep' © John Carey, reproduced with kind permission of Lotus PR and The Fat Duck

Figure 5.6: Reproduced with kind permission of Marcel Buerkle

Figure 6.2: © Space Copenhagen, 2012

Figure 6.3: © Cornell University, 1999

Figure 7.1: *Lonely* © Jon Krause

Figure 7.2: Mella Jaarsma, *I Eat You Eat Me* (2000). Performed at 'Feast: Radical Hospitality in Contemporary Art', Smart Museum, Chicago, 2012. Photography: Smart Museum. Courtesy of the artist

Figure 7.3: Marije Vogelzang, *Sharing Dinner* (Tokyo, 2008). Photography: Kenji Masunaga. Reproduced by kind permission of the artist

Figure 8.1: © The SAS Museum, Oslo Airport, Norway

Figure 8.2: © The SAS Museum, Oslo Airport, Norway

Figure 9.1: Menu map copyright © Dave McKean, reproduced with kind permission of Lotus PR and The Fat Duck

Figure 10.1: © The Coca-Cola Company, 2017

Figure 10.2: © *Chicago Tribune*, 2012. All rights reserved. Distributed by Tribune Content Agency. Photography: Scott Strazzante

Figure 10.3: 'Sweet Shop' © John Carey. Reproduced with kind permission of Lotus PR and The Fat Duck

Figure 11.1 © Alex Lentati

Figure 11.2: Underwater restaurant © Crown Company PVT Ltd trading as Conrad Maldives Rangali Island, 2013; Dinner in the Sky, Toronto © Dinner in the Sky

Figure 11.3: © David Ramos/Getty Images

Figure 11.4: © Liz Ligon

Figure 11.5: Barbara Smith, *Ritual Meal* (1969). Excerpt from 16mm film by William Ransom and Smith of a performance event in Brentwood, California. Lent by the artist

Figure 12.1: © Food Ink, 2016

Figure 12.2: © Charles Spence and Piqueras-Fiszman; licensee BioMed Central Ltd, 2013

Figure 12.3: 'Sound of the Sea' © Ashley Palmer-Watts, reproduced with kind permission of Lotus PR and The Fat Duck

Figure 12.4: © Association for Computing Machinery, Inc., 2012

Figure 12.5: © Intellect Limited

Figure 12.6: © Association for Computing Machinery, Inc., 2011

Figure 12.7: © REUTERS/Sheng Li

Figure 13.1: 'The Futurist Table', c.1931 (Filippo Tommaso Marinetti Papers, Beinecke Rare Book & Manuscript Library, Yale University) © DACS Author photograph © akg-images/MPortfolio/Electra

Figure 13.3: © akg-images/MPortfolio/Electra

While every effort has been made to trace copyright holders, the publishers will be happy to correct any errors of omission or commission at the earliest opportunity

Index

KU-084-298

LIVE   INSTITUTE
OF HIG   DUCATION

LIBRARY

WOOLTON ROAD,
LIVERPOOL, L16 8ND

RELIGIOUS EDUCATION
AND
YOUNG ADULTS

LIHE

RELIGIOUS EDUCATION AND YOUNG ADULTS

A ground-plan for religious education
in the senior forms of
the Church secondary school

Edited by Donal O'Leary

 St Paul Publications

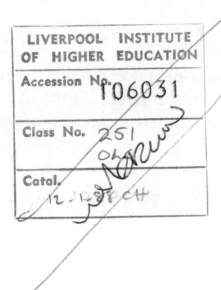

LIVERPOOL INSTITUTE
OF HIGHER EDUCATION

Accession No. 106031

Class No. 251

Catal.

St Paul Publications
Middlegreen, Slough SL3 6BT, England

Copyright © St Paul Publications 1983
First published October 1983
Printed by the Society of St Paul, Slough
ISBN 085439 229 7

St Paul Publications is an activity of the priests and
brothers of the Society of St Paul who promote the christian
message through the mass media

CONTENTS

MAIN CONTRIBUTORS

Mr John Brennan (Cambridge University)
Unemployment and religious education

Mr Brian Davies (General Secretary, CAFOD, London)
Development education in the secondary school

Rev. John Glen (Religious Education Centre, Northampton)
Principles and practices of liturgy and worship

Mrs Margaret Grimer (Catholic Marriage Advisory Council, London)
Education for personal relationships in the sixth form

Rev. Edward Matthews (Allen Hall, Chelsea)
Principles and practices of liturgy and worship

Mr Duncan MacPherson (St Mary's College, Twickenham)
World religions in Christian sixth form R.E.

Mrs. Teresa O'Donovan (St Mary's College, Twickenham)
'Multi-cultural' religious education

Rev. Donal O'Leary (St Mary's College, Twickenham)
Theology and teaching
The ministry of teaching in the Church school
Away-days, residential courses and community experiences

Miss Theresa Sallnow (St Mary's College, Twickenham)
Catechesis or religious education?
Religious education and the creative arts

Dr Marion Smith (Roehampton Institute of Higher Education)
Contemporary theories of psychological growth

Dr Geoffrey Turner (Trinity & All Saints College, Leeds)
Theology in the sixth form

Mr Patrick Walsh (London University Institute of Education)
The Church secondary school and its curriculum

Westminster Diocese Working Party members:
Miss Janice Burn, Deputy Headteacher, St Thomas More School, Chelsea; Mr David Evans, Headteacher, Cardinal Bourne School, Herts; Miss Veronica Melia, Deputy Headteacher, St Mark's School, Hounslow; Mr Michael Murphy, Head of R.E. Dept, Douai Martyrs School, Ickenham.
The status and structure of sixth form R.E.

To my brother Joseph
"a sign of contradiction"

FOREWORD

Over the past couple of decades a considerable number of books on religious education for the lower forms of the Church Secondary School have appeared, but very little has been written either for the sixth former or for the sixth form R.E. teacher. For all sorts of reasons educationalists seem to have fought shy of this particular field.

Undoubtedly the sixth form is a difficult group to provide for. It is a far less homogeneous group than it used to be. A significant proportion of those now staying on into sixth form are not pursuing academic courses of study. On this score alone it is therefore necessary for a school to offer its sixth formers a variety of courses in R.E. This in turn demands from sixth form teachers an increasingly wide knowledge, at a time when, following the Second Vatican Council, there has been a very considerable development in Church teaching.

Such factors make the appearance of this book most welcome. It is a tentative production. It does not attempt to offer final and polished answers, but rather aims to stimulate teachers of R.E. in the senior classes of Church Secondary Schools to tackle the curriculum and syllabus planning for their own schools. From my own experience of sixth form R.E. teaching I am convinced that this has to be the approach. It goes without saying that experienced teachers and writers can help prepare detailed courses on given topics, but in the end each class teacher (and in particular each sixth form teacher) has to decide both what the pupils need and what he/she is able to give them. Much of the material that follows will prove invaluable in planning future programmes.

It is interesting to read what a group of sixth form pupils once said about the qualities they would like to see in their R.E. teacher:

"The teacher should realise that he is talking to adults

(even if young adults); he should be able to express an opinion without suggesting that anyone who does not agree with him is a fool; he should be open-minded in the sense that he is able to change his mind; he should be well-informed; he should be completely sincere; he should be a good teacher."

Although these comments were made more than ten years ago, doubtless they still stand. I take a certain encouragement from the fact that the students who made these comments were sufficiently interested and concerned to answer questions about their R.E. teaching thoughtfully and honestly. There is no reason to think that today's students are in that respect any different. Apart from their comments giving all of us something to aim at, I believe they should also give us great hope.

✝ David Konstant
Bishop in Central London   September 1983

PREFACE

During the past decade Christian educators in Britain have indicated their deep concern regarding the nature, aims and content of religious education in the Church secondary school.

By way of response to such concern — concern which is shared by all Christian denominations — this book is offered as a basis for future curriculum and syllabus planning in the senior forms of post-primary schools with occasional special reference to sixth form R.E. From the many issues at stake in this vast area we have selected a basic, coherent and topical list for consideration.

While the contributors are drawn mainly from a Roman Catholic background, their contributions will be seen as emanating from a thoroughly sympathetic understanding of the difficulties facing all Christian denominations in their educational endeavours for the future.

For that reason, this ecumenical venture will be of interest to all religious education teachers in Church schools, heads of R.E. departmentas, headteachers and lecturers in colleges of further and higher education, who are responsible for the training of teachers in the Christian tradition.

It is offered also as a ground-plan to dioceses and local education authority bodies involved in devising new guidelines and agreed syllabuses for Church schools from a Christian perspective.

All contributors were approached because of their experience and expertise, their sensitivity to renewal and tradition and their commitment to excellence in religious education, both in its theory and practice. This is the common bond which lies at the heart of the book.

As well as the main contributors who are listed separately, I wish to convey my thanks to all those people who, over the past three years of working parties and consultations, have offered their advice and amendments regarding various sections of the

1

book. Among these I would like to mention Miss Ann Barton, Mr Jim Conroy, Rev. Perry Gildea, C.M., Rev. Peter Hackett, S.J., Rev. Paddy McCrohan, Sr Margaret Walsh, R.S.C.J., and many of my colleagues on the academic, secretarial and library staff at St Mary's College. My thanks also to Bishop David Konstant who initially commissioned the project and to Theresa Sallnow who co-ordinated and edited a great deal of the material.

Finally I am indebted to Rev. Kevin Raerty, C.M., and the Vincentian Community at Strawberry Hill who generously supported and facilitated the writing of many sections of this book.

D.J.O'L.

CHAPTER 1
EDUCATIONAL ISSUES

THE CHURCH SECONDARY SCHOOL AND ITS CURRICULUM

The good Church school

From earliest Christian times it has been a principle of theology that everything in our lives which has been lived through by Jesus, God incarnate and a man like us in everything except sin, has been in principle by that very fact blessed, redeemed, divinised. 'Everything' includes our birth and our death and, among the many things between these points, our development towards adulthood, our learning, our natural desire for achievement, our teaching, our love of the young and our concern for their development and achievement. Apply this to Catholic schools! We want them, of course, to be 'good schools', ones that stand comparison with the best around, ones in which our young will get 'a good education'. And we want them to be generally Christian places that provide a solid religious education for our children. Only we need not — and I think should not — see this religious element as an obligatory 'extra', as a Christian icing on the cake of a pre-baked education. It is more like a secret ingredient *in* that cake, the special yeast of the Gospel parable. Rather than looking to our schools for a good education *plus* a significant contribution to the Christian upbringing of our young we would do better to *stop* at asking them for a good education — but one that is good in a Christian way.

Schools that provided this would be likely to strike the unprejudiced outsider also as good, elusively 'different' indeed, but in a way that was at once strangely appealing and strangely challenging. They would meet the Gospel requirement of being 'a light to the world'. To say this, however, is not to imply that they would be the only light in the firmament. It is not to deny that Church schools have, and would continue to have, much to learn from their secular counterparts. The Holy Spirit fights his uphill battles everywhere — not just in Christian hearts and institutions. As Christians can discern and be shamed by the virtues of their non-Christian friends so can the Church school be humbled by the virtues of its secular neighbour. To that we might add that secular schools have quantities of committed Christians working in them as well as a broad Christian heritage on which they still draw. All I am claiming and all I would claim here (though it is indeed quite a lot) can be expressed in three propositions. First: Take the Christian who is properly apprised of the parables of the leaven and the mustard seed, who recognises that the dynamic of unfolding faith is *via* the impregnation of human life, concerns, culture and education. To this Christian a good school must in the last analysis be one in which this leavening dynamism is *in fact* at work, in which therefore pupils can grow in *real* faith. We could say that it must be a Christian school in fact, whether or not it recognises itself as such. Secondly, it is ordinarily the case that a school will be helped to become and remain such a school if it deliberately aims at it and draws explicitly upon Christian sources and resources, i.e. if it does call itself a Christian school. Additionally, the flourishing of schools in which everything is thus made explicit can itself work as a leaven in a wider and secular school-system — provided, of course, that what is made rhetorically explicit here actually governs the practice of these schools. Thirdly, the professedly Christian school in holding itself open to being in its turn instructed by secular schools should keep at the ready Christian criteria of discernment. For it would be all too easy here to take the wrong lessons and leave the right ones. (Arguably, it is thus that we have often 'learnt' in the past!) It is above all to signs (of which there are many) that some secular schools are in certain respects more 'Christian' than it itself is that the good Church school will be alert.

But what makes the good Church school? An enquiry with aspirations to completeness might find it convenient here to distinguish (but also to relate) three broad areas of the life of a school: its relations with externs — families crucially, but also the neighbourhood, parishes, other schools, and so on; its own community life — the best context perhaps in which to consider government, discipline, the pastoral system, the liturgical life, the 'hidden curriculum' — among other things; and its curriculum in a narrower sense, including its R.E. This very incomplete enquiry will confine itself to the third of these areas.

The curriculum

The idea is probably as old as Church schools themselves that their profession of Christianity should make a difference not only in R.E. but across the curriculum — and this suggestion is often accompanied by talk of Christian culture or Christian humanism. But what sort of difference? Though we have this hunch we remain perhaps a bit vague about it and for lack of clarity achieve less in practice than we might.

(i) Christian culture, Christian studies?

If black studies for black children then, it might be asked, why not Christian studies for Christian children (black and white)? If in schools where there is a strong concentration of black children it is a good idea to lace the curriculum with black literature, black history, black music forms, etc., then why not something analogous where there is a strong concentration of Catholic or Christian children? There is a rich enough heritage to draw on in the arts and humanities side of the curriculum: an ongoing tradition of sacred music and of art with Christian motifs; a body of literature, still being added to, about Catholic or Christian experience and/or by Catholic or Christian writers; and a history syllabus could be designed to pay more rather than less attention to periods, institutions and figures that were inspired by Christian idealism. Let us insert a couple of cautions here. One: the art, music or literature must be good *as* art, music or literature and the history must remain critical. In other words, these subjects must maintain their own identities, must not be absorbed or manipulated by theology. Two: the diet should fall some considerable way short of being exclusively Christian in the above ways. Granted these provisos it seems to me a valid enough proposition that the Christian slant of our schools can reasonably be reflected in the choice of content for the arts and humanities areas. But this observation does not get us to the heart of the matter. More radically, we should enquire about the *point* which is to be assigned to the curriculum as a whole and in its parts, and consequently as to the *spirit* in which it is to be conducted in the Christian school.

We have been enjoined by our Lord and Master to love not only God but each other, humankind at large, and indeed (implicitly at any rate) the whole physical universe — that universe which Genesis represents God the workman as surveying with satisfaction and pronouncing good in each of its parts on the evenings of the days of creation and which is, in our understanding of creation, sustained moment-by-moment in existence by God's own love for it. This love of created things is but the other side of the coin of our love of God. It is indeed a necessary dimension of our love of God even when we do not — and perhaps should not — think explicitly of God in connection with it. Let us start from that fundamental premiss and ask what it implies for our teaching and our learning about humankind and the universe. *What would it mean to conduct the teaching and the learning that comprise the curriculum in love?*[1]

First, let us distinguish between the more academic and theoretical side of the curriculum and its practical side, with broadly speaking on one side the sciences and the humanities and an emphasis on the value of knowledge for its own sake, and on the other side the arts, crafts, design and technology, as well as moral and political education. In the latter case the main point is to *make* something, whether good products or good decisions. Correspondingly, love has two sides to it, the contemplative and the practical. There are on the one hand times of fascinated, wondering, affectionate resting in the beloved — on the other hand, times of concern and work for the beloved. The love itself holds these two aspects together and it is important that they should feed off each other, but it is convenient to discuss each in turn. At this stage I will focus on the contemplative side of love, and on the opportunities in the curriculum for developing *a contemplative regard for creation*. And I shall do this by dwelling briefly on two examples: science and history.

Science as food for the soul. Suppose we ask why science should be on the curriculum. We can envisage different kinds of answer. One will stress the '*pay off*', the contribution of science mediated by technology to the creation of wealth and power, and the need, therefore, to produce scientists and technologists and also a scientifically literate public to support them. That answer would be, I think, the government's one. Another will emphasise the intellectual *fecundity* of science, its endlessness, the opportunities it provides for developing and stretching young minds, for continuing intellectual challenge and adventure. That is an answer which will spring readily to the lips of most educationists. A third will insist instead on the *rationality* of science, on its embodiment of qualities like critical objectivity and a passion for truth, ethical qualities with which, hopefully, it infects those young minds which are exposed to it. Broadly speaking this might be taken as the philosopher's answer to the question: why science? Like the previous one it has elements of the answer I am looking for but is not yet it. A fourth viewpoint is that all these answers, while being all right as far as they go, don't get to the heart of the matter. For the fundamental value of science is one that comes to it from the *object* which it studies, that is, from the order of the universe in general and in its parts. This order invites attention and enquiry (and an education in these) simply by being itself awesome, wonderful, marvellous, beautiful — and thus fit *food for the soul*. And this, by the way, was Einstein's answer. As a young scientist he wrote of a longing to escape "from one's noisy cramped surroundings into the silence of the high mountains where the eye ranges freely and fondly traces out the restful countours apparently built for eternity". And he added that the state of mind which enables one to approach science in this manner was "akin to that of the lover and religious worshipper". The Christian could well say that it was not only akin to but actually a part of worship. And so, I would suggest, the good Christian school is one which in its science teaching is indeed concerned to turn out the odd scientist for the public good, to make science fun and to make pupils critical, but *above all* it will be at pains to nourish rather than stifle, to preserve rather than extinguish, the child's sense of wonder and admiration at our universe and God's. It would be an additional value of this approach to science-education that it would connaturally develop in pupils an ecological consciousness. For to such a consciousness reverence is an even more basic ingredient than the calculation of long-term human utility.

8

History as piety. In the case of history the analogue to wondering admiration at the universe would be love of the human past. By this I mean, quite simply, love of the human beings and human worlds that are dead and gone, a critical respect and regard for these as they were in themselves, a sense of solidarity with their sufferings and their liberations, a readiness to draw inspiration from them — in a word a love that *relates* us to them. This of course is what the word 'piety' originally meant, and the Christian can give to it a firmer and deeper reality by connecting it with the doctrine of the communion of saints. This piety has a logic similar to that of charity, of which indeed it is one dimension. Like charity it would begin at home with the history of one's own society, but would not jingoistically end there. Like charity it would sometimes be critical and indignant, but also inclined to forgiveness and to being in the first instance self-critical (i.e. it would not simply assume that our times stand on some kind of ethical pinnacle in relation to past times). Like charity it would have a special eye for the poor and oppressed of times gone, more particularly where time has yielded a people not a final victory but a final destruction. I suggest that the good Church school would, in its history teaching, be concerned to make the past live for pupils in such a way that they could actually relate thus *personally* to some parts of it (while holding themselves open to relating thus to further parts) and that such a sense of personal relationship with the past would be likely to become a permanent part of their lives.

(iii) *A concerned curriculum*

Let me turn now to the practical side of love for humankind and the world as it might impinge upon our general curriculum. Here I shall step over arts, crafts and design and take my illustration from what might broadly be considered social and political education. The proposition I want to advance here is, I think, relatively simple. There exists a whole range of causes and concerns which are of more or less critical importance for the future that our pupils (and their children) will live in, matters which are all profoundly ones of *justice*. I am thinking of issues that are usually captured in labels like the following: world-hunger and North-South relationships; disarmament and the bomb; ecology, energy-conservation and nuclear power; racism

9

B

and the multi-cultural society; women's rights — among others. These typically are matters of some considerable complexity, ones that require patient and sustained reflection towards a multi-disciplinary understanding (part scientific, part historical, part economic, etc.). They are also typically matters in connection with which there are battles for the support of public opinion, in which propaganda and counter-propaganda are the orders of the day. Now, first, if the basic information and concepts of many different kinds that relate to these issues (as well, of course, as a basic concern about them) is not imparted in school, it is unlikely that many people will go to the considerable trouble of seeking it out in later life. I would propose that a substantial start be made on these matters in the secondary school. Second, there are in fact already curriculum projects and proposals that relate to such issues, ones that go under titles like peace-studies, black-studies, women's studies, energy studies. I am not aware that our Church schools have been to the forefront in pioneering such work. But the forefront is where we should surely be, just because these are areas in which learning relates directly to matters of justice and love. Third, it might be that our greater degree of independence vis-a-vis the state in comparison with wholly secular schools will increase our responsibility here. For the introduction of some of these projects is being met with some 'official' nervousness and discouragement. Fourth, the task is not one of indoctrination along some party-lines or other, but the much more difficult one of a) provoking a sustained and critical *concern* about issues and b) imparting the requisite information and concepts for *understanding* issues. Fifth, what is required in the long run is likely to be less a matter of introducing *ad hoc* extra courses that compete with existing ones than of reinterpreting existing courses, of opening them out towards these issues, of, we might say, the critical pursuit of justice *across* the curriculum.

(iv) *A prophetic and nurturing R.E.*

Curricular isolation — whether of the reverential or the dismissive kind — does not become R.E. in the Church school. First the dominant mode of Church school R.E. is now (in theory at least) 'experience-centred'. Its aim is to exhibit to the pupil

10

the illuminative power of the Word of God (in particular, but not exclusively, the Christian Word) in regard to his or her life-experience. But school-experience, indeed curriculum-experience, is a considerable fraction of that life-experience. It is also (in theory at least) formative of his or her future experience. Hence the 'relevance' that is to be exhibited in the Word of God must include, and be mediated through, the relevance it has to the various 'words' of human culture and the general curriculum. Secondly, a curriculum that was conducted in the spirit described earlier would already embody an attitude of faith, if only implicitly and 'anonymously'. For its spirit is one of contemplative and practical love of creation and, therefore, implicitly of God. And charity presupposes faith. In that case the Word of God and the word of culture would have a special resonance for each other and R.E. would move towards becoming something like the prophetic interpreter of the meanings of the curriculum in general. Its relationship with other subjects would thus become a more positive one than that which is evoked by such current R.E. themes as 'religion and science', 'religion and progress' or 'religion and the arts' — which tend to be approached in R.E. class as exercises in the delineation of boundaries. There would still be disputatiousness of course — if only because the Word of God always comes to us in the words of particular cultures, including ones that are often esoteric to us — but it would now be what Plato once called the benevolent disputatiousness of comrades.

But this is to assume the legitimacy of R.E.'s taking upon itself a nurturing-by-explication role in regard to faith. To this assumption we must now turn.

Faith-nurture and indoctrination. R.E. has been a statutory requirement on *all* British schools since the 1944 Education Act, but the approach to it in Catholic schools is of course highly distinctive — *more* distinctive nowadays indeed than it used to be. Not so long ago pretty well all schools in the country acknowledged at least some responsibility in fostering a Christian faith in their pupils — an allegiance if not to some particular Christian denomination at any rate to a common Christian set of beliefs. In those days the Catholic school was distinctive only in its particular orientation to Catholic beliefs. But nowadays, what with the greater secularization of society and the much greater number of pupils in schools of non-Christian religious background and culture, R.E. is quite generally conceived in state schools as education *about* religions — where indeed it is not further re-interpreted as 'social studies'. In the better departments this education about religions is seen as not just a matter of providing information about religions but of giving also some imaginative understanding of them and of their appeal, of communicating a little of what it *feels* like to be a Moslem, a Hindu — or a believing Christian. But in general schools other than Catholic ones no longer see it as part of their belief to nurture Christian *faith*. And we should add here that by and large the other Christian Churches are quite happy with this state of affairs and with this kind of distinction between religious education and religious nurture. For them religious education (interpreted as education about religions) is the school's job; religious nurture on the other hand is the job of the family, of the parish and of that long-standing British institution — the Sunday school.[2] Now, of course, Catholics and their schools have been by no means unaffected by the more secular and multicultural atmosphere in which they too live, but it still remains true that they would not here distinguish — or at any rate distinguish so sharply — education and nurture (or catechesis). And, I suggest, if ever we who are Catholics did come to accept a *sharp* form of this distinction we would have lost a large part of our rationale for having Catholic schools. The hunch or intuition on which we proceed is rather something like this: that a genuine open-minded nurturing in Christian truth and relationships is about the most profound educational experience one could have! That being so, isn't it better to go on working at it through the week rather than just on Sundays, better to rely on teams of profes-sionally trained teachers rather than a few volunteers, and, incidentally, shrewder (as well as being perfectly legitimate) to continue funding the enterprise by that considerable part of our own taxes which go to pay teachers than to rely exclusively on Church monies? To that we can add the potential value of the mutually enriching relationships between a nurturing R.E. and a general curriculum that would be conducted in the spirit des-cribed earlier.

I would suggest then that the good Church school will be

quite simple and straightforward about having as its ideal the nurturing in its pupils of a faith in God and in Christ that is explicit, clear, intelligent, articulate, open-minded and sincere. It will not be too defensive or embarrassed about this. (Of course, the particular role of the R.E. class in even the Church school is centrally the development of understanding. But here it is an understanding of the kind that presupposes faith and tends to the promotion of a fuller faith.)

By taking this line the Catholic school will, of course, continue to be almost the number one target for accusations of indoctrination. So let me say two words about that. The first is that our success in convincing *others* that we are not in fact indoctrinating is never likely to be complete. Somebody who is convinced that the main Christian beliefs are irrational, incredible and irrelevant to anything that matters is going to remain suspicious of us no matter how open and honest our R.E. methods seem. For how else, he wonders, can you account for so many people being wedded to crazy beliefs if not by indoctrination? (And note that he may appeal here to our own distinction between faith and reason to convict us out of our own mouths. For do we not ourselves admit that the beliefs are beyond reason, a leap in the dark and so on? This suggests that we need to be more careful and clearer than perhaps we are about what we are doing with this distinction — but this is too complex a matter to go into here!) To change his mind it would not be enough to bring him to our classrooms. We should also have to begin to show him that our beliefs were not so crazy after all! And of course we should be ready to try this — ready, in St Peter's words, "to explain the faith that is in us".

My second remark, however, about the charge of indoctrination is that we must be careful not to deserve it. If we cannot realistically hope to persuade everyone of our innocence we must at least try to be sure of it ourselves! Earlier I spoke of the promotion of an explicit, intelligent, sincere belief as the ideal. Ideals are of the utmost importance but only if we relate them to messy reality, in this case the reality of children who are in many cases from only nominally believing backgrounds, or who have no obvious faith even if their parents have, or who have quite natural doubts, difficulties and problems even if they still think of themselves as believers. The position of such pupils

13

must be respected, and even protected. The good Church school will respect the unbelievers and searchers in its own midst — and that goes also for teachers in this position. To extend this respect is not a departure from the earlier ideal. On the contrary it is required by it. For real faith is always a personal and free response to God and will be hampered, humanly speaking, by an atmosphere of compulsion and fear — by any attempt at thought-control. We want to convey to our pupils a picture of the life of faith as adventurous, searching, big with intellectual honesty — as a living by the truth. Any indoctrinatory methods we were temped to employ would most surely convey an opposite message. A remark of Simone Weil's captures the thing well, I think. Faced, she said, with what seemed to be a choice between Jesus Christ and the truth she would choose truth, and only thus might she hope to find Jesus Christ.

But how do we ensure in practice that the unbelievers in our midst, real or apparent, are respected? There are, I think, a few touchstones here. One is freedom of speech. Is the atmosphere in R.E. class such that everyone can feel free to express and to share *their* experience, *their* view-point, *their* objections and difficulties? Furthermore: for those pupils who are happier — whether at a given stage of their development or even for their whole careers in school R.E. — to take a non-committal stance like junior anthropologists of religion, this approach should be available. This, of course, is to impose on the teacher a duty of considerable flexibility as regards methodology. She has somehow to contrive an environment in which all (the believer, the searcher and the detachedly curious spectator) feel at home with their teacher and with each other, and in which all feel that they are learning something worthwhile from each other, from the common course-materials and from the small-group and individual projects that would be necessary elements. Furthermore, she has to pull this off while keeping alive, in herself and in her class, a sense that the *norm* remains the pursuit of an understanding of faith within a community and a tradition of faith. Though all this requires skill, no little imagination and a multi-layered kind of awareness, it is by no means an impossible ideal — nor indeed one that has never been realised in practice!

14

The R.E. teaching team. What, more generally, is to be said of the R.E. teachers and their department in the good Church school? Much that is uncontroversial on paper is not always found in practice. They will be reckoned by pupils and colleagues to be among the best teachers on the staff. There will be enough of them. They will be allotted enough time for the teaching of their subject. They will be equipped with an adequate supply of books, files, project-materials, etc., and perhaps with an R.E. room or study-centre in which these can be readily available for their own and their pupils' uses. As a department they will hold their own in relation to other departments, but in such a way that they are supported rather than resented by the rest of the staff. They will regard an on-going, school-based curriculum development — which includes intelligent response to outside initiatives — as a professional 'must'. They will aspire to supplement their own (good) initial training by participation in some of the inservice courses that are everywhere mushrooming, and in this they will be supported by the school. Thus — as well as by their private reading — they will keep in touch with, and feel the excitement of, that continuing theological ferment which lies back of the dramatic developments in their subject, and some of them may indeed be modestly active contributors to this ferment. All these things would be so many indicators that a Church school was properly serious about its R.E.

Let me, however, draw attention to two further indicators that are not, perhaps, so obvious and uncontroversial. We have implied that there would be not simply good relations but an active commerce between the R.E. and the other departments. Members of these others would be drawn into the R.E. teaching as auxiliaries, occasional participants, joint promoters of interdisciplinary units and so forth, and the R.E. teachers would return the compliment. This follows, of course, from the idea canvassed earlier of a prophetic role for R.E. vis-a-vis the general culture and curriculum. And it in its turn implies that the R.E. teachers, in addition to their obligation to their own discipline narrowly conceived, would feel it incumbent upon themselves to command — at any rate collectively as a department — *a wide and deep culture.*

Secondly, there is the question of the teachers' own faith-commitments, delicate but unavoidable once R.E. is being taken

15

as having not only an academic but a faith-nurturing function. If it be true of the teacher of any subject that his teaching carries real depth and resonance only if he is *personally* involved with his subject, if the good literature or history teacher is one for whom history or literature will be a significant part of his life outside the classroom and school, then this is especially true, I suggest, of the Christian R.E. teacher. Christianity will have marked him inwardly. It will have penetrated his sense of who he is — or might be, and of what he is — or might be — called to become. It will insinuate itself into his everyday reading of everyday events. In a word it will *matter* to him. His pupils will sense the personal involvement behind his teaching and be challenged by it. But may we take the further step of expecting him to be 'committed'? There are dangers here — even after we have allowed that to be committed is not necessarily to be a saint, nor one who has never a doubt, nor one whose orthodoxy could never be impugned by anyone, but one, rather, who is always restarting on the business of discipleship. For there are still issues of freedom of conscience and employment rights, as well as our obligation of hospitality to the 'searcher'. And yet the faith-commitment of teachers is clearly not an irrelevance to a faith-nurturing R.E.! Perhaps what we are entitled to look for here is a *preponderance* of quietly committed Christians in the R.E. department as a whole.

One thing that all this implies is that anything which builds up and nourishes the spiritual lives of R.E. teachers contributes to their effectiveness as teachers. Now this is, I imagine, an endless topic and in any case one much better handled by somebody else. But I cannot resist one brief observation. That is that we would do well to fight against the inclination to think of school-liturgies, class-prayers and so forth only as things put on 'for the children'. Somehow, I feel, teachers in general have to contrive to be Mary as well as Martha here, to be, and to be seen as, not only conducting prayers but praying, not only ushers and organisers but worshippers. And those who design and prepare the spiritual interludes that punctuate the daily and weekly timetable of the good Church school (often the R.E. teachers) should do it, I think, with themselves and their colleagues in mind as well as their pupils.

An ecumenical R.E. Finally, the R.E. in a good Church school will surely be strongly ecumenical. By this I mean in the first place that it will be open-hearted and sympathetic rather than suspiciously defensive in its handling of other Christian Churches and traditions, of non-Christian world religions and indeed of noble non-religious philosophies of life. (This does not mean, however, that it should eschew criticism. For that matter there is much in our own tradition to be critical of!) But secondly, 'ecumenical' means in regard to other Christian Churches that the aspiration to Christian unity should be deliberately fostered. We should be turning out pupils who want Christian unity and will work for it. So they will appreciate the importance that our own Church and others attach to this goal, echoing in this the clear prayer and command of Jesus himself. They will learn to be troubled by the scandal of a divided Christianity and to be possessed of some glimmers of a vision of how much truer and saner, how much holier and more attractive a reunited Christianity would be than perhaps any of its still splintered parts can be on its own. (Speaking personally, it was a few short stays at the ecumenical monastery of Taizé, which has discovered in its unity an extraordinary power to draw to itself the youth of Europe and beyond, that brought this home to me.) They will also have some realistic grasp of the obstacles that stand in the way of unity, the legacies of old suspicions and bitterness, the differences of doctrine and so forth, but will not in the end be deterred by these, for in this matter, they will realise, we can be quite certain that we have God with us.

Now I have some acquaintance with one of the handful of new ecumenical schools, a joint Catholic and Church of England secondary school. Are such schools the shape of things to come? Personally, I would love to think so. For in such schools the ecumenical ideal I have sketched would — it seems obvious — be in principle much easier to realise, and in the one I know it is coming to be realised.[3] But, of course, it is not only for Catholics to decide about this and I would imagine that other Churches might find it a problem here that they have not hitherto committed themselves theologically or financially — not at any rate to anything like the extent that Catholics have — to the separate Church school. But whether ecumenical schools become the norm or remain exceptional witnesses and laboratories from which ideas and materials are disseminated throughout our schools, there can be no doubt that ecumenism should be a high priority of Church school R.E. in our times.

To look back and to summarise. We began by deriving from the incarnational principle that Christianity is rather leaven than icing the twin formal ideas of an education and a school that would *as wholes* stand up to Christian criteria of the good. Focussing then on the curriculum of such a school-education I argued that it might well allow itself to draw significantly, though not exclusively, on the heritage and ongoing traditions of Christian culture in the arts and humanities areas, but, more fundamentally, that it would develop in pupils (i) (especially in such theoretical subjects as science and history) that contemplative regard for creation and humankind which is one dimension of our worship of the creator, and (ii) (especially in more practical curriculum areas) an informed, intelligent and active concern for creation and humankind. Finally, and within this wider curriculum-context, we focussed more narrowly on R.E. in the Church school. Here four conclusions were argued for: (i) that it had a prophetic role in regard to the curriculum at large as interpreter of the point or meaning of the variety of activities that comprise this curriculum; (ii) that it should be unafraid to take the nurturing of faith as its norm, while yet being scrupulous to avoid indoctrination and while providing alternatively for those who would not wish to accommodate themselves to this norm; (iii) that, therefore, its teachers — in addition to their professional obligations more narrowly conceived — need both to be immersed in general culture and (in the main) to be themselves Christians whose commitment is generous and constantly renewed; (iv) and that it should be pervasively and decidedly ecumenical — whether or not the school be formally set-up as an ecumenical one.

This paper has grown out of an address to the recent Brentwood Diocesan Education Day. It is deeply indebted to the teachers and parents, representing every Catholic school in the diocese, who listened and then made me listen, to Dr Phil Dineen, my inspiring partner in the day's lecture-team, and to Mr Frank Murphy of the diocese's R.E. Commission who inivited me in the first place.

References

1. The few recent philosophers to whom this question has not been a stranger include A. N. Whitehead, R. S. Peters (in some of his writings) and — most notably — Simone Weil. Before them there was Plato!
2. Cf. the British Council of Churches publication *The Child in the Church*.
3. St Bede's Secondary School, Reigate, Surrey.

LIVERPOOL INSTITUTE OF
HIGHER EDUCATION
THE BECK LIBRARY

CATECHESIS
OR RELIGIOUS EDUCATION?

The problem

Probably the dominant question in the minds of many Church school teachers of religion today is 'what exactly am I doing?' This is not the kind of question that arises necessarily from inefficiency or a lack of competence in the subject. It is not the same order of question that might be asked, for instance, by a surgeon when requested to survey the foundations of an office-block, or by an engineer when faced with transplanting a heart. In the case of the teacher of religion, the dilemma arises from the nature of the subject itself. There is an implicit confusion in the activity of communicating religious meanings. For here we are concerned with a realm which is ultimately mysterious; there is, in the long run, an immeasurability about the teaching of religion which no amount of analysis or definition of aims can encompass. Those involved in the communication of religious dimensions and meanings in life must face up to this quality of mystery inherent in their subject.

As individuals committed to a specific faith-system, namely Christianity, the problem is particularly acute. In the Church school, we feel comfortable neither with a purely confessional approach, nor with a phenomenological approach. Nevertheless, the *via media* between these two frequently lacks both clarity of character and aims and consistency of method. The teacher is left with the difficulty of exercising commitment on the one hand and neutrality on the other. Little wonder then that confusion abounds.

Some way to a solution has been attempted in recent years by the debate on catechesis and R.E. Discussions have tried to focus on the nature of the enterprise in school and parish contexts and the respective aims of the activities undertaken in each. Before proceeding to examine these areas, it should be noted that the terms 'catechesis' and 'R.E.' do not correspond necessarily to the 'confessional' and 'phenomenological' approaches. Rather, each enterprise properly aims at the balance of personal experience and objective content appropriate to its context and to the students within it. Thus, both catechist and teacher are involved in equally valid educational tasks in the Christian community.

Chiefly, the problem occurs when the R.E. teacher in the classroom understands his or her responsibility to be one of catechising, that is, of encouraging in explicit terms a deepening of a Christian faith assumed to be already present in the pupils. It is this assumption that needs to be investigated with great thoroughness, since the experience of most teachers today would seem to indicate that a high proportion of pupils come from nominally Christian backgrounds with little experience of Church attendance or any other form of commitment. This is especially evident at the secondary level. So, the catechising approach, hitherto largely unquestioned as a classroom method in the Church school, does not tally with the real situation. This has produced a crisis of identity for the teacher.

The nature of the enterprise

Is there a difference, then, between the nature and aims of catechesis and those of R.E.? It is worth exploring some of the recent writing on this subject. Thomas Groome in *Christian Religious Education*[1] appears to draw a distinction of character between the two in the following way: catechesis is a deliberate and intentional attending with pupils to the activity of God in our present, to the story of the Christian faith-community, and to the vision of God's Kingdom, the seeds of which are already among us. Catechesis therefore concerns itself with the pupils Christian faith-experience in relation to a past and a future, and should evoke a threefold commitment of believing, trusting and doing on the part of the catechised. As an activity, catechesis is a species of religious education, which may be defined, according to Groome, as a deliberate attending to the transcendent dimension of life, by which a conscious relationship to an ultimate ground of being is promoted and enabled to come to expression. This should in fact, he says, characterise all good education. A

Victor Frankl "Man's search for meaning"

21

problem with Groome's analysis is that he employs the term 'Christian religious education' rather than catechesis, which rather confuses the issues. It might be argued that R.E. can be Christian without necessarily being catechetical. Further, where he refers to R.E. in the general sense, it seems to be a synonym for good education as a whole — certainly interesting, but not developed adequately by Groome.

Paul Hirst, in his article 'Education, Catechesis and the Church School',[2] suggests that catechesis is the activity of presenting religious belief and practice from the stance of faith. This faith stance can nevertheless only take over where reason leaves off, for reason must always be maintained as prerequisite of faith if we are to avoid the pitfalls of indoctrination. This is another way of saying that R.E. prepares the ground for catechesis; Hirst calls R.E. a form of pre-catechesis. R.E., he says, presents religious belief and practice from the stance of reason, observing the objective status of the information and appropriate justification of the evidence. This is a fairly acceptable analysis on the face of it, as long as it implies no qualitative distinction betwen catechesis and R.E., such that the *aim* of R.E. becomes one of getting the pupil over into the catechetical camp. To see any enterprise as a form of 'pre-catechesis' is to beg the question and to damage the openness of the shared task of teacher and students, as well as the freedom of the individual pupil.

In the past two years, two official documents have been produced on the issue of Christian education. *Understanding Christian Nurture*[3] emerged from the British Council of Churches and refers in the main to the Church of England community, but offers many valuable insights to the debate on Christian education in general. The document suggests that catechesis — or nurture — is an activity conducted with critical openness, which assumes that teacher and child are *inside* the Christian faith. A child can be catechised or nurtured, therefore, only in his *own* tradition. One cannot, for example, catechise a Muslim child in Christianity. One can, however, educate him/her in Christianity. Religious education, observes the British Council, is an activity, again conducted with critical openness, but this time inviting the pupil to *imagine* what it would be like to be inside a faith — or indeed, if it is the child's own faith that is the topic of the lesson, what it would be like to be outside it. In other words, what is required by R.E. is a certain suspension of belief or dis-belief, for the purposes of learning beyond one's own faith or commitment. A child can thus be educated in several other tradi-tions as well as his/her own. This document is definitely deserving of close study. Perhaps its most significant contribution is its stress on the necessity for 'critical openness' in both the cate-chetical and the religious educational contexts. We have tended to reserve such terms for the objective study of religious pheno-mena, assuming an implicit 'closed' approach in the realm of catechetics. By insisting on critical openness in catechesis as well as R.E., the document encourages an honest respect for the freedom of committed and uncommitted alike.

Signposts and Homecomings: The educative task of the Catholic community[4] was addressed in 1981 to the bishops of England and Wales, under the chairmanship of Bishop David Konstant. It contains some discerning comments on psychology and sociology in relation to young people in modern Britain, and also some interesting theological insights which are worth developing. As its subtitle indicates, however, this document is mainly concerned with the Catholic — that is to say, the Roman Catholic — community. For this reason, it suffers from rather narrow definitions. In the first place, the nature of R.E. is seen to derive from distinctively Catholic education: it starts from the explicit belief in Jesus Christ as universal saviour, as con-cretised in the proclamation, especially the gospels, and the doctrine of the Catholic Church. There are no real distinctions drawn between catechesis and R.E., for all such activities proceed by way of induction of pupils into a way of life — by implication, the Catholic Christian way of life. This process must observe, nonetheless, the universal educational canons of rational judge-ment and personal autonomy. No attempt is made, however, to clarify how this relationship between catechetical induction and educational principles might be validly forged within the context of a single activity. Consequently, the contribution of this docu-ment to the catechesis/R.E. debate is somewhat limited.

The aims of the enterprise

With regard to the actual aims of catechesis and R.E., the same selection of writings may be usefully surveyed. Groome asserts that the aims of catechesis are to enable pupils to appropriate the Christian faith tradition as their own, by nurturing their spiritual development, fostering their relationship with God in Jesus Christ, and promoting goodwill towards their fellow human beings. Catechesis focuses on sponsoring pupils to live lives of unity between what they profess to believe and how they actually engage in the world. The ultimate objective is a lived Christian faith. Religious education, in common with all good education, aims to *empower* pupils in their quest for the transcendent and the absolute ground of existence. There is a broadly 'spiritual' character to all authentic educational enterprises. This cannot be disputed, but it fails to account for the specific task of R.E. that every school teacher is confronted with, and moreover must tackle in the face of various levels of non-commitment, hostility and indifference. The problem cannot be solved by such abstractions, which themselves draw heavily upon a catechetical vocabulary.

For Hirst, catechesis aims at a free response in decision and faith by the pupils, which will result in the development of the committed Christian life.

In R.E. the aim is to develop in pupils their natural reasoning and autonomy, without in any way attempting to advocate one set of beliefs as true, or trying to induce commitment. The distinctive contribution of Hirst in delineating these aims, is that he maintains the possibility of pursuing both activities of catechesis and R.E. within the school environment, precisely as separate enterprise, rather than transferring catechesis totally to the parish setting. Moreover, he contends that the two activities should be conducted by two distinct sets of staff, so as to preclude any suspicion in the minds of the pupils and to accord equal respectability to both catechesis and R.E.

For the British Council of Churches, catechesis aims at the intentional building up of faith and a deepening of personal commitment. A spirit of critical openness must be sustained at all times in fulfilling these aims, so that it should be quite possible for a pupil to opt out of explicit commitment and the catechetical environment if he so chooses. R.E. aims at deepening this critical openness in the pupil, in relation to a wide range of religious and non-religious attitudes and world-views, including Christianity. It is worth pointing out that some catechetical aims might well be fulfilled in R.E., but only incidentally, not intentionally. The indispensable character of critical openness may prompt a pupil to reflect on the level of personal faith, and so motivate him/her towards specific commitment. In this case, catechesis becomes appropriate for the pupil, *in addition to* continuing religious education. One ventures to suggest that all children should be religiously educated, but by no means are all children ready for simultaneous catechesis.

It is difficult to discern any real distinction of aims in *Signposts and Homecomings*, since, as has already been said, all definitions tend to subsumed under the umbrella phrase of 'Catholic education'. Thus, in the context of Catholic education in general, R.E. is seen to serve catechetical ends. It is suggested that catechesis occurs implicitly through the 'hidden curriculum' of the Church school. Undoubtedly, we hope that a child will assimilate certain values and attitudes from the overall ethos of the school, and as Christian teachers we must make every effort to create a favourable environment for this to take place. Nevertheless, it could be argued that Christian values education need not entail personal commitment to the Christian faith. Many people today live their lives along broadly Christian principles without adhering to systematic Christianity. If catechesis is an intentional activity, then the ethical influence of the 'hidden curriculum', important as it is, cannot strictly be classified as catechetical — or at best it is only incidentally so. According to the document, explicit catechesis takes place through the R.E. lesson. *& Liturgy*

In short, the aims of R.E. are considered to be twofold. First, to enable pupils to experience through the structures, teaching, worship and relationships within the class and in the school as a whole, a believing and integrated community, which itself belongs to the local Church community. Second, to develop in pupils a sense of personal responsibility, autonomy and integrity, essential as a preparation for proper citizenship and for deliberate acceptance of their membership of the Church. While attractive and visionary in one way, the generality of these aims offers cold comfort for the teacher caught between the two stools of catechesis and R.E. For a start, the problem of corporate worship is very often heightened today, not only by the presence of 'non-practising' Christian pupils (and perhaps staff) but also by the

25

increasing presence of pupils from non-Christian backgrounds within Catholic schools. Furthermore, the notion of a 'believing community' grows ever harder to define in realistic terms, especially with regard to the school. While pupils should be prepared to live decent and caring lives, the relationship of the caring life to Church membership is not at all obvious to many young people. Unless we clarify the activities of school R.E. on the one hand, and catechesis, whether parochial or school-based, on the other, then the practical task will remain dogged by a lack of realism.

Content reflects requirements

The need to distinguish aims arises from a fundamental difference in the requirements of the young person in the catechetical and educational contexts. Groome observes that a young person's faith emanates from his/her self-identity, which is determined largely by the socio-cultural environment. For the catechetical process, he or she therefore requires socialising interaction with a Christian faith-community capable of forming him or her in such faith. The child to be catechised, whether as parishioner or scholar, needs to encounter the appropriate symbols, role models, world-view, and value-system that can be interiorised as his or her own Christian self-identity. This is to say, the child requires a shared Christian praxis. Specific instruction in doctrine and traditions is an integral aspect of this process, as is the need for personal participation in various sacramental celebrations and liturgies.

The young person's self-identity is determined too by the wider socio-cultural environment, beyond his or her immediate faith-community — if indeed he has one. In this more wide-ranging sphere, the pupil in R.E. needs to encounter religious phenomena and attitudes in a controversial light, by means of rational and objective investigation, as Hirst has suggested. In this process, a young person can exercise autonomy without suspecting an explicit intention on the part of the teacher of promoting commitment. Indeed, *Signposts and Homecomings* points out — though it remains rather an inconsistent observation in relation to the major tone of the document — that the non-assumption of faith in R.E. allows for different degrees of commitment, in

26

that those who have given up sacramental practice may not necessarily have lost faith.

Pupils experience a different kind of freedom when they undertake exploration independent of truth-claims. Such an enterprise allows for the possibility of opting out for a time (or indeed permanently) from faith-commitment, while continuing to take seriously the meaning of religious faith for others. In the modern world particularly, the skills of empathy are essential for genuine community, and never more so than in a multi-cultural and multi-faith society. When compulsory formal worship is not considered a prerequisite for sympathetic understanding of various religious beliefs, a young person may well be more inclined to approach religion as a whole with greater openness and seriousness.

In turn, the needs and requirements of young people within the catechetical or R.E. contexts determine the actual content of the respective programmes. The content of catechesis will be necessarily denominational, for it assumes a specific faith, placing heavy emphasis on the doctrinal element of Christian belief. World religions and other faith positions are viewed from inside the Christian faith-community, with the primary aim of reinforcing personal commitment. Such content is necessarily presented in a comparative way, because truth criteria are assumed by definition to be Christian. There must also be a stress on sacramental preparation and practice, and special training in specifically Christian spirituality and prayer-life. It is essential that the content of catechesis be focussed and specialised in this way, for without it a child's faith cannot be nurtured and deepened into a mature Christian commitment.

The content of R.E., however, should reflect the ecumenical and multi-faith dimensions demanded by a multi-cultural society in which every Christian child must learn to live. These broader dimensions are perhaps also demanded by the overall definition of the Christian school, with its character of tolerance and openness, and its commitment to the quest for all truth manifested in a variety of forms. Moreover, there is a need for thoroughgoing educational approaches to content-areas which are essential to proper human — and thus Christian — development. Examples of such areas might be: justice and peace studies, political, social and sexual ethics, third world and development studies, and ecology and conservation education. Constraints of priority and time would render a thorough study of these secondary in a catechetical programme, yet a proper grounding and understanding in these areas is an integral facet of Christian maturity. Only in the context of an integrated R.E. programme, which draws upon the expertise of other disciplines within a school, can education of this kind be adequately achieved.

In summary, then, a valid distinction can be made between the nature and aims of catechesis on the one hand, and the nature and aims of R.E. on the other — a distinction founded on the specialised, faith-assuming character of the former, with Christ as the focus of belief; and on the broader, investigative character of the latter, with multifarious religious phenomena as the arena of exploration. This distinction takes account of the differing requirements of the young Christian in his or her developmental process, and is reflected in the modified content of each activity.

Structure reflects theology

However, the dilemma is not resolved simply by recognising these distinctions. Recognition clears the ground, but the educational task demands that we address ourselves to certain pragmatic considerations. These will vary, depending on the type of educational establishment, its location, its catchment, and so on. Nevertheless, what follows may provide some general guidelines for teachers and catechists alike. It always remains, of course, the duty of the practitioner to test out all theories against his or her situation.

In the first instance, should the whole debate revolve around content as it tends to do, or should we attend to something more fundamental about the Church school, which after all belongs at least in part to the public sector? It has already been suggested that the 'hidden curriculum' cannot properly be categorised as intentional catechesis. A Christian theory of the curriculum in Church schools does not necessitate only catechetical content, nor does it imply that the primary aim of the Church school is *de facto* only a furthering of personal faith-commitment. This is an assumption that must be challenged, for it represents the hub of the problem. Traditionally, the existence of the Church school has been rationalised on this basis, by contrast with the county or state school which claims no particular religious orientation. Yet, in this day and age, the claim of the Church school to be the 'radical alternative' in education can hardly be sustained, if it operates within the relatively narrow parameters of selecting Catholic children to produce Catholic adults by means of 'pure' Catholic content. The notion of the 'radical alternative' must surely imply much more than this.

The great danger is in placing too much emphasis on *Catholic content* at the expense of *Christian structure*; in constructing a (Catholic) Christian content-framework into which our students' experience must fit, rather than providing a flexible, methodological structure for an honest observation and interpretation of experience. For example, is a primary objective to ensure frequent attendance at Mass, rather than to evoke a basic respect for the worth and values of others? Simply to observe the rites, as we know, far from indicates the real presence of community. Or is obedience to ecclesiastical authority given precedence over informed responses of freely deciding individuals? Obedience which lacks reflection is too frequently founded on fear and stunts the development of a human being's own sense of responsibility. This could have far-reaching consequences for young people preparing to live in a pluriform world.

What is needed is a holistic vision. For far too long 'Christian education' has been taken to imply a separate compartment of educational experience. This is not surprising. The Church itself has been inclined to place itself alongside the rest of the world, and adopt rather a superior and monopolistic attitude towards its concerns. With a deepening awareness of human solidarity, however, this is changing. Our understanding of the meaning and purpose of Christian education and the Church school should therefore follow suit. The British Council of Churches suggests that we can speak of 'Christian education' in the sense of "a Christian rationale for the processes of learning".[5] It argues that this rationale is completely non-catechetical, and arises purely from the 'internal logic' of Christian faith.

This notion of 'internal logic' deserves careful examination, for it really provides us with the theological justification for maintaining a distinction between R.E. and catechesis. The reality of the incarnation stands central to Christian faith: God so loved the world that he became man and united himself with it forever.

The logical imperative of this all-encompassing belief is an unequivocal ecumenicity, which demands that the Christian treat with the utmost reverence and seriousness the total spectrum of religious and humanistic beliefs that are manifest in the world. More than this, that every effort at discovery and self-understanding at the creation of community and loving relationships, on the part of human beings, be considered as indispensable dimensions of the relationship between God and man, between Christ and his world. Implicit in the kenotic character of Christ is the kenosis of the Christian religion.

What this means is that authentic Christian faith precludes a sectarian approach. By definition, Christianity possesses a universal orientation; its truth-claims are not closed and systematised, but derive from the infinite possibilities of the world itself, in all its variety; a world which, irreversibly, has divinity-within-humanity as its heart. Believing this to be so, Christian faith-commitment requires total commitment to all truth *in its own right,* not by assimilation, but by non-dogmatic recognition and affirmation of all that is best in humanity.

In terms of general Christian education, *Signposts and Homecomings* asserts that we should seek "to establish transcendental curricular principles . . . on which the curriculum might permanently rest".[6] By insisting on attending to the development of the whole child and not just to discrete areas of his intellect, emotions or even soul, and by affirming the ultimate unity of truth through the employment of correspondingly integrated methods, the Church school might begin to realise the meaning of 'radical alternative'. In terms of R.E., a pupil who is not committed is rarely convinced by the particular if he is denied access to the general; he cannot realise the meaning of Christianity if he has not apprehended the meaning of human questing. Conversely, a pupil who is committed cannot possibly grow in faith and indeed may lapse into complacency, if he never has the opportunity to explore and absorb the great insights of humanity which extend beyond the Christian tradition. The young Christian must learn the value of an insect, a tree, a stone even, from the Indian tradition; he must realise the meaning of endurance from the Jewish tradition, and the totality of God from Islam. He must gain a deeper understanding of human celebration from the worship of other faiths, and intensify his awareness of birth, life, suffering and death through their beliefs and ceremonies.

This is not an exercise in charity on the part of the Christian believer, neither is it merely a 'good thing' from an educational point of view. It is theologically and, therefore, humanly imperative: Christian faith *demands* the mutual enrichment of the world's diversity in order to become more authentically, more fully itself. In this process, the world's variety in its quest for truth, must in turn be allowed its autonomy. Any relationship of love entails as **much**.

One ventures to suggest, then, that if we get the Christian structure of the school and the curriculum right, the content of the R.E. programmes will reflect, naturally, openness and tolerance in this way. Educational philosophy and theological perspective contribute equally in determining the pedagogical content of Church school R.E.

The essential dialectic

Notwithstanding the distinctions between catechesis and R.E., have we polarised the terms of reference in the debate and thereby lost a sense of the whole? This is always a temptation. The truth is that both activities of school-based R.E. and parish or school-based catechesis are equally necessary to maintain a genuine dialectic of human experience for the young Christian.[7] In the R.E. enterprise, general human experience challenges and interprets Christian faith as well as other religious faiths; in the catechetical enterprise, specifically Christian faith challenges and interprets human experience. R.E. should be seen not as an activity which takes second place to catechesis, nor simply as a form of 'pre-catechesis', as Hirst would put it. It is an activity essential to a holistic vision of man and perhaps especially of the child. There must be maintained a functional dialectic between catechesis and R.E., consonant with the dialectical character of human experience in general and of Christian faith in particular. Authentic Christian faith must never lose sight of this. The total process is one of 'ongoing synthesis' which is the more enriching to a person's life the more he or she realises that these experiences are not mutually contradictory, but equally valid aspects of the whole.

Young adults today face a difficult and challenging world.

With long-term unemployment as the prospect for the majority, the ever-present threat of nuclear war, technological revolution, increasing terrorism, ecological destruction and disintegrating relationships, the future looks bleak enough. As Christian educators, we cannot afford monopolistically to ignore the richest insights offered by the faiths and philosophies of the world's cultures. Young Christians of today will shape tomorrow's earth and they will need all the integrity that comes from a well-founded apprehension of this existential dialectic. Above all, they will need imagination. To hold convictions, yet remain open; to achieve fulfilment, while recognising ever more potential; to grasp truth and know that the mystery is enexhaustible; to be confident in the Christian faith and give credence to all possible faiths; to share in community and be true to one's own individual quest — this is the integration which, whether as teachers or catechists, we should aim at achieving in Christian youth. In short, a healthy Christian must be a healthy human being, possessing faith in the particular, only because of his or her commitment to the universal.

Many people — young and old alike — would seem to fear the consequences of allowing to both catechesis and R.E. their full value in Christian education. From a catechetical perspective, the committed Christian fears in R.E. a *loss* — of certainty, truth, essence, purity of faith. R.E. poses a threat, the beginning of something vague and insecure. From the religious educational perspective, the uncommitted Christian fears in catechesis a *possibility* — of pressure to conform, exclusiveness, limitations to personal questing. Catechesis poses a threat, the end of something desirable and free. To those of us who claim to be committed equally to the universal human quest for truth, and to the undying image of God in humankind, as epitomised in Christ, there is nothing to fear from either enterprise and everything to gain from both. A rounded Christian education should accommodate itself prophetically to the various attitudes and beliefs of young people, by discerning what is appropriate and when is the right time.

References

1. Thomas Groome, *Christian Religious Education* (Herder & Herder 1980).
2. Paul Hirst, *Education, Catechesis and the Church School* in British Journal of Religious Education (Spring 1981).
3. British Council of Churches, *Understanding Christian Nurture* (BCC 1981).
4. D. Konstant et al., *Signposts and Homecomings: The educative task of the Catholic community* (St Paul Publications 1981).
5. *Understanding Christian Nurture*, p. 26.
6. *Signposts and Homecomings*, p. 69.
7. Graham M. Rossiter employs the phrase 'creative divorce' in his article entitled *The Need for a 'Creative Divorce' between Catechesis and Religious Education in Catholic Schools* in Religious Education, Vol. 77 No. 1, Jan.–Feb. 1982. I prefer the term 'dialectic'. Rossiter's article is well worth reading.

P.69. "teaching as a ministry is concerned with the a most immediate and specialised way with the liberation and humanising of others".

RELIGIOUS EDUCATION
AND THE CREATIVE ARTS

The business of senior form R.E. is a problematic one at the best of times. When we come to discuss the place of the creative arts within an R.E. programme, the difficulties seem to be compounded. To begin with, few teachers feel happy about areas with which they are perhaps relatively unfamiliar and in which they are probably totally untrained. "I haven't done art since 'O' level"; or " 'To be or not to be' is the only literature I recall", might be the fearful complaint. Furthermore, the teacher who does attempt a more adventurous approach to R.E. with senior forms is frequently disappointed with an unenthusiastic response It is not easy for young people, unaccustomed in their previous R.E. classes to dancing or writing poetry, suddenly to begin to share themselves let alone express themselves through creative media of this sort. People cannot be forced to throw off their inhibitions or communicate intimately; their vision must be fostered so that these activities become natural and personal modes of exploration. Similarly, no-one can be coerced into 'liking' a piece of music or sculpture; we can be encouraged in a way of listening and a way of seeing with the heart, but our response must be all our own.

It is the contention of this section, therefore, that any investigation into the possible role to be played by the creative arts in upper school R.E. must take into account a larger number of factors than would initially seem apparent. First must be examined some pragmatic factors which have resulted in the misuse of the creative arts in education. I suggest the real need for educational balance throughout a person's school life and cite one or two problems arising from this. Second, the specific needs of the senior student are to be considered in planning a general R.E. programme. Third, we must explore the possibility of a theological justification for including the creative arts in the

R.E. curriculum, as an essential factor in the maturing process of a person. Fourth, I suggest that a sense of appreciation and a capacity for expression are central to human development, which must take place through a young person's own selection and synthesis of image-experiences, many of which will necessarily be ambiguous at different ages and stages. Fifth, it is contended that religious interpretation and the creative arts are closely related, and that R.E. at senior level, whether of a general nature or specifically cetechetical, should aim to encourage mature integration of the paradoxical character of human existence with which both religion and creative activity are centrally concerned. Constraints of time and space have caused me to restrict my remarks somewhat to the Christian tradition. I do not believe that even in the Church school the actual content of a general R.E. programme should include only the Christian or western artistic traditions. Appreciative maturity must extend beyond this both aesthetically and religiously; the practical responsibility for content lies with the R.E. teacher.

Maintaining the balance in religious education

Christian education aims at the mature integration of all aspects of the human personality. The tendency to overstress the left side of the brain, responsible for cognitive and reasoning activity in the individual, has led us to neglect the right brain which triggers our capacities for feeling and emotion. Recent writing and research in this field insists that the totality of the child's personal growth does not reside in cognition alone, but demands the nurturing of his imaginative and affective faculties as well. This recovery of an holistic approach has had an impact on all disciplines, not least on R.E. Yet, enthusiasm for allowing greater creative freedom to the child frequently causes our methodologies to run wild. We have sometimes burnt the textbooks and replaced them by paint-pots, demolished the blackboard and substituted the projector, removed the desks and gone overboard on open-plan.

As a consequence, many teachers at both primary and secondary levels, have experienced loss of control over the educational situation — pupils, content, method, aims, overall purpose. The R.E. teacher in particular is susceptible, for he or she very often possesses a heightened awareness of the shortcomings of traditional 'input' approaches and realises the pressing need for child-centred emphases if R.E. is to succeed. It may be, however, that in our attempt to redress the balance, we have tended to regard two pedagogical approaches as absolutely alternative methodologies. There is really no sound educational theory to sustain this tendency; it is rooted rather in the natural need for simple certainty and practical clarity. Unfortunately, this is an elusive ideal in the teaching profession and should never blind us to the reality of the classroom context, which always demands a good deal of variety in both content and method.

On the other hand, creative activity in R.E. has also been employed as a crutch for the failing strength of a more traditional method, with the aim of recovering the interest of the pupils. Too often this represents an abuse of both the creative medium and the lesson material. The relief of boredom can never properly be an end in itself, but only an integral result of good, well-founded teaching.

Religious education and the needs of the student

In the primary school, under the present system, there is perhaps more scope for achieving a degree of methodological integration. The persistent emphasis at the secondary level on the autonomy of the various disciplines, however, contracts this scope. Examinations become an important, if not essential, factor for all abilities, and pupils are encouraged to make options from the age of fourteen or so. It is not uncommon to find amongst many teenagers and students, those who 'opted' for sciences in adolescence and henceforth encountered nothing of literature, art, music or any of the so-called arts. The converse is also regrettably true; scientific education ceases if, for instance, it becomes obvious that a youngster is not strong on mathematics.

We cannot all be specialists in all subjects, but the imposition of this sort of experiential selectivity on a youngster at a tender stage of development could produce in them certain vacua not easy to remedy when they move into the pluriform world of adulthood. A person might suffer on two counts. First, the personal — the lack of maturity in a fundamental dimension of his or her humanity, whether in discerning the mystery of the empirical or the mystery of the aesthetic; second, the social — he or she must learn to live in a rapidly changing world, which requires some acquaintance with the progress of science and of appropriate ways of utilising increased leisure time. This dimension of education has nothing to do with careers, technical expertise, artistic talent or intellectual ability.

By the time a student reaches the sixth form, his or her specialisations are completely determined. The increased recognition of R.E. as a respectable academic discipline and its consequent place alongside other subjects in the curriculum, is greatly to be welcomed; but this has, in one way, proved a dubious virtue for the Church school. Viewed as one amongst many possible choices on offer — such as French, chemistry, history, biology — R.E., it seems, must face with other disciplines the possibility of straight rejection as a specialist subject. The Church school, which already finds itself in the throes of a crisis in respect of its catechetical/educational functions, seeks to ensure that every student continue to pursue the subject nevertheless, under the terms of 'General R.E.' Most schools structure their own general R.E. programme, which is organised and taught with varying degrees of success. In some instances it may be a highly exciting and educational programme: in others, the lesson-time may be no more than a timetabled opportunity for marking and homework.

Clearly, most upper school students will not be content with a type of veiled catechesis which re-presents to them much of what they have probably already studied throughout their school R.E. Neither will they be happy with an R.E programme that makes too many demands on them in terms of time, reading or preparation — they already have the pressures of their chosen specialisms. The majority, too, will be largely unresponsive to a re-hash of the so-called social topics — drugs, race, war, sex, abortion and so on. Those who wish to deepen a personal faith through the detailed study of doctrine and scripture and liturgical participation, should be allowed a different context from the timetabled general R.E. This should constitute an activity catering for a wide range of commitment or non-commitment, presenting material which is potentially of interest to all, primarily in terms of their common humanity rather than of their faith-profession. A student should not feel oppressed or threatened, even where sixth form R.E. is compulsory. It should be a time of liberation and discovery, of challenge and relaxation, which permits the student — albeit in small or limited ways — to grow in seeking out new forms of creativity and self-expression, consonant with his or her passage into mature adulthood. This will undoubtedly involve discussion and verbal expression of ideas, but it also requires the exploration into a whole spectrum of symbols and communicative media if we are to remain true to the profound nature of the human person. For this reason, the creative arts fulfil a particularly significant role in senior form R.E.

A theology for the creative arts

Scholarly explorations into the philosophical, psychological and theological grounds for integrating the creative arts into the R.E. curriculum, are offered by Philip Phenix in his *Education and the Worship of God*,[1] by a number of eminent thinkers in *Aesthetic Dimensions of Religious Education* edited by Gloria Durka and Joanmarie Smith,[2] and by members of the *Christian Education Movement* and the *Religious Experience Research Unit*.[3]

Here, I intend to argue that the creative arts are not simply a useful educational method, but are indispensable in the light of the twofold character of all creative activity: appreciation and expression. Let us examine these qualities from a Christian theological perspective. The focal theme of Christian belief is love. Love is not considered to be merely an attribute of God, but as constituting his very nature; as St John puts it, God *is* love. All love is necessarily relational and demands both appreciation and expression, which qualities must be in constant mutual interaction if love is to be clarified, to deepen, intensify and multiply. God expressed his loving impulse in creation, and especially in humanity. So profoundly did he appreciate the significance of this fragile creation of his, that he himself entered historically into it, in the person of Jesus Christ, as a further expression of his creative love, thus clarifying, deepening, intensifying and multiplying his original impulse. Christ then is the expressed appreciation of God for man — Love incarnate.

But the story does not end there. In turn, every man, in the image of this loving God, and after the paradigmatic figure of

Jesus, bears the crucial responsibility of appreciation and expression, if he is to grow into the fullness of his humanity. These qualities must be actualised in love for other persons, and also in a loving relationship with reality as a whole. True creativity resides in the ability to discern the equal worth of both the smallest and greatest. Many of the great painters afforded as much attention to the dog in the corner of the canvas as to the saint at the centre! Love cannot simply appreciate with eternal passivity; it must express itself. Similarly, the expressive dimension of love becomes vacuous, meaningless, a lie even, if divorced from a genuine appreciation of its object. This is as true of artistic creativity as of the creativity of human relationships.

The creative arts and human development

Let us consider now these qualities of appreciation and expression from a Christian educational perspective. A young person must learn the art of appreciation through the development of his or her emotive and responsive faculties; he or she must also learn how to express that appreciation through the medium of his or her own creativity. This, one might say, constitutes the core of the educational enterprise. While, in the integrated experience of the person, the appreciative and expressive senses cannot properly be classified discretely, it is nonetheless helpful to glance at each of these dimensions in turn in our attempt to establish the peculiar place of the creative arts in the process of human development.

(i) *Learning to appreciate*

From babyhood, the modern young person is exposed to a vast number of images, many of which are aesthetically indifferent. The advent of television and video especially, has made available a whole spectrum of visual images and symbols which in many households are assimilated indiscriminately by the child. Discrimination and appreciation must be learnt through careful education, if long-term damage to the child's sensibilities is to be avoided. This process should not necessarily be a moralistic one which places unreasonable compulsions and restrictions on the freedom of a normal child; but it should encourage him to test the images he sees against his own feelings, fears and inclinations. Genuine distinctions between appropriate and inappropriate images must grow out of a young person's own capacity to discern the congruence of certain images with himself, and the lack of congruence of other images. Such discernment, at various stages of childhood and adolescent development, may not always accord with adult views, yet in the context of a loving education at home and in school, we should respect with trust and confidence the emerging power of self-determination reflected in their choice of images. I personally recall the endless parental patience when, for a number of months, I would rush home from primary school and promptly disappear to a neighbour's house to watch Popeye on a coveted T.V. Through gentle discussion of the programme's merits, and encouragement in infinitely more interesting diversions, boredom with Popeye soon set in.

I well remember the horror I felt from the story of Baba-Yaga the witch, yet with a strong fascination I returned to read it again and again. I recall, too, an old News Chronicle annual containing comic-strips about a family of anthropomorphised rabbits. Among a host of dislikeable characters was a sadistic recluse by the name of Miss Sheep, to whom Baby Rabbit had the misfortune to be sent for a holiday. After a nightmarish sequence of being imprisoned in his cot, locked in the stairs-cupboard and all manner of deprivations, Baby Rabbit is reclaimed by his family in a happy ending. The last line of the story was: "And as for Miss Sheep, we must just cut her dead!" Together with my brother and sisters, after each successive reading, we would ritually score the misfortunate ewe with a pair of scissors.

Contrary to certain theories, especially the Christian concept of sin, some forms of aggression are part and parcel of healthy psychological development. Similarly, aggressive response to aggressive images, within a controlled context, can sometimes aid the growth of appreciative sensitivity in a young child. A parent or teacher, then, cannot dictate in this process of image-selection, but only *relate*. A sincere relationship of love and trust between adult and child is the firmest foundation for the development of the appreciative sense in a young person. Where this may be lacking in some homes, the teacher bears the brunt of the responsibility for this relationship, being for many children the only other consistent adult presence. While all Christian teachers should be committed in this way, the function of the R.E. teacher would seem to demand that specific attention be given to this dimension of child growth. Its concerns are *directly* those of the child and in particular of his or her capabilities for loving as a mature human person. A child who is unable to discriminate between resonant and dehumanising images, and whose general sense of appreciation has atrophied, is unlikely to form happy and committed relationships in adult life. What we are trying to do as educators — whether parental or professional — is to encourage young people to discover the *meaning* of the images around them, through reflection and relationship. Life is not a bed of roses. We must face reality through the images that mediate it — trivial or profound, gentle or violent, reassuring or fearful, life-giving or death-dealing. Appreciation is to do with the overcoming of fear, the daring to love and the emergence of a healthy self-image.

(ii) *Learning to express*

A human being starved since childhood of adequate visual, aural and tactile stimulation, will in most instances lack imagination. From the very beginning, images of facial expression, voice, sound, movement, colour, shape and texture, determine an individual's responses and lay important foundations for personality-building. As the child grows, he should be permitted to explore an expansive variety of images, both of his own choosing as we have suggested and also those that education might offer him for his free acceptance or rejection. Art, music, poetry, literature, dance, drama, should be integral aspects of the learning situation from the first day to the last of a young person's school career.

The expressive dimension of an individual's relationship to reality, can only mature out of a rich, well-nurtured imagination. A child's earliest drawings and models, often fashioned with great love and care as a gift for a parent, indicate the inseparability of creative expression from the power of appreciation, and this universal impulse to make and create should not be neglected as a young person matures. Perhaps a big mistake in education has been to assume the necessity for obvious talent in some sphere of the creative arts. Moreover, this talent has generally been

41

assessed, especially in the secondary school, on the basis of a pupil's ability to *copy* accurately, either from life or from some great masterpiece. Similarly, in the area of dance and movement, a pupil might be praised for his or her ability to execute with skill a sequence choreographed by the teacher. While there is, of course, a place for definite skill of this sort, it should never be a priority for the R.E. class. Appreciation of and participation in the creative arts in R.E have the function of deepening self-understanding in relation to the mystery that reality is; our primary aim is not the development of technical expertise. In this respect, the encouragement of *honesty* in self-expression is extremely important; we are not looking for the 'clever' poem or painting but for a creation which communicates uniquely and personally a response, an emotion, a realisation, an awareness. Above all else, the person that the student is must be allowed to blossom, whether through expressions of joy or sorrow, humour or tragedy, certainty or doubt, life or death, fullness or emptiness, beauty or ugliness. Our task as educators is to aid in the integration of all these strangely inescapable facets of existence, the ever-present paradox of humanity-in-the-world.

Through the medium of the creative arts, it is possible to say with the students: this paradox is eternal, in the images both terrible and wonderful in life around me; but it is also within. What I see and what I feel reflect one another, not as mirror-images but in a mysteriously transformed manner, so that my expressions do not simply copy reality but *interpret* it.

Religion itself interprets reality in this way. From the Christian perspective, one can discover some of the most profound images, which themselves have given rise through the ages to outstanding artistic creativity in painting, sculpture, poetry, music, dance. Creation, incarnation, trinity, transfiguration, resurrection, ascension, judgement . . . these doctrines, rich expressions of the Christian apprehension of reality, reflect humanity's deepest capacity — actually to *name* experience and interpret it creatively. All the great religious traditions indicate this, in their scriptures and mythologies, in their magnificent visual symbols and imagery, and in their attempts to *influence* reality through their rituals and practices.

An essential factor shared by the creative arts and a religious interpretation of life is — community. There is always the danger that the artist, like the fanatic, privatises his art or uses it as a medium for the destruction of what is truly human. Art, like religion, should be prophetic; even in its statements of horror and death, it can be redemptive. Of course, it is far from easy to make judgements of this sort, when each individual's relationship to a painting or a poem will be, like religious faith, largely personal. But the communitarian dimension of appreciation and expression is as indispensable to education in the creative arts as it is to R.E. itself. When we attempt to marry these two in an integrated programme, we must be ever more aware of this. Autonomy and community, personal and social, receiving and giving, must be maintained in creative tension in the educator's efforts to nurture appreciative and expressive maturity as an integral feature of mature religious understanding and an adult faith.

Catechesis and the creative arts

In the context of general R.E., we should avoid the tendency to overt catechesis. This would be inappropriate if we are to assume that any given group of, for instance, sixth form students would consist of a proportion of non-committed or perhaps non-Christian people. As we have already suggested, provision should be made for those students wishing to study and pray within an assumed-faith setting. It would be a gross mistake, however, to suppose that the creative artistic dimension should be subordinated in catechesis to the didactic dimension. The synthesis of cognitive and affective, reasoning and feeling, intellect and imagination, memory and emotions, logic and intuition, is as essential here as anywhere in the educational enterprise. A mature and open Christian faith demands such a synthesis. The creative arts, therefore, occupy a central place in the exploration of the believer into his or her faith, both in terms of history and tradition, and in terms of personal commitment. The role of the creative arts in catechesis should be developed within a twofold process, involving a universal 'thread' and a specific 'thread'!

It has been the contention of this paper that appreciation and expression are the qualities of true humanity. Thus, by integrating the creative arts into catechesis, the catechist's concern is on the one hand to encourage appreciation of, and participation in, aesthetic experience as universal — as general expressions of the human spirit. On the other hand, the concern is also to evoke appreciation of, and participation in, aesthetic experience as specific — as a particular Christian meaning expressed through the same creative media.

Let me explain what I mean. Probing the general creative expressions of humankind through art, music, dance, drama, literature and poetry, should render a young person more sensitive to the creative expressions of his or her own tradition. This can happen through the emergence in the individual of new modes of seeing and listening beyond the immediate and superficial. The visual symbols of Christianity: the cross, church architecture, icons, colour, can be looked at through new eyes if the power of art is apprehended. Liturgy with its feelings of celebration, thanksgiving, the evocation of memory, the experience of awe, can be enhanced by an appreciation of the language of music. The significance of gesture, such as blessing, genuflection, the laying on of hands and the sign of peace, can assume a deeper significance through an appreciation of dance and drama. Scripture, with its living communities, its stories and literary devices, and its profound inspiration to generations, can be better appreciated and understood through an awareness of great literature. Doctrinal formulations and credal symbolism can take on new and undreamt-of meaning in the light of the truth of poetry.

Moreover, for young people to participate personally in creative activity (through their own paintings, poems and dances) incarnating imagination is to develop the power of response which is a prerequisite for adult Christian faith and for free participation in the faith-community. To be able to respond to life and to others in a spirit of faith and commitment is perhaps the most important sign of healthy Christian development.

Conclusion

In short, only an authentic human being can be an authentic Christian — one who acknowledges that the cross and resurrection of Jesus Christ do not abstract us from reality, but plunge us ever more deeply into it. For Christian revelation is the ultimate creative statement, arising from and addressing itself to the endless creative potential of humanity. This being so, we do a grave disservice to young people, particularly those on the brink of adulthood, whether they be committed Christians or not, if we deprive them of the opportunity to develop this dimension of creativity. To appreciate and express is to love, and the creative arts are unique in their capacity to foster and stimulate this deepest quality of human existence.

Senior form non-examination R.E., therefore, frequently bears a heavy responsibility to provide students with creative artistic experiences — perhaps the only such experiences in a student's school life — which make a positive contribution to adult maturity. While this does not require technical expertise, it does demand adequate research and a good deal of sensitivity on the part of the teacher in nurturing a suitable atmosphere for sharing and creating. To enlist the support of other relevant departments is enriching and indeed essential to the success of a truly integrated programme. The R.E. teacher should not feel that he or she should be able to 'go it alone', since R.E. is by its very nature a comprehensive discipline, drawing upon and interpreting out of the multifarious sources that go to make up life itself; this is what renders it all the more exciting for both teacher and students.

Finally, it must be emphasised that the creative arts should not be used as a substitute for so-called 'religious' content in R.E., but should be considered as integral and imaginative modes for the exploration, interpretation and communication of the religious sensitivities of humanity; that is, they are the means for expressing humankind's relationship to mystery itself.

> 'Sculpture, the first of arts, delights a taste still strong and sound: each act, each limb, each bone are given life and, lo, man's body is raised. Breathing, alive, in wax or clay or stone. But oh, if time's inclement rage should waste, or maim, the statue that man builds alone, its beauty still remains, and can be traced back to the source that claims it as its own'.[4]
>
> — Michelangelo

References

1. Philip Phenix, *Education and the Worship of God* (Westminster Press, Philadelphia 1966).
2. Gloria Durka & Joanmarie Smith, eds., *Aesthetic Dimensions of Religious Education* (Paulist Press 1980).
3. Brenda Lealman & Edward Robinson, *The Image of Life* (CEM 1980); and *Knowing and Unknowing* (CEM 1981).
4. Joseph Tusiani (transl.) *The Complete Poems of Michelangelo* (Peter Owen Ltd. 1961) p. 77 No. 85.

CHAPTER 2
THEOLOGICAL ISSUES

THEOLOGY IN THE SIXTH FORM

In this paper I make the assumption that it is indeed desirable to teach some form of religious education in the sixth form. On that point there need be no argument here. The question I want to deal with is what form religious education should take in non-examination courses in the sixth form, both as far as method is concerned and content.

The prevailing orthodoxies suggest that *catechesis* is appropriate for Roman Catholic schools and *phenomenology of religion* for non-confessional schools. Both these methods of dealing with religious education represent a considerable advance on what had been done previously in each type of school, though unfortunately it would seem that some schools have still not caught up with these recent developments. However, a number of people in the religious education business are beginning to realise that there are problems with both catechesis and phenomenology and that some further model might be required for explaining what should be done in religious education in schools. It is one such model that I should like to outline in this paper.

Historically, Catholic schools were created to protect the children of Catholic parents from an aggressively Protestant world and later from an uncomprehending and sometimes cynical secular world so that they could be educated in an environment which fostered a Catholic culture with certain specifiable social and religious characteristics. It was also the intention that the children should be given religious instruction, though the enthusiasm with which this was done varied from the negligent to the fanatical. The primary model for religious education in Catholic schools, then, was that of 'instruction' and out of this catechesis has developed in recent years as a method of religious education which is designed to preserve, foster and develop the specifically Catholic faith of pupils, while at the same time avoiding the charge of indoctrination.

In the early Church catechesis was the instruction given to

48

adults who had asked to be baptised and a number of catechetical sermons have survived from the fourth century,[1] those of Cyril of Jerusalem for example. Its most debased form came much later with the catechism presented to children to be learnt parrot-fashion, and with about as much understanding as a parrot. The modern version of catechesis, however, is an attempt to revive and adapt the earlier tradition and is seen as a form of education which developes and matures the faith of believing Christians. Or, to put it another way, one may say that the aim of catechesis is to deepen the understanding of those who have faith. The nature of catechesis has been admirably described by Kevin Nichols in the opening pages of *Cornerstone*.[2] He makes it clear, as does Pope John Paul II's *Catechesi Tradendae* and the *General Catechetical Directory* (21) that catechesis is "a dialogue between believers" which encourages people to grow up in the Christian faith.

There can be no doubt that catechesis is an admirable activity and cannot be neglected in the life of the Christian community. The limitation of catechesis is that it can only take place among those who explicitly profess belief in Jesus Christ, or who have, to use Kevin Nichols' phrase, "at least the spark of faith in them". The problem is whether catechesis can be used in confessional secondary schools and more specifically in sixth forms.

The presumption of those who advocate the adoption of catechesis as an appropriate method of religious education in confessional schools is that all the pupils of a class awaiting catechisation will be believing Christians who require further development of their faith. But anyone with recent experience of teaching in a Catholic secondary school will know that this is a false assumption. Virtually any class, and this is especially true of sixth form classes, will be made up of students with a wide range of religious allegiance from the committed Christian to the staunch atheist, with a variety of positions in between: the Christian who wants to stand back, as it were, and examine the religious luggage he has been carrying for the previous ten years; the student who is genuinely puzzled by questions of religious belief; the one who was baptised sixteen years before but who may have become completely secularised; and those belonging to a non-Christian religion. As Catholic schools are increasingly coming under pressure to broaden their intake of pupils to fill up the school at a time of falling rolls and to widen its racial make-up, the more likely it is that only a proportion of our students will in any serious sense 'have the faith'. Increased pressure on pupils from the mass media and their peers in our predominantly secular culture guarantees this, and even the *General Catechetical Directory* recognises adolescence as a time of searching for meaning, of religious crisis and conflict. It is a stage in personal development when traditional authorities and earlier convictions are systematically questioned. These factors have been recognised and well described in *Signposts and Homecomings*,[3] and the point need hardly be made that we do our students no service if we close our eyes to these difficulties. It is precisely these factors, among others, which must determine our method for teaching religious education, and it is these factors which make catechesis ineffective, inappropriate and, indeed, improper for use in a class at which attendance is compulsory.

In the lower part of a secondary school and probably in the sixth form also in a confessional school it is most likely that religious education lessons will be compulsory. If we are to respect the freedom of our pupils as was most forcibly recommended by the Second Vatican Council document on Religious Freedom, we cannot catechise them so long as catechesis presupposes that its participants 'have faith'. Enforced catechesis, like enforced worship, is a nonsense and an offensive nonsense. Moreover if we want to teach efficiently we will not even attempt to catechise pupils in the classroom because it will inevitably come across as a kind of preaching, evangelisation in the Billy Graham sense, and it will be met with firm consumer resistance. Catechesis is normally better suited to adult life or to voluntary groups in school who feel the need for catechesis. At any rate it should be firmly excluded from normal time-tabled religious education lessons.

Teachers who recognise the inadequacy of catechesis for classroom work and especially sixth form work might be tempted to adopt the approach common in state schools, the phenomenology of religion pioneered at Lancaster University in the last two decades. There is a sophisticated and a crude version of the phenomenological method. My concern is with the crude version, though I am not sure that my criticisms do not in some respects apply to the sophisticated version as well. This method advocates a quasi-scientific study of religion in which the student disregards any commitment he might have in order to examine the phenomena of religion as a neutral observer. At its worst this develops into a Cook's Tour of religion around the world; a cross between a religious geography, a religious sociology and a religious anthropology. The main focus of attention is often festivals and religious customs, while beliefs are looked at from a distance, as it were. The generic term 'religion' is often used in an undifferentiated way as though all religions approximated to a quasi-Platonic ideal of religion. In fact, one only encounters concrete religions with specific beliefs and different histories and cultures and in different societies. We do no service to any religion, least of all Christianity, if we abstract it from its historical development, if we refuse to examine differences as well as similarities, and if we ignore doctrinal, theological questions which are at the core of religions. Some advocates of the Lancaster approach would insist that what actually happens in schools is a debasement of what they advocate, but it would clearly be a false turning for Catholic schools to become committed to this style of the phenomenological method. Questions of Christian theology are going to have to be central in the religious education of a confessional school. As one member of the Christian Education Movement said to me recently, "State schools seem to be missing the point of religious education".

The context of religious education in the sixth form makes it particularly problematical what is to be taught and how it is to be taught. These students no longer have a legal obligation to be in school; they could, if they so chose, be in employment or at a college of further education. They are likely to see their own educational needs largely in terms of acquiring qualifications for higher education or employment and they are notoriously reluctant to attend courses, voluntary or compulsory, which are not immediately useful for the advancement of their future careers. How we are to teach them will be determined by our perception of their intellectual and emotional needs and their reaction to what we offer them. One thing is clear, what we offer them and how we teach them must be as professional and academically respectable as the rest of the work they are doing in the sixth form. The level of thought we demand from our students should not be less than what is expected in other subjects, and in terms of their sensitivity to new ideas and their ability to think critically and form personal judgements we may expect more than do some other teachers. Much of this is supported by the *General Catechetical Directory* (83) which says that in young adulthood "The method which seems most desirable is that of treating fundamental problems and problems of most concern to this age with the serious, scholarly apparatus of the theological and human sciences, using at the same time a suitable group-discussion method".

In the light of all this, I suggest that what we should be doing in the sixth form, as elsewhere in the school, is *theology*. By this I am proposing a method of teaching; I am not thereby suggesting that the content should be any more taxing than a student can actually cope with. Theology should be as academic as is suitable for the people one is teaching. It should be professional, rational, open-minded and open-ended, but dealing specifically with issues and problems of religious belief and theological meaning.

There are, of course, different forms of theology. What I would advocate is the theological method associated with Clement of Alexandria, St Augustine, Peter Abelard, St Thomas, Newman and Rahner among others which uses reason and culture in an attempt to grasp meaning from the world of experience. I can see little place in school sixth forms for the theology of Tertullian, St Bernard, Kierkegaard and Karl Barth which spurns reason and culture to focus on an authoritarian claim for a revelation of God or which regards devotional insight as its primary aim. Abelard's maxim is particularly pertinent to the place of theology in schools: "Fides non vi, sed ratione, venit", faith grows from reason, not compulsion.

Theology must be taught in an environment where students can feel free to speak openly about their convictions, frustrations and doubts. In theology we must be concerned to critically analyse the traditional language of religion, in our case this will be primarily Christianity, to see whether it conveys meaning which illuminates our own experience. Above all the student must be encouraged to ask himself/herself where he/she stands with regard to claims to truth made by Christianity, or any other religion. The advocates of neutrality in religious education, splendidly criticised by Edward Hulmes, are 'copping out', if I

52

may so put it. The importance of doing theology is that it concerns each individual and should be related to the practicality of life, to the pattern of life to which we commit ourselves.

Unlike catechesis, a student does not need to have faith to do theology. A student of theology may or may not have faith. For most of the readers of this paper, *fides quaerens intellectum*, faith searching for understanding, will be the model of theology they adopt. But if we accept that there are important questions of meaning, knowledge and truth in theology, we will acknowledge that in some sense theology is a science which can and should be pursued by all students. We may re-define theology as *homo quaerens intellectum de Deo*, man searching for understanding about God. It is such a model of theology that I would recommend for use in both confessional and state schools because it is flexible enough to meet the needs of all schools while also pinpointing the central issue of religious education, the exploration of truth and understanding in religion. The content of a theology course in any school will be determined by the cultural and religious traditions of the members of that school, but the comprehensive meaningfulness and truth of those traditions cannot be presupposed, they have to justify themselves. For many Catholic teachers the teaching of theology to sixth formers as to younger pupils also will involve a 'letting go' which they may find threatening. But it is children that we teach, not doctrine, and we have to teach them as we find them.

It follows that the content of non-exam theology courses will vary to a greater or lesser extent from school to school depending on the ability and interests of teachers and students alike. A number of preliminary points need to be made, however. First, the promotion of theological enquiry among students who voluntarily stay on at school to do A-levels or to avoid unemployment is a very demanding task that should be done wherever possible by graduate theologians, or at any rate by professionally trained theology teachers. It is not a job for amateurs or the reluctant. Secondly, it is eminently desirable that there should be a small group of A-level theologians in the sixth form to act as a point of focus for and give academic respectability to non-examination theological work. And thirdly, national and diocesan syllabuses which predetermine in detail what should be taught are not helpful and are doomed to failure because they cannot meet the

HIGHER EDUCATION
THE BECK LIBRARY

needs of a particular school. Flexibility is the order of the day not standardisation. On the other hand guidelines about what might be done could be very helpful.

The method of teaching theology, then, is to get students to think through issues of religious belief. This will involve a good deal of discussion. Information of a suitably demanding nature must be fed into the group by way of books, duplicated material, films or videotaped television programmes. The content of courses, as I have said, will vary from school to school and group to group and a teacher's plans will have to be flexible enough to meet the needs of the students, even to the extent of altering course in the middle of a term. It is desirable that a number of options are available. Students of any age and especially sixth form students can express great resentment at having to do compulsory non-exam courses, especially one which touches them so personally as theology. Having some choice about which course they follow can often defuse much of this resentment.

The content of any course must be determined by each teacher in consultation with his colleagues, but one crucial principle must be preserved: that theology should not evade central issues of religious belief and practice. The content of what is to be done will also depend on what has already been taught lower down the school. At the school at which I have been working until recently we agreed that there was a range of theological issues which it would be appropriate to examine in the sixth form, at the centre of which was 'the God problem'. Pupils lower down the school, in my experience, find it very difficult to handle the concepts involved in talking about God, so the lower sixth appears to be the best place to tackle this.

My colleagues and I have not been too concerned to cover everything in the time available; we have been more interested in developing intellectual skills than feeding information. We work within a predetermined 'field' but we do not cover every bit of ground in the field and we enter it by different 'gates'. We usually have three teachers working at the same time and each teacher takes a different starting-point so that students can join a group of their own choice. One group may begin with some general questions of philosophy of science, deliberately designed to attract A-level science students who often have positivistic ideas about 'truth', 'knowledge', 'proof' and 'facts'. Such pre-

Popperian attitudes have to be cleared up before they can begin to do any theology. Another possible starting-point would be an exploration of different theoretical views of man (philosophical anthropology) — Christian, Buddhist, Marxist, Humanist, materialist, etc. — and their practical consequences for ethics and politics. Alternatively the philosophy of religion provides a suitable way into theological investigation. Ethics and politics are other possibilities which are valuable areas of investigation in themselves and which can be used as a way into fundamental questions of theology.

Other groups in other schools may want to study part of the Bible (I have just finished reading Galatians and Romans with one group); or another religion and its relation to Christianity; or Christology. The study of a suitable religious or theological book might provide a good peg on which to hang discussions of religious issues. There is a very wide range of possibilities. My only concern is that whatever is chosen should have a theological justification.

One distinguishing characteristic of theology is that it is critical. It does not presuppose what is true Christian teaching, but is an open enquiry into truth and understanding. It takes traditional authorities seriously — the Bible, the early Fathers, Popes and Councils — but does not slavishly defer to them. The style of theology I am recommending is prepared to criticise the substance of the Bible and statements of orthodoxy where this is demanded (what the Germans call *Sachkritik*). As I have already said elsewhere,[4] such a theology is not primarily at the service of the Church; it is an open search for truth which, we may believe, is ultimately to be identified with God. "Yet such a theology would ultimately be of service to the Church because it would give intellectual rigour to the questioning already taking place among young people inside and in many cases outside the Church. Such a questioning in our schools, far from being dangerous, is invigorating; the only thing which is dangerous and depressing is an unquestioning and mindless acceptance or rejection of Christianity".

References

1. Cf. E. Yarnold, *The Awe-inspiring Rites of Initiation*: Baptismal Homilies of the Fourth Century (St Paul Publications 1981).
2. K. Nichols, *Cornerstone* (St Paul Publcations 1978), 6–9, p. 15f.
3. D. Konstant et al., *Signposts and Homecomings* (St Paul Publications 1981).
4. Cf. *The Tablet*, 30 August 1980 and 28 February 1981.

THEOLOGY AND TEACHING

This section is about religious education in the wide sense. It endeavours to point out that while essential doctrines and truths can never change, the teaching Church, at any given time, may be moved by the Spirit to draw the attention of the faithful to rich and challenging aspects of the Christian faith. Some of these aspects, concerning Catholic theology and teachings, which may have been temporarily set aside in the past, giving way to the need for other more pressing emphases, have the potential, we feel, to reach the hearts and minds of mankind in the 80's. Since theology grew, in the first place, out of the prayerful thoughts and formulations of the early faith-communities, we hope that a brief theological explanation might be offered to our teachers today, as a basis for new effort towards more effective dialogue.

The place of theology in education

Pope John Paul II has written of the "irreplaceable mission of theology in the service of faith" and religious education, especially during this "important but hazardous time of theological research".[1] In the same paragraph we find ". . . every stirring in the field of theology also has repercussions in that of catechesis" The implications of the re-emergence of an incarnational theology of revelation for religious education and catechesis in home, parish and school, for an understanding of the liturgy, for renewal of homiletics and for adult education are beginning to appear quite significant. Part of the task facing us is (a) to clarify and accurately express this theology in the interests of avoiding confusions and error where God's people are concerned,[2] e.g. to provide an over-all and unifying view of the central mysteries and truths of the Church and (b) to provide readily-understandable guide-lines for religious education, catechesis and liturgical celebration at all levels and places, ages and stages,

E

but in the present context, with special reference to young adults.

The dignity of man

As the world enters a new decade at the close of the second millennium the Christian message is addressed to man in his humanity; it is addressed to him previsely as a human being wherever he works, creates, loves, suffers, sins or doubts. Religious education testifies to the fact that Christianity affirms man in the very essence of his manhood, emphasising his unique dignity and challenging him to become ever more fully the truly human image of the living God who dwells in the depths and the heights of the human, questing spirit. The precious and "surpassing dignity of man"[3] is forever established in the incarnation. Man is a being who seeks. His whole life confirms it. The history of each one bears witness to it. Man searches for meaning.[4] Equally man's life is a search for love. "Man cannot live without love. He remains a being that is incomprehensible for himself, his life is senseless, if love is not revealed to him, if he does not encounter love, if he does not experience it and make it his own".[5] The incarnation of God is the fulfilment of this search. In it the true nature of man is revealed. His life is given its full meaning. "Christmas", Pope John Paul wrote, "is the feast of man". Christ is the answer to man's search. "God became man so that man could become God".[6] This astounding truth is the message of the Church for all its members and for all mankind. The unique destiny of man is the good news for today's world. "Through the incarnation God gave human life the dimension that he intended man to have from his first beginning. . .".[7] A new and official document on education which calls for the harnessing of all human questing after love and meaning, among young and old, as central to the work of the Spirit in continuing revelation and in the redemption of the world, can only bring about a transformation of the apathy that often colours the attitude of people to their role in the Church. "The 'divine destiny' of man is advancing, in spite of all the enigmas, the unsolved riddles, the twists and turns of 'human' destiny in the world of time".[8]

A traditional incarnational theology of the Church sees the whole of creation as Incarnation writ large. Nothing and no-one are beyond the absolute presence of the risen Christ. All that men know and experience, in potency and in actuality, is intrinsically related to the movement towards the full humanity for which Christ is the paradigm. An incarnational theology — which can be authenticated in Scripture (e.g. prologue to John's gospel), the Fathers of the Church (e.g. Irenaeus, Gregory Nazianzus), the Vatican Council Documents,[9] and in the writings of Pope John Paul II — speaks to people, particularly the young, in a way that has meaning and truth. (The extrinsic approach of much post-Tridentine theology can often have a disturbingly negative effect on the young and old of this generation, as recent surveys and enquiries have shown.) Much careful work remains to be carried out in the study and presentation of the truths of the faith as seen in this renewed light.

Christ — the First Man

Truly incarnational theology, with its positive and affirming approach to humanity, allows Christ the full value of his humanity and deals with the enterprises and endeavours of every person in the light of it. Wherever people are loving, sharing, forgiving, seeking truth and meaning, overcoming sorrow or difficulties — that is precisely where the Spirit is present. The name 'Christian' signifies, among other things, our awareness of this fact. Resurrection guarantees the nature and destiny of man. Incarnation provides the last Word on creation. "The God of creation is revealed as the God of love".[10] From within graced man, of the line of David, God emerges, revealing love and meaning at the heart of all life. Religious education sees Christianity as the specific expression of the reality of this new creation and vision that is already established and active universally. When we talk of the Christian vocation of all men today we are really talking about the potential of true human vocation; when we explore the meaning of Christian community we are exploring the full possibilities of human community. Revelation is not first and foremost about special 'mysteries' and secrets left behind for man concerning a remote God, nor even is it primarily about an institution called Christian but about graced mankind as a whole. From all time, in God's initial and loving design, man was created so that in Christ God's self-communicating love and man's original and graced capacity to identify with that love could find their full unity and final expression.[11] Incarnation is then seen in terms of the completion and fulfilment of God's first plan for his beloved creatures as well as of the redeeming and healing that is required by the ever-present counterforce of original sin.[12] Perhaps this is the emphasis that will once again bestow on teaching and preaching a colour and an excitement which is long overdue.

Eucharist and community

The Christian faith is radically communitarian. Salvation is intrinsically bound up with the way in which we live out our relationships with others: "I tell you solemnly, in so far as you did this to one of the least of these brothers of mine, you did it to me".[13] Contemporary Church documents on religious education continue to remind us of this often neglected truth. "Man's journey to God is a growth achieved in community with others. . . Man grows through his relationships with others".[14] The essential need for community was exemplified most strongly in the Last Supper which Jesus shared with his disciples. Here was the perfect example of community and, in the washing of feet, a total offer of service.

Christian education will always see the Eucharist in terms of God's self-sacrificing people offering themselves with Christ in mutual service as "a living sacrifice of praise",[15] where the communal celebration is "a sacrament of love, a sign of unity and a bond of charity".[16]

The Christian community demonstrates again the full salvific value of all true fellowship in the sacrament of the Eucharist. It is in the Eucharist that the amazing reality of the incarnation is celebrated — that in the life, death and resurrection of Christ the potential of all men has been determined; that the condition for achieving this full humanity is to "love one another as I have loved you". "Eucharist", Pope John Paul II explains, "constitutes the soul of all Christian life, which is expressed in the fulfilling of the greatest commandment — in the love of God and neighbour". He stresses that "the authentic meaning of the Eucharist becomes of itself the school of active love for our neighbour. If our Eucharistic worship is authentic, it must make us grow in awareness of the dignity of each person. . . The sense of the Eucharistic mystery leads us to a love for our neighbour and for every human being".[17]

This paper emphasises the fact that Eucharistic celebration can never be divorced from Eucharistic living. In the Eucharist God's people celebrate the real presence of his incarnate and self-emptying love both in the world and also in the experience of that love in their own communities, e.g. at home, in the parish, in school and in the wider contexts of their lives.

The power of the young

"Tell everyone that the Pope believes in you and that he counts on you. You, the young, are the strength of the Church and of the world".[18]

Now is the time for a positive world-view to be addressed to its people by the Church. Having made the world, God saw that it was very good. Without in any way playing down the sharp edge of evil or the awful daily evidence of sin in the world, people today, especially the young, are waiting for a confirmation of the worth and value of creation in general and of themselves in particular, and an affirmation of their place in and contribution towards a world in sore need of redemption. They seek for a recognition of their talents and ideas by a Church that often seems to have little place for authentic lay participation. They believe in their power and vision and the future they can build, "making life ever more human" in a community of love and freedom[19] but they are sometimes confused as to how this desire for commitment can function as a life-source of the community called Church.

A theology of Revelation which transcends the sharp dichotomy between sacred and secular, which sees all life as already permeated by God's grace and which holds out for a continuing revelation within human experience[20] can only revitalise the latent spiritual power — "the power that permeates everything that is human" — within today's youth. It sees all creation, made by God in love and assumed by him in Christ as the very mode of his being, as gift rather than threat. God is not lessened when man is ennobled and identified in his creatureliness with Christ, the Son of God. Such a theology "is concerned with the ultimate meaning of life and it illumines the whole of life with the light of the Gospel . . ." since revelation must not be "isolated from life or juxtaposed to it".[21] Perhaps this is the time to reformulate a 'new' apologetics for a young, questioning and restless world.

The challenge of Christian education

Christian education is concerned with providing the circumstances, the atmosphere and the inspiration for the emergence and actualising of the God-implanted destiny of each person with his own unique gifts and talents. To facilitate the attainment of one's true identity and to assist the full development of each one's particular personality, conforming ever more closely to that of Christ the first man, is the task of all Christian educators, be they parents, teachers or priests. This growth, however, in awareness and knowledge demanded by such a challenge can only happen in community.[22]

A central aim of Christian education must then be that people develop a fundamental humanity and sensitivity towards all others and a general sense of responsibility to life and tasks as they serve one another in the spirit of the Master. Jacques Maritain, the Catholic philosopher and friend of many popes, observes that education is the process of "becoming our true selves", of realising the fullness of our redeemed humanity in mutual sharing.[23] Perhaps the radical contribution of Christian education to a 'secular' world is that it concerns itself, in spite of all persistent influences to the contrary, with the total person and not just his/her usefulness to society or intellectual capacities. In the light of Christ's refusal to dissect people into a mind/heart or body/soul categories, and of his persistence in seeing each person as a whole entity, whose total response to God is found in the Spirit-filled love of others, it would follow that all Christian education is directed towards the emergence and strengthening of that 'divine spark'[24] wherein man is established in the image of God. The learning of the apostles and disciples, for instance, bore fruit only in their deepened awareness, through and in Christ, of community service and their own unique and privileged place in living out that total demand.

The revelation of Christ then, as reflected in the writings of Scripture, in the celebration of liturgy, in the formulated wisdom of doctrines, in the history and witness of the Church-community and in many other ways, is seen as the key to the true nature and development of every person. The hope of the Christian educator, regarding each member of God's family, is that the Father may give "the power through his Spirit for your hidden self to grow strong . . . until, knowing the love of Christ, which is beyond all knowledge, you are filled with the utter fullness of God".[25] Church school religious education and catechesis, therefore, sees grace and salvation in terms of the human experience of people and in their personal relationships with God and with one another.[26] 'Handing on the faith' must always be a balanced commitment. It must never become a merely cerebral activity.[27]

Knowledge alone is insufficient. It must be internalised into wisdom. Wisdom is knowledge laced with love. We must become what we know. The Christian message enables man to actualise and realise who he potentially is. "The man who wishes to understand himself thoroughly . . . must come to Christ with his unrest and uncertainty, and even his weakness and sinfulness, his life and his death. He must, so to speak, enter into Christ with all his own self, he must appropriate Christ and assimilate the whole reality of the incarnation and redemption in order to find himself".[28]

General aims of Christian education

As well as knowledge and understanding, Kevin Nichols singles out *insight* as central to the over-all aim in Christian education. "I have chosen this word", he writes, "because it goes beyond knowledge and understanding, beyond the cognitive. It includes sensitivity and feeling. . . An insight into, say, the Church's doctrine implies that I am able to see something of its inner unity and its relevance to world affairs and to my own life".[29]

It is generally agreed that the scriptures, liturgical celebration and the doctrines of the Church play the major part in the achievement of this objective. While much remains to be examined in this whole area, a few brief remarks on the vital role of each of these channels of Revelation may be helpful here.

(a) It seems vital that the Scripture should be so presented as to provide insight into the meaning of each person's experience. Scripture is the living word that transforms the very heart of man, overcoming blindness and ignorance and sin through the spirit of wisdom and love. This evokes a response of the total person. There was agony and ecstacy on the road to Emmaus when Scripture was explained. "Good teaching should enable us to feel our way into the Bible text as well as knowing its meaning; to respond to it emotionally as well as mentally; and to respond in faith".[30] The gospels express the emerging insight of the first Christians that the story of the life, death and resurrection of Jesus provides an answer to the mystery of all human life and death.

(b) Liturgy, too, must be seen in terms of the affirmation and celebration of the Christian life in the Spirit-filled community. The awareness and experience of redeemed human life lead to a heartfelt response in liturgical celebration, in sign, symbol and ritual. Liturgy springs from the joy, gratitude and needs of the children of God in the new creation. And, like Scripture, it must involve the whole person. "It is closely connected with the education of the emotions".[31]

While providing the means of expressing the deep awe and wonder of man in the face of the paschal mystery of Christ, it is important, too, to preserve as far as possible a basic element of spontaneity and creativity at the heart of all liturgical celebration. This is man's response to the good news that, because of this Easter mystery, Christ is alive in all men, setting them free to conform, in their new humanity, ever more closely to the image and likeness of God.

(c) "Doctrines are not the object of faith. . . They have a secondary or subsidiary function. . . It is important to remember this truth, both in accepting and in teaching doctrines. If they are themselves elevated into objects of faith, they can easily become idols which block our road into the realities of faith rather than opening it up".[32] Nevertheless, as Nichols points out, they are central to the faith, preserving as they do in a most important way, the accumulated wisdom and truth revealed by the Holy Spirit, within the Christian community. This truth is not about an abstract God removed from this life but about the Word made flesh, incarnate in the very heart of each person. Doctrines endeavour to reveal the significance of human life, the potential and destiny of all men, the meaning of suffering and death.

They grew out of the Spirit-filled experiences of the first Christian communities and they enrich the quality of life of God's people today. "Christian faith is, among things, a view of the meaning of the world and of life, and doctrine formulates this, making explicit and clear many things which are expressed . . . in more implicit ways".[33] The question at issue here, as with Scripture and liturgy (above), concerns the fact that only in Christ, the First Man, do we find the full and final revelation of the dignity and destiny of all men. Scripture, liturgy and doctrines, therefore, must always stand under the primacy of the mystery of the God-Man Christ. This fact will in turn influence the manner in which these three channels of Revelation are presented and considered in the handing on of the faith; i.e. are they seen as ends in themselves, divorced from people's daily experience of the human condition, or as essential interpretations of God's indwelling in the world he created and "so loved that he sent his only Son" to save and complete it?

"Christ is risen so that man can live his own human life more fully".[34]

Postscript on contemporary ethical issues

The task of ethics is to describe the Christian life, which is a mature response to the Father who approaches us with a loving invitation to fellowship in and through Christ. On the personal level, the response asked for by Jesus is conversion — a radical change of heart — that has as its goal the standard of the Father's love, which Jesus incarnated in his life and death. We are called to love one another as he has loved us — to be perfect as our heavenly Father is perfect. This call to discipleship and conversion is experienced within the community of the Church. The sacramental initiation into the life of Christ is the beginning of a process of growth which is the Christian's life task. It is the function of ethics to reflect upon and describe in practical terms the implications of this invitation to grow in Christ.

While it is within, and with, the community of Christians that each individual must accept personal responsibility for 'putting on Christ', as St Paul expressed it, this cannot be done in isolation from the wider human family. In recent years Christian ethics has been considering its role vis-a-vis the larger human community. Since the days of *Rerum Novarum*, at least, the Church has addressed itself with increasing awareness to the quality of human life. In the recent renewal of ethics the theological underpinnings of this concern have been more clearly perceived, and the implication for all who are called to be disciples of Christ is more apparent. In Jesus, the Father has reconciled an estranged world to himself, and this reconciliation is to become a reality, not only in the lives of individuals but in all human history. The Church and the individual Christian look not only to their own personal growth, but realise that authentic growth in Christ involves taking responsibility for the world also. For man the world is a blessed gift. If he recognises it as God's creation renewed in Christ, then his relationship with the world is decisive for his relationship with God. We must be convinced that this world has God's promise and love; that it is offered to the Christian in solidarity with Christ, as the forum of loving endeavour. We are called to be the leaven, to change ourselves by changing the world. Since the time of Pope John XXIII the Church, with increasing confidence, addresses all men. The Second Vatican Council speaks of the Christian's solidarity with the whole human family.[35]

It is precisely in the realm of responsibility for this world of ours that ethics faces its most difficult challenge. The prodigious advance in scientific knowledge and technology allow us to control much of the world around us and many aspects of our own existence. These and future possibilities for control and management raise the question of limit of the manipulation we may exercise, or aspire to exercise, over our very existence. This is especially apparent in the field of medical ethics: transplant procedures raise the question of what is death; pre-natal screening the question of the acceptance of handicapped; medical techniques the question of the necessary and unnecessary prolongation of life. Christian ethics with its conviction of the positive value of the individual life has a responsibility to engage in the debate. This is however relatively familiar ground for Christian ethics. If the Christian has a responsibility to and for the world, and by this one means the whole human family in its environment, then all the major problems facing this family are of pressing concern.

There are however many other equally pressing problems that

must be recognised as demanding consideration. These may be global, for example, the situation of the inhabitants of the disadvantaged third or southern world — we have a responsibility to them as members of the same human family. They may be national or local, for example, discrimination against our neighbour for a variety of reasons. The central point is that all human problems are matters of concern for the Christian and especially for the business of Christian ethics. Previously a major concern of Catholic morality was sexual ethics. This is still of importance, but it must now be seen as a small corner of a much larger canvas. Christian ethics is now ready to struggle with a much wider range of subjects because these are the problems of a world loved, and offered healing, by Christ.

Some of these are new issues, others newly perceived. But Christian ethics must work towards the answers in dialogue with all concerned parties, and with a certain humility. Consider the situation created by developments in genetic engineering. Not only are new horizons opening but they are extending with amazing speed. Today's possibility is tomorrow's achievement.

One must respect the possibilities of human ingenuity, listen to the scientists, strive to understand what they propose to do, and discuss this in terms of the essential responsibility to, and reverence for, human life. The scientist, too, has a responsibility to listen to those who are concerned about the human meaning of life itself. The point is that a development like 'in vitro' fertilisation poses new questions. Old answers are not sufficient. This is not a cause for dismay. If we look at the Acts and the epistles of the N.T. we see a similar situation. The early Christians had to grapple with the problems of their worlds and their commitment to being true disciples engaged in loving Christ and realising his love for the world. That is the role and task of Christian ethics today. It is not easy, neither is it impossible, but it is urgent.

References

1. Catechesi Tradendae, no. 61. John Paul II, 1979.
2. ibid.
3. Redemptor Hominis (C.T.S. 1980) no. 10.
4. Catechesi Tradendae, no. 60 (op. cit.).

5. Redemptor Hominis (op. cit.) no. 10.
6. Irenaeus.
7. Redemptor Hominis (op. cit.) cf. ch. 2.
8. Redemptor Hominis (op. cit.) no. 15.
9. Gaudium et Spes, para. 12, 22, 34, 38, 39, etc. Vatican II Documents.
10. Redemptor Hominis (op. cit.) no. 19.
11. Dei Verbum para. 4 (Vatican II Documents).
12. Redemptor Hominis (op. cit.) no. 12.
13. Matt 25 : 40.
14. Westminster Diocesan Pastoral Policy on Education 2.2; 2.3.
15. Eucharistic Prayer IV.
16. Sacrosanctum Concilium, no. 47 (Vatican II Documents).
17. Eucharist, Mystery and Worship, Feb. 1980. John Paul II. (C.T.S. 1980).
18. Pope John Paul's Address to Youth, Sept. 1979.
19. Redemptor Hominis (op. cit.) no. 14.
20. Gaudium et Spes, ch. 3 (op. cit.).
21. Gaudium et Spes, ch. 3 (op. cit.).
22. Westminster Diocesan Pastoral Policy on Education, para. 2.
23. Westminster Diocesan Pastoral Policy on Education, para. 2, 4.
24. Gaudium et Spes, no. 12, 22 (op. cit.).
25. Eph 3 : 16, 19.
26. General Catechetical Directory, no. 74 (C.T.S. 1973).
27. Catechesi Tradendae, para. 22 (op. cit.).
28. ibid. no. 61.
29. Art. Building on Cornerstone, no. 6 C.T.F.
30. Cornerstone, no. 138 (St Paul Publications 1978).
31. Building on Cornerstone, no. 7.
32. Cornerstone, no. 141 (op. cit.).
33. Building on Cornerstone, no. 7.
34. Pope John Paul's Easter Message, Urbi et Orbi, 1980.
35. Gaudium et Spes, no. 1–10 (op. cit.).

THE MINISTRY OF TEACHING
IN THE CHURCH SCHOOL

In past ages the Church was often a pioneer of community-care and social services, sensing the needs of each generation before those needs were generally recognised. At a very vital moment St Benedict founded his monasteries which became centres of education. The Church continued to be the only provider of education through the dark ages and the middle ages, and the chief agent of education right into our own century. Similarly, with medicine and travel. "Today those services are rendered by secular bodies, whose titles have about them a curiously religious ring — the Ministry of Health, the Ministry of Transport".[1] This word 'ministry' needs a brief explanation. In the early Christian centuries the term referred to the variety of operations within the Church community, all being necessary for the harmonious tenor of its life. Gradually, however, "the ministry began to be set over against the Church, with the idea of function subordinated to that of office and privilege so that the people of God became dominated, instead of guided, by the clergy. This deformation of the ministry into an exercise of dominant authority is now being challenged strongly, as the idea of the ministry of the laity is gradually recovered".[2] It is in this latter sense that the term is employed here. Teaching has always been regarded as a profound activity by the Church and this regard has secured for it a privileged place among the Christian ministries.

When applied to teaching, the word 'ministry' implies a view of this profession and vocation which categorises it as being concerned in a most immediate and specialised way with the liberation and humanising of others. This indicates the sacredness of the teaching activity, calling for an exceptional degree of commitment and dedication on the part of the teacher. In 1873 the French bishop of Orleans wrote: "While there shall still remain on earth a creature of this race, of whom God has said, 'Let us make man to our own image and likeness', the education of man will be the greatest of works, a providential and sacred labour, a kind of priesthood. While there shall remain on earth intelligences which God has created, capable of knowledge and wisdom, capable of truth and light, capable of thought and memory, capable of science and genius, it will be praise-worthy; it will be divine, to labour for the education of such great creatures".[3]

Under the following three headings an effort will be made to single out some central but often neglected aspects of the teaching profession which a Christian vision of education would urge us to consider carefully. There is always the danger that this vision may become clouded by the pressing demands of the 'establishment' with its urgent priorities of academic success and prescribed achievements. The Christian teacher can easily resemble the man on the horse in the following incident. There was a man on a horse galloping swiftly along the road. An old farmer, standing in the fields, seeing him pass by, called out: "Hey, rider, where are you going?". The rider turned around and shouted back: "Don't ask me, just ask my horse!" Many Christian teachers have become passive victims of a powerful system; they are no longer masters of their own professional lives. Like the rider above, they are carried away, out of control: the horse is in the saddle! We are often in short supply of the prophetic when it comes to discerning the subtle sins of 'secular' principles of education.

A. CONCERN FOR THE NATURE OF EDUCATION

Because of his vision of man in his totality and of the unity between knowledge and the knower, the Christian teacher might well be forgiven his anxieties in the face of many current approaches to the phenomenon of learning. Few would deny that in the currently accepted canons of the theory and practice of education, the cerebral is emphasised at the expense of the emotional, knowledge at the expense of meaning, the cognitive at the expense of the affective and the analytical at the expense of the imaginative. Christian teachers and educators will be aware of this imbalance in the more or less established view of education in this country and many are eager to rectify this imbalance in their own teaching and to contribute to research and movements that concern themselves with this issue. Philosophical, psychological and theological factors will play their part in this debate. There are many complex and sometimes conflicting issues to be reckoned with in the pursuit of an educational theory that respects the wholeness of the learner and the mysterious qualities of each creature. In general terms the search is for an approach to pedagogy which sees the child as an integrated personality with capacities for feelings, reason and imagination.[4] The discussion will also call for some careful sifting of the conceptional distinctions between P. Hirst's celebrated 'forms of knowledge' and P. Phenix's 'realms of meaning'.[5]

One starting-point would be "to take the meaning/knowledge relation and examine it, not from two different sets of assumptios, but from a single premise: meaning is knowledge taken up with wisdom and interiorised with love. This premise derives from an incarnational understanding of man himself, which holds any form of dualistic anthropology to be a contradiction of the divine image in which man is lovingly created. "The reciprocity and inner union of knowledge and love reflect not only the primordial ground of man's unity, but the concrete history of his personal unity as well. . . As a man loves more truly, in a way proper to the object of his love, his knowledge becomes more profound".[6] Cognition remains detached from humanity unless taken up by the subject into himself and interpreted wisely. "Subjectivity", Maritain asserts, "this essentially dynamic, living and open centre, both receives and gives. It receives through the intellect, by superexisting in knowledge. It gives through the will by superexisting in love".[7] There can thus be diverse levels of what might be termed 'cognitive meaning' from the briefest insight to the most profound awareness or expertise.[8]

Looking at the contemporary scene, we would like to point out here that the currently orthodox notions of initiation, rationality, objectivity and autonomy are often less than adequate as the basic criteria for the total evaluation of an educational approach. Kevin Nichols wonders whether a substantially deepened understanding of education could be constructed around wider pivotal ideas. He selects the notions of commitment, search and dialogue which "chime in a different philosophical register" to the demanding and rigorous style of the English philosophical tradition. "In giving such a key place to the words commitment, search and dialogue, I am not dismissing the ideas of initiation, rationality and autonomy. I am arguing that they are not on the first level and should not have the first priority. So reason is not rejected in favour of some esoteric alternative. Rationality, like autonomy, is assumed into another paradigm. In the interplay of commitment, dialogue and search, it is assumed that rationality in the sense of civilised discourse and the rules of evidence will be honoured. But there are different kinds of rationality and each should have only its appropriate place. In matters of religion we ought to recognise not only the function but also the limits of reason".[9]

B. CONCERN FOR A COMMUNITY-SETTING IN EDUCATION*

A theological basis

It seems to me that theology is determinative in all areas of evaluation. Our views of man, incarnation, revelation, the relation of Church and world, are decisive for any conception of what Christian community is, and consequently of what an academic institution should be. This is so fundamental that any discussion concerning the character of a Christian school or college either operates at cross-purposes or remains on a somewhat platitudinous level of abstraction, unless the theological ground is cleared beforehand. If the horse is to pull the cart effectively, then we should try to determine just what sort of animal he is.

The kind of theology to which I would subscribe is that which sees the whole of creation as Incarnation writ large. Nothing and no-one are beyond the absolute presence of the risen Christ. All that men know and experience, in potency and in actuality, is intrinsically related to the movement towards full humanity for which Christ is our paradigm. It seems to me that an incarnational theology, which can be authenticated in Scripture, the Fathers of the Church and many Vatican documents, speaks to people, particularly the young, in a way which has

---

*Some Material for this section is drawn from 'The Christian School: A Caring Community' (unpublished paper by T. Sallnow).

72

meaning and truth. It is a theology which allows to Christ the full value of his humanity and deals with the humanity of every person in hte light of it. The point I am re-emphasising here about the imminent presence of the Spirit has already been introduced in the previous section (cf. p. 58). The name 'Christian' signifies our awareness of this fact, but the existence of the believing Church does not effect that presence — it is already guaranteed by the resurrection. Christianity is the specific expression of a reality that is already established and active universally. When we talk of Christian vocation we are really talking of the potential of true human vocation; when we explore the meaning of Christian community we are exploring the full possibilities of human community. Revelation is not first and foremost about a community called Christian, but about mankind as a whole. Any Christian community must be a micro-model of the community called Church. One cannot, therefore, establish a Christian academic model of community until a quest has been made for a satisfactory ecclesiological model. From among the various models of the Church with which we have now become familiar the option here is made for the 'sacrament-servant' model. We cannot, therefore, avoid examining the ecclesiological concept of mission which is central to an understanding of Christian community. This task cannot be pursued here. But, in brief, it may be said that a vital aspect of the school or college community would be to reflect, in its educational milieu, the values of true humanity, fellowship and love, to the wider local and global community.

Lived knowledge

Maritain observes that education is the process of "becoming who we are", of realising the fullness of our humanity — a humanity which essentially constitutes the living image of God and is exemplified supremely in the person of Christ. Christian education should facilitate the development and maturation of the 'divine spark' in each person; that powerful gift of the Spirit which enables us really to live out the one command "love one another as I have loved you". Christ's primary concern was to open the hearts of those who encountered him. His disciples were not learned men; they grew in knowledge as they grew in

F

love, actualising their true human potential in the challenges and demands of daily life in community. What they knew in cognitive terms had continually to be made meaningful and to be given depth through loving service to others.

The overall aim of the Christian school or college, therefore, should be that students enter into wider society with a fundamental humanity and sensitivity towards all people and a general sense of responsibility to life and its tasks, whatever their profession or intellectual capacities. All that takes place in a Christian community should operate within the context of this primary aim. The implications of liberation theology, too, might well be explored in this connection. Political connotations of liberation-thinking, important as they are, should not blind us to the fundamentally personal meaning of the term itself. In the education sphere we are concerned first and foremost with the liberation of persons.

There is another point here. Our future world promises to be a strangely transformed one — the age of the silicon chip. Employment will be increasingly hard to come by in most sectors. We will need to be constantly on our guard against the dangers of dehumanisation and loneliness. Two of the most significant areas of an education programme for this kind of future will be leisure and community concern. The humanisation of our world will depend a great deal on the kinds of leisure activities which people indulge in, and on the rendering of mutual service on the part of all persons.

The pursuit of 'intellectual excellence' is, of course, essential for any academic institution, but the development of new curriculum structures and courses involving pastoral and community studies, and opportunities for students to pursue 'appreciation' courses in the arts and sciences, may well widen the concept and perhaps 'earth' it a little more. Intellectual excellence, of itself, has limited value, unless it is seen as a function serving the human excellence of the whole person.

One does not have to be able to read music in order to appreciate Mozart or Berg; nor does one require the ability to paint for the appreciation of great art; one need not be a writer to enjoy literature and poetry. The same applies to dance and drama. Man's fundamental gift is that he has the capacity to transcend his own particularity, to go beyond the limitations of himself and actually to experience, relate to and love that which seems beyond him. Everyone has a natural flair for loving — it is with this intuitive, foundational characteristic of all who study or work in a Christian school or college that the business of Christian community should in the first place concern itself.

The academic community

The process of acquiring knowledge in a Christian setting cannot be a purely cognitive one. If the 'learning community' is fundamentally a 'caring community', then the learning process is more than simply a cerebral activity. It is an activity involving the whole person — heart as well as intellect. Education then aims, as Maritain suggests, not so much at the student gaining 'knowledge about' things but primarily at developing a sensitive 'knowledge into' the human significance of his subject(s). An incarnational theology sees the wise interiorisation of knowledge as essential for our growth towards full humanity.

A body of material which remains on an abstract level in both its content and the method we employ for teaching it, cannot find any real and loving response in people. In a sense we become what we know. Unless abstraction can be existentially anchored in ways appropriate to the students' self-awareness, then ultimately they cannot be assisted in their quest for meaning. In the sphere of Christian theology, for instance — revelation tells us as much about man as it does about God; as a discipline therefore theology should deal with the human predicament and the activity of God within it, in a manner which is living and meaningful for young human beings. It is not that we are looking for a 'point of contact' between Christianity and education, or an 'overlap' area. Christianity reveals what we are capable of becoming; education, in the spirit of that revelation, is the process of that becoming. Whether we are concerned with the cognitive, the affective or the creative/imaginative fields of a young person's life, we are involved in the integral development of his humanity in the light of Christian awareness. It is for this reason that our principal model should be that of a caring community lived out in sacramental witness and service. "If he is to fulfil his mission in the world, the mission for which he was baptised, the Christian must not build his own imitation world out of Catholic material and take refuge in it, he must manifest the kind of adult concern for the quality of life that is characteristic of the intelligent humanist".[10] An incarnational theology takes the humanity of Christ to be the living reality of his divinity. "The kenosis of the Redeemer, his surrender of that which he might have held, will then be the perfect manifestation of the kenosis of God. The 'emptiness' of the Redeemer, in the poverty and humility of his historical existence will point to the 'emptiness' of God in and through his eternal activity: and the kenosis of Christ, so far from impairing the fullness of his disclosure of God, will in fact contain the very heart and substance of that disclosure".[11] All that is human has been redeemed and consecrated in Christ. Our starting point, therefore, must be the real experience of young people, and our aim should be to deepen that experience and illuminate it in the light of the gospel.

C. CONCERN FOR HUMANNESS IN EDUCATION

In his book on 'Creative Ministry' Henri Nouwen reminds us that education is ministry not because of what is taught but because of the nature of the educational process itself. Perhaps we have paid too much attention to the content of teaching, he suggests, without realising that the teaching relationship is the most important factor in the ministry of teaching. Most of the remainder of this section is devoted to an examination of what he describes as two basic models of teaching — the violent model and the redemptive model. As long as teaching takes place in a context where knowledge is seen as a product to be possessed and used in an individualistic fashion, it is doomed to be a violent process. There will be those who reject the term 'violent' as being an accurate description of any aspect of our educational system. Life is a kind of warfare, they will point out, and knowledge is a powerful weapon. There is little room for trust. You keep the system going and the system will keep you going. And so you outwit and outsmart the 'other' who is the enemy. You must remain one step ahead. And so you are not free.

On the other hand, the introduction of the term 'redemption' already indicates the unique point of departure of Christian education which distinguishes it from ideological indoctrination, general education and functional learning. "Christian education is not in the business of promoting a body of knowledge, theories, doctrines or creeds. It is concerned with redemption and with the proper locus of trust. It is concerned with where in a world of untrustworthiness, amid judgement and criticism, there is a trustworthy, affirming, forgiving experience. Here the Christian community turns to its unique individual and corporate experience of Jesus of Nazareth as the forgiving friend of sinners, as a liberating and trustworthy locus of trust. This experience is liberating because it frees from dependence on authority and ideology for support, and frees for servanthood. For now, the 'other' is not the enemy, the judge and critic, but a part of our humanity. This is the kind of radical alternative needed to overcome the critical and judgemental dilemma".[12] The basic ideas in this section, and the terminology for the six sub-section headings are drawn largely from the work of Henri Nouwen.

1. *Violent teaching*

Among the characteristics of teaching as a violent process, one can number the competitive spirit, the unilateral nature of the process and the alienating propensity which results from these.

a) *Competitive*

While the spirit of competition is identifiable in most pursuits and justifiable in many, its dehumanising power is often overlooked. The quality of relationships within the learning community whether in school or college is silently eroded when subconscious rivalry reaches a certain point. The fruit of study, whether through memorised data or written notes, is protected and defended. Grades and marks are significant, not because they may happen to be good in themselves but because they are better than the others. Information is stored only to be unpacked on the occasion of the exam, much as a telephone number is remembered only until the call is made. As Doomsday draws near, anxiety grows and peace of mind lessens. The troubled spirit then begins to be afraid. "This fear makes many students oversensitive to the reaction of their friends and teachers. This fear makes them extremely self-conscious, highly defensive in their relationships with others, constantly concerned about the possibility of failure, and very hesitant to take any risks or do anything unexpected. Often this fear becomes the unaccepted ruler over everything they write, say, or even think. Through this fear, competition has become the great preventive in a student's free development of his total personality".[13]

b) *Unilateral*

The unilateral nature of many modes of teaching, in spite of the presence of educational and pedagogical theories which would seriously question this very practice, is another aspect of the educational milieu in which we operate, that demands careful scrutiny. Much as we may care to deny it, the teacher is still the one who knows and the pupil or student is the empty vessel waiting to be filled. The fact that there necessarily is a certain amount of truth in this cannot justify the fairly common practice of the unilateral procedure adopted in contemporary teaching. The teacher has a difficult and thankless job and it is easy to criticise. To survive in the profession, and to do so successfully, one must accommodate oneself to the 'system'. But the whole point of this paper is to question the system, and search for ways and means of 'redeeming' it. It seems an impossible task but even within the inherited structure there may be avenues and attitudes that can be developed and explored. If, for instance, there are teachers who believe the following passage to be true, then surely there is room for further discussion. "Underneath many methods of teaching is still the prevailing supposition that someone is competent and that someone else is not, and the whole game is to try to make the one just as, or nearly as, competent as the other. When this ideal is realised, the teacher is no longer considered as a teacher and the student as a student and both can depart with not much more accomplished than the ability to tell stories about each other as an entertainment in their later years. In this context the teacher is strong: he knows and should know. The student, however, is weak: he does not know and should want to know. The whole movement, therefore, is from teacher to student, from the strong to the weak, from one who knows to one who does not yet know. It is basically a unilateral process. Many teachers as well as many students still operate under the faulty supposition that it is better to give than to receive. Teachers want to give something to students — an idea, an opinion, a specific skill, advice, or any other thing they think students are waiting for — and students, in turn, value their teachers according to what they have to give".[14]

c) *Alienating*

We turn now briefly to what we have called the alienating propensity of the way in which we sometimes endeavour to teach pupils and older students. Just as we often regard the child merely as one who will one day be a man — as though this were the only goal of his life — so, too, we often see education as a preparation for the future, detached from the present life of the student. His learning will lie in cold storage until the moment when the fruits of his labours will be revealed for one reason or another, as the need arises. The desire to know, and to know for its own sake, whereby man's grace of self-trascendence is forever testified to, is often brutally distorted by 'forced-feeding and front-loading' where learning is concerned. Because of insensitive cramming in result-hungry institutions, innumerable adults have been almost totally closed to the healing beauty of the work of many great writers, artists and musicians. What Violet Madge here writes concerning children may be equally applied to the college student, whether sixth former or under-graduate. "If we are to help children towards a personal discovery of truth we must be content to wait on inner growth, yet at the same time provide opportunities which will foster development, anticipating, but not imposing, the formulation of conceptions. We must be sensitive to seize the moment when instruction or suggestion from us will promote further understanding. A. J. Tessimond described the teacher's function when he suggested that —

> Man can be taught perhaps only
> That which he already knows,
> For only in soil that is ready
> Grows the mind's obstinate rose:
> The right word at the wrong time
> Is wind-caught, blown away;
> And the most that the ages' sages'

Wisdom and wit can say,
Is no more to the quickest pupil
Than a midwife's delicate steady
Fingers aiding and easing
The thought half-born already".[15]

Just as bad R.E. teaching, based solely on criteria of instant conversion or religious practice, can dull the readiness of the human soul for the sensitising and spiritualising graces of God, perhaps for ever, so, too, the functional and hurried treatment of the world's works of beauty in an educational setting can wither at source the small plants of the spirit that need time and space and nourishment to grow. "If we accept the need for growth from within, we shall encourage children to develop their sense perception and shall, at the same time, recognise their emotional needs. Thus, they may not only be helped towards the intellectual and analytical aspects of scientific enquiry, but may sense something of that mystical and unifying vision which is an experience of personal revelation.

Man is born with the religious capacity as with every other. If only his sense for the profoundest depths of his own nature is not crushed out, if only all fellowship between himself and the Primal Source is not quite shut off, religion would, after its own fashion, infallibly be developed".[16]

Henri Nouwen expresses his opinion in a somewhat similar fashion: "The violent process of teaching is alienating because the eyes of the student are directed outwards, away from himself and his direct relationships into the future where the real things are supposed to happen to him. School, then, comes to be seen as only a preparation for later life, for the 'real' life. One day the classroom will at last be left behind, the books will be closed, the teacher forgotten, and life can begin. School is just an indoor training, a dry-swim, a quasi-life. It is not surprising, therefore, that many students are bored and tired during class and are killing time by anxiously waiting until the bell rings and they can start doing their own thing. Nor is it so strange that many say they have nothing or little to do with what happens at school and must go by blind faith that one day they will be thankful for the knowledge they received. It is not so strange, then, that many teachers are looked upon as belonging to a world that is not the world of the students, and that a hidden hostility often grows from this, expressing itself in a total lack of thankfulness towards those who have given much of their time, energy, and concern to prepare them for society".[17]

2. *Redemptive teaching*

a) *Evocative*

Just as we are prepared to learn from the etymology of the word education (*educare*) as a leading out and a drawing forth, so, too, the term 'evocative' (*evocare*) indicates a mutual calling into being, an invitation to realise one's own abilities, to express, even though only partially, something of one's true self. This is a redemptive experience. Students and teachers can discover and find themselves in mutual searching and sharing. There is an identity that emerges in this kind of exchange and achievement. This identity can only emerge where there is no divisive competition and rivalry where there is the kind of mutual respect and sensitivity that Christian community is supposed to foster. There should be no air of unreality about these remarks. The plea for a strong communal presence, in terms of respect and sensitivity, at all educational encounters, especially where the Christian institution is concerned, is totally consonant with the highest degree of professionalism and skill in the exciting cut and thrust that the process of education often is. There is no lowering of standards here, no lessening of the demands on teacher and pupil alike, no blunting of the challenge to grow in wisdom and knowledge. But what is ruled out is any element of 'scoring' over another, of accentuating 'losers and winners' attitudes, and of comparisons that compel competition, placing 'success' in seminar groups, essays and exams above the feelings and unique sensitivities of each individual student.

We have all experienced teaching and learning situations where discussions have become evocative forms of learning, and personal experiences were offered to teachers and fellow-students in order to facilitate a clearer understanding. There is no threat in that kind of learning experience and the absence of fear will often free the atmosphere for deep and lasting sharing and communication. "This all suggests that he who wants someone else to grow — that is to discover his potential and capacities,

to experience that he has something to live and work for —
should first of all be able to recognise that person's gifts and
be willing to receive them. For man only becomes fully man
when he is received and accepted. In this way, many students
could be better students than they actually are if there were
someone who could make them recognise their capacities and
could accept these as a real gift to him. Students grow during
those moments in which they discover they have offered something new to their teachers without making them feel threatened
but, rather, thankful. And teachers could be much better
teachers if students were willing to draw the best out of them
and show their acceptance by thankfulness and creative work.
Too many people cling to their own talents and leave them
untouched because they are afraid there is nobody who is really
interested. They then regress into their own fantasies and suffer
from a growing loss of self-esteem".[18]

b) *Bilateral*

When we say that the second characteristic of a redemptive
model of the teaching relationship is that it is bilateral, more
than a mere sharing of words is envisaged. It means that the
teacher is also cast in the role of the student, searching with
the others for what is true and valid, giving to each other the
chance to play each other's roles. It is not so much, then, the
intellectual superiority of the teacher that counts as it is his
maturity, in facing the unknown, discovering and setting aside
his defences, and a willingness to leave unanswerable questions
unanswered. There are many pressures such as course content
and course requirements with limited time to cover all, which
make suggestions such as this appear quite challenging and even
threatening. And so they are generally and popularly resisted
or even ridiculed. But the greatest pressure and the most
powerful resistance comes from within. Because it is very difficult
to truly share. "The final and most powerful resistance against
learning, however, is much deeper and much more profound. It
is the resistance against a conversion that calls for a 'kenotic'
self-encounter. We will only be able to be creatively receptive
and break through the imprisoning strings of academic con-
formity when we can squarely face our fundamental human

condition and fully experience it as the foundation of all learning in which both students and teachers are involved. It is the experience that teacher and student are both sharing the same reality — that is, they are both naked, powerless, destined to die, and, in the final analysis, totally alone and unable to save each other or anyone else. It is the embarrassing discovery of solidarity in weakness and of a desperate need to be liberated from slavery. It is the confession that they both live in a world filled with unrealities and that they allow themselves to be driven by the most trivial desires and the most distasteful ambitions.

Only if students and teachers are willing to face this painful reality can they free themselves for real learning. For only in the depths of his loneliness, when he has nothing to lose any more and does not cling any longer to life as to an inalienable property, can a man become sensitive to what really is happening in his world and be able to approach it without fear".[19]

c) *Actualising*

The plea here is for a manner of teaching which will endeavour to link the present education of the student with his very person in all its variety and capacities. It is to anchor 'purely academic' preparation and training for the future, with the feeling and spiritual side of each student's life now. Many would say that there can never be an area of study which stands 'in itself'; that the days of pure, rarefied academia are virtually over. Courses should be designed, it is maintained, and taught in a way that encourages the practical application of the principles of those disciplines in the actual lives of the students. Only in this way can they grow towards meaning and wholeness in terms of both intellect and heart. Since the actual quantity of knowledge may mean very little, the main aim of all Christian education will be that the truly educated student can apply and live out the little or much that he knows. A case in point would be the manner in which theology, for instance, is taught to a pupil or student who intends to become a teacher. The presentation would differ, many will argue, from the manner of presenting theology to someone who wishes to have a mastery of theology for reasons other than that of communicating it. As an example, let us take the preparation of a religious studies student following the B.Ed. course in a college of further or higher education. The suggestion would not mean that the theological content of such a degree would be less 'academic' or trivialised in any way, but that the mode of teaching that content would actually communicate to the student preparing to teach, the methods by which he or she should themselves communicate content to his or her pupils. To imagine that the course could be structured in terms of a 'content' component and a separate 'method' component (pedagogical) is to drive the very wedge into the teaching of R.E. which we are attempting to remove, and which seems to present so many difficulties for teachers of R.E.

Henri Nouwen makes the same point from a slightly different perspective. When schools are places where community can be experienced, where people can live together without fear of each other, and learning can be based on a creative exchange of experiences and ideas, then there is a chance that those who come from them will have an increasing desire to bring about in the world what they experienced during their years of formation. In this sense, schools are not training camps to prepare people to enter into a violent society but places where redemptive forms of society can be experimented with and offered to the modern world as alternative styles of life. Teaching can then become a way of creating a new life-style in which people are able to relate to each other in a basically non-violent way. And the teacher himself, in trying to live this way, will discover that learning itself is a way of life that goes far beyond the classroom situation, that it creates new relationships that do not finish when students leave, that it is a process that asks for continuation and is not limited by grades and degrees, and that it is a challenge to renewal of one's own style of life.[20]

Postscript

"Jesus can be called Teacher in the fullest sense of the word precisely because he did not cling to his prerogatives but became one of the many who have to learn. His life makes it clear to us that we do not need weapons, that we do not need to hide ourselves or play competitive games with each other. Only he who is not afraid to show his weaknesses and who allows himself to be touched by the tender hand of the Teacher will be able to be a real student. For if education is meant to challenge the world, it is Christ himself who challenges teachers as well as students to give up their defences and to become available for real growth. In order to come to this conversion we might be thrown from our horses and be blind for a while, but in the end we will be brought to an entirely new insight, which might well bring about a new man in a new world".[21]

References

1. S. Verney, *Peoples and Cities* (Collins 1969) p. 36.
2. A. Richardson (ed.), *A Dictionary of Christian Theology* (SCM 1969) p. 216.
3. Mgr. Dupanloup, *The Child: A Work on Education* (McGlashan and Gill 1873) p. 264.
4. Cf. E. Robinson, *The Original Vision* (RERU 1977); K. Nichols, *Orientations* (St Paul Publications 1979).
5. Cf. P. Hirst, *Knowledge and the Curriculum* (Routledge and Kegan Paul 1966); P. Phoenix, *Realms of Meaning* (McGraw-Hill 1964).
6. G. Moran, *Theology of Revelation* (Herder & Herder 1966) pp. 150, 157.
7. J. Maritain, *Human Knowledge and Metaphysics* (Yale University Press 1940) p. 91.
8. D. J. O'Leary and T. Sallnow, *Love and Meaning in Religious Education* (OUP 1982) p. 72.
9. K. Nichols (ed.), *Voice of the Hidden Waterfall* (St Paul Publications 1980) p. 117.
10. H. McCabe, art. in *The Meaning of the Church*, ed. D. Flanagan (Gill & Son 1966).
11. W. P. Vanstone, *Love's Endeavour, Love's Expense* (DLT 1978) cf. ch. 4.
12. A. Reuter, art. on *Ideology, Criticism and Trust*, in Religious Education, Vol. 76, no. 1, p. 82).
13. H. Nouwen, *Creative Ministry* (Doubleday & Co. 1971) p. 6.
14. ibid., pp. 8, 10.
15. V. Madge, *Children in Search of Meaning* (SCM 1965) p. 102.
16. ibid., p. 102.
17. H. Nouwen, op. cit., p. 8.
18. ibid., p. 17.
19. ibid., pp. 18, 19.
20. ibid., pp. 13, 14.
21. ibid., p. 20.

CHAPTER 3
PSYCHOLOGICAL ISSUES

CONTEMPORARY THEORIES
OF PSYCHOLOGICAL GROWTH

The theories to be discussed in this chapter are all concerned with human development over a life-time. The usefulness of these theories for the teacher is not restricted to what is said about the earlier stages as such, because a perspective which takes account of the whole life-span may well affect our vision of the parts within the whole. Furthermore, a conception of human development which carries on well past school-leaving age enables us to hope that what has not been attained by sixteen or eighteen years may yet be accomplished, especially if the foundations are secure.

The first part of the chapter mentions several theories of personality and its development. Although each theory has distinctive features, there is a common approach to human beings and one which is compatible with the Christian view of man, for each has regard for the uniqueness of the individual and for the importance of studying his experience in its totality. In the second part of the chapter there is a description of the development of moral reasoning as observed in the research of Lawrence Kohlberg, and the third part is devoted to the very recent work of James Fowler on the stages in development of faith. All of this work merits the consideration of those involved in religious education, because the effectiveness of teaching or of pastoral care depends very much upon the match between what is offered and what can be utilised and understood.

The development of personality

In contemporary psychology, there are three types of theory of personality and its development. *Freudian theory* assumes that man can never attain fulfilment because he is involved in the constant conflict between inner biological pressures and the external pressures of society. By seeing man as the victim of circumstance, Freudian theory diminishes his freedom and responsibility. *Behaviourist theory* also has a mechanical view of man in its assumption that personality is the outcome of external stimuli. All behaviour, it is said, results from positive or negative reinforcements, from rewards or punishments. No attention is paid to anything which may go on 'inside' a person, and concern is concentrated upon the external, the measurable and the modifiable. These vignettes of two important theoretical positions scarcely do justice to their complexity, or to their usefulness. However, for religious education, *humanistic theories* have more to offer, because they attend to those characteristics, such as the search for meaning and the capacity for choice, which distinguish men from animals.

The personalist theory of *Gordon Allport* is based upon the analysis of a number of personal documents: diaries, letters, autobiographies, interviews or other reports of an individual and largely spontaneous character. Allport is concerned with the normal development of a mature person. Such a person is able to coordinate and direct all the elements of temperament, physique and intelligence which make up a human being. Maturity is obviously a relative term and it grows gradually over a life-time. The characterising features of Allport's mature person resemble those of a mature religious person. For Allport, such a person has a critical and reflective attitude with differentiation among the range of experiences. Religion is an active power in his life and a consistent set of moral values develops. The mature religious person is concerned about the most important issues of human existence and all the main aspects of life are integrated into religious beliefs. Yet rigidity is avoided and the mature person is always ready to move on to a more adequate expression of faith.

Abraham Maslow, like Allport, collects his data from the observation of ordinary people, or rather, from those exceptional people who, in Maslow's judgement, had fully realised their potential as human beings. He calls this 'self-actualisation' and the seventeen headings in this list of characterising features itemise what we should expect in a highly competent and successful individual. Such a person, for example, accepts reality, is self-reliant, enjoys life, establishes sound relationships, is egalitarian, has a proper sense of values, can resist the pressures of society, and is well-integrated. Maslow is an optimist who implies that self-actualisation can proceed almost automatically providing that there are no major hindrances: self-actualisation is merely the unfolding of human potential. His 'theory' of personality is a description of possible goals in life rather than a fully worked out account of human development.

Carl Rogers' person-centred theory is derived from his observations as a psychotherapist so that it has a firmer foundation in research than that of Maslow, though both theories are alike in a number of ways. Rogers claims that it is a part of man's genetic make-up to attempt to develop all his capacities to the full in a process of self-actualisation like that described by Maslow. From his clinical experience, Rogers perceived that failure in self-actualisation arises from the sacrifice of personal evaluation to win the approval of significant others. In the conditions of psychotherapy, the client is encouraged to accept his own experiences, to trust in his personal evaluation and so to become responsible. In applying his theory to teaching and learning, Rogers takes the teacher's role to be one of 'facilitating' self-directed learning in both the cognitive and emotional domains by offering the resources necessary to the student's personal development. Rogers claims that as a person comes to accept what he actually is, he also becomes ready to accept others as they are.

There are other theories of personality which, although they incorporate much of what Allport, Maslow and Rogers say, attach greater importance to the human characteristic of searching for meaning. These theories are grounded upon existential philosophy which considers man as he is 'emerging' or 'becoming' in the conditions of his world. The strenuous activity of decision making, that is, of finding meaning in all that confronts us, is the way in which we construct our personality.

This decision making accepts the unalterable features of our world but strives always for what is still a possibility for us. The chief limitation on our freedom is the certainty of death, but throughout life also there are the lesser deaths of disappointment and frustration. Rooted in the nature of man — and not thereby faulty upbringing — are the emotions of guilt and anxiety. Guilt arises from failure to change or to move away from our past so as to create new meaning for ourselves. The experience of guilt has the function of prompting change, but this inevitably leads to anxiety, for it cannot be known ahead whether it will be possible for us to survive in the future we choose. This is where, in Tillich's phrase, 'the courage to be' is needed, with faith in oneself as being able to manage in the unknown future, and also with the conviction that taking such a risk is the only way to find meaning and to exercise our human capacity for freedom.

For the existential psychologist (and philosopher) the goal is to become an 'authentic' person. Psychotherapy has been the background of much of the exposition of theory and we can understand how the contrast between authentic and inauthentic has been noticed in the treatment of clients. The description of the authentic person resembles Maslow's and Rogers' self-actualised person, or Allport's mature person. There is distinct individuality, defined tastes and preferences, reflectiveness, and willingness to change. The authentic person is active in bringing about change in social institutions, is willing to enter into close relations with others and to respect their individuality. Such characteristics give the authentic person a greater measure of freedom and control over his existence. At each point the in-authentic person is at the opposite extreme — vague, with stereotyped or conventional preferences, less conscious or aware of his existence and with no clear impressions about the meaning of his life. He is passive and acquiescent, superficial in human contacts, 'playing' social roles but incapable of entering deeply into them, and tending to go on in a routine manner, missing opportunities for growth or for richness of experience. Such a life fails to realise the potential for being human, but for existentialists this true fulfilment demands discipline and involves stress.

Among such psychologists, *Rollo May*, for example, emphasises the place of will and decision, and the necessity to become aware that there is always the possibility of some personal response towards our circumstances even when we appear to be their victim. It is through meeting whatever conflicts are presented to us that we achieve our individuality. A similar approach is to be found in *Viktor Frankl* and his technique which is known as 'logotherapy'. In this people are encouraged to search out whatever has meaning for them in a seemingly indifferent and meaningless world, even if nothing else is possible for them but consciousness of what it is they experience. His work is influenced by the experiences of the concentration camp and the extremes of deprivation and limitation within which some people still managed to *be* human.

The description of the authentic person refers primarily to the adult, but there are obvious ways in which parents and teachers can help or hinder the development of personality. Existential psychologists attribute decision making ability to the three characteristics of symbolisation, imagination and judgement, and these can be encouraged at home and at school. Symbolisation is the faculty of abstracting categories or ideas from concrete experience, and, in imagination, these categories or ideas are combined in new ways to effect changes in thinking. Judgement involves the assessment of experience and hence decision about values and preferences. Exercises in these three areas foster self-reliance and independence in decision making but it is also important that clear limits should be set, for it is not true of life that there are no limitations. Meeting failures and frustrations with the courage to try again is the way to achieve individuality. Failures, at any time of life, can bring the information necessary to re-evaluate goals, to reformulate plans and to try again. Creativity can come from suffering, conflict and frustration, and this work of constructing the self continues, or can continue, over the experiences of a life-time.

Kohlberg: moral reasoning

Lawrence Kohlberg's research into the development of moral reasoning is impressive because he has made a rare longitudinal study. Over a period of at least twenty-five years, the original participants in his research have been re-interviewed, at three-yearly intervals, with the same test material. The responses have been preserved and after detailed analysis a manual has been produced which defines six stages of development and offers an interview and scoring technique for those who wish to carry out their own researches. Underlying Kohlberg's account of moral reasoning development, therefore, there is the evidence of changes recorded as children grew into adolescence, took jobs, married and became parents. There have also been numerous replications of Kohlberg's work using subjects from other cultures, and examining the effects of variables such as sex or social class.

Kohlberg's attention has been directed to the reasons people give when they talk about moral issues, and this reasoning is a function of their intellectual development. His theoretical basis is, therefore, cognitive-developmental. Accepting the description of the structures of logical thought which Piaget has elaborated over a lifetime's investigation, Kohlberg has shown that reasoning about moral issues follows a similar but distinct path. It is not simply that egocentric thinking about the physical world is matched by egocentric thinking about moral issues, or that concrete thinking about the physical world is matched by concrete thinking about moral issues, or that abstract thinking in the one sphere matches abstract thinking in the other. It has been shown that intellectual development may advance *without* corresponding development in the sphere of moral reasoning. That is, although moral reasoning of a given kind can occur only when that same level of intellectual development has been reached, the converse is not necessarily the case. This means that moral reasoning may lag behind intellectual development, and, in some instances, may never catch up, so that someone may have the capacity to reason in the abstract, formal or hypothetical manner with respect to the physical world without being able to do so about moral issues.

It has also been reliably shown that levels of moral reasoning are dependent upon the development of social perspective. Here the work of Robert Selman has supplemented that of Kohlberg. Selman's account of the development of social perspective, i.e. the developing capacity to take account of other people's thoughts, feelings and needs, is also in the cognitive-developmental tradition. He has found that certain levels of intellectual development are needed before corresponding levels of social perspective-taking are to be found, but the presence of intellectual development of a given kind does not inevitably mean that this will be used in the social sphere; again there is a time lag, and again the level in social development corresponding to that in intellectual development may not be attained.

What we have, therefore, is an account of moral reasoning

which at any level demands the prior attainment of a corresponding-ing level of social perspective which, in its turn, depends upon a specified level of cognitive development. However, we are not to assume that anyone with adequate intellectual development will necessarily apply that quality of reasoning to moral issues. In many ways the realisation that the development of moral reasoning is a distinct dimension of personal growth is as important as detailed knowledge of the stages through which it normally passes. There is certainly a relation between development in these three spheres (intellectual, social, moral) and there is also in each case a relation with age, but whereas among younger children there may not be significant differences in the pace of development in each dimension, by adolescence and among adults there may be marked discrepancies. With some, moral reasoning will follow on social and intellectual development quite smoothly, whereas in other cases, even among the intellectually gifted, moral reasoning levels may not have moved much beyond the most rudimentary. There will, therefore, probably be wide differences in moral reasoning ability among the members of a sixth form even though their intellectual attainments are similar.

A feature of Kohlberg's work, and one which newcomers to it sometimes find confusing, is that he describes kinds of arguments, and the forms of expression used in discussing moral issues; he is not, except at the very highest level, concerned with content. His interview procedure helps to explain this. He has selected a number of stories which present a moral dilemma. In any one of these stories there is a conflict between two moral considerations, such as the value of life set against respect for property, or obedience set against trust, or theft set against fraud. Probe questions elicit comment upon many aspects of these issues, and, in the evaluation of the type of reasoning employed, what counts is not a decision about right or wrong but the justification for it. For example, is the reason primarily to do with personal benefit, or the good of society, or the upholding of some principle?

Because he is concerned with the way in which people reason about moral issues, Kohlberg is not specifically investigating moral behaviour. There cannot be any reliable empirical test to determine a relation between what people say and what people do, especially what people do under pressure. However, the theory cannot be dismissed as having no practical significance. While it is certainly true that moral development and a moral life are much more complex than reasoning about moral dilemmas, the practice of reflecting upon issues is highly important in the formation of a morally responsible adult. In civilised society, conventional attitudes carry us through many of our perplexities, but society changes rapidly. New problems confront us which our parents and grandparents could never have envisaged, and the increased mobility of life and, very often, the experience of other life styles put us in the position of having to decide our moral stance on rational grounds which can be defended.

The six stages of development which Kohlberg has proposed are at three levels; the two stages at each level have a common structure though they display progression within this structure. At successive levels, the considerations underlying a moral decision widen. The first level is characterised by concern for what is best for oneself (in avoiding punishment, at stage 1, or in winning favour, at stage 2); the second level is characterised by concern for maintaining social structures (whether by looking for the approval of individuals in one's own circle, at stage 3, or by upholding the rules and values of society at large, at stage 4). This is known as the 'conventional' level and decisions take account of what others consider fair and just. The third level Kohlberg has called 'autonomous' or 'principled' and it is characterised by a willingness to be pioneering in moral sensitivity. Frequently there is a widening of the range of application of the values society already holds, e.g. a respect for human rights which is to be extended to those of a different race or nation. This is perhaps another way of saying that 'society' comes to mean all mankind. At stage 5 there is an attempt to change others' opinions so that the majority agree to alter laws in favour of the new understanding of justice, and social cohesion is maintained. Sometimes, however, an individual, on grounds of conscience, believes he has to stand out against the morality he thinks is limited, at the cost of personal ignomiy and perhaps of death. Examples of this stage 6 level of moral reasoning are rare, but, over the centuries, these have usually been the people who have had a profound and civilising influence on humanity.

If we assume that the development of moral reasoning is as Kohlberg has described, those concerned with religious education can utilise his work in several ways. The use of dilemmas to encourage reasoning about moral issues normally arouses interest and the activity of weighing the various arguments against one another impresses the notion that moral judgements are complex. Christian responsibility is exercised in careful reflection upon the courses of action which are open to us and in making decisions according to our understanding of the fundamental principles of our faith. Practice in thinking round the issues and in considering a range of viewpoints (whether they are those of the characters in a narrative or the others in the group) is likely to be helpful when a new or unexpected situation arises. At the very least, such discussion will make it easier for someone to give considered reasons for what is done and will promote a sense of responsibility about choices.

The role of the teacher in encouraging the discussion of moral issues is not, of course, to attempt to push the pupils on too fast, but rather to be aware of the kind of reasoning which is being employed and to introduce arguments at the next higher stage. Such intervention will help to effect transition to the next stage and this is particularly desirable where it is evident that intellectual and social development are more advanced than the type of reasoning used in the consideration of moral questions. In relating moral questions to a Christian view of life, it is important to encourage a level of thought which matches development in other dimensions. For otherwise if an immature understanding of God persists, it is possible that religion itself will be cast aside when someone perceives the mismatch between his notion of God and his understanding of the world and of his relations with other people. At each of Kohlberg's three levels there is an underlying 'authority' for moral choice, and this can be expressed in religious terms. At level one, self-interest matters, and God is understood as one who punishes wrong doing (stage 1) or as one who will give us good things if we obey his wishes (stage 2). At level two, the majority view matters, and God is understood as guiding moral decision through the institution of the Church, which may serve as the group of people among whom we may find approval and support (stage 3) and whose rules we try to uphold and maintain (stage

4). At level three, where the autonomous conscience matters, God is understood as inviting his pilgrim Church to move on to new ways (stage 5) and each Christian to take a personal responsibility for his moral decisions (stage 6).

Supposing a teacher has given a good deal of attention to the discussion of moral issues and has been skilful in providing arguments at a stage just ahead of what he hears his pupils using, how far can he expect them to have moved along Kohlberg's sequence? Probably school leavers will not have gone beyond stages 3 and 4, according to accounts of programmes which have been very carefully designed to effect progression, and, if this kind of moral reasoning has been consolidated, the school will have done well. Perhaps the most important contribution a teacher can make is to leave the way open for further development, and for a deeper understanding of moral judgements within a Christian perspective of life.

Fowler: stages of faith

Fowler has built upon the findings of Kohlberg and adopted the same theoretical model, but, in the elaboration of his account of faith stages, he has also made use of the work of other psychologists, such as Erikson, in the Freudian tradition. Fowler's definition of 'faith' has also drawn upon the work of several theologians and philosophers. We are, therefore, faced with a variety of positions which have to be evaluated separately as well as an overall account to be considered as a theory of faith development. Nevertheless, while proceeding with caution, we can acknowledge the value of an attempt to describe development of this complexity and including so many facets of human nature.

Fowler's conception of the term 'faith' is not restricted to what we commonly think of as 'religious' faith, that is, of faith expressed in doctrines, rituals and institutions. For him, faith is an apprehension of the ultimate environment, it is an active mode-of-being-in-relation to it, and it is the primary motivating power in life. This is a very brief summary of the descriptions of faith Fowler has expounded in his writings, in which he draws upon the insights of theologians and philosophers who have influenced him, particularly H. Richard Niebuhr. In religious language, 'ultimate environment' would mean the kingdom of God, or Le Divin Milieu; in ordinary terms it means the central values which command our loyalty. The dynamic quality of faith emphasises its function as permeating the chief activities of our life; it signifies our heart's commitment and the apprehension of that centrality of value in our lives involves our powers of knowing, constructing and composing. Fowler's own writings must be studied before judgement is passed upon the adequacy of his concept of faith. At least it has the merit of refusing to use the term without an attempt at its definition, and also the merit of distinguishing the experience of faith from what may be called its 'content'.

Fowler's theoretical position is 'structural-developmental'. His research has led him to postulate a sequence of six stages of faith. These occur always in the same order and they have internally consistent characteristics of thought and expression, distinct at each stage, and with the later ones built upon and incorporating features of the earlier. In this he follows Piaget's and Kohlberg's conception of 'stage', and accepts that intellectual development is one of the key factors in faith development. Alongside this cognitive-developmental approach is the admitted influence of Erikson's scheme of psychosocial stages. This work is grounded in Freud and a particular feature of it is the account of 'predictable crises' over a life-cycle. These are the challenges of life which everyone must face. Here are to be found the tensions and conflicts the resolution of which will effect transition to the next stage: examples of such crises are the adolescent's shift to the adult responsibility of earning a living, or the adult's acceptance of the demands of parenthood, or the older person's recognition of ageing and the prospect of death. As with all cognitive-developmental theories there is no guarantee that every individual will achieve the transitions in moral or faith development compatible with his intellectual capacity, even though he cannot hold off the crises of his human progression through the years.

The evidence for the detailed description of the faith stages has been collected from a number of interviews (359) conducted across the full age-range (4-80 years). The interview has a series of questions grouped under four headings. First there are questions to give a general picture, a 'life-review', and these include factual details as well as information about the way the individual sees the pattern or the turning-points of his life. In the second part, there are questions about relationships, particularly within the family, and with those figures of the past who have shaped values and interpretations, though, of course, these are all described as they are seen in the present. The third part is concerned with current values and commitments, and takes up questions about the purpose and meaning of life: personal, for society, and for mankind as a whole. The fourth part has questions using religious terminology and answers are used from both the avowedly religious and also the non-religious.

The data has been analysed (though not as yet completely) under seven aspects. These may be listed here to give some indication of the range of features which make up the description of each stage.

(A) The form of logic: in Piagetian terms of pre-operational, concrete or formal operational.

(B) Perspective taking: in Selman's scheme of modes of taking account of others' thoughts, feelings and needs.

(C) Form of moral judgement: in Kohlberg's scheme of stages.

(D) Bounds of social awareness: the extent of inclusiveness of groups to which loyalty is felt.

(E) Locus of authority: persons, ideas, institutions, individual judgement as the ground of validation or legitimation.

(F) Form of world coherence: how and to what extent the meaning of separate events and experiences are related.

(G) Symbolic function: the use and understanding of symbols as representations of ultimate meaning.

It is evident from this list of aspects that a full account of the characteristics of each of Fowler's stages of faith would take up more space than is available in this chapter. He gives labels to each of the six stages but these are not particularly helpful without a fuller description of what he means by them. Details of his projected stages are to be found in the accounts listed at the end of this chapter.

Fowler's interest is not merely with the theory; he has an active concern for its application to the work of the ministry and to religious education. Since some of the most significant developments of faith occur in adult life, he is eager to see appropriate adult opportunities for growth made available, and, while acknowledging that someone may be a Christian all his life, he introduces the imperative that the *way* of being a Christian should deepen as the years go by. In the course of interviewing, he discovered that adults appreciated the chance to talk of such matters, and that the effort of articulation not only helped them to clarify and consolidate their position but even, at times, to make a transition to a further stage. He is careful to say that there is no requirement to push anyone towards a higher stage in a way that sets an evaluation on the stages, for each has a value of its own at the appropriate point in a lifetime. His account of the stages shows the importance of the language, rituals and symbols which express a religious tradition and which need to be made available for the expression of faith. There must also be an attitude to language, ritual and symbol which allows them to be the vehicles of deepening meaning over the years. The scope of Fowler's conception of faith prevents any artificial separation of faith from everyday life. He notes that growth is not attained without pain and disruption, confusion and a sense of loss, but, nevertheless, when there is a conviction of some further position to move into, where faith can be reconstructed, the sense of anxiety is reduced.

Fowler's theory assumes that there is an in-built desire in human beings always to move on to a more adequate position, or apprehension of our being-in-the-world. Intellectual endowment, temperament, events and circumstances combine to affect the way we develop from stage to stage. The theory of progression through ordered stages is valuable in helping us to recognise where another person stands and hence to assess what may be most appropriate for his immediate needs, but there is always a danger of over-ready categorisation and of forgetting the inconsistencies or of regressions under pressure. Furthermore, we need to remember that Fowler and his colleagues have not as yet lived out their own life-span and it is an accepted feature of cognitive-developmental theory that there is only partial understanding of the stage ahead of one's own, and almost nothing of what may lie beyond that again. As he says, stage 6 is exceedingly rare and in his sample of 359 he found one instance in the 60-plus age group. Stage 5 appeared from the age of about thirty years but was found in only 7 per cent of his total sample. Even those not statistically minded can recognise that we are only at the beginning of the collection and analysis of data. Much further investigation is required but this work is likely to be well worth our future attention.

Conclusion

Developmental theories are always unsatisfactory in one all-important respect. Maybe we do not want to push people ahead faster than their natural pace, yet we cannot avoid curiosity as to *how* transition from one stage to another comes about. This is a little like watching to see the plant actually growing. The theories described all assume that there is an inbuilt, i.e. genetic, tendency towards development, but observation has also shown that sometimes there are hindrances to a smooth transition, or, according to other theorists, that succumbing to a hindrance is in some sense a lack of courage or a failure of will. Kohlberg and Fowler also suggest that transition to a further stage of development is prompted when the even tenor of life is somehow disrupted, not necessarily by disaster, but often by some event such as a change of job, or a move to another town, or by new responsibilities.

A final remark about developmental theories concerns the limitation of language which suggests an evaluation of the stages. It is common to speak of stages as higher or lower, or of people as mature or immature, and the words sometimes take on an evaluative connotation not intended by the psychologists. What is important is that each successive stage is appreciated as it appears in the history of any one person, and that the teacher should recognise what is appropriate to meet the needs of that stage, and to prepare for what is to follow. The educator has the delicate task of allowing each stage to be fully expressed while not hindering anyone from shifting to the next at the proper moment for him. There is an open-endedness about developmental theories which reaffirms our belief in life as pilgrimage.

Reading list

Allport, G. *The individual and his religion* (Macmillan, N.Y. 1950).

Fowler, J. & Vergote, A. (eds.) *Toward moral and religious maturity* (Silver Burdett Co, Morristown N.J. 1980). This is a useful collection of articles, including ones by J. Fowler and L. Kohlberg.

Berryman, J. *Life maps* (Word Books, Waco Tx. 1978). J. Fowler contributes a main chapter which describes his work in detail.

Duska, R. & Whelan, M. *Moral development*: a guide to Piaget and Kohlberg (Gill & Macmillan, Dublin 1977). This includes some of the dilemmas used in Kohlberg's interview, but without the probe questions. It specifically considers moral development from a Christian perspective.

Duska, R. & Whelan, M. *The nature of prejudice* (Addison-Wesley, Reading, Mass. 1954).

Erikson, E. *Childhood and society* (Norton, N.Y. 1950).

Erikson, E. *Young man Luther* (Norton, N.Y. 1958).

Erikson, E. *Identity: youth and crisis* (Norton, N.Y. 1968).

Erikson, E. *Gandhi's truth* (Norton, N.Y. 1969).

Frankl, V. *Man's search for meaning*: an introduction to logotherapy (Hodder, London 1964).

Frankl, V. *The will to meaning*: foundations and applications of logotherapy (Souvenir Press, London 1971).

Frankl, V. *The doctor and the soul from psychotherapy to logotherapy* (Penguin 1973).

Maslow, A. *Toward a psychology of being* (Van Nostrand, Princeton 1968).

Maslow, A. *Motivation and personality* (Harper, N.Y. 1970).

May, R. *Psychology and the human dilemma* (Van Nostrand, Princeton 1969).

May, R. *Existential psychology* (Random House, N.Y. 1969).

Rogers, C. *Client centered therapy* (Houghton Mifflin, Boston 1965).

Rogers, C. *Freedom to learn* (Merrill, Columbus Oh. 1969).

Rogers, C. *A way of being* (Houghton Mifflin, Boston 1980).

Selman, R. *The growth of interpersonal understanding* (Academic Press, N.Y. 1980).

Gruber, H. & Vonèche, J. *The essential Piaget*: an interpretative reference and guide (RKP, London 1982). This is an anthology covering the whole of Piaget's works.

EDUCATION FOR
PERSONAL RELATIONSHIPS
IN THE SIXTH FORM

There is no doubt that courses in personal relationships are popular with sixth formers. At one Catholic comprehensive school a relationships course was offered recently as one option among seven, and it turned out to be the first choice of over half of the sixth formers.

Education for personal relationships is not, of course, synonymous with religious education but there are many reasons why the R.E. department should be concerned with relationships. Any worthwhile religious education is rooted in love, love of God and love of one's neighbour, rooted therefore in relationship. It is then appropriate that any education for relationships within a school should draw on the R.E. department as one of its resources. Furthermore, if it is intended to run a course in personal relationships available to all sixth formers and making use of the expertise of several different departments within a school, the contributions from each department will need co-ordinating. The R.E. department is often well placed to do this work since it already offers a general course of study unrelated to examinations and available to every sixth former.

It is worthwhile to look a little more closely at the task facing an R.E. department responsible for offering a sixth form course in relationships in order to clarify the aim and objectives of such a course and to consider what steps can be taken to ensure that the course is interesting, valuable and competently taught. This has indeed been attempted at some length in the booklet *Sixth Sense* and what follows is a summary of the proposals made in that booklet, which is published by the service for teachers of the Catholic Marriage Advisory Council.

It would be possible to have the cognitive aim of teaching

about relationships. However, sixth formers are likely to see their own needs in a more practical way and to be more concerned with understanding the feelings and behaviour of themselves and of others and with learning to cope with their relationships better. It is, therefore, suggested that the aim of a sixth form relationships course should be to increase the sixth formers' ability to relate in a satisfying way.

Three objectives will help sixth formers to achieve this aim: experiencing good relationships during the course, learning the skills of self-development and of relating, and exploring many different lifestyles so as to adopt a lifestyle and behaviour which reflects their own values and convictions.

There are two further objectives which help to make explicit for sixth formers the Christian dimension of relationship: recognising the Church's potential for understanding and celebrating the milestones of life and for helping people to love; and developing sixth formers' sense of the mystery at the heart of creation and seeing here the presence of God. It is useful to take each of these objectives in turn and to show how an R.E. department might organise itself to achieve them.

Unless sixth formers *experience good relationships* during their relationship course it is probable that they will derive from it no lasting benefit. The best chance of ensuring this experience is when sixth formers meet regularly in fairly small groups, and it is worth making a great deal of effort to provide them with this opportunity. Sometimes a large teaching group is suitable, as is obviously the case when a film is shown or a visiting lecturer is invited, but if the sixth formers are to participate honestly and with any degree of self-disclosure small groups are essential.

Leading a small discussion group is a skilled task. The teacher who does this well allows plenty of time for the group members to get to know each other, helps them to tell the group about their attitudes, feelings and experiences, protects them from loss of face, shows interest and concern for them, helps them to confront those they disagree with without being unduly aggressive or submissive and directs the group gently towards the intended learning. This is quite a challenge for teachers who are often untrained in small group leadership, and there is much scope for in-service training and for teachers' workshops in these skills.

104

A second objective for a relationships course is for sixth formers *to learn relationships skills.* Relating well requires goodwill, unselfishness, generosity and a sense of responsibility: these come from within the individual and cannot easily be taught. But relating well also requires skill. Many relationship skills have been analysed; there are exercises which practise them and it is perfectly possible to teach them. Thus sixth formers benefit from exercises which aim at self-development and self-awareness, at building up their sense of their own worth and at clarifying their own deeply-held values. Other valuable exercises are those which help sixth formers to admit to their feelings, opinions or experiences and not to be fearful of sharing these with others when appropriate. Some exercises show pupils how to put themselves in another person's shoes while others help them to understand the extent of their dependence on other people and to practise their ability to co-operate. Further exercises help to develop ability to listen well or show the extent of a person's prejudice or aggressiveness. There are exercises which challenge unthinking stereotyping of male and female characteristics, those which can show sixth formers how to challenge and confront people with whom they disagree without being unduly aggressive or submissive, those which teach the best way of conveying understanding, of avoiding vagueness and abstraction, and of asking clearly for what one wants.

All of these exercises teach the skills of relating in a direct and practical way and can be enormously valuable to sixth formers in helping them to improve their ability to relate well. Such exercises are most effective if teachers themselves have tried them and appreciate their value from personal experience, and if teachers are willing to adapt those exercises designed for the American cultural and educational setting to their own very different circumstances and pupils. Relationships exercises work best when approached with a sense of discovery, a sense of adventure and a sense of fun.

A third useful objective for a relationships course is for sixth formers to *explore many different lifestyles.* Through the mass media they hear of ways of living totally at variance with their own and those of their families, and even in times of unemployment they have possibilities of choice in their personal and sexual lives which are perhaps wider than any other genera-

H

tion has yet faced.

There are many lifestyle topics concerned with social relationships: there are those which show the need for honesty, and deal with such matters as stealing, lying, cheating and shoplifting; those which show the need for peace, and consider such matters as domestic violence, civil unrest, law and order, dangerous driving, war, nuclear weapons, and the limits of legitimate defence; those which point to the need for justice, and reflect on matters like inequality, unemployment, famine, overpopulation and racial and religious prejudice; those which have to do with moderation, and consider the use and abuse of alcohol, gambling, drugs and tobacco; those which point to the need for community and study urban society, loneliness, mental illness or handicap, old age, suicide and death. Other lifestyle topics have to do with sexual relationships: there are those concerned with the changing patterns of marriage, divorce, family size and the roles of men and women; there are those which consider sexual choices and reflect on such matters as boy/girl relationships, pre-marital sex, choosing a partner, preparing for marriage, marriage and commitment, the meaning of sexual intercourse, marital breakdown, sexually transmitted diseases, the single life, whether lay or religious, and homosexuality; and there are topics which are concerned with sexuality and children and consider such themes as conception control, abortion, the sexual education of children and one-parent families.

It is obvious that only a few lifestyle topics can be considered in any one sixth form course. Indeed, if there is to be real reflection at any depth it is essential to avoid rushing from one topic to another to satisfy some overall scheme. In this lifestyle study we ask sixth formers to examine their feelings, attitudes and values when confronted by differing ways of living. We ask them to learn a few facts, to discuss and compare their attitudes with others in their group, to try to sympathise with the suffering inherent in many life situations, to consider the guidance offered by religious and other authorities and finally to make a judgement about the principles which they hope will govern their behaviour.

These three objectives of experiencing good relationships, learning relationship skills and exploring different lifestyles will take sixth formers a long way towards the aim of increasing their ability to relate well. There are occasions where we can go further, where teachers and sixth formers explicitly share a common membership of the Christian community and it is appropriate to take as a further objective that sixth formers should be able *to recognise the Church's potential for understanding and celebrating the milestones of life and for helping people to love.*

This potential springs from the Church as a community, from the good news which that community preaches and from the liturgy and sacraments that mark and celebrate the milestones of life and give significance to daily experience.

Sixth formers often see the Church solely as an authority, a policeman, and are apt to resent the claims it makes on their obedience; it can be helpful for the teacher instead to present Christ's community as a resource drawing on two thousand years' experience of helping people to live well but still developing in response to the living situations that community members find themselves facing.

It is important for teachers not to assume a commitment to the Church that sixth formers do not have. If for the moment they are standing apart from the Christian community and see it as having nothing to offer them that they can use, teachers can help them to work out a morality to which they can for the time being give assent, to become conscious of the values to which they really do feel a sense of commitment. When the difficulty of doing this single-handed becomes apparent they may be more amenable to looking again at the teaching and inspiration to be found within the Christian community, and in that case the teacher again has the opportunity to offer it to them. Always the teacher has the opportunity of showing the individual pupil an acceptance, interest, concern and love which reflects the love of Christ for us all.

One final objective for a sixth form relationships course is *to recognise the mystery at the heart of relationships, and to sense here the presence of God.* As teachers we do our best to teach relationship skills clearly, to discuss sexuality plainly and to consider behaviour logically. Yet we must admit that much about relationships is deeply mysterious. Birth, death, uniting in sexual intercourse, begetting a child, love, sacrifice, commitment: these are the deep things of life. These people recognise

107

as having meaning and significance and reality. Here individuals look far into their own consciousness. Here they experience something profound. This is intensely personal, yet it is linked with the universal sense of some profound meaning and significance and reality which transcends time and space and is present behind the whole of creation.

Perhaps in our own time people are less conscious of this sense of the 'mystery deep down things'. It has no place in the scientific, rational mode of thought. Urban society, far removed from the rhythm of the seasons, is not attuned to mystery. It does not flourish in a world of man-made objects where people work for money and everything has its price. Yet the sense of mystery is not easy to suppress. It shines through the joy and pain of giving birth, the anguish of lovers parted, the happiness of lovers reunited. There is awe and wonder to be found in microphotographs of pre-natal life, in the union of ovum and sperm and the slow development of a new human being in the space capsule of the uterus. There is a profundity in the best human relationships that causes individuals to realise for themselves that love is stronger than death.

Perhaps teachers should not be too hasty to give this sense of mystery the name of God. The slow revelation to each individual of God incarnate in creation can be hampered when teachers attempt to confine God within their own definitions, but where sixth formers catch a glimpse of the mystery the teacher can recognise it and help them as they struggle to express it. We can put them in touch with that literature, poetry, music, art and drama which express more certainly what they are groping to articulate.

The person of faith can have no doubt that this sense of mystery does indeed point to God: what is important is to have confidence that God really does speak to each of our pupils in the depths of their being. If we ourselves are open to this voice we will also catch its echoes in our pupils' hearts.

Much planning remains for a religious education department before it is ready to offer an adequate relationships course for the sixth form. It is challenging to select staff from every relevant discipline within a school, to work out with them the best methods for this type of course and to offer them in-service training in those methods, and to secure adequate facilities and timetable space. Often the resulting course will not seem to have much to do with religion, concentrating as it does on acquiring relationships skills and studying different possible life-styles. Our critics will certainly be quick to ask, "Where's the religion in that?"

Yet relationship is central to our belief. At the heart of all things there are three Persons in relationship. Every loving thought, word and action that passes between two human beings bears the likeness of that love which shines out from the centre of all things. From this belief comes the conviction that helping sixth formers to relate as well as they can is one of the most valuable services teachers can offer to prepare them for life here and hereafter. Relating is God's work. It reflects God's life. That's where the religion is in education for personal relation-ships.

CHAPTER 4

METHODOLOGICAL ISSUES

THE STATUS AND STRUCTURE
OF SIXTH FORM R.E.

Introduction

In what follows we have made the assumption that we are considering a general sixth form R.E. course — one in which all members of the sixth form participate — and that those pupils who wish to follow an examination course in R.E. will be catered for at some other time in the week. We have also assumed that we are talking about a Church school because that is where our experience lies.

We are very aware that over the past twenty years the teaching of R.E. has become more and more fraught with difficulty and the teaching of R.E. in the sixth form perhaps doubly so. We know well that there are times when the going is difficult but we have also had enough positive reaction — often coming in the most indirect way — to make us more optimistic than pessimistic.

The status of the subject

The status of all religious education in the school and especially sixth form R.E. will reflect the attitude of the head, the senior staff and the governors to the subject. The status of the head of department (e.g. scale post) will also be significant to the general esteem given to religious education.

The head with the staff will determine the nature of the syllabus and must ensure that what is taught and the way it is presented stretches all the students intellectually, spiritually and is related to their life style and future aspirations. Opportunity must be offered to the student to make a response not only within the classroom but also in action within the parish, the community and, most important, in prayer and liturgy. (The essential question concerning the role and interaction of the school regarding the wider community and the implications of this for R.E. is discussed in ch. 1.)

The effectiveness of the RE. programme in the sixth form will depend upon the status given to this essential aspect of the curriculum by managerial function of the school and it is in the structuring of the whole curriculum of the sixth form that the senior staff can either facilitate the work of staff and students in the classroom or can undermine the efforts being made by the head of department and the teachers and students.

There must be an adequate time allocation to this work and positive efforts must be made to place the R.E. sessions in a favourable place in the day or week. Early in the morning and in the earlier part of the week make for a more receptive response on the part of staff and pupils. Last period on Friday afternoons could render abortive the most imaginative of teaching. This then will need to be a positive criteria for the timetabler early in the composition of the timetable — if it is left until the end, the session will only be squeezed into the spaces left.

Adequate staffing must be made available both in quality and quantity. Much of the work will involve discussion and questioning of belief and, therefore, a staff/pupil ratio (of 1–15 at the maximum) must be provided to enable this type of session to function effectively. Obviously, if there is a guest speaker, a film, or an outside visit or even some liturgical or para-liturgical ceremony, then a higher ratio would suit. The possibility of team-teaching or flexible staff grouping is available.

The staff who teach R.E. at any level need to have integrity, empathy and sincerity and, of course, need to be committed to the faith. It is an old saying that faith is caught not taught. Children can see through a person who is not sincere very quickly. All those who teach the faith inside or outside the classroom would ideally try to present the Christian faith to pupils through their own lives and personalities as well as by the formal contact of teaching matter. This is doubly necessary for those teaching sixth form students who are in a difficult phase of their lives. They are emerging from adolescence, becoming independent adults who will rightly question their own beliefs and relationship to the Christian commitment. This can be seen by the teacher as a challenge to their own deeper-held convictions and who must be able to be objective in their relationships with the sixth former. It is also necessary to consider the experience of the teacher involved in this type of programme and, if possible, create a situation where inexperienced teachers working with one who is experienced can develop their expertise in this difficult field. It should also be remembered that senior staff involvement in the R.E. programme does give status and importance to the subject. If the sixth former can see that R.E. is important to the headteacher and the deputy then they are more likely to consider it important to them. For similar reasons the male teacher very often removes from the sub-conscious of the pupil the notion that R.E. is only for the few.

Resources are essential to any teaching enterprise at an adequate level of expenditure and are fundamentally important to the success of the sixth form R.E. programme. Text books, reference books, film material and TV material must be available together with the facilities to use these media in the teaching programme. The rooms allocated must be of a size suitable for the group to allow for discussion, to enable a serious yet relaxed atmosphere to be created together with black-out furnishing for A.V. material. Again, an important consideration for the timetabler and an indication of the 'pecking-order' afforded to R.E. in the school.

The students

When planning the sixth form R.E. course our first consideration must be the students who are taking part in it. We must look at who they are, why they are there and what their needs are.

It is no longer the case in most schools that the majority of sixth formers are there because they wish to follow advanced level examination courses and go on to further education. Many are there because they wish to improve on low grades gained in the fifth year; others because they want to try new subjects or want to embark on vocational courses, some because they do not feel themselves mature enough to leave the sheltered environment of school and a growing number because, in the face of increasing unemployment, they feel there is nowhere else for them to go. A number will be suffering a sense of disappointment and failure and it should be an essential part of the course to give them some sense of their own value. All are young adults concerned about themselves and their place in the outside world and seeking independence.

114

We must always remember that the student is in the sixth form because, for whatever reason, he has chosen to be there, and that the R.E. course is frequently the only compulsory part of his timetable. There may well be an element in the group which is dismissive of R.E. seeing it as a 'waste of time' and 'not of any use in the future'. It is clear, however, from discussion with many individual sixth formers that for the most part they feel that there should be an R.E. course in the sixth form.

This does not always prevent them feeling some resentment and antagonism when faced with the actuality of the R.E. lesson. It is important that they be made to feel that R.E. time is in a special sense theirs, that they are at liberty to express opinions and ask questions, and that their views will be listened to and respected even when the teacher disagrees with them. The organisation of R.E. time must aim to motivate the students so that they do not just 'put up with' R.E. lessons but participate in, and grow through, them.

The group

To be really effective and to allow for real personal development the group needs to be small; if at all possible no more than twelve. Relationships both between pupil and pupil, and pupil and teacher, are of vital importance and it is impossible for any relationship to grow in a situation where the group is so large that free expression is inhibited and the teacher is pushed into a situation where he is lecturing. There should be some element of self-selection in the grouping but, in order to ensure that no group is seen to be an elite and that all members of the group are really given the opportunity to learn from each other and to grow as individuals, each group should be made up of a cross-section of the sixth form — able, less-able, boys, girls and — where the sixth form as a whole is timetabled for R.E. — upper and lower. One possibility might be for students to give in their names in groups of three or four and for the final grouping to be left to the leader of the teaching team who would, hopefully, know the pupils and staff and be able to take into account staff/pupil relationships.

The place in which the group meets is of importance. In the best of all possible worlds this should not be a classroom but a room with some comfort and a feeling of informality — a common

115

room, library, social area, recreational area. However, in most schools this is unlikely to be a possibility for all groups, therefore it is important that the teacher makes every effort to create a relaxed informal atmosphere as unlike classroom teaching as possible in whatever spaces are available.

The language

One aspect of our teaching which should and almost inevitably will be affected by the small group is the language used. This should be non-academic and not specifically 'religious', so that neither pupil nor teacher can hide behind it. Too often in our R.E. lessons throughout the school we use language which remains meaningless to the pupil and which, therefore, prevents any response from him. This can be equally if not more true in sixth form R.E. But if the situation of a small group in a non-classroom atmosphere has been achieved and students are given a real opportunity to express their own ideas the language will be everyday language and there will be real communication between members of the group and between the teacher and the group.

Time

The time given to R.E. on sixth form timetables varies greatly as does the way in which the allocation of that time is arranged. Our feeling is that a double lesson per fortnight is the very minimum which should be allocated. We will come back to this topic again when we discuss the role of the head of department. Here we would like to make two points. First, we cannot state strongly enough our belief that it is better to do a little well than a great deal badly. More time will not necessarily result in more being achieved. The quality of staff and the size of the group should never be sacrificed in order to gain more time. Second, we feel that in addition to the regular time-tabled period some provision be made during the school year for full days either in the form of 'away-days' or total days in school off normal time-table with perhaps outside speakers involved. There should also be the possibility of residential weekends during which students would have the opportunity to explore in greater depth some of the topics touched on in

116

the course and to experience living together as a worshipping community (cf. following section).

The staff

In the organisation of the sixth form R.E. course we feel that the choice of staff is crucial. We see the staff involved in R.E. as a team; and in order of importance, as we see them, we would look for the following 'qualifications' in a teacher forming part of that team:

1. Empathy with and commitment to the pupils;
2. Personal commitment to Christianity;
3. An openness to the views of others and an acceptance of the value of other people's opinions;
5. An awareness of current developments in the Church and a willingness to learn about them.

In putting empathy with pupils before personal commitment to Christianity we would like to make two points. Not everyone, not even every teacher, no matter how strong his own faith, is able to communicate on a personal level with young adults and it may be that there is a non-Christian member of staff who is able as a member of the sixth form R.E. team to contribute a great deal to the content of the course and to the personal growth of the pupils.

We have put academic qualifications low on the list not because we do not think that the input to sixth form R.E. is important but because we assume that members of the R.E. department will form part of the team and be available to help those staff who are not so well qualified academically.

It is of vital importance that staff involved in sixth form R.E. meet together regularly, work together as a team, are prepared to be open with one another and provide each other with support. This last point will be of particular importance to any new member of the team.

The programme

When it comes to the head of department drawing up the programme it is important that this is done in consultation with the other staff who are going to share in the course. This is not only to

HIGHER EDUCATION
THE BECK LIBRARY

117

ensure their understanding of the content of the course, but also that they recognise they have something worthwhile to contribute.

What guidelines are there as to what is both appropriate and relevant for students of this age?

Cardinal Hume, in commenting on the conclusion reached by the bishops at the end of their General Synod, emphasises:

1. the need for greater understanding of prayer;
2. the importance of the dignity of the human person;
3. the dignity of all men — so that Christians have a deeper understanding of the need to work for justice as part of their Christian commitment;
4. to appreciate the value of the Christian Community in the home, the parish and the school.

Following this, Kevin Nichols, in his book *Cornerstone*, reminds us that the content should be concerned mainly with:

Scripture
Doctrine
Liturgy
Experience

Nichols' ideas and suggestions have been taken up and developed in various ways by many educationalists in recent years.

To avoid the course becoming too set or dated in any way, it can be useful to devise a new theme each year — while at the same time including the same basic elements.

For example, such a course might include:

the dignity of human life and the moral responsibilities involved;
the development of human relationships and the Christian sacrament of marriage;
the need to work for justice;
our responsibility to the underprivileged;
the Word of God in scripture;
the expression of worship in the liturgy;
the teaching authority and universality of the Christian Church.

The approach to these subjects through what is often described as 'life themes' is again following the guidelines stated in the *General Catechetical Directory* of the Roman Catholic Church; "Any divorce of faith from life represents a grave risk for the Christian". "Anyone wishing to have an effective discussion about

118

God with the man of today, must have human problems as his starting point, and he must always bear them in mind as he communicates his message".

Pope Paul VI also expresses similar ideas in his document on evangelisation when he says: "Evangelisation would not be complete if it did not take account of the incessant interplay between the Gospel and man's everyday life. This involves an explicit message adapted to different situations about the rights and duties of every human being; family life; life in society; international life; peace; justice; development. This message today is about liberation".

To implement a programme involving such topics requires careful forward planning:

— **either an experienced teacher or specially invited speaker** is needed to introduce each topic;
— to allow the subject to be explored in reasonable depth the 'blocking' session is essential;
— the staff who will be involved in group activities needs to be properly prepared, allowing for individuality of style.

The method

In considering methods of teaching it is important to consider very seriously the fact that we live in an age of mass communication. It must be realised what a fundamental change this has brought about in human beings.

Desmond Brennan, of the West London Institute of Higher Education, describes this transformation by saying: "We have passed from a linear civilisation to an audio-visual world".

If young people grow up in a world of sophisticated communication, it is extremely important that we are not responsible for failing to communicate fully, because we are relying solely on the written or spoken word. We should ask seriously how much use we are making of radio, video, TV, film, tape, to illustrate more richly the message we are attempting to communicate.

Our schools are often full of expensive equipment but what use is the R.E. department making of the overhead projector, the film and slide projectors, radio, TV, video programmes, tape recorders, music centres, etc.

Through the incarnation, God chose to communicate fully

with man — but are we, for our part, making full use of man's gifts and senses to share in this communication?

For example, to what extent do we use art, music or drama to communicate those mysteries which are difficult to express in language alone?

To what extent do we make liturgy an integral part of the course so that it becomes possible for our students to realise that worship lies at the heart of man's relationship with God?

The structure and presentation

Many students will be members of the sixth form for only one year. Where lessons are concerned this may well mean less than one year if, as is often the case, the timetable ceases to run once the public examinations begin. Others will be there for almost two years. The structure of the course needs to take account of this.

The course must have built into it three important ingredients — choice, action and variety. It is highly beneficial if the students themselves can be involved in the planning of the course. (This may be a fruitful exercise for sixth formers remaining in school at the end of the summer term.)

The course content should allow the individual student a substantive element of choice as to the topics which he is going to explore. The range of topics should ensure variety and have a strong practical bias built in. Local conditions and circumstances will influence the final selection of topics as will the particular expertise of the members of the R.E. team.

Our experience suggests that students gain most if they pursue their chosen topic for at least two terms. We here suggest a list of topics for consideration:

*Community service
*Justice and peace
*Sixth form catechists course
  Liturgy group
*Charities
  Morality today
  Ecumenism
*Self awareness group
  Theology
  Comparative religions

120

Church history
Philosophy of religion
Prayer
Other religions
Most popular topics

The head of department

Any team will need a leader and we have assumed that usually the leader will be the head of the R.E. department. His role is crucial to the effectiveness of the teaching of R.E. in the sixth form. We assume that he will be doing in the normal course of events what other heads of department do in pressing for adequate capitation and time for his subject and in building up his department throughout the school. As leader of the sixth form R.E. team he needs to be aware of the difficulties staff may have and of the need to build a supportive group. He must be sensitive to the extent to which people are being asked to lay themselves open, to the ease with which people may be damaged and to take all this into consideration when asking for staff to be included in his team.

It happens perhaps more often than we would like to admit that the difficulties of R.E. in the sixth form are compounded by the attitude of those outside the R.E. team. Students may, for example, have activities or outings arranged for them during R.E. time because "you will only be missing R.E." The R.E. teacher is then left with the dubious choice of either letting the student go, thereby seeming to agree that R.E. time is not important, or of refusing to let him go and then creating resentment. In this the head of department is of great importance as far as non-R.E. staff is concerned. They will look to him, consciously or unconsciously, for his attitude to people. He needs to be seen as a good professional. His philosophy of religious education must be consistent, balanced and clearly apparent to all staff.

This is asking a great deal of the head of department and it cannot be said too strongly that his task with regard to R.E. in the school in general and the sixth form in particular will be an impossible one unless the school as a whole values R.E. and its central place in the curriculum. Religious education in its widest sense needs to be seen as the business of the whole school not just the R.E. department. The timetable must reflect this concern for R.E.

I

Interdepartmental work

Interdepartmental work is advantageous to religious education not least in the six form because

1. It shows that religion is concerned with, and affects, all areas of life;
2. It shows in practice the openness of the truly religious person;
3. It underlines the belief that we need to grow as human beings if we are to grow as Christians;
4. It involves the wider school community in the religious education of the pupils which should not be regarded as the task of the R.E. department alone so much as the purpose of the whole school. Too often a Christian school actually means a school with a Christian R.E. department.

We need to look at interdepartmental work from two angles, that of the subject and that of personnel. We involve other members of staff because of their particular subject area and the contribution that subject can make to a particular topic. A group concerned with justice and peace would be helped by contributions from the geography, sociology and economics departments. Indeed it is difficult to see how they could study this topic at any depth without some input from these departments. A liturgy group would be enriched by help from the music, art, drama and English departments; an ethics group by the science and English departments. Such contributions from other departments help pupils to see that religious and moral questions are at the heart of many areas of the school curriculum. They show that R.E. is concerned with all areas of life. To be really effective R.E. needs other subjects. R.E. without interdepartmental work may be seriously impoverished and dangerously isolated from the mainstream of the pupil's life

But we need to involve members of staff from outside the R.E. department not just because of their expertise in other areas but also because of the kind of people they are, and this for two reasons. First, as we said earlier when discussing staffing, the personnel involved in sixth form R.E. needs to be sensitive, emphatic and able to build excellent relationships with the pupils — this is an essential quality and obviously will be found as readily outside the R.E. department staff as inside. Second, the example of the personal faith and commitment of an exemplary Christian from another department can be a very powerful influence in showing the pupils the value of Christian belief. We think for example of a member of a science department involved with a liturgy and prayer group. Such involvement of staff from other subject areas can be particularly valuable on an 'away-day' or 'away-weekend' when there is time for relationships to develop and for an individual's own commitment to show clearly.

Celebration

For the most part we have spoken of sixth form R.E. in terms of small groups. We feel, however, that it is important for the small groups to come together as a large group from time to time particularly for the purpose of celebration. Joy, peace and happiness should be the keynotes of the Christian faith. We must show this visibly and powerfully in our sixth form R.E. programme. In practice most human celebrations revolve around food, drink, music, etc., so the R.E. in the sixth form should involve such activities as socials explicitly to celebrate the joy we believe Christ brings us. We need also to build into our programme carefully planned liturgical celebrations but unless the students understand 'celebration' they will surely not understand liturgical celebration.

Christmas and the end of the academic year in May or July are two clear occasions for organising some sort of R.E. social. We should not be afraid of organising something which is pure 'fun'. It would be a happy result if the students' last memory of their Church school's R.E. programme was a joyful celebration.

Reference

1. e.g. cf. O'Leary, D. J. & Sallnow, T. *Love and Meaning in Religious Education* (O.U.P., 1982).

AWAY-DAYS, RESIDENTIAL COURSES AND COMMUNITY EXPERIENCES

Away-days are essentially about Good News. The Good News for the Christian is that God loves him, whom he has made, no matter what. For young people who often lack the deeper relationships they long for, this Good News becomes real when they share with each other, and with understanding adults, their deepest needs and aspirations. Away-days provide the atmosphere and the support that helps young people to overcome the fears and inhibitions that prevent them from being open and taking the risk of getting hurt in their relationships with each other, and with their families, and thus with God. To experience acceptance and genuine friendship, and to know this has something to do with God and the Church, brings great healing and a sense of reconciliation and new hope. To love becomes possible.

The experience of living together, and being free to think about common problems, can give a new insight into the aim and purpose of life in general and the Christian life in particular, through deeper awareness of relationships within the group, the family, the Church, the world and with God. Knowing that all are formed as human beings, not by being fed information but by living and learning with others right from the first moments of life, one cannot doubt that one of the most effective ways to deepen Christian experience is by the interaction of individuals in a real Christian community atmosphere. This happens at three levels.

First, the youngster is given time to consider his own self and his deepest hopes, the relationship between his inner and outer life, and the working of God's spirit in him. Then, through the community, he can experience his own Christian reality finding expression in the loving group of people who are as much a part of him as he of them. This does not arise out of anything the leadership team may do or say, but flows from the love which the individual experiences and to which he contributes. It is the Christ-identity which arises in any community of love.

Ultimately, he can begin to see that, as a Christian, he is called to go out to others, family, friends, and beyond, carrying Christ to the world.

These are in fact the three ways the incarnation is realised: building up the body of Christ in each Christian in the worshipping communty, and in all mankind.

The preparation

A group of 15, or at most 20, young people gather at a residential centre, away from the ordinary routine of school, together with a leader and some of their teachers. There is plenty of time to talk and discuss the more important aspects of life today without being hampered by time-table and the type-casting which results from the formal structure of schools. Here, there will be new pressures, but they may be more stimulating for growth and change than those we are familiar with in school R.E.

It is important that the leader and other adults relate to the young people as friends and fellow-pilgrims committed to the whole experience.

The first essential is to choose a leader who is likely to prove sympathetic to that particular group, and to prepare for the course by arranging a day's 'talk-in' with students and staff to find out what they want to discuss. This should involve a genuine personal contact some weeks before the course is due to take place. Talking over the suggestions with the young people should lead without difficulty to the acceptance of a unifying theme, as most youngsters are much more ready to accept and appreciate guidance than is often supposed. The next thing is to work out a rough time-table and send it to the school in advance of the course. The final time-table — light-hearted and illustrated if possible — is handed out during the opening session of the course itself, when it will be received with lively interest as a guide to the unfamiliar situation in which they find themselves.

The length of the course will vary according to circumstances and financial constraints, but it is usually harder to assemble a group at weekends, because of family commitments and so on. It is advisable that, even if some form of subsidy is available, the young people themselves should make some contribution to the costs. An element of sacrifice enhances the value of such a project. It is important, too, that these 'away-days' and residential courses should have an ecumenical dimension.

Experience suggests that it is advisable to keep within a single year group, at least up to sixth form level. Though boys and girls do not develop at the same rate, there are questions of status involved which are likely to affect the way the 'peer group' information responds to the demands made on it. Whether a single-sex group can be as lively and open as a mixed one may be doubted; but fourth year groups may not be mature enough to 'be themselves' in the presence of the opposite sex. Probably fourth year is the earliest useful time for this kind of course.

Ideally, attendance should be voluntary, but some encouragement is essential, and occasionally even a little moral pressure, though this risk of generating resentment should be kept in mind. Situations vary from school to school. Difficult groups could have the course presented to them as a normal part of the curriculum; they change their attitude once they have tried it. Generally, a class that has had experience of a course will do a lot to encourage others to participate, the word gets round that 'it was great'. Clearly it is vital to have the enthusiastic backing of the school staff for the project.

The action

Once the course is under way, the time-table should allow for plenty of free time with adults available for chats and discussions. In the mornings, there may be a talk — no longer than twenty minutes — or perhaps a short film or other kind of presentation — followed by small group discussion (6 in a group) and a general feed-back. The aim of the chairman of the feed-back session is not to tie everything up Q.E.D. fashion. If a matter has been clarified that has value in itself. Adults are inclined to talk too much and take the drawing of conclusions out of the young people's hand, happily unaware that they are not only failing to convince their audience but leaving them cynical and resentful of adult incomprehension.

Afternoon are generally left free for group excursions, and working sessions begin again after tea until supper time. Evenings are usually quieter, with films, music and socials. Team games, though sounding 'old hat', are immensely popular if organised properly. Above all there should be room for spontaneity and the flexibility to disregard the time-table when the occasion calls for it. This is particularly true of 'lights out'. Often an apparently superficial youngster will pose some very deep questions at 2 a.m. in the morning. The drawback is that the adults involved may not get much sleep, but the reward is more than worth the effort made.

Once the youngsters are committed, in some sense, to the group as a community of people trying to relate more closely to God, the focal point of the experience will naturally be the evening Eucharist, which will embody in living and symbolic form all that has transpired and been experienced during the day. There is a fertile tension between what happens in talks, discussions, and games, and what happens to the community through what it does at the Table.

However, it may sometimes be advisable to leave attendance at the Eucharist in the beginning of a course lasting several days to the choice of the youngsters, so that they can discover a need for it.

A talk about prayer may be followed by the opportunity for a meeting for guided prayer in the quiet of a softly-lit homely chapel or oratory. As we know from the experience of Taizé, many young people are hungry for silence and a meditative interval in the presence of God.

There will certainly be the opportunity for them to prepare a special liturgy to express their new insights.

The place

As for the setting, a religious house is often more satisfactory than an L.E.A. Centre, which may be subject to rules and regulations that cannot easily be lifted and lacks the atmosphere of a house with a chapel which can play a vital part in the day.

The opportunity to experience the peace and quiet of the country is especially valuable for children from the inner cities, who will appreciate being able to roam and explore, and opportunities for vigorous physical exercise provided by a swimming pool, gymnasium, or simply a field to kick a ball around in.

Some of the houses that have been found particularly helpful are Allington Castle in Kent, the Cenacle at Grayshott, London Colney Pastoral Centre, Park Place Pastoral Centre in Hants,

Broom Hall in Surrey, St Cassian's Centre in Kintbury, Maryvale Centre at Guildford, Sandrock Hall self-catering lodge, Spode House, and even Caldey Island; also Damascus House in Mill Hill, on the edge of the Green Belt.

A journey to a distant centre can be educative in itself, with the picknicking on the way and dropping off to see some building or place of interest. Rough sleeping accommodation, with camp beds and sleeping bags, can be an advantage as it encourages a do-it-yourself atmosphere in which a sense of community develops very quickly. It may sometimes be better to look for a self-catering house rather than an official pastoral centre for this reason.

Teachers who come on residential courses find that the friendly informal atmosphere of a course enables them to develop warmer relationships with their pupils which arc of help when they are back in school. They may incidentally decide that a residential for staff could help them see their own problems and difficulties and to be a more united community. The more the school has already realised its identity as a Christian community, the more the teachers will be at ease on the course, and the more they will be able to give and to receive.

To our knowledge, no one who has tried these courses has anything but enthusiasm for them. We have learned a lot, and made a lot of mistakes in this sphere of work, and new approaches are being explored. For instance in the formation of 'search groups', week-ends led by young people, springing from sixth form encounter days, and parish confirmation residential groups, which stimulate the parish too. There seems to be something in this kind of experience which particularly appeals to our contemporaries as they try to work out their personal destiny and grope for a deeper awareness of the meaning of life and the presence of God.

PRINCIPLES AND PRACTICES
OF LITURGY AND WORSHIP

While some of the suggestions in this section have relevance for all Church schools its main thrust applies to the Roman Catholic confessional system of education. Because of the wide-ranging nature of this topic, we will not be considering the issue of the school assembly here.

There are many Christian countries where religious worship in state schools is banned. Britain, by contrast, has a law obliging all schools in the state system to a daily act of worship. Whether such a sensitive matter is susceptible to legislation is a debatable point, but given the ease with which such a law can be circumvented or ignored it is an impressive fact that a large number of schools remain faithful to the law in varying degrees. Church schools have a strong tradition on the point. Even before the 1944 Education Act prayers and hymns at assembly and before each class were a regular feature of school life, not to mention in R.C. schools events like the rosary, benediction and stations of the cross.

The Eucharist was more rare, largely because it required a church for its celebration. All that has changed. The Eucharist has penetrated the very classroom, and boys and girls frequently go away on residential retreats, or 'youth gathers', during which elaborate celebrations of the Eucharist play a prominent part.

Only the embittered and the change-resistant would complain about this new state of affairs. Nevertheless, a Eucharist for-every-occasion-policy may not be the best for children, teachers or school in general. Indeed, the question is often asked "why celebrate Eucharist in school"; why celebrate any act of worship in a school? Answering that double-barrelled question is important because it provides a foundation, a policy, for school worship, particularly in the secondary school system where, because of the complication of time-tabling and general organisation, any act of worship can appear to be mere icing on the cake — good to have but not essential.

The Eucharist is the Christian liturgy par excellence, but it is not the only one. Unfortunately, in recent years the normal parish programme has left one with the impression that there's nothing but the Eucharist and this has sometimes tainted another approach to liturgy.

First, schools (and parishes) need a variety of liturgies because within the schools there is a variety of needs, of occasions, not always adequately met by the Eucharist. For example, it may be that a teacher is about to be married and wants her class to pray about it and celebrate with her. The Eucharist might be too 'heavy', too formal. Much more appropriate would be a specially planned celebration based upon discussions the class would have had about marriage and upon the special relationship of teacher and pupils. Similarly with the birth and baptism of teacher's children and brothers and sisters of the pupils.

Apart from such special occasions, there are normal occasions for short, informal celebrations marking the various stages in the R.E. course. Recalling what we said earlier, it is important that the content of R.E. be prayed and celebrated. It is not being claimed that such an approach will bring about the instant and dramatic conversion of the boys and girls, but it does move the ideas being taught into the living presence of God. In addition it provides boys, girls and teachers with the regular opportunity of praying. Many need the chance of melting their own weak, halting prayers with those of the group where they are supported, supplemented, by the combined prayers of all — not to mention those of Christ.

Second, the Eucharist is a somewhat sophisticated form of liturgical celebration. It is a big jump from private prayer to a liturgy with a history of over two thousand years; therefore, it is useful practice to have intermediate celebrations, less formal and intense than the Eucharist, which act as stepping stones to it. Hopefully, the impression would also be gained that worshipping together is a necessary part of the Christian way of life, not merely an adjunct to it.

Third, there are more sacraments than just the Eucharist. Obviously apart from the sacrament of reconciliation (and even that can be a problem) the other sacraments would not normally be celebrated in a school. Yet there are many opportunities for using the ideas, symbols and texts of all the sacraments in non-sacramental celebrations. Marriage was mentioned above; Holy Orders might perhaps be linked to discussion and prayers for vocations; Anointing of the sick linked to the needs of a particular pupil or teacher, and so on.

Reconciliation deserves more attention than can be given here. More obviously than most of the other sacraments, Reconciliation is to do with relationships — with neighbour, with God, with personal conscience. For teenagers especially, relationships are fraught with difficulties, and frequently healing is required. A sensitive approach to occasional services of Reconciliation provides the young adult with opportunities for reflection, for renewed resolution and a sense of solidarity with others in their struggle towards maturity. Above all, even in a non-sacramental celebration, Christ is present with his healing, helping grace.

Group celebrations may lead to individual celebrations of Reconciliation. How many teenagers are there who desperately want to talk out their problems but do not have the opportunity? A warm-hearted, compassionate and loving priest or vicar can do untold good in bringing the peace of Christ to their lives. However, like the Eucharist, it's not easy to jump straight into the intense, high-level celebration of the sacrament: intermediate celebrations need to be part of the school's way of life.

D.I.Y. liturgy

How do we set about constructing a non-sacramental celebration? There is no one easy answer because situations differ so much. Frequently it is a matter of instinct: the teacher recognising the need for a celebration in the same instant as he or she sees the materials in the idea currently being discussed.

There is one good foundation for all liturgies, official and unofficial alike: call and response. In the liturgy God speaks to us and we reply. An obvious example of that idea can be seen in the Eucharist: the liturgy of the Word (God's call, or voice to us) and the liturgy of the Eucharist (our reply through, with and in Christ). But even then there is no absolutely clear division between the two halves — bits of one are found in the other, and vice versa.

The call-response pattern is good because that is exactly what goes on in the liturgy. It's neither God exerting his power on a passive, inert audience, nor an excuse for a group of care-free individuals to do their own thing.

Symbols also play a part in good liturgy. The besetting sin of modern liturgy is wordiness. Young people are not simply intellects, or disembodied voices. They have five senses, all of which can be, and must be, put to the service of the worship of God. The sacramental symbols of water, eating and drinking, anointing, the laying on of hands, have a great many non-sacramental uses. There are, too, many other symbols which can be called into service. For example, a group faced with the challenge of a length of rope developed a moving celebration lasting about five minutes in which the rope was seen as a sign of strength, of binding up the shattered, of uniting all who held it, of a means of pulling together. . .

Dance, mime, movement, gesture, are all part and parcel of the same principle. Christians believe that all people are God's creation, body and soul; they must worship him with all that they have and are. It was once assumed that only girls were open to a more physical and demonstrative form of celebration, yet experience proves that fourteen and fifteen year old boys are sometimes prepared to give themselves more fully to such celebrations.

However, the teacher needs a certain degree of sensitivity when leading liturgy-planning, especially with the older teenager. For silence is equally important. There are times when what is required is quiet, meditative reflection. Obviously, this is as true of the official liturgies as of the unofficial.

Music

Like silence, singing and instrument playing are at home in the specially devised celebrations and in the Eucharist. Music plays a big part in the lives of young people, from the basic transistor blaring out its diet of pop and rock to the sensitive and skilled young musicians common in so many of our schools. What is surprising is how often this fact is ignored in the school's liturgy — the music of religious celebration bearing little relation to the actual skills and tastes of the young people concerned.

Possibly the problem lies partly in the fact that individual teachers are themselves not very musical and are not sure how to get the best musically out of a situation. Another problem arises when the thin partition walls in some schools make one class's celebration another's aggravation. Nevertheless, there still remain many possibilities; and one has taken part in prayerful celebrations which have featured a rock group, the school orchestra, as well as the ubiquitous guitar.

Celebrating the Eucharist in school

Not many secondary schools are able to gather all pupils and teachers together for a single celebration because few schools have the architectural facilities. But it is a good idea to do so, if at all possible. The reason for this will be clear from what has been said above. An all-school Eucharist once a term certainly confirms and celebrates the school's raison d'être.

Meanwhile, the more frequent form of the celebration will be in smaller groups — year, class, interest or subject group. The advantage will be obvious — greater intimacy, more personal involvement, ease of planning. For many youngsters and teachers the group Eucharist is a welcome change from the hustle and bustle of a frequently anonymous Sunday parish Eucharist.

How often such a celebration? Local circumstances usually dictate the answer to this question, but once a term for each pupil in the school seems reasonable. Certainly, even if a chaplain is freely available, more than twice a term might be excessive under normal circumstances.

How do we celebrate a school Eucharist?

R.C. secondary school children are in a liturgical limbo — officially at least. For adults there is the standard Roman Missal; for pre-adolescents there is the *Directory for Children's Masses*, published by the Vatican in 1973. The aware teacher or R.E. department will be able to strike a balance between the two.

Indeed the Directory is nothing more or less than the

application of basic educational and liturgical principles to the Roman Missal. Therefore, in the secondary school it is simply a matter of adapting the Eucharist, where it is adaptable, to the needs of the particular group. In practice, this will not require a great deal of work because the older the pupil the greater his or her need for an adult form of celebration. A typical sixth form celebration is frequently a low-key, rather meditative affair.

Two things are essential: flexibility on the part of the priest, and a good working knowledge of the parts of the Eucharist by both priest *and* teacher. The priest becomes a 'problem' when he brings to the celebration an overly legalistic and 'churchy' attitude. In fact, experience shows that many priests who object to so-called liberties taken with the Mass in school frequently betray a lack of knowledge of the permissions and flexibilities built into the rubrics of the Roman Missal.

Here is an outline list of possible adaptations:

— Pupil-written and delivered introduction following the priest's greeting;
— Specially composed penitential rite (by pupils);
— Specially selected readings; the reduction in number, but retaining a gospel; mime, dramatisation of readings; use of visuals with readings — slides, film, posters; dialogue homily;
— Prepared or spontaneous bidding prayers;
— Use of one of the children's Eucharistic prayers or, taking those prayers as a model, the inclusion of more acclamations at different points throughout the adult prayers;
— Preparation and baking of eucharistic bread by pupils; simplification of communion rite.

In addition, it is presumed that, as far as possible, what is meant to be sung is sung, though celebrations do differ in mood, some requiring less music than others.

The golden rule in adaptation is: let the Eucharist be the celebration of *this* particular group within the framework of the true meaning of the liturgy.

'Theme Masses?'

Liturgy and life should come together in a harmonious unity. Therefore, boys and girls should be encouraged to bring to the Eucharist their own needs, hopes, experiences. Maybe that is what is missing from some Sunday parish Eucharists. Allow them to become totally 'other-worldly', and we create a situation where many will drift away because of their seeming irrelevance.

However, the Eucharist is equally the celebration of the Church's faith. We have said something on this at the beginning of this chapter. The relevance of the Eucharist is not created solely by the secular ideas brought to it by the boys and girls but also by the events of faith made present by Christ, especially his death and resurrection. It is an admirable idea to celebrate such themes as birthdays, sport, parents, dating, third world concern and so on, but they should always be celebrated within the context of the living and redeeming Christ.

Preparation

It is important to involve pupils as much as possible in the preparation. After all, it is theirs. Just as important is the involvement of the priest-celebrant. It is difficult for both pupil and priest to understand, let alone become enthused about a celebration which is presented as a pre-packaged affair. This sometimes happens when a teacher is seized by an obsessive idea which he or she is determined to implement. At the opposite end of the scale is the one who does not bother with any preparation at all "because the Mass is the Mass and it always works".

Questions that crop up

Much depends upon the priest. Some schools are lucky to have a full-time chaplain who invariably has a special ability for chaplaincy work, in which case he will be fully involved in the liturgy of the school at all stages. But what if such a gem does not exist? Many schools invite pupils' own parochial clergy to celebrate an occasional Eucharist or penitential service — a practice which certainly involves priests in an important sector of their parishioners' lives. Yet it has some disadvantages: for individual priests the contact may not be frequent enough to lead to familiarity with and understanding of the school liturgy.

Another system is to invite clergy from the school's feeder parishes to become responsible for an entire year. In some cases the priests move up the school with their year, thus establishing a long and regular relationship with one group.

Should non-practising pupils take part in school Eucharists especially at communion? That is a question that can be fully answered only in each individual case. Some boys and girls have rejected the Sunday Eucharist because of a conscious decision; that decision should be respected and they should not be forced to participate in what they do not believe. However, one then wonders why they attend a Catholic school and what requirements should be made clear to pupils by the head when selecting new entries. A chat with the parents might also be necessary.

Most pupils who are not regular church-goers are such because of a rather casual, unthought-out attitude. Without wishing to lay down a rigid rule, perhaps teacher and chaplain should remember that the Mass is not a reward for good behaviour but the celebration of a very diverse and far from perfect community. Christ came to call lost sheep. And it could be that together with the encouragement by teacher and priest the school Eucharist will be a way back to the parish Eucharist.

Should the school Eucharist be so different from the normal parish Eucharist? In one sense not, because the basic structure of the Eucharist is the same in both cases, no matter how varied the form of celebration. Equally, they *must* be different because the celebrating communities are different. The Eucharist should always be adapted to the needs of real people. However, this question arises almost exclusively when the parish liturgy is inadequate — for the adults as well as youngsters. What is then required is the renewal of the parish liturgy, not the watering down of the better-planned, better-celebrated school Eucharist.

Yet the school has a responsibility to educate its pupils in an understanding of, and fuller participation in, their parish Eucharist. There is no future in building a Berlin Wall between parish and school. Celebrations have as one of their aims a fuller, more conscious and fruitful patricipation by pupils in the adult celebration.

Liturgical celebration in the sixth form

Why is it important to celebrate liturgically with some frequency with sixth formers? There are two reasons. First, pupils of this age can all too readily identify their Christian faith with concepts, doctrines and logical argument. The resurrection of Christ, it is true, has been and always will be the subject of speculation, formulations, translation into various categories of thought. Nevertheless, the resurrection of Christ, as the immediate Apostolic reaction testifies, was first of all one of *welcome* before being categorised by the understanding. The Christian faith centres on a person, Jesus Christ, and persons are primarily met, encountered, before they can be said to be understood. In the liturgy the risen Christ stands in our midst, among doubters like Thomas, among weak in faith like Peter, among the strong in love like John. In a sense, liturgy planners need only prepare the room for the coming of the Lord. What takes place between Christ and the individual sixth former in the intimacy of the heart will always ultimately escape detection. But pupils who have occasion to return to this 'source moment' testify that they experience, as a result of the celebration, something of their essential human and Christian solidarity and the power of the love of the risen Christ.

Second, the celebration of the liturgy discloses the intrinsic triangularity of human relationships. There is God, myself, my neighbour. In the liturgy what we celebrate first and foremost is what *God* has done *for me*. Christ came not as just a super-teacher, the moral guide of all time. He stands as the embodiment of God's love for us, and in the liturgy when he holds us together it is in the embrace of his love not by the fear of the sword or the fear of hell. In a sense that moment sums up the vision of the Christian life. It celebrates what is possible with the risen Christ. Indeed, the liturgical celebration not only states the vision but, because the risen Christ himself is present in this moment, the celebration is the activating point in the long process of humanisation and divinisation of man and woman. Liturgy continually, in the image given us by St Ignatius of Antioch, rechords not just records Christ's presence among us, challenging us to live as Christians in harmony, concord and charity. When through the pressure of life we find, like guitars, that we are out of tune with each other, that we are quarreling, speaking badly of others, breaking faith with each other, then liturgy becomes more, not less, urgent in our lives since it is at this point that everything needs rechording on Christ. Liturgy

137

J

and Christian living go hand in hand; what God has opened up to me, in Christ Jesus, is also what God extends to my neighbour and the vision of it all is stated and activated in the liturgy.

Opportunities to celebrate together

One can ask when, given all the other commitments of staff and pupils, does one find time to celebrate liturgically with the sixth form? Three occasions spring to mind.

First, if what liturgy proclaims is true, namely, that Christ stands in our midst *now*, whenever we celebrate in the name of the Church, then every 'teachable moment' (and we are assured by educationalists that such moments do occur) is at the same time potentially a 'kairotic moment'. 'Teachable moments' surface throughout the school year. For instance, September seems tailor-made for small group celebrations based on the hopes of all for the year ahead. Second year sixth form pupils throughout the months of February, March and April begin the process of serious revision. A liturgy during this Lenten period centred on a 'revision of life' can help them appreciate their relationship with God more profoundly. May and June (examination time) lends itself to services of intercessory prayer! Exams, like shipwrecks, have a habit of concentrating the mind! Early July, before the great diaspora, can be a moment of real nostalgia, a moment of thanksgiving to God for the whole of school life. Within this general framework, there will be the usual dramatic events of life — a death in the family or in the school, someone's mother will have a baby, there will be illness, some unforeseen incident. There will be Christmas and Easter and other occasions when a group of sixth formers might like to celebrate as a group.

In an increasing number of schools, I am finding that opportunities are provided for first year sixth form pupils, in the course of the year, to have a few days together, either in the school or in some residential centre. These open up wonderful opportunities for Christian celebration of life. In some schools, sixth form tutors encourage pupils to become actively involved in parish education programmes. Other pupils help in the preparation of, and the celebration of, the liturgy of the Word with infants and small children in the parishes on Sundays. A few schools encourage some sixth formers to help pupils lower down in the school who have reading problems. Frequently, such sixth formers are invited to their class Masses. The opportunities for celebration undoubtedly exist. The problem is selection.

Second, I have found in my experience that ninety per cent of the fun and long term benefit of any liturgical celebration lies in its preparation by all involved. As a journey can be said to begin with the first step so, too, can the celebration of the liturgy. The choice of readings to suit a celebration can be both constructive and revealing. Who, for instance, would have considered as a first reading on the theme of hospitality, the Mamre incident (cf. Gen 18)? It surfaced in a small group liturgy. Again spending time aiding pupils to decorate the place of worship creatively is well worth the effort. Each group finds new permutations of light and darkness, music and silence, gesture and prayer formulae. The time spent in this way is invaluable for two reasons. First, needs of groups change, and liturgy needs to keep apace of *real* not *past* needs. For this reason, no two liturgical celebrations ever quite manage to be the same. Second, care in liturgy preparation is a witness that today's liturgy demands the pooling of our total resources. One purpose of liturgy planners is to enable all those resources to be mobilised in order that in sign and word sixth formers might express their authentic religious experience. So liturgy planners contribute most when they prompt and co-ordinate, not when they dominate a sixth form liturgical celebration.

Lastly, my experience with sixth form liturgies has led me to celebrate in a variety of quite unlikely places. This is bad news for the kingdom of darkness. There is no longer a sacred place that the devil can call his own. There is no hiding place from the liturgy of Christ. I have found that some sixth formers celebrate together before school begins, during lunch time, after school has formally closed. The obvious advantage of using the school premises is the ease of access for everyone. But at other times, some sixth formers prefer to meet and pray together in their own, or the school chaplain's, or the parish clergy's home. (Cardinal Hume, for example, meets young people and prays with them in his home each week.) One advantage of this lies in the way the celebration absorbs a more intimate, domestic atmosphere.

I think that it is the experience of most priests, and indeed most sixth form staff, that liturgical celebrations with the sixth form are uniquely rewarding. Sixth form liturgy, as with every other school activity such as exams, concerts, sports, has a way of creating its own standards, its own momentum, its own expectations. These expectations are relatively new still. We are far from an established liturgical tradition in our sixth forms. To achieve that, much more critical reflection on our common experiences will be necessary. This essay, we hope, will encourage others to lay bare their secret thoughts on the matter.

CHAPTER 5

SOCIOLOGICAL ISSUES

WORLD RELIGIONS
IN CHRISTIAN SIXTH FORM R.E.

Few would question the proposition that the modern sixth former should have a sound introduction to the main non-Christian religious traditions. Without some understanding of Islam, for example, the sixth former will have great difficulty in evaluating the politics of the Middle East. Without some insights into Hinduism he or she will only be able to develop a very limited appreciation of the art and architecture of South Asia. All this is as true for the sixth former in a Christian school as in any other. However, what I wish to argue in this article is that world religions are not simply important for general education but that they also provide an essential part of the context for Christian religious education.

World religions in religious education

In county schools R.E. has begun to appear in a less exclusively Christian context. In an increasingly pluralistic and post-Christian society the importance of teaching about non-Christian religions and ideologies is quite apparent. The county school does not seek to provide the Christian initiation and formation which is properly seen as the function of the Christian family or the local Church. In denominational schools, on the other hand, Christian initiation and formation is often the primary purpose.

The denominational school is viewed as connected with the Christian home and of the Christian Church. However, liberally interpreted, religious education is provided within an ethos of Christian commitment and of a specific Christian tradition. In such circumstances, the teaching of world religions is often either ignored or tacked incongruously and briefly at the end of the course. Since sixth form religious education is the end of the course for many, world religions often tends to be brought in at this stage. However, it is important on both educationanl and theological grounds that world religions should not only be approached in a serious academic manner but also that they should be seen as integral to Christian R.E.

On educational grounds it has become impossible to ignore the wider social framework of Christian education. However specific the religious tradition of the school, it is still concerned to prepare its pupils for life in a pluralistic society — no matter how much some Christian believers might wish it to be otherwise. Christian faith is only one option among a number of competing religious and ideological commitments. When the rival truth-claims of these other traditions are ignored, the pupils, particularly the older pupils, begin to suspect that the omission is a deliberate attempt to keep them in ignorance.

On theological grounds, the omission is equally indefensible. Increasingly, Christian theologians have been obliged to come to terms with the existence of other world religions. If the religions of the world were simply rival systems of belief it would be possible to deal with the problems by recourse to exclusivist apologetic. But they do not merely compete. They also converge and overlap to an astonishing degree. To account for this convergence, theologians, and in particular Roman Catholic theologians, have developed a theology of world religions, often seen as one of the most significant growth areas in modern Christian thought.

According to this new theological perspective, the old distinction between natural theology and revealed theology is regarded as inadequate to account for truths which Christians perceive in religions outside their own traditions. According to the older, largely discarded theory, the truths in non-Christian religions are a product of man's natural reason rather than of God's revelation. Cut off from communion *with* his creator, fallen man nevertheless has access to certain truths *about God*. ". . . Holy Mother Church holds and teaches that God, the beginning and end of all things, may be certainly known by the natural light of human reason, by means of created things".[1] The possession of these truths does not bring about his salvation but merely underlines man's own implicit longing for the one authentic *Christian* revelation.

Inadequate and exclusive as it seems, this traditional Roman Catholic view does at least take a more positive approach to non-Christian religions than the radical Protestant position. Radical Protestantism rejects the premise that fallen man's natural reason has any access to divine truth whatsoever. Utterly corrupted by sin, fallen man has no access to God other than by repentance and conversion. From Calvin to Karl Barth, the problem of world religions is simply not seen as a problem — except for the millions of unregenerate souls who adhere to them. Other non-Barthian Protestant theologians have placed great emphasis upon the positive character of non-Christian religions. Rudolph Otto and Paul Tillich spring immediately to mind. John Hick stands out as a Christian who comes close to denying any unique character to Christian revelation.

In contrast to all these viewpoints the new Roman Catholic theology of world religions sees God's self-revelation as universal as well as historically specific. The history of Israel and the divine incarnation in Jesus Christ are seen as referring to the same fundamental realities experienced by mankind as a whole. Christianity retains its uniquely authorititive position but the biblical revelation is not reducible to the partial and largely European categories possessed by Christian theology until now.

All human cultural intellectual religious experiences must be tested and transformed by their contact with the gospel of Jesus Christ. Catholicism is seen not so much as an empirical phenomenon but more as an ultimate future ideal. Hindus, Sikhs, Muslims, Buddhists already encounter the unknown Christ in their beliefs and practices. As they come to know him explicitly, they will illuminate areas of truth about him not available to western Christianity. In this way, the Church will approximate more closely to the universalism to which it is called.

This approach commends itself to many as a theological basis for the inclusion of world religions teaching in the Christian school. It appears to combine a principled commitment to traditional Christian orthodoxy with tolerance and openness to learn from other religions. Nevertheless, it presents its own dangers and difficulties. The most obvious difficulty was highlighted in a recent debate between Hans Küng and Karl Rahner. Küng criticised Rahner's concept of the anonymous Christian and asked him what Christians would say if they were called condescendingly 'anonymous Buddhists'. Rahner's reply that he would regard it as a compliment to be called an 'anonymous Buddhist' fails to meet the force of Küng's question. It seems better to treat people's beliefs on their own terms rather than begin from the perspective that such beliefs are merely a distorted or, more kindly, a complementary version of one's own.

Indeed, despite the sincere protestations of tolerance and dialogue evinced on the part of Christian theologians, anonymous Buddhists and others may, understandably, suspect that all roads still lead to Rome — even if it is to be a renewed and historically abstract Rome. Secondly, and more seriously, it leads itself to the procrustean tendency to reduce unfamiliar ideas to the familiar, thus robbing them of their intrinsic meaning. For example, if the Buddhist does not talk about a Supreme Being then we are tempted to discover such a being in the concept of nirvana; or if the Muslim denies the Christian doctrine of the uncreated Word of God, a comparable doctrine must be concealed in the notion of the eternal uncreated Qur'an. With this attitude goes the related tendency towards an evolutionist overview of world religions. Whatever their unique contribution, all world religions are represented as finding their fulfilment in Christianity and we find ourselves imposing an evolutionary pattern where there may be none.

The fact that nineteenth century rationalism played the same evolutionist game with religion, albeit with a different jumping off point, should give us pause. The findings of modern anthropologists have largely invalidated this way of dealing with phenomena so diverse as religions. But even if they had not, we are still left with the fundamental question of why anyone should accept that the evolutionary process should find its fulfilment in Christ rather than in, say, theosophist universalism or dialectical materialism.

If religion is not one thing but many, is it possible to sustain any position which sees all religion as tending towards the same reality? Far from being a key to dialogue, the new theology of world religions may actually preclude the possibility of understanding any religion on its own terms.

The other objection comes from the exponents of religionless Christianity and political theology. Armed with their characteristic insight that the Gospel is about *this world*, such Christians

145

view with horror the attempt to shore up the crumbling meta-physical, mystical and sacralist elements in Christianity through recourse to a positive appreciation of just such elements in the religions of the East.

It is, of course, possible to some extent to avoid these dangers by taking a sociological approach to world religions. Nor is it necessary to have some grand evolutionist overview in order to admit that there is a thematic relationship between Christianity and the other world religions. Perhaps world religions do not all address themselves to the same fundamental reality but they do inevitably refer themselves to some common insights and problems. It is this thematic relationship which provides the most useful basis for the teaching of world religions in the Christian school. However secularised or political our theology may become, historical Christianity has and does share with other religions a common vocabulary of religious language, ceremony and ritual.

General principles in the classroom

When it comes to the actual teaching of world religions in denominational schools I would like to advance some general principles based on the argument I have outlined:

1. The way in which world religions are taught must not *either* reduce other religions to distorted expressions of Christianity *or* present Christianity as though it were one version, among others, of a substantially identical faith.

2. Christian formation must take place against the background of a basic general knowledge of world religion. This is to be effected in two ways:

(a) by specifically Christian denominational teaching thematically linked with comparable beliefs and pratices in other world religions;

(b) by offering a general objective account of each world religion *within its own terms of reference*. This is particularly appropriate to the sixth form where world religions could be given serious treatment as part of the general sixth form course. Those sixth formers taking religious studies 'A' level can engage

146

in more specialised study of world religions. Papers are available in non-Christian faiths in the syllabuses of the University of London, the Welsh Joint Board and the Joint Matriculation Board. The Associated Examination Board offers an option which includes comparative treatments of Christianity and other religions.

It is possible to illustrate the thematic links between the teaching of world religions and of confessional Christianity but, apart from this serious specialist approach to world religions, it is possible to link up with Christian symbolism over a large number of areas. These areas would include prayer and meditation, festivals and fasts, holy books and holy people. Most of them centre upon the liturgical life of the Church and the expressions of faith included in its creeds and doctrinal formulae. Denominational R.E. is also very much concerned with the formative education of children within a specific Church community. In theological terms, the sacraments are constitutive of the Church itself. Sixth formers in Church schools have usually been through a religious formation which has included preparation of children for first communion, confirmation and, in the case of Roman Catholic schools, first confession and, although the majority of Christians are still baptised during infancy, preparation and instruction for any of the other sacraments presupposes a basic grasp of the sacrament of baptism also.

Modern catechetics approach the teaching of all the sacraments thematically. In the case of Baptism, the symbolism of water is explored at numerous levels through discussion and project work.

In most cases our sixth formers will not have been encouraged to see any world religious dimension to these themes. With serious treatment of world religions, however, the common religious language of water can be seen in most belief-systems, e.g. Hindus bathe in the waters of the sacred river Ganges, Muslims wash before entering the mosque, etc. Although other religions employ varying symbols of ritual initiation, some of the basic symbolism of water is universal, reflecting as it does man's continual dependence upon water for life.

In the case of Confirmation reference points in world religions abound. The son of a Hindu Brahmin receives the sacred thread to wear upon his shoulder as a sign of being ready to assume the religious duties of his caste. In some Buddhist countries young boys imitate the great renunciation of the Buddha. The young boy is dressed up as a prince in order to exchange this garb for the robe of a monk. Then, just as the Buddha abandoned the delights of the palace for an ascetic quest after truth, the young Buddhist leaves home on a donkey for a brief stay in the local monastery. Although this is a religious coming of age ceremony, some aspects of the practice provide a closer parallel with Baptism than with Confirmation. The identification of the individual with the historical renunciation of the founder of Buddhism is analogous with the Christian identification with Christ in his death and resurrection through Baptism.

The Christian Eucharist, uniting the ideas of sacrifice and fellowship meal, has parallels in most world religions. If we interpret sacrifice in the weaker sense, as a setting aside of gifts to honour a divine or exemplary being, then there is no major religion which does not have some form of sacrifice. Even Buddhists offer gifts of incense and flowers to the statues of Buddha. Taking sacrifice in the stronger sense as involving the death, or commemoration of the death, of a victim, there are examples in some surprising places. Thus Islam, perhaps the least priestly or sacral of all religions, includes the sacrificial slaughter of goats during the great Meccan pilgrimage.

Here, as elsewhere, it is dangerous to assume that the common symbols of sacrifice all refer to similar ideas. I remember an African student preparing lesson plans for his teaching practice. He proposed teaching the Eucharist with reference to a traditional African rite for placating angry ghosts. We finally agreed that the only point of comparison was a negative one. The notion that the sacrifice of the Mass might be offered in order to placate an angry ghost — in this case God the Father — was clearly unacceptable theology.

On the other hand, the common symbol of the fellowship meal provides less difficulty. The sikh 'communion meal' of karah parshad, like the Christian Eucharist, symbolises not only the unity of the believers, but the wider unity of all mankind in the sight of God. But the Eucharist is more than, and different from, any of the analogies provided in world religions. Only in Judaism, with the passover meal, do we find any close parallels, hardly surprising in view of the historical connections between these two religions.

In nearly all the examples given, similarities tend to highlight differences. The sand at Cannes may remind us of the sand at Weymouth but the seaside cafes are of a different type. Frenchmen and Germans both speak of their countries as the 'fatherland' but the character of their patriotism varies in each case. So too with religions. It is illuminating and of vital importance for Church children to know that many of the beliefs and practices of their faith are analogous with those of others but it is important also to give time to the teaching of world religions in their own right. In this way it may be possible to prevent other religions from being seen simply as distorted versions of our own.

This chapter is an adaptation of an article on 'World Religions in Denominational Schools' which first appeared in *Perpectives on World Religions* ed. R. Jackson (Extramural Division of the School of Oriental and African Studies, University of London, 1978).

References

1. Vatican I, Dogmatic Constitution on the Faith, see Denzinger, Enchiridion, Editio XXXII (Roma 1963) para. 3004; ET (ed.) K. Rahner, *The Teachng of the Catholic Church* (Cork 1966).
2. See H. R. Schlette, *Towards a Theology of Religions* (Burns and Oates, London 1966).

'MULTI-CULTURAL'
RELIGIOUS EDUCATION

1. *Education for a multi-cultural society*

For the purposes of this paper, instead of the popular term 'multi-cultural education' I would like to use 'education for a multi-cultural society' to describe not only the education of racial, ethnic, cultural, religious or whatever minority groups in our society but, in addition, the education of *all* children, brown, white as well as black, for living in our changed and ever-changing multi-racial, multi-cultural, multi-faith or pluralist society. If this is a political interpretation of 'multi-cultural education' then so be it. The political concept employed here has little to do with party politics; it has everything to do with changing our educational system to further the development of justice, equality and peace in society. And it is closely connected to the development of the individual person in that society.

Attempts to reach an understanding of 'multi-cultural education' have often been based on the assumption that 'it' is something to be considered in addition to education, a programme or set of courses to be 'tacked on' to the curriculum of schools and as such an optional extra. This, I feel, is to miss the whole concept of multi-culturalism in education and educators who see 'multi-cultural education' as some form of activity in addition to, or apart from, the mainstream curriculum are doing *education* a great disservice. I say this because 'multi-cultural education' and 'education' must be seen as synonymous in our pluralist society. 'Education' is not optional in Britain and neither is 'multi-cultural education' which is the education suitable for individuals who are changing themselves and being changed all the time and for a society whose circumstances are altering almost daily. All must receive this education, otherwise the concept of equality of opportunity on which our educational system rests has no meaning in reality. Multi-cultural education is a total educational approach not an extra subject or area of knowledge; it is a curriculum development process not a method only, although method is very important as indeed is content.

The word 'culture' as used in the term 'multi-cultural education' needs analysis, since it is open to so many different interpretations. The word is used most commonly to refer to the accumulation of knowledge and experience deriving from history, myth, language, religion, family mores, dietary habits, art forms, which tend to differ between national and/or ethnic groups. It has been argued that teachers need a thorough knowledge of the cultural backgrounds of all their pupils in order to teach effectively and to help pupils to understand each other's backgrounds.[1] According to this, teachers would need to assimilate a vast quantity of information and to be all things to all persons and this would seem to be unrealistic to say the least. Teachers must not remain ignorant of other cultures, of course, but neither must they become walking polymaths to answer the needs of multi-cultural education. I prefer the interpretation which regards 'culture' not so much as group experience distilled, but as the kaleidoscope of thoughts, feelings, ideas and influences which produce an individual. This would be too complex for any anthropologist to classify or for any teacher to commit to memory. Pierre Bourdieu[2] calls this the 'cultural capital' which children bring to education. And he suggests that this 'individual' capital is constantly under-valued, in fact viewed as deficient in schools, especially so in the case of ethnic minority group children. Basil Bernstein corroborates this view: ". . . all that informs the child, that gives meaning and purpose to her/him outside the school ceases to be valid or accorded significance and opportunity for enhancement within the school".[3] Life as lived has never been considered 'the proper stuff' of education in the past nor indeed has the religion of the child, the family or the group.

A major aim of all education must be to encourage in children a critical understanding of those influences which are forming them and, therefore, of the culture of which they are a part. Such an understanding is necessary if they are to make sense of themselves and of the world, if they are to make moral choices, if they are to confront issues such as prejudice and racism which form part of the fabric of their lives, if they are to make decisions about their own future, and be able to accept responsibility for those decisions, if they are to learn 'to respect and to reverence' human beings and all life around them. If education takes the 'cultural capital' of children into account this could prove a significant step in the direction of helping them to become autonomous adults and not just members of their own particular groups, important as this might be in their lives. It should also aid them in developing skills and abilities to analyse those values, both manifest and hidden, which our education system transmits. Religion must surely be one of the most powerful influences in this 'culture capital' which the child brings to education. For this reason alone religious education must have a central role in education for the multi-cultural society in all its connotations, yes, even in the political sense described above. But it must, in general, become a much more open pursuit than it has been in the past. There should be no need for secrecy when it can be easily demonstrated that religious education is relevant to the whole education of the individual for life in the community and for life in a wider multi-cultural world. Multi-cultural education is education for justice, a theme which must surely pervade the whole of religious education today.

2. *Religious education for a multi-cultural society*

Religion still holds a central position in the British educational system. The Education Act of 1944 remains without amendment and still requires the teaching of Christianity in all schools. There can be little doubt that religious practice has changed in the past forty years: there has been a sharp decline in Christian commitment, in church attendance, during the very period in which the 1944 Act has been on the statute book. And Britain has become increasingly a multi-cultural and multi-faith society; bringing, inter alia, a heightened interest in world religions in education. The role of religious education in schools has become crucial in this new situation, but as yet remains insufficiently clarified. It has become obvious that its former aim of nurture in one particular Christian tradition can no longer be defended outright. What then should its aims be if it is to be relevant to these changing times? Should it try to stay 'neutral' in the multi-cultural, multi-faith situation and restrict itself to a comparative and objective study of religion? The short answer to the last question is 'no!' If it is Christian religious education it can never be neutral.

Recent research has shown that secondary school pupils are apathetic about religion in general and display negative attitudes towards the religious education they are receiving especially in the fifth and sixth forms. The favourable attitudes towards religion detected among primary pupils at Church schools had vanished by the time these same pupils had reached their fifth year in the secondary school.[4] Reasons for this abound, from the presentation of doctrines and practices as timeless and permanent, part of a fixed, unchanging culture and language, to methods of teaching which were too academic and which allowed for little or no participation on the part of pupils. Certainly, Christian teachers have failed to stress sufficiently that religion is about life, about human beings, about pupils and students, about 'the possibilities intrinsic to their humanity'. Religious education is about values, attitudes and behaviour based on the quite revolutionary commitment to love which the Christian ethic embodies. It is about concern in practice for human beings 'of every description' and herein lies its main relevance for the multi-cultural society. "Our love is not to be just words or mere talk, but something real and active".[5]

The Christian is a witness and an apostle of justice: he/she must not look away or stand aside from any form of inequality, injustice or oppression in local or global society. Multi-cultural education, like development education is closely concerned with justice. From this perspective religious education cannot abdicate social and personal responsibility even if it wished to be academic. And to remain neutral is certainly not adequate, for neutrality involves a hidden option. It is negative not positive; in its 'no risk' stance it inclines to death rather than to life. The taking of risks can promote new life in the Church. For educators, particularly for religious educators, there is no room for neutrality.

Church leaders of all denominations in Britain have condemned injustice, racism and the unequal treatment of minorities in recent years. It was clearly right that they spoke out on social, political and economic issues as diverse as the Nationality Act, the problems of inner-city areas like those of Liverpool, Bristol

153

K

and London, on multi-racial education, on the housing crisis, etc. For the Churches to remain unconcerned with these issues would be tantamount to a declaration that the Gospel was irrelevant to the life of the community.[6] Alas, too often there it rests. Bishops' statements, essential as they are, can provide an excuse for lack of action on the part of members of their Churches. The Church has spoken; our duty is done! Unless timely statements by Christian leaders in Britain are followed up by participation on the part of the people, their effectiveness will wane. And schools, no less than any other type of institution, reflect this syndrome. In my recent experience of surveying some Catholic schools throughout the county for evidence of curriculum action on multi-cultural education, I noted a very wide knowledge and a concern for truth in the eloquent policy statements of headteachers which were not matched by practice in the schools. The hierarchical 'trickle down' appeared to be far too slow and there were few signs of real participation on the part of staff or pupils in education for life in a multi-cultural society. Even R.E. teachers were unconcerned, but, it must be said, more from lack of guidance than from any inherent hostility to change. Sadly, there were examples of strong racial and ethnic prejudice among sixth form pupils. The task for religious education was clarified for me by these experiences.

3. *Religious education and racism*

When I use the term 'religious education' it is in the broad sense of education which is 'informed' by religion and which speaks to the whole human condition. By 'racism' I mean not just a set of irrational beliefs and attitudes, such as race prejudice towards 'outgroups' in society, but the actions which stem from these. For racism goes beyond thoughts and feelings to include discriminatory practices and insulting behaviour. It has been defined as "prejudice plus the power to activate that prejudice" and power is certainly a factor adhering to it — the power exercised by the ingroup over the outgroup, by those deemed superior in society's terms over those deemed inferior who become 'marginalised' in the process.

Racism is most often connected with 'colour' prejudice but it may be directed towards any group considered inferior by societies differentiation procedures. Irish people living in Britain, Polish, Spanish, Italian immigrants can also become its victims. Slavs, Poles and Gypsies perished alongside Jews in Nazi gas chambers on account of their alleged racial inferiority although skin colour was not a factor. However, it would be true to say that in Britain today it is black people who are the main victims of race prejudice and discrimination. Over the past twenty years three Race Relations Acts have been passed in Britain and bearing in mind that legislation comes about only when it is strictly required, it had obviously become necessary to make race discrimination an offence punishable by law in our society in this period.

Statements condemning racism as a serious violation of human rights have been issued worldwide from Church and state in recent years. Significantly for our present topic, the National Pastoral Congress of the Catholic Church at Liverpool in 1980 produced specific recommendations about race education: "Education policy must encourage the multi-racial/multi-cultural nature of our society"[7] and listed five major points condemning any form of race discrimination in education generally and in Church schools specifically. It is the responsibility of teachers to ensure that these recommendations do not remain as mere policy but are implemented through school curricula. But then the Church's teaching on racism has always been clear and direct. Every human being is made in God's image and we are all one in Jesus Christ. Christianity disregards differences, including physical differences, and regards all human beings as belonging to one 'family'. The New Testament makes several anti-racist statements, which could become the basis of a whole programme of R.E. lessons for the fifth and sixth form!

The Christian Churches in Britain have commissioned many bodies, often ecumenical in composition, to examine the multi-ethnic, multi-faith society and the phenomena of race prejudice and discrimination which arise even within their own Churches. Condemnation of racism, in theory and in practice, emerged from these deliberations. Invariably education to counter racism has been suggested as the solution so the mandate is clearly stated.

It would be naive indeed to think that our educational system has not been partly responsible up to the present time for the

LIVERPOOL INSTITUTE OF HIGHER EDUCATION THE BECK LIBRARY

maintenance of racism in society. I suggest also that our *religious education* must take some of the blame for the racism of individuals and of institutions in our society, if only because many programmes of religious education have failed to make explicit the Gospel vision of a just society where all human beings are equal and included, without exception, 'in the scope of Christ's redemptive love'. The report *Signposts and Homecomings* in pointing to "evidence of considerable racial prejudice and discrimination in British society", was moved to declare strongly that "it is clear that any form of racialism is totally unacceptable".[8]

I am suggesting strongly that racism is one area of specific interest to religious educators in schools. Almost invariably 'social education' and the teaching of 'race relations' become the province of R.E. teachers in L.E.A. schools. Church schools, by definition, should be havens of peace and good human relations but the evidence too often suggests otherwise. White parents frequently choose to send their children to Church schools, even when their links with Christian practice are tenuous to say the least, because the school is all or predominantly white. In a recent survey nine out of thirteen headteachers of Church schools said that 'race' was a factor in parental choice of schools.[9] This could become a scandal in educational and human terms. Sadly, there is evidence of racist behaviour in Church schools in the same measure as in secular schools. My suggestion here is that examples of racist name-calling, slights and insults should be noted and discussed openly with the pupils. They should never be ignored as this might lead pupils to believe that silence means consent to their racism. A policy of ignoring such behaviour is adopted by teachers in some Church schools of my acquaintance. Instead I suggest this anti-social, anti-human behaviour should become the focus of R.E. lessons. There is ample Scriptural reference to call upon, a plethora of documents from Vatican II and other contemporary source material in the form of Christian writings which are anti-racist in spirit and in content. I believe the direct approach is the correct one in this context.

In educating against prejudice and racism the R.E. teacher is called to be a witness to justice.

4. *A curriculum for religious education in the multi-cultural society*

What kind of R.E. curriculum would answer the requirement of preparing *all* pupils for life in our multi-cultural society and for the task of building "a world where every man, no matter what his race, religion or nationality can live a fully human life"[10] in justice and in peace? It would need to be an 'interventionist' curriculum since in education things do not occur by chance. Positive steps need to be taken by Christian educators to produce a curriculum based more firmly on the Gospel vision of man, on this radical vision which inevitably is at odds with a materialistic society which accepts racism and ethnocentrism and discriminatory practices to operate against some 'out-groups' within it.

The study of world religions is most important if only because today in parts of Britain the multi-cultural classroom could contain a mixture of Muslim, Hindu, Sikh, Jain, Jewish and Christian pupils, certainly in L.E.A. schools. Anglican schools are no longer mono-religious or mono-cultural and even Roman Catholic schools, which have always been multi-cultural, are not all mono-religious nowadays, according to last year's survey sponsored jointly by the Runnymede Trust and the British Council of Churches. But the study of world religions is concerned, of course, not just with pluralism, but with the way in which religious convictions and institutions affect and are affected by society, a crucial area indeed for R.E. This attempt to see the interplay of religion and daily life in religions other than one's own, for example to note how Sikhs in Southall or Buddhists in Birmingham react to life around them, could be very fascinating, but also very academic. It could fail to acknowledge the presence and importance of feeling, it could emphasise 'knowing at the expense of feeling', and R.E. is about feeling as well as understanding.

The teaching of world religions in secondary schools may mean no more than a sketchy account, against the background of a pluralist society in Britain, of the distinguishing features of certain non-Christian religions. What is even more serious is that the work done, because of its very superficiality, may not even suggest that other religious systems deserve study as such. Frequently, the treatment of other religions may even imply that, when they are not Christian, they are somehow inferior, or merely 'different' or 'quaint'. The student becomes almost a 'voyeur' of other religions and certainly not a participator. Yet in some secondary schools the teaching of 'word religions' has been put forward as the schools' answer to the religious pluralism of our society, while in others it has been made commensurate with the total religious education of non-examination pupils in the fifth and sixth forms when it should form but a part, a valuable part, of it.

I suggest that world religions should be taught in multi-cultural schools in order to promote understanding and tolerance leading to empathy with others; further, I would suggest that they should be taught in *all* schools as part of the preparation for life in a multi-cultural society, particularly where they are 'living' religions which can be experienced at first hand. The 'global' dimension of the study of world religions needs stressing in addition if international understanding is to be promoted through education. The study of cultures, religions and world perspectives other than our own must form part of the school curriculum.

The curriculum for R.E. in the senior classes of the secondary school could draw on a whole variety of areas of relevance to the multi-cultural society. One possible theme could be the significance of the sharing of food and drink as found in different cultures. Kevin Nichols, when he was National Adviser for Religious Education, observed that the only real multi-cultural education he found was taking place in domestic science classes! Here the treatment of food across cultures, the diets, the eating habits, the custom of sharing food with others, all formed part of the pupils' education and gave valid opportunities for the expression of similarities and differences between them. Surely there is scope here for the R.E. teacher because of the deep significance adhering to the preparation and sharing of food present in so many religions.

The R.E. curriculum must not avoid the more global issues of conflict and justice. I am struck by the unwillingness of religious educators to engage with unjust structures, to examine issues of racism and inter-racial relationships in inner cities, for example. These matters are left to the sociologists and the journalists and if they are 'tackled' at all in R.E. they become additional to the main concerns and are often left to external speakers and other do-gooders in a kind of cosmetic exercise. Yet throughout the world there are conflicts which have been attributed to religion, rightly or wrongly, which would merit study from a religious perspective. Close to home we have the struggle in parts of Ireland where Christianity, in two different forms, has been put forward, quite erroneously, as the root cause of political, social and economic ills: where religious divisions and conflicts have been increased by mutual lack of understanding of one group's beliefs and practices by the other. Dare we hope that a more 'open' and broadly-based religious education of school pupils might serve to heal the differences and dispel the hatred?

Religion as an area of knowledge can be a great 'bridge' not only between people, but with other school subjects in the study of human issues and problems. Then questions of values and morals are not confined to the R.E. lesson, but will permeate the whole curriculum. The relevance of this bridge for the study of race and racism, for issues of human rights and of world development are obvious. The contribution of religious education to the whole process of education for living has not yet been fully realised through the curriculum. There may well be a profound practical truth in Whitehead's statement that "The essence of education is that it be religious".

References

1. The Rampton Report (HMSO 1981).
2. P. Bourdieu, 'The School as Conservative Force', in *Schooling and Capitalism* (RKP 1971).
3. B. Bernstein, 'Education Cannot Compensate for Society', 1971.
4. L. Francis and M. Carter, *Church Aided Secondary Schools: Pupil Attitudes Towards Religion*, in British Journal of Educational Psychology, Nov. 1980.
5. 1 John 3 : 18.
6. Catholic Commission for Racial Justice, Notes and Reports, No. 11.
7. *Liverpool 1980*. Official Report of the National Pastoral Congress (St Paul Publications 1981), p. 264.
8. *Signposts and Homecomings:* the educative task of the Catholic community (St Paul Publications 1981).
9. A. Dummett and J. McNeal, *Race and Church Schools*, p. 16.
10. Pope Paul VI, *Populorum Progressio*, 47.

DEVELOPMENT EDUCATION
IN THE SECONDARY SCHOOL

The evidence for widespread poverty and oppression in the world is overwhelming. It is estimated that 800 million people suffer from hunger and malnutrition; that 12 million babies die each year before reaching their first birthday and 870 million adults cannot read and write. Nearly half of the world's people do not have safe water to drink and 500 million people have no means of earning a livelihood. There is a vague awareness of these facts and at the same time a general feeling that, since the problem is getting worse, there is very little that can be done about it.

What is not so widely understood is the extent to which Britain as a rich country is part of the problem. In this country we import half of our food and two-thirds of the raw materials used in industry. Arguably we are more dependent on the outside world than any other rich country. This can be expressed in another way: poor countries are dependent on what we pay them for their commodities. In 1980, the real price of the raw materials exported by them was at its lowest level for thirty years.

It is possible to give endless statistics to illustrate the poverty of the poor south comprising 75% of the world's population, but the important point to grasp is that the countries of the rich north, where only 25% of the people live, have immense economic power in the rest of the world. The north controls 90% of the world's wealth, 80% of the world's protein and probably over 90% of the world's technological capacity. In practical terms this means that the lives of the majority of the human race are entirely in the hands of the minority who happen to live in the north.

The implications of all this for the school were well acknowledged in the D.E.S. publication *Education in Schools* (1977):

"We live in a complex, interdependent world and many of the problems in Britain require international solutions . . . our children need to be educated in international understanding".

The Brandt Report gave strong support to the same view. Yet a survey commissioned by the Overseas Development Ministry, gave a depressing picture: "of two thirds of the nation with parochial and introverted attitudes, unsympathetic to a world perspective, clinging to the past and untutored to approach the future constructively. Attitudes towards the underdeveloped countries in particular are confused by stereotype images, post-colonial guilt, racial and cultural prejudices, limited, unbalanced knowledge, concern about future domestic employment, the belief that overseas development is synonymous only with aid and that aid is motivated only by charity. It also shows a near obsessive concern with the present economic situation in the U.K.".

Theological basis

(a) *Scripture*

The Bible has a message about social justice, but we certainly do not understand it by trying to find proof-texts to justify a particular system. Our concern must rather be to locate themes which run throughout Scripture and which have given rise to values which people of faith have tried to understand and live throughout the history of salvation.

From the very beginning people set themselves against God and sought to find fulfilment apart from his design for the world. As society developed, the experience of alienation and violence gained ground. Yet the Bible also traces the people's many attempts to be reconciled with a loving and merciful God by living in justice within the community. Justice demanded that they treat their enemies and the poor in a special manner (Exod 23:1–7); social responsibility was to be hallmark of the people of God (Lev 19:15).

The story of the exodus was about God's deliverance of his people from slavery to the freedom of the promised land. That experience of liberation remained central in the Jewish understanding of their history, relived each year in the celebration of Passover.

It is especially in the teaching of the prophets in the face of injustice that the highest ideals of the community are expressed. They called the people to return to the social responsibility they had known in the desert. They condemned the comfort and luxury of urban life in their own day (Amos 3 : 15; 6 : 8). The social values lived during the exodus were seen as part of the plan of salvation. The reconciliation of the community to God could only be achieved through personal and social conversion. It would have to wait until the time of the promised one (Isa 11) who would bring true justice to the nations (Isa 42 : 1).

From the very outset of his public ministry Jesus identifies himself as the one who was to come:

> "The spirit of the Lord has been given to me,
> for he has anointed me.
> He has sent me to bring the good news to the poor,
> to proclaim liberty to captives,
> and to the blind new sight,
> to set the downtrodden free,
> to proclaim the Lord's year of favour".
>
> (Lk 4 : 18–19, quoting Isa 61 : 1–2)

The gospels give as a key-sign of Jesus's messianic mission his special concern for the poor and oppressed (Mt 11 : 5; 19 : 21; Mk 10 : 21; Lk 18 : 22; Mk 12 : 42–3; Lk 14 : 13). He teaches that through the poor he is present among us: what we do to the least of our brethren we do to Christ (Mt 25 : 31–40). Throughout his life Jesus constantly associates with the outcasts, with those society considers marginal and of no account.

The beatitudes present a vision of the kingdom in which all social classes are united (Mt 5 : 3–10). The first four beatitudes are for the poor and wretched of the world: those who have experienced injustice and oppression will have a special place in the kingdom. The second four beatitudes are for the protectors and prophets of the poor: those who work for peace and suffer because they work for justice. It is in Christ that the fullness of justice and peace will be achieved in creation, so that "those who hunger and thirst for justice will be satisfied"

(Mt 5:6) and "those who suffer persecution for justice sake will have the kingdom of heaven" (Mt 5:10).

One particular point about the biblical idea of justice needs to be properly understood. The parables of the Kingdom made it plain that for Jesus justice is not at all the same as fairness. The stories of the labourer in the vineyard and the prodigal son are not at all fair: a different framework of values is being used in which preference is given to the dispossessed, the outcast and the powerless. It is the poor who are the inheritors of the kingdom; they are really the rich ones. The value-system of the gospels turns worldly priorities on their head. Christian justice supercedes fairness: it is of a different order.

(b) *The Church's tradition*

From the beginning the Christian community understood that faith in Jesus meant a conversion that was social as well as individual and involved a voluntary sharing of material goods. Their way of life is described in Acts: "The whole group of believers was united, heart and soul; no one claimed for his own use anything that he had, as everything they owned was held in common. The apostles continued to testify to the resurrection of the Lord Jesus with great power, and they were all given great respect. None of their members was ever in want, as all those who owned land or houses would sell them, and bring the money from them, to present it to the apostles; it was then distributed to any members who might be in need" (Acts 4: 32–35).

St Paul has some sharp words for the community at Corinth. It seems that when they came together to celebrate the Lord's Supper some were so busy eating their own food that they ignored the needs of the poor. This was to pretend there was unity in one Body expressed in the Eucharist when in fact they were divided. "A person who eats and drinks without recognising the Body is eating and drinking his own condemnation" (1 Cor 11: 29).

The Church Fathers addressed themselves to the great economic inequalities of their time. They held that material possessions were essentially common property belonging to all people. St Ambrose in his preaching condemned the accumulation of great wealth: "It is not from your own possessions that you are bestowing alms on the poor, you are but restoring to them what is theirs by right. For what was given to everyone for the use of all, you have taken for your exclusive use. The earth belongs not to the rich, but to everyone. Thus, far from giving lavishly, you are but paying part of your debt".

Much more recently the social message of the Gospel has been presented in the encyclicals of the Popes. *The Progress of Peoples* (Paul VI, 1967), in particular, is a most radical document. Among other things it teaches that individuals must be willing to pay higher taxes and wealthy nations must trade on equal terms with the poor (n. 56–61). Development is the new name for peace: so much of the violence in the world finds its roots in economic injustice (n. 76–79).

The same Pope went on to stress the importance of political action to obtain justice because so much about social life is decided by politics. His letter, *A Call to Action* (Octogesima Adveniens, 1971), outlines our political vocation as Christians, and criticises the political ideologies of both the left (Marxism) and the right (Capitalism).

In 1971, the Synod of Bishops met in Rome and produced the document *Justice in the World*. The message has shaped much of the renewed action for justice in the Church. The following is a summary of what it has to say about the way justice enters into education.

Education for Justice implies the following:

1.   Teaching people how to base the whole of their lives on the Gospel. (Schools often tend to be too taken up with the 'establishment' order of things where Gospel values so easily become blurred.)

2.   Preparing people for a way of life that is genuinely and utterly human. (It is meant to awaken a critical sense towards society, towards the way men and women live and the values they adopt.)

3.   Helping people resist manipulation by mass media or by political powers. (Education should enable people to take control of their own destinies, creating communities which are genuinely human.)

4.   Practical steps to get everyone involved in action for justice. (This works by having direct contact with injustices.)

5.   A start at home, within the family. (This is where we learn

respect for other human beings.)

6. Exploring what the liturgy has to teach us about justice. (The Eucharist, above all, forms the community and leads it to the service of mankind.)

(c) *The voice of the poor*

If we accept that the poor are the inheritors of the kingdom and that Christ speaks to us through 'the least of his brethren', then there is good reason to take heed of what the Church in the third world is saying with increasing insistence. We are being reminded that an understanding of justice does not come from the abstract reflections of theologians but from the lived-experience of those who suffer oppression and injustice. The priorities of the poor are a better indication of the values of the kingdom than the concerns of the respectable and powerful of our own society. The poor have a privileged access to a God who reveals himself as liberator.

In terms of the kingdom our worldly order of things is completely reversed. It is a matter of "upsetting, through the power of the Gospel, mankind's criteria of judgement, determining values, points of interest, lines of thought, sources of inspiration, and models of life, which are in contrast with the Word of God and the plan of salvation" (Paul VI: Evangelisation in the modern world, n. 19).

It is only in economic terms that we are rich; in human values so often we are the poor. We need to learn and receive from the shanty-dwellers who share what little they have with their neighbour, and from the rural peasants who care for the children and the aged within their village as members of their own family. Mother Teresa of Calcutta on a visit to this country reminded us of where we stand: "But you in Britain have a different kind of poverty, a poverty of loneliness and being unwanted, a poverty of spirit. And that is the worst disease in the world today".

What is development education?

By now the general area of concern in this article is clear but the problem is that the label 'development education' is not widely understood. There is for example the possibility of confusion with curriculum development and even with Piaget's stages of psychological development. Education about development really had its origin in the late 1960s, with the work of the aid agencies, the Churches and the U.N. regarding 'third world' development. This sponsorship has tended, at times, to restrict the scope of the subject to providing information to support fund-raising activities; or to considering issues like food or water that are politically safe.

(a) *Development*

'Developed' has so often been applied to the rich countries of the north, and 'underdeveloped' to the poor countries of the south with the clear implication that progress for the latter consisted in pursuing paths taken by the former. In fact, all countries are developing and there are other criteria for progress than the economic. A league table which attempted to evaluate moral or spiritual development might be difficult to compile but would certainly give a very different result to that of the World Bank.

'Development' has been so controlled by the rich countries of the north that the very concept is seen by many in the south as simply another means by which the rich continue to exploit the poor. Charles Elliott, Director of Christian Aid, has shown that 15 years of 'development' has actually meant a transfer of 60% of the wealth of the poor south to the rich north. It is largely a matter of perception: from the north, the problem is seen to be poverty, explained mainly in terms of natural causes such as drought, disease and ignorance. From the south, the perception is not so much of poverty, understood in static terms, but increasingly of oppression, understood in terms of structures, and ultimately of people with the power to keep the poor powerless. This is not the place to argue the whole case: it is enough to indicate the ambiguous nature of 'development' and why liberation is the more appropriate response from the south. The Asian Ecumenical Conference for Development put the matter succinctly: "Development is not simply food, medicine, education, population planning or a just wage for a day's work. Development is liberation".

(b) *Education*

If we focus for a moment on the second part of the title — 'Education' — we shall discover that there are different percept-

166

tions here as well. For some, education is mainly about passing on the wisdom and culture of a society to the next generation. Starting from ignorance, the student gradually becomes better informed through a process of induction. The assumption is that society's values and priorities are beyond reproach.

On the other hand, those who speak on behalf of the poor and oppressed would suggest a different starting point: the real problem is not ignorance; it is indoctrination. The illiterate and inadequate poor of the shanty towns are conditioned by generations of failure to expect very little. Paulo Freire has shown that, if they can begin to realise that the causes of their poverty are not in their failure as people but in the structures of their society, then change and progress become thinkable. The changed perception is liberating and leads on to action. Freire refers to this process which promotes critical awareness as conscientisation. (See his 'Pedagogy of the Oppressed'). The example could be much nearer home. If the rate of unemployment among blacks in some of our cities is three or four times higher than among whites, then as a society we are apparently saying something about their relative worth. Society's values need to be questioned.

It is interesting to note that the two views of education that have just been suggested are to be found, in perhaps a less obvious form, in recent documents concerned with the school curriculum. Something akin to the former view has been proposed by the D.E.S. along with a greater degree of central control over the curriculum, whereas a much more enlightened approach has been propounded for a number of years by Her Majesty's Inspectorate and the Schools Council.

After starting out as a programme to mobilise public opinion towards aid for developing countries, development education has undergone considerable growth. Many definitions have been given, but perhaps the one that best expresses the scope of the subject and its orientation towards action is that provided by the U.N.:

"The objective of development education is to enable people to participate in the development of their community, their nation and the world as a whole. Such participation implies a critical awareness of local, national and international situations based on an understanding of the social, economic and political processes . . . Development education is concerned with issues of human rights, dignity, self-reliance and social justice in both developed and developing countries. It is concerned with the causes of under-development, of how different countries go about undertaking development, and of the reasons for and ways of achieving a new international economic and social order".

On this basis, development education is about personal change, political awareness, and action here at home as much as 'third world' development.

(c) *Development education in the curriculum*

The case for development education in our schools is a strong one. Yet the attention it has received has often depended on the interest of particular teachers of geography or R.E. and the choice by pupils of such optional projects as 'world hunger' or 'the third world'. There certainly does not seem to have been a widespread conviction in schools that this area is of critical importance.

There are a few schools around the country that have made world studies or development studies part of the basic curriculum of the school. (The World Studies Journal produced at Croby Community College provides evidence of the work that has been done in evolving such courses.) But the educational climate has changed: what was done at a time of expansion in the mid '70s is no longer possible with falling rolls and a constricting curriculum. The emphasis now is on a core curriculum of subjects such as maths, science and English and this automatically tends to devalue other subjects and areas of interest.

It would have been a challenge for some Catholic schools to produce a syllabus for, say, 'O' level world studies, but we are late in the field and must take the lead from others. The alternative procedure is to recognise development education, peace education, world studies and multi-cultural education, as a 'closely-related family of concern' which can be grouped together as 'education for justice', and which should be an essential dimension of most, if not all, the existing subjects in the curriculum. Two booklets, 'The Changing World and RE' and 'The Changing World and Geography' (from the Centre for World Development Education) were an attempt to show what could be done about development education in two particular subject areas, within the scope offered by existing syllabuses. In fact, very much more can be done than is generally appreciated.

168

(d) *The question of justice*

Education for justice makes a number of assumptions. They need to be given a central place in the whole rationale for Catholic schools, particularly with regard to the hidden curriculum. We should be asking questions about a bias towards the poor and disadvantaged. In other words, is the school demonstrably part of a Church at the service of the poor or not?

The assumptions made with regard to education for justice are the following:

1. Social justice is an essential part of the life of the Church;
2. The Church has a right and a duty to preach and teach about society;
3. World peace cannot be achieved without justice in the world;
4. There exists a legitimate pluralism within the options offered to achieve justice;
5. Before we can teach about justice, we must be perceived as just;
6. No economic or political system exists in a pure form; therefore, structural analysis of each system is required;
7. The poor must have a special place in our concern for society.

In two separate reports from the National Pastoral Congress held at Liverpool in 1980 there were explicit recommendations about development education. If the above ideas were taken seriously, we would be taking the necessary steps, as far as the schools are concerned, to bring about the required change. "We stress the critical need for the Church in this country to commit far more resources to raising the awareness of our people about issues of international justice and human rights, through work by national groups, in dioceses and in parishes, schools and colleges".

Conclusion

The theological basis in Scripture, the Fathers and Church teaching, as well as the living witness of her members, are the traditional ways of exploring an area of faith. It was important to reflect on ideas about justice and development at some length since there has been a tendency in the Churches of the rich world to neglect this area of belief — or, at any rate, to see it as a concern

169

that can be left to be pursued by fringe groups who are interested in that sort of thing. We should be clear that such a view is no longer tenable: concern for justice is a central and constitutive element of the Gospel. It is not possible for a school to call itself Christian and to neglect it.

Bibliography

Brian Wren, *Education for Justice* (S.C.M. 1977)
Susan George, *How the other half dies* (Penguin 1977)
The Brandt Report, *North-South: A Programme for Survival* (Pan 1980)
World Studies Project, *Learning for Change in World Society*
World Studies Project, *Debate and Decision*
World Studies Project, *Ideas in Action: A Handbook for Teachers*
    (World Studies Project items available from CAFOD or from One World Trust)
Paulo Freire, *Pedagogy of the Oppressed* (Penguin 1972)
Barbara Ward, *Peace and Justice in the World* (CAFOD 1981)
James Pitt, *Good News to All* (CAFOD and CIIR)
Brian Davies (Ed.), *The Changing World and R.E.* (CWDE 1980)
Populorum Progressio (simplified form, *This is Progress*, CIIR)
Justice & Peace Handbook (CAFOD, CIIR & Pax Christi)
    Church in the World series (15 already published)
    Statement by Bishops' Conferences in the Developing World (CAFOD and CIIR)

UNEMPLOYMENT
AND RELIGIOUS EDUCATION

(TOWARDS A CONTINUING PARTNERSHIP IN 16–19 EDUCATION)

The teaching and caring ministries of the Church cannot be allowed to cease at 16 because so many of young people may have finished their years of formal schooling. The Church's commitment to all its young people needs to be manifest in the strategies adopted by the Church to ensure the continuing presence of the Church in the day-to-day lives of its youngsters. The Church must be seen to be where the youngsters will be, and as such its easiest task lies in working with those who stay on into the sixth form, whether for a traditional sixth form course or as members of the 'new sixth'. Much of the rest of this publication is concerned, and rightly so, with many of the problems associated with that work, and it is of vital importance that we clarify our objectives in a rational and realistic way about this sixth form work. It would be unfortunate if these immediate discussions were to be carried on in isolation from fundamental debate about the general secondary R.E. curriculum, as there is usually an unfortunate snap in the continuum of both content and method between fifth and sixth form.

The implications of any willingness to alter substantially the school pattern of religious education are governed by a lack of clear, systematically-gathered evidence about what is done with all sixth formers at the moment. The lack of this information reflects undoubtedly the lack of consensus amongst teachers about the type of work which is appropriate to the 16–19 year old age group. Perhaps a concentrated study of Dr Leslie Francis' recent study of the religious notions of young adults might help to guide us a little: his study clearly begs a number of questions about the general lack of evaluative research activity into religious education in Church schools.[1] Is it unreasonable to have expected that the

Church would have gathered such information through its diocesan R.E. advisory teams? We have a real need to redirect some of our energies into the collection of appropriate data about the actual classroom reading which goes on in our schools: only in such a systematic way do we unearth our centres of excellence from which the rest of us can learn. In the absence of really clear evidence about what we are doing with the 16–19 age group, perhaps we can only look at some of the social factors which ought to influence any curriculum which is concerned with the development of values in our present society.

There are perhaps three areas of significant social change which are linked directly to the process of establishing an individual identity: in the past we have assumed that our young people would lead a busy working life in an established community where leisure was seen as a direct counterpoint to work. Now, we can look forward to a society in which work will no longer provide the major key to a young person's identity: where he/she will live in a multi-racial, multi-faith community, faced, possibly, with a tremendous problem of time. These considerations should be making us ask many basic questions about our formal curriculum and its domination by irrelevant academic examinations. The 'we' in this context is all those of us with a concern about the provision of opportunities in formal and informal education: parents, teachers, religious, youth workers, residential centre staff, and R.E. development teams. Perhaps a developed programme for sixth formers in school could involve a contribution from all of us in a process which reflected some notion of the community educating its youngsters into an adult faith which can survive the challenges of intermittent employment and urban tension. Despite some of these problems, and many others mentioned in these papers, I reiterate my viewpoint that the easiest work in the 16–19 year old bracket is going to be with those in school sixth forms. That represents the challenge to the schools: the challenge to the Church is in devising structures, initiatives, developments and modifications which enable the Church's commitment to all its young people to be manifest. After the success of harnessing resources for the visit of the Pope, it is no longer acceptable to be told that the Church does not have the resources: the question is whether the Church has the will and the knowledge to respond.

The fastest growing sector of the public service at the moment is that which deals with the government's alternatives to employment. This should not surprise us when we stop to examine the range of the problem: there has been a significant drop in the numbers of young people, even those in work, who can obtain apprenticeships. In the 1960s, 40% of 16 year old boys obtained apprenticeships: by 1980, this was about 18%; apprenticeships dropped from 236,000 in 1968, to under 150,000 in 1980 and are thought now to be under 100,000. At the same time the number of young people in manufacturing industries receiving any formal training fell from 210,000 in 1968 to 90,000 in 1980 and is fewer every year. This awesome drop in real training for real jobs has been accompanied by a quadrupling of unemployment in just seven years, 1975–82; whilst overall unemployment quadrupled, unemployment amongst under 18's increased 5 times over. The Manpower Services Commission estimates that 57% of 16 year olds and 48% of the 17 year olds in the labour market will be unemployed in September 1984. The Youth Opportunities programme for the year 1982–83 catered for 630,000 entrants, with a budget of £730 million, an increase of 77% over 1981–82. For the 16 year old school leaver in September 1983, the following figures are projected:

Total age group	Employed	Unemployed	Schools	Further education
889,500	210,600	297,300	270,400	115,400

Such figures have prompted the government to accept in total the proposals of the Manpower Services Commission Task Force Group Report (April 1982) and widespread reform of post-16 provision is about to take place. I want to examine this area of development under three linked headings:

A: The government's proposal to build a permanent bridge between school and work, both through their New Training Initiatives and community programmes

B: Some aspects of the personal crisis faced by families and young people in areas of significant long-term unemployment

C: A programme of response for the Church, using all its physical and membership resources in co-operation with the overall strategy.

A: The government's proposals: the government is now committed to providing a permanent bridge between school and work, and this commitment is broadly based upon the following objectives:

a^1 to develop skill training in such a way as to enable young people entering at different ages and with different educational attainments to acquire agreed standards of skill appropriate to the jobs available, and to provide them with a basis for progression through further learning;

a^2 to move towards a position where all young people under 18 have the opportunity of continuing in full-time education or of entering a period of planned work experience combined with work related training and education;

a^3 to open up widespread opportunities for adults to update their skills and knowledge;

The programme to be financed by the government will bring all training work with young people together under the New Training Initiative proposals, although pilot schemes in some colleges are being advertised as Youth Training Services courses. The government hopes to see programmes of at least one year, of high quality, including a minimum of 3 months 'off the job training' and/or 'relevant further education'. Schemes will be mounted locally, and require the co-operation of employers, local education authorities, colleges and voluntary organisations. All young people taking part in the courses are to be volunteers and to be regarded as 'trainees': as trainees they will receive an allowance of £25 per week and will be reimbursed their travelling expenses in excess of £4 per week. Courses will be sponsored, within a local area, by 'managing agencies', and voluntary organisations are encouraged to take on such a role. It may be necessary for the sponsors to negotiate with other training or placement agencies (e.g. the Careers Service) to implement part of the overall programme for their trainees, but there is clearly an opportunity for any committed organisation to make a contribution of significance: "any kind of organisation might apply to be a managing agency . . . new organisations might be formed to act as agencies". All the detailed requirements of the training programmes and of the managing agencies can be found in the appropriate report, as can the very simple financial arrangements. It is important to stress that one of the ways of providing a recognised programme is through a community project and it is also relevant to stress that the entire programme can be funded by the Manpower Services Commission.[2]

B: Some aspects of the problems faced by families and young people affected by the prospect of long-term unemployment: in a recent conference paper,[3] Mr Colin Leicester, Associate Fellow, Centre for Employment Studies, Management College, Henley, projected a view of the employment scene which effectively countered any notion that we are experiencing a short-term loss of employment. Not only could he see a continued unemployed pool of 2 millions until 1990 but he predicted that the prospect of long-term unemployment for 16–20 year olds would increase by a factor of 20, and for 20–24 year olds by a factor of 25: there would be long-term 'stacking' of unemployed youngsters, and this was without us knowing anything about the real impact of technological change. He argued for an emphasis on self-employment; on new, maverick, innovatory, self-reliant entrepreneurs; on equipping youngsters with a fundamental survival kit — all of these being needed from a system of schooling which he labelled as 'isolated and inert'.

Dr Bill Law of the National Institute for Careers Education and Counselling drew attention to what people had customarily gained from work: achievement, stimulus, esteem, company, structure and goods; work provides both a formal structure and companionship; if we deny people the opportunity to work, from where are they going to experience these features of life? He thought perhaps the fact that parents and children were experiencing unemployment, and realising that it was not their fault, was perhaps bringing families together. Unemployment, however, formed its own ghetto where talk of work, knowledge of possible jobs, were excluded; there was a clear need for those unemployed to be kept in regular, friendly, supportive touch with their own community. We need, therefore, to accept the long-term nature of the present problem, and to look for hope within our own local community. These features were recognised in the Task Force's report which stressed the importance of local involvement and it is towards this that I now want to turn.

C: A programme of response for the Church, using all its resources in co-operation with the overall strategy.

A unique feature of the Church in Britain has been its partnership with the state in the provision of denominational education. The survival of this co-operative system has been due, in very large measure, to the sensible adaptation of the Church's provision to that of the local education authorities: where appropriate we have recognised with sixth form colleges, upper schools, 11–18 schools, 11–16 schools, middle schools, junior schools, first schools, in a recognition of both local need and political reality. Where we have failed to do that, I suspect our final solutions will be an unfortunate compromise of the undesirable with the minimal. We are now faced with other issues:

a) does the Church accept the need for a continuing presence in the education and training of those young people who are no longer at school?
b) does it have either the appropriate structures for this or the willingness to develop them?
c) can it work alongside the national plan in a continuation of the partnership which has been so widespread for children from 5 to 16 or 18?

If the answer to a) is no, then we must accuse the Church of continuing to discriminate on grounds of ability, because those in universities and sixth forms will be provided for, at least in part, whilst those at colleges of further education, in training workshops, on community projects, or unemployed will be denied resources, and those resources can be made available at minimal cost to the Church. Of course, any presence of the Church in this post-16 work will raise urgent questions of aims, objectives and methods, just as the same questions are being energetically debated by all those already involved in existing training programmes. As flexibility is the aim in an unpredictable scenario, can I make a series of tentative suggestions which might become a ground plan for development:

1) Every Catholic secondary school should consider whether it can make a contribution to the programmes offered at post 16 within its area; this contribution could take the following terms:

176

1.1 Producing a package of work, perhaps in the area of social and life skills, which could be part of an integrated programme for 16+ youngsters: these youngsters would return to their original school for perhaps a day and a half per week for this work.

1.2 Receiving youngsters into the school as part of their traineeship on some local training programme: this may involve work with groundsmen, caretaking staff, school meals staff, or working as an ancillary in a department which could offer some vocational training, e.g. craft and design, home economics, science and physical education.

2) Individual schools or groups of schools should consider establishing themselves as managing agencies for training youngsters within their locality. This would require at least the following initiatives:

2.1 Schools must become familiar with the details of the national plans for youth training.

2.2 They should ensure that they are clear about the arrangements being made by local boards (see initial report), and that they can identify the key people within the local Manpower Services team and the local education authority, especially within the Youth Service and Careers Service.

2.3 They should identify members of staff with interest and expertise in informal ways of working with young people, and encourage such staff to be made available to the local training programmes, both in initial course planning, execution and evaluation.

2.4 Schools, either individually or in groups, should consider whether, in co-operation with local parishes, employers, and diocesan teams, they could take on the function of a managing agency, initially perhaps for a group of 15–20 young people, remembering that the young people do need to be volunteers for training.

2.5 Schools should have detailed and sustained discussion across their catchment area with all representative Church groupings, to identify community needs: subsequently, they can arrange to meet these needs through a community project. This initiative by the schools gives the Church's local, parish communities a tremendous opportunity to embark on a really responsible programme, by which I mean a programme which gives young people real responsibility. Perhaps the outlines of an example would be more significant than continuing prose:

Stage 1   Schools/parish representatives meet in catchment area/deanery meeting to discuss the contribution which 'the Church' can make to unemployment/employment.

Stage 2   Hopefully, a decision emerges to adopt a community programme approach; two developments are needed
a) the identification of likely community tasks;
b) early discussion with the local board of Manpower Services Commission to establish their willingness to consider the proposal.

Stage 3   The creation of a small, *working* group to act as the managing agency to sponsor a training programme; such a group might look something like this:
1. Headteacher of secondary school
2. Representative head of junior schools
3. Chairman/nominee of P.T.A.
4. Clergy representative: someone active in youth work
5. Parish representative: preferably employers or those active in community work
6. Diocesan R.E./Youth adviser
7. Secondary school staff representative
8. L.E.A./M.S.C. representative
   The group should not exceed 7–8 people and should create the minimum bureaucracy necessary; it is almost certainly going to need to be serviced by secondary school.

Stage 4   The working group refines the community project approach, ensuring that there is a year's work available for the number of young people involved, and gives consideration to the following essential features of the programme:

a) the final design and management of the programme;
b) the recruitment of young people, in co-operation probably with school staff and careers service;
c) the supervision/monitoring of young people during the programme;
d) the maintenance of quality and standards within the programme;
e) recording and certifying progress during the programme;
f) ensuring that trainees have access to support and advice during the programme.

It is important here to stress that the planning group would have needed to make links with other agencies as it refined its proposal to see what part they would be willing to play in the programme: these agencies would include:

the careers service
the youth service
the local college of further education
community service volunteers
diocesan organisations, e.g. social workers, etc.

However, I would hope that the school would see itself as being central to the programme by providing:

i) the initial impetus;

ii) an induction programme for the trainees;

iii) a substantial part of an accompanying programme to the practical work, for example one and a half day per week programme of continuing education;

iv) playing an active part in the monitoring of the programme; many schools already have large numbers of pupils doing community service; some resources could be switched to the specific programme; and many parents would be willing to volunteer to this work;

v) perhaps providing a physical base in which the programme could be housed; with the dramatic fall in roll in urban areas especially, this should be no problem. (Indeed, some secondary schools will have enough space on a permanent basis to offer to house a training workshop on their premises.)

3) Diocesan authorities should consider in particular whether there is an opportunity within the diocese for all young people involved in trainee programmes to experience at least one week of residential community life: the integration of such an experience into the overall programme would be an essential part of the programmes' structure: ideally two or even three residential spells should figure in the programme. Diocesan R.E./Youth service staff should be involved significantly in the joint planning of the residential blocs. These should aim at giving young people most of the responsibility for actively managing the week, for learning by experiencing together and reflecting together on the project work they have been undertaking. This calls for very skilled work on behalf of the adults involved, and for an appropriate environment in which the youngsters can meet, remembering all the time that they are volunteers in the first place. Any diocese without a youth work specialist, with a local network also in operation, is unlikely to be very significant in this process and should do something about it; certainly all dioceses should, as an absolute priority, have reviewed their permanent commitment to young people in the light of the present national crisis; youth advisers appointed ten or fifteen years ago to do a different low-key job may wish to respond energetically, having discovered their mission at last, or they may wish to step aside.[4]

You may say, having read all this that it is impossible. Well, it may look that way, but the fantastic growth in Youth Opportunities programmes showed that people could respond, although in the main the Church was insignificant in responding to those opportunities. All over the country, hard pressed careers service, youth service and manpower services staff are already running pilot 'youth training schemes'. The tremendous expansion of staff working within this sector is obvious, and the amount of curriculum work being done on programmes for 16–18 years olds is very significant.[5] New ways of assessment are being looked at seriously and will undoubtedly be used throughout these courses; they will be used by local groups managing local courses, and at last we may see employers coming to terms with assessments other than formal examinations.[6]

To summarise a somewhat complex scene: the government hopes to provide youth training schemes for all young people who are willing to participate in them; it envisages 300,000 places being provided on schemes sponsored by employers, and 160,000 places provided by training workshops and community projects. The Government is prepared to provide £1,850 per trainee, of which £1,450 will be paid to the trainees as allowances; any agency acting as a management agency will receive an additional £100 per trainee, so there is no real shortage of finance. For our young people, they may be a number of alternatives, including perhaps a year in school studying for the 17+ qualification, the certificate of prevocational training (C.P.V.E.). The temptation is to say that our schools should develop that as their only provision for any youngsters interested in continuing education: certainly, it should be considered, but alongside, if possible, other provisions: the training schemes and the pre-vocational courses can have a lot in common: they could even be timetabled together and joint arrangements made with other agencies to support the work of both initiatives. At a conservative estimate, for many areas, we are talking about 50% of youngsters not wanting traditional school or further education and looking for alternative programmes. Just as the government provides a national system of schooling in which the Church is a negotiated partner, so it is now proposing a national youth training scheme; the submission in this article is that the Church should at least try to become a willing partner in this national programme. This requires more than anything local initiative, and, possibly, a commitment of some resources; some successful practice would, however, really extend our notion of school and community, and give some real definition of 'the Church' to young people already deprived of so many sources of identity and significance.

References

1. Leslie Francis, *Experience of Adulthood* (Gower Press, 1982).
2. Manpower Service Commission, *Youth Task Group Report*.

3. *Integrated Approach to 14–18 Education,* Conference Report, Cambridge Institute of Education.
4. Report on the future of the Youth Service, *The Thompson Report* (H.M.S.O.).
5. *Vocational Preparation* Themes and Development (Coombe Lodge Publications).
6. *Profiling and Profile Reporting* (Coombe Lodge Publications).

251081

This book is to be returned on or before
the last date stamped below.

- 4 NOV 1996

1 5 MAR 2006
- 5 NOV 2009

LIBREX

LIVERPOOL HOPE
THE BECK LIBRARY
HOPE PARK, LIVERPOOL, L16 9JD